HANDBOOK OF TERRESTRIAL HEAT-FLOW DENSITY DETERMINATION

SOLID EARTH SCIENCES LIBRARY

HANDBOOK OF TERRESTRIAL HEAT-FLOW DENSITY DETERMINATION

with Guidelines and Recommendations of the International Heat Flow Commission

Edited by

R. HAENEL

Department of Geophysics, Geological Survey of Lower Saxony, Hannover, FRG

L. RYBACH

Institute of Geophysics, ETH Zürich, Switzerland

and

L. STEGENA

Department of Environmental Physics, Roland Eötvös University, Budapest, Hungary

KLUWER ACADEMIC PUBLISHERS

DORDRECHT / BOSTON / LONDON

Library of Congress Cataloging in Publication Data CIP

International Heat Flow Commission.
Handbook of terrestrial heat-flow density determination: guidelines and recommendations of the International Heat Flow Commission / edited by R. Haenel, L. Rybach, and L. Stegena. p. cm. — (Solid earth sciences library)
Bibliography: p.
Includes Index.

ISBN-13: 978-94-010-7780-4 e-ISBN-13: 978-94-009-2847-3
DOI: 10.1007/978-94-009-2847-3

1. Terrestrial heat flow. I. Haenel, Ralph. II. Rybach, Ladislaus. III. Stegena, Lajos. IV. Title. V. Series.
QE509.I575 1988 87—32078
551.1′4—dc19 CIP

Published by Kluwer Academic Publishers,
P.O. Box 17, 3300 AA Dordrecht, Holland.

Kluwer Academic Publishers incorporates the publishing programmes of D. Reidel, Martinus Nijhoff, Dr W. Junk and MTP Press.

Sold and distributed in the U.S.A. and Canada
by Kluwer Academic Publishers,
101 Philip Drive, Norwell, Ma 02061, U.S.A.

In all other countries, sold and distributed
by Kluwer Academic Publishers Group,
P.O. Box 322, 3300 AH Dordrecht, Holland.

TABLE OF CONTENTS

LIST OF CONTRIBUTORS

N. Balling	Laboratoriet for Geofysik, Geologisk Institut, Aarhus Universitet, Finlandgade 6, DK – 8200 Aarhus N, Denmark.
A. E. Beck	Department of Geophysics, The University of Western Ontario, London N6A 5B7, Canada.
V. Čermák	Geofysikalni Ustav, Československa Akademie Ved, Praha 4 – Sporilov, CSSR.
D.S. Chapman	Department of Geology and Geophysics, University of Utah, Salt Lake City, Utah 84 112, USA.
C. Clauser	Institute of Applied Geophysics, TU Berlin, Ackerstraße 71–76, D – 1000 Berlin, Federal Republic of Germany.
E. E. Davis	Pacific Geoscience Centre, Energy, Mines and Resources, Sidney, B. C. V8L 4B2, Canada.
G. Delisle	Department of Geophysics, Federal Institute for Geosciences and Natural Resources, Stilleweg 2, D – 3000 Hannover, Federal Republic of Germany.
R. Haenel	Department of Geophysics, Geological Survey of Lower Saxony, Stilleweg 2, D – 3000 Hannover, Federal Republic of Germany.
F. Mongelli	Dipartimento di Geologia e Geofisica, Università di Bari, I – 70122 Bari, Italy.
H. N. Pollack	Department of Geological Sciences, The University of Michigan, 1006 C. C. Little Building, Ann Arbor, MI 48109, USA.
W. G. Powell	Department of Geology and Geophysics, University of Utah, Salt Lake City, Utah 84112, USA.
L. Rybach	Institute of Geophysics, ETH Zürich, Hönggerberg, CH – 8093 Zürich, Switzerland.
J. H. Sass	United States Department of the Interior, Geological Survey, 2255 North Gemini Drive, Flagstaff, Arizona 86001, USA.

L. Stegena — Department of Environmental Physics, Roland Eötvös University, Kun Bela tér 2, H – 1083 Budapest, Hungary.

D. Stromeyer — Central Institute for Physics of the Earth, Telegrafenberg, DDR – 1500 Potsdam, German Democratic Republic.

S. Uyeda — Earthquake Research Institute, University of Tokyo, Bunkyo-Ku, Tokyo 113, Japan; and Department of Geophysics Texas A & M University, College Station, TX 77843, USA.

G. Zoth — Department of Geophysics, Geological Survey of Lower Saxony, Stilleweg 2, D – 3000 Hannover, Federal Republic of Germany.

FOREWORD

There comes a time in the affairs of every organization when we have to sit down and take stock of where we are and where we want to go. When the International Heat Flow Committee (as it was first called), IHFC, was formed in 1963 at the San Francisco International Union of Geodesy and Geophysics with Francis Birch as its first Chairman, the principal purpose was to stimulate work in the basic aspects of geothermics, particularly the measurement of terrestrial heat-flow density (HFD) in what were then the 'geothermally underdeveloped' areas of the world. In this, the IHFC was remarkably successful.

By the beginning of the second decade of our existence, interest in the economic aspects of geothermics was increasing at a rapid pace and the IHFC served as a conduit for all aspects of geothermics and, moreover, became the group responsible for collecting data on all types of HFD measurements. In all the tasks that are undertaken, the IHFC relies on the enthusiasm of its members and colleagues who devote much of their time to the important but unglamorous and personally unrewarding tasks that were asked of them, and we are fortunate that our parent institutions are usually quite tolerant of the time spent by their employees on IHFC work.

As with any youthful discipline, individuals had their own ideas on what should and should not be done and reasonable assumptions were made which later proved questionnable if not untenable. For example, for many years we felt results would be free of 'surface' noise sources if we confined our measurements to depths greater than about 30 m, the depth below which perturbation due to variations in the annual surface temperature became negligible; we now know that in many cases there are longer period variations to be concerned about. We now know that slow subsurface fluid flows may have far more influence than previously thought on the reliability of the results. Therefore, how deep should we go in order to obtain a reliable HFD value? What do we really mean by a reliable HFD value? Having obtained a reliable HFD value, over how wide a region does it pertain? Should we differentiate between local and regional HFD values? What characteristics should we look for to define an economic geothermal resource? What is the best method for exploring for such resources, called reserves?

Every person who has been involved in any aspect of geothermal work, has had to ask questions like these and produce some sort of answer. That means that there is a large body of experience to draw upon; we know many (but certainly not all) of the traps and tricks of the trade and one of the purposes of this handbook is to discuss the advantages and disadvantages of the most used techniques.

To cover all aspects of geothermics in depth would require several volumes. What we have tried to do is cover geothermics in sufficient depth to please the experts in one area who wish to have a better understanding of the problems and practices of another area, and to give newcomers sufficient information to start a programme without having to spend too much time on a literature search in a region where library facilities may be relatively poor; naturally, as the programme develops extensive further reading becomes necessary.

Inevitably, there is some overlap between the various sections and just as inevitably it will be discovered that some authors might hold different opinions from their colleagues. This is not a bad thing; it would be terrible if the day arrived when we thought we knew everything and we all agreed with each other. Nevertheless, there is general agreement on such things as techniques that yield the best results in most cases (however good a general technique is there will always arise the special situation when something new has to be devised) and on the types of dubious situations one should look out for (but having identified a tricky situation it might not be nearly so easy to decide what should be done about it).

Therefore, although we have tried to make this handbook a reasonable stand alone reference manual, each section is copiously referenced and if the readers are uncertain about any statements made they should refer to the original literature.

Other problems of considerable proportions were those of decisions on terminology and units. For example, some contributors do not like the term Heat-Flow Density (HFD) as a replacement for the older Heat Flow or Heat Flux; but it is logical and correct and has the advantage of becoming an unambiguous acronym. Again we can hope to obtain universal acceptance of the recommended way of writing the values of the geothermal units in SI; for example, should we use $mm^2\ s^{-1}$ or $10^{-6}\ m^2\ s^{-1}$ for thermal diffusivity. Someone has to take the bit between the teeth and make a recommendation; those of the IHFC appear in the text.

In view to these and other problems the editors faced a difficult task in undertaking such a monumental project with so many international contributors; anyone who has undertaken an editing job of even half this size will understand the problems. This book is a tribute to their energies.

Let me return to where I started with history of the IHFC; the late Dr. Elena Lubimova was a founding member and the second president who lead us for 12 years; her unexpected and too early death in 1985 left members of the heat flow community in a state of shock. This manual is dedicated to her memory.

A. E. BECK
Chairman, IHFC
London, Canada

July 1987

1. INTRODUCTION

R. HAENEL
Department of Geophysics, Geological Survey of Lower Saxony, Hannover, FRG

L. RYBACH
Institute of Geophysics, ETH Zürich, Switzerland

L. STEGENA
Department of Environmental Physics, Roland Eötvös University, Budapest, Hungary

The study of the Earth's internal temperature has a long history stemming from the time of Plato to the present. It has been described by Bullard (1965), completet and summarized in the Table given below. In contrast terrestrial heat-flow density determinations, which give more exact geothermal information about the Earth, were first actually measured by Benfield (1939) about the middle of this century. Since that time worldwide temperature and heat-flow density determinations have increased slowly, but continuously. A status report on the methodology and results of heat-flow density on the worldwide scale was given by Lee (1965) in the Sixties.

Interpretation of continental heat-flow density measurements was significantly assisted by three developments in the 1960's. They include: the fundamental work about conduction of heat in solids (Carslaw and Jaeger, 1965), the recognition of the heat-flow density/age relationship (Lee and Uyeda, 1965; Polyak and Smirnov, 1968), and the establishment of the concept of continental heat-flow provinces on the basis of a heat-flow density/radioactive heat generation relationships (Roy *et al.*, 1968).

The late 1960's also signaled a rapid development of the concept of plate tectonics. It was recognized from the very beginning that the related geodynamic processes have heat-flow density signatures and that heat-flow density values thus represent strict boundary conditions for mathematical modelling of the processes in question. In this context, only a few basic studies shall be listed here:

- oceanic heat-flow density/age relationship, and the related matter of sea bottom topography caused by sea floor spreading and lithospheric cooling (McKenzie, 1967; Francheteau *et al.*, 1970; Sclater *et al.*, 1970);
- thermal effects of lithospheric subduction (Bird *et al.*, 1975) and of continental collision (Minear and Toksöz, 1970);
- lithospheric thickness/heat-flow density variations (oceanic: Parker and Oldenburg, 1973; continental: Crough and Thompson, 1977; Chapman and Pollack, 1977);
- heat-flow density variations at continental margins and in extensional basins (Sleep, 1971; Stegena and Horvàth, 1975; McKenzie, 1978);

R. Haenel, L. Rybach, and L. Stegena (eds.)
Handbook of Terrestrial Heat-Flow Density Determination, 1—7.

– heat-flow density in continental rifts and at transcurrent faults (Lachenbruch, 1978).

Triggered and motivated by all these studies, the number of heat-flow density determinations grew rapidly. For example for a continent such as Europe the summary of measurements presented by Čermák and Rybach (1979) includes about 3100 measurements.

During this period a separate development also gave impetus to further investigations in geothermics. During the 1970's it became clear that the world's known nonrenewable energy resources are decreasing rapidly and may be exhausted within the foreseeable future. This and the sudden rise of oil price in 1973 generated increased interest in alternative sources of energy, including the field of geothermal energy. Geothermal prospecting aims at the evaluation of the temperature field and the mode of heat transfer in the subsurface because from these data the geothermal potential of the area under study can be assessed. Heat-flow density is a key property to understand the thermal conditions and processes at depth.

A worldwide search for geothermal energy began and research into possibilities of its utilization was started. Published summaries of geothermal research include Kappelmeyer and Haenel (1974); U. N. Symposium (1975); Kestin *et al.* (1980); Rybach and Muffler (1981); Edwards *et al.* (1982). The interest in thermal aspects of geodynamic processes as well as the activity in geothermal research both lead to an increase in the number of scientists were involved in geothermics. Correspondingly measuring equipment, measurement techniques, correction procedures, interpretation methods, and representation of geothermal data have undergone refinement.

Simultaneously, some uncertainty has arisen with regard to the most suitable equipment, measuring methods, correction procedures, the minimum borehole depth for a reliable heat-flow density determination, etc. The difference in quality and reliability of presented temperature and heat-flow density data are sometimes considerable. The comparison of different temperature maps and also of different heat-flow density maps are often difficult or impossible. To avoid misunderstanding, certain terms had to be reconsidered and to be defined more clearly. These include the terms geothermal potential, geothermal resources, geothermal reserves, and also the heat flow in cal cm^{-2} s^{-1} itself have been transformed into heat-flow density in mW m^{-2}, in accordance with the International Standards (ISO Standards, 1982).

In view of these developments in the field of geothermics, it has been felt for quite some time to reconsider certain terms and to define them more clearly, to prepare correction procedures and to recommend standardization of measurements, correction of geothermal data and forms for data presentation. After previous attempts (Haenel, 1979; Balling *et al.*, 1981), a first workshop on 'Standards in Geothermics' was organized by the International Heat-Flow Commission (IHFC), during the 21st General Assembly of the International Association of the Seismology and Physics of the Earth's Interior (IASPEI) in London, Ontario (Canada), 1981 (Haenel and Gupta, 1983).

Notwithstanding the wide range of topics and the large number of ideas and views presented and discussed in this workshop, there remained some conspicuous gaps. It was felt that further efforts are necessary to effectively cover the entire field of geothermics, including geothermal resources, to prepare guidelines and recommend standardization of terms and methods etc. Thus, the IHFC acted to form a Working Group on 'Standards in Geothermics'. A second workshop was held in Liblice (CSSR), 1982, and a third one in Hamburg (FRG), 1983. The results are incorporated in this book.

The objective of the present Handbook is to up-date the theory and practice of heat-flow density determination techniques. This concerns guidelines for standardized use of terms and units, for recommended equipment, measuring techniques, correction procedures, and comparable data representation. Moreover, the accumulated knowledge and experience of researchers in heat-flow density determination and interpretation are made available. This book tries to cover all phases which are necessary for heat-flow density determination and topics which are strongly connected to these phases.

In Chapter 2. the fundamental equations are presented as a basis for a better understanding of the following chapters. With regard to the increasing interest in understanding the influence of convective heat transport on the conductive heat-flow density, and the increasing interest simulating these processes by numerical models, both of these topics have been stressed. The following eight chapters cover the methods for determination of temperature and temperature gradient, the geothermal parameters (thermal conductivity, thermal diffusivity, specific heat capacity, heat production rate, opacity), the heat-flow density on continents and oceans, the representation of geothermal data (temperature, temperature gradient, heat-flow density, resources, reserves), the crust and mantle geothermics, and the already mentioned implications (geodynamics, exploration, palaeogeothermics, engineering applications).

In the past an attempt was made to gather and codify the geothermal information by the establishment of the International Geothermal Information Exchange Programme (IGIEP) for the exchange and dissemination of all published articles and data. Two computerized centres were established – one in Pisa, using the combined facilities of the International Institute of Geothermal Research (CNR) and the Centro Nazionale Universitario Calcolo Elettronico (CNUCE), and the other in the USA using the facilities of the Lawrence Berkeley Laboratory, California, and the US Geological Survey in Reston, Virginia. In 1974, the co-ordinating group of IGIEP initiated the following activities:

- GRID a computerized bibliography of geothermal literature at the Lawrence Berkeley Laboratory, California.
- GEOTHERM a computerized data file on geothermal fields, wells, and other geothermal topics compiled by the US Geological Survey.
- Implementation of GEOTHERM at the computer centre of the International Centre for Geothermal Research, Pisa, and at the Lawrence Berkeley Laboratory.
- Co-ordination of activities between the United States, New Zealand, Italy and other countries, and between the US Geological Survey and the Lawrence Berkeley Laboratory.

The agreement between the Italian and American centres operated for five years until 1979, but was not renewed thereafter. There is no longer a single world centre for published geothermal data, so the student of geothermal energy undoubtedly has a problem in keeping up-to-date with the latest developments. For heat-flow density values can be contacted the World Data Center A, for solid earth geophysics, Boulder, Colorado (USA).

The Handbook of Terrestrial Heat-Flow Density Determination has been initiated by the International Heat Flow Commission, chaired at that time by L. Stegena.

TABLE

A short history which leads to geothermics; (1): from Meyer (1913).

Plato, (1)	427–347:	Explains heat and cold by atoms
Aristoteles, (1)	384–322:	Nature consists of four elements: air, fire, water, and Earth
Pliny	23–79:	Reports on volcanic eruptions, thermal springs, fumaroles, etc.
al Bîrûnî, (1)	1000:	Discusses the ideas colder and hotter
Agricola	1530:	Describes the higher temperatures in mines down to 1000 m depth
Galilei, (Burckhardt)	1600:	The first thermometer; Galilei's thermoscope
Bacon	1620:	Introduces a theory of heat in his opus Novum Organum Scientiarum
Sanctorius	1626:	Describes seven thermometers
Ferdinand II of Toscana, (1)	1641:	Constructs thermometers similar to present types
Boyle	1665:	History of cold
Amontons	1702:	Inventor of the air thermometer
Fahrenheit, (Boerhaave)	1709:	Fahrenheit thermometer scale
Rèaumur	1730:	Rèaumur thermometer scale
Celsius	1742:	Celsius thermometer scale
Black, (1)	1728–1799:	Heat of fusion, heat of evaporation, heat content, etc.
Crawford	1778:	Determination of specific heat capacity of different materials
Fourier, (1)	1768–1830:	Theory of heat conduction
Gay-Lussac, (1)	1778–1850:	Gas laws
Kelvin, (1)	1824–1907:	Introduction of absolute temperature, thermoelements, heat regime of the Earth
d'Aubuisson	1802:	Temperature measurements in mines of Germany
De La Rue, Marcet	1832:	Temperature measurements in a borehole at Pregny near Geneva (Switzerland)
Reich	1830/32:	Attempt of topographic correction
Mayer	1842:	First law of thermodynamics
Kirchhoff	1860:	Radiation law
Clausius	1865:	Second law of thermodynamics
Stefan	1875:	Kinetic gas theory, radiation law
Stapff	1883:	First estimate of heat-flow density for continental Europe 77 mW m^{-2}
Boltzmann	1884:	Kinetic gas theory, radiation law
Unnsteinsson	1888:	Soil is heated by hot water to grow potatoes in Iceland

Table 1 (Continued)

Chun	1898: Temperature measurement up to 4700 m water depth from the research vessel Valdivia
Wells	1890: Thermal water (77 °C) was used for heating purposes in Boise, Idaho (U.S.A.)
ENEL	1897: Natural steam heats a boiler producing pure steam for a reciprocating engine in Larderello, Italy
Planck	1900: Radiation law
ENEL	1904: Electric power from natural steam is used for electric lamps in Larderello, Italy
Parsons	1904: Proposed a twelve mile deep shaft to meet 500 °C, The Hellfire Exploration Project
Nernst	1906: Third law of thermodynamics
Joly	1909: Correlation of heat-flow density and local rock radioactivity
ENEL	1912: The first turbine generator was installed with 0.25 MW in Larderello, Italy
Benfield	1939: First heat-flow density calculation from temperature measurements in a borehole and thermal conductivity determination in the laboratory
Pettersson	1949: Heat-flow density determination in the deep sea
Beck *et al.*	1956: Thermal conductivity *in-situ* measurement
Berteaux *et al.*	1980: Heat-flow density *in-situ* measurement

REFERENCES

Balling, N., Haenel, R., Ungemach, P., Vasseur, G., and Wheildon, J.: 1981, *Preliminary Guidelines for Heat Flow Density Determination*, Dir.-Gen. for Research, Science and Education Brussels, EUR 7360 EN, 32 p.

Benfield, A. E.: 1939, *The Terrestrial Heat Flow in Great Britain*, Proc. Roy. Soc., A, London, 1973, 428–450.

Bird, P., Toksöz, M.N., and Sleep, N.H.: 1975, 'Thermal and Mechanical Models of Continent – Continent Convergence Zones', *J. Geophys. Res.* **80**, 4405–4416.

Bullard, E. C.: 1965, 'Historical Introduction to Terrestrial Heat Flow', in Lee, W. H. K. (ed.), *Terrestrial Heat Flow* Geophys. Monogr. 8, Port City Press, Inc. Baltimore, Maryland, 1–5.

Carslaw, H. S. and Jaeger, J. C.: 1979, *Conduction of Heat in Solids*, 2nd edit., Clarendon, Oxford, 510 p.

Čermák, V. and Rybach, L. (ed.): 1979, *Terrestrial Heat Flow in Europe*, Springer Verlag, 328 p.

Chapman, D. S. and Pollak, H. N.: 1977, 'Regional Geotherms and Lithospheric Thickness', Geology, 5, Boulder Col., 265–268.

Crough, S.T. and Thomson, G. A.: 1977, 'Upper mantle origin of Sierra Nevada uplift', Geology, 5, 7, Boulder Col., 396–399.

Edwards, L. M., Chilingar, G. V., Rieke III, H. H., and Fertl, W. H. (ed.): 1982, *Handbook of Geothermal Energy*, Gulf Publ. Co., Book Div., Houston USA, 613 p.

Francheteau, J., Sclater, J.G. and Menhard, H.: 1970, 'Pattern of Relative Motion from Fracture Zone and Spreading Rate Data in the North-Eastern Pacific', *Nature* **226**, 746–748.

Haenel, R.: 1979, 'A Critical Review of Heat Flow Measurements in Sea and Lake Bottom Sediments', in Čermák, V. and Rybach, L. (eds.), *Terrestrial Heat Flow in Europe*, Springer Verlag, 49–73.

Haenel, R. and Gupta, L. (ed.): 1983, 'Results of the First Workshop on Standards in geothermics', *Zbl. Geol. Paläont.*, Teil I, 1/2, Stuttgart, 184 p.

ISO Standards: 1982, *Units of Measurements*; ISO *Standards Handbook 2*, ISO Central Secretariat, CH–1211 Genève 20, Switzerland, 258 p.

Kappelmeyer, O. and Haenel, R.: 1974, *Geothermics with Special Reference to Application*, Geoexpl. Monogr. **1**, No. 4, Borntraeger, Berlin, 238 p.

Kestin, J. (ed. in chief), Di Pippo, R., Khalifa, H.E., and Ryley, D. J. (ed.): 1980, '*Sourcebook on the Production of Electricity from Geothermal Energy*', US Dept. of Energy, Washington, D.C. 20402, Contract No. EY-76-S-4051.A002, 997 p.

Lachenbruch, A. H.: 1978, 'Heat Flow in the Basin and Range Province and Thermal Effects of Tectonic Extension', *Pure and Appl. Geophys.* **117**, 34–50.

Lee, W. H. K. (ed.): 1965, *Terrestrial Heat Flow*, Geophys. Monogr. 8, Amer. Geophys. Un., Washington D.C., 276 p.

Lee, W. H. K. and Uyeda S.: 1965, 'Review of Heat Flow Data', in Lee, W. H. K. (ed.) *Terrestrial Heat Flow*, Geophys. Monogr. 8, Amer. Geophys. Un., Washington D.C., 27–190.

McKenzie, D.P.: 1967, 'Some Remarks on Heat Flow and Gravity Anomalies', *J. Geophys. Res.* **72**, 6261–6273.

McKenzie, D.P.: 1978, 'Some Remarks on the Development of Sedimentary Basins', *Earth and Planet, Sci. Letters* **40**, 25–32.

Minear, J. W. and Toksöz, N. M.: 1970, 'Thermal Regime of Downgoing Slab', *Tectonophysics* **10**, 367–390.

Parker, R.L. and Oldenburg, D.W.: 1973, 'Thermal Model of Ocean Ridges', *Nature Phys. Sci.* **242**, 137–139.

Polyak, B.G. and Smirnov, Ya.B.: 1968, *Relationship between Terrestrial Heat Flow and the Tectonics of Continents*, Geotectonics, Acad. Sci. USSR, 4, Washington 1969, 205–213.

Roy, R. F., Blackwell, D. D., and Birch, F.: 1968, 'Heat Generation of Plutonic Rocks and Continental Heat Flow Provinces', *Earth and Planet. Sci. Letters* **5**, 1–12.

Rybach, L. and Muffler, L. J. P. (ed.): 1981, *Geothermal Systems – Principles and Case Histories*, Wiley & Sons, Ltd., 359 p.

Sclater, J.G., Jonas E.W., and Miller S.P.: 1970, 'The relationship of Heat Flow, Bottom Topography and Basement Relief', *Tectonophysics* **10**, 1–3.

Sleep, N. H.: 1971, 'Thermal Effect of the Formation of Atlantic Continental Margins by Continental Break Up', *Geophys. J. Roy. Astron. Soc.* **24**, 325–350.

Stegena, L. and Horvàth, F.: 1975, *Ensimatic and Ensialic Interarc Basins*, J. G. R. 80.2., 281–283.

U.N. Symposium: 1975, *Proceedings, 2. U.N. Symposium on the Development and Use of Geothermal Resources*, San Francisco, California, USA, 20–29 May 1975, Washington, Vol. 1, 2, and 3, 2466 p.

Table References (The year of edition must not be the same as given in the Table):

Agricola, G.: 1530, Bermanus, Froben, Basel.

Amontons: 1702, in Mém. de l'Académie Royale, Paris.

Aubuisson, d' J.: 1801, *Journal des Mines* **11**, Paris, p. 517–528.

Bacon, F.: 1620, 'Novum Organum Scientiarum', in Devey, J. (ed.) *The Physical and Metaphysical works of Lord Bacon*, London 1910.

Beck, A. E., Jaeger, J. C., and Newstead, G.: 1956, 'The Measurement of the Thermal Conductivities of Rocks by Observations in Boreholes', *Austr. J. Phys.* **9**, 286–296.

Benfield, A.E.: 1939, *The Terrestrial Heat Flow in Great Britain*, Proc. Roy. Soc., A, London, 1973, p. 428–450.

Bertaux, M. G., Bienfait, G., Jolivet, J., and Vasseur, G.: 1980, 'Direct Heat Flow Measurement Probe Project', in Strub, A. S. and Ungemach, P. *Advances in European Geothermal Research*, D. Reidel Publ. Co., Dordrecht. Holland, 545–555.

Boerhaave, H.: 1732, Elementa chemiae. Lugduni Batavorum (*Elements of Chemistry*, Trans. Timethy Dallowe, M.D., Vol. 1, London 1935).

Boltzmann, L.: 1884, *Wied. Ann.* **22**.

Boyle, R.: 1665, *New Experiments and Observations Toaching Cold or an Experimental History of Cold*, London.

Boyle, R.: 1671, 'Of the Temperature of the Subterraneal Regions as to Heat and Cold', in *Tracts Written by the Honourable Robert Boyle etc.* (usually called Cosmical Qualities), Davis, Oxford.

Burckhardt, F.: 1867, *Die Erfindung des Thermometers*, Reinhardt, Basel (1902).

Celsius: 1742, in *Vetensk. Akadem. Handl.*, Aar.

Chun, K.: 1899, *Berichte des Leiters der Expedition Prof. Dr. Chun an das Reichsamt des Innern*, VII. Int. Geographen-Kongress; die Deutsche Tiefsee-Expedition auf dem Schiff Valdivia, Jena.

Clausius, R.: 1867, *Der zweite Hauptsatz der Wärmelehre*, F. Vieweg und Sohn, Braunschweig, 1–17.

Crawford: 1778, *Experiments and Observations on Animal Heat*, London.

ENEL: 1976, *Geothermoelectric Power Plants of Larderello and Monte Amiata*, Series Corandi Impianti, 3, Public Relation and Press Office, Milan, Italy.

Kirchhoff, G.: 1860, *Poggendorf Annalen der Physik*, 109, Leipzig.

Joly, J.: 1909, *Radioactivity and Geology*, Archibald Constable & Co, London.

Mayer, R. J.: 1867, *Die Mechanik der Wärme*. Hg. v. J. J. Weyrauch.

Meyer, K.: 1913, *Die Entwicklung des Temperaturbegriffs im Laufe der Zeiten*, Wissenschaft, 48, F. Vieweg u. Sohn, Braunschweig, 160 p.

Nernst, W.: 1918, *Die theoretischen und experimentellen Grundlagen des neuen Wärmesatzes*, Knapp, Halle, VII, 218 p.

Parson, Ch., 1904: *The Hellfire Exploration Project*, see: Armstead, H., 1983, *Geothermal Energy*, 2nd ed., E. & F. N. Spon, London, New York, 404 p.

Planck, M.: 1900, *Über eine Verbesserung der Wienschen Spektralgleichung*, Verh. d. Deutsch. Phys. Ges., 2, Braunschweig, 202–204.

Plinus, E.: 1840, *Naturgeschichte*, Bd. 1, Buch 2, in deutsch von H. Külb, Röm. Prosaiker, Metzlersche Buchhandlung, Stuttgart.

Pettersson, H.: 1949, 'Exploring the Bed of the Ocean', *Nature* **164**, 468–470.

Réaumur, R.: 1730, in Hist. et Mém. de l'Academie des Sciences, Paris.

Reich: 1830/32, in Stapff, F. M., 1893: *Ein paar Worte über Bodentemperatur und artesische Strömung*, Z. prakt. Geologie, Springer Verlag, 381–385.

Rue, A. de la and Marcet, F.: 1834: *Quelques observations de physique terrestre*, Mém. Soc. Phys., Genève.

Sanctorius: 1626, 'De Temperamentis' in *Die Schwankungen und Pendelungen des Klimas Europa seit dem Beginn der regelmäßigen Instrumentenbeobachtungen (1670)*, F. Vieweg u. Sohn, Braunschweig 1967.

Stefan, J.: 1875, *Sitzungsberichte der Kaiserlichen Akademie der Wissenschaften*, math.-nat. Classe, Abt.II, Bd.65, Wien.

Stapff, F. M.; 1883, 'Some Results of Observations on Underground Temperature During the Constructions of the St. Gotthard Tunnel', *Proc. North Engl. Inst. Min. Mech. Eng.* **32**, 14–34.

Unnsteinsson, G. J.: 1978, *The Development of Landscape Gardening and Greenhouse Cultivation in Iceland, their Present Position and Future Outlook, and the Utilization of the Hot Springs and Steam for the Purpose of Heating*, State Horticultural College, Reykjum, Ölfusi, Iceland, 1–10.

Wells, M.: 1971, 'Early Development of Western Geothermal Resources', *Journal of the West*, **X**, Los Angeles, 53–71.

2. FUNDAMENTALS OF GEOTHERMICS

R. HAENEL
Department of Geophysics, Geological Survey of Lower Saxony, Hannover, FRG

L. RYBACH
Institute of Geophysics, ETH Zürich, Switzerland

and

L. STEGENA
Department of Environmental Physics, Roland Eötvös University, Budapest, Hungary

2.1. DEFINITIONS, STANDARDS, CONVERSIONS

The expression 'geothermics of the Earth' is understood to be restricted to the solid Earth and is usually shortened to geothermics. Hence, the field of geothermics starts as soon as the solid Earth has been entered. Geothermal phenomena above the Earth's solid surface (geothermics of the hydrosphere and atmosphere) are considered only if they affect the subsurface conditions in terms of boundary conditions, such as palaeoclimatic temperature changes, annual air temperature variations, and incoming solar radiation.

Table 2.1 gives the activities covered by geothermics, and thus its scope. Sometimes, geothermics is subdivided into applied geothermics and pure geothermics. An attempt is made to consider this subdivision in Table 2.1, but one should

TABLE 2.1
The scope and tasks of geothermics.

1. *Pure geothermics*

1.1. Geothermal field of the Earth
    Temperature and heat-flow density determinations, their representation and interpretation
    Determination of thermal properties of rocks, including temperature and pressure dependence
    Downward continuation of geothermal data (crust-mantle geothermics)
    Heat sources, heat sinks, and their distribution

1.2. Thermal history of the Earth
    Thermal regime of the Earth and planets during the last 4.5×10^9 a
    Geodynamic studies involving plate tectonics, see floor spreading, mountain forming and basin evolution

1.3. Development of equipment
    Instruments for measuring in boreholes on land
    Instruments for deep-sea measurements
    Instruments for determination of thermal properties

1.4. Development of methods
    Geothermal modelling
    Correction procedures for elimination of the effect of topography, palaeoclimate, uplift, denudation, water movements, etc.
    Standardisation

R. Haenel, L. Rybach, and L. Stegena (eds.)
Handbook of Terrestrial Heat-Flow Density Determination, 9–57.

Table 2.1 (Continued)

2. *Applied geothermics*

2.1. Geothermal energy
- Exploration for geothermal resources and reserves
- Development, charge and recharge of geothermal reservoirs
- Development of methods for extracting heat from hot dry rock
- Study of groundwater heat balance for utilisation of heat by means of heat pumps

2.2. Hydrogeology
- Exploration of thermal water, drinking and industrial water
- Subsurface hydraulic studies
- Soil studies: evaporation, water tension, permafrost

2.3. Volcanology, tectonics
- Delimitation of hot intrusive bodies
- Monitoring of volcanic areas
- Location of faults, horst structures, salt domes, diapirs

2.4. Oil and gas prospection
- Hydrocarbon genesis studies
- Palaeogeothermics
- Cementation
- Gas release

2.5. Prospecting for mineral deposits
- Salt diapirs, anticlines, synclines
- Sulphidic ores
- Radioactive minerals
- Hot brines in deep sea

2.6. Mining
- Freezing shaft problems
- Temperature distribution in mines
- Cooling of galleries for working, prediction of temperatures

2.6. Environmental protection
- Monitoring of drinking, mineral and thermal water districts
- Surveying of waste (hot) water from power plants and industry in boreholes, rivers, lakes, etc.

keep in mind that a separation into applied and pure geothermics is more or less artificial, and most of the results may be used for both.

Many parameters are used in geothermics, and these parameters, by convention, are associated with certain symbols and units. The *S*ystème *I*nternational d'Unités (SI) is fast becoming the international technical language. This system is the result of international agreements reached at the General Conferences of Weights and Measures and is strongly recommended for utilization. In Table 2.2, all important quantities, their symbols, definitions and internationally used units are listed. In compliance with the principles of SI, only these should be used in geothermics, too, in the future.

It is recommended that the units and symbolism be as follows: K or °C and T for temperature, Km^{-1} (or $\mathrm{mK\ m}^{-1}$) and Γ for geothermal gradient, $\mathrm{W\ m}^{-1}\ \mathrm{K}^{-1}$ and λ for the thermal conductivity, $\mathrm{mW\ m}^{-2}$ and q or HFD for heat-flow density, and $\mathrm{W\ m}^{-3}$ and H for heat production rate.

TABLE 2.2

Quantities and units of heat (ISO Standards, 1982). The table has been supplemented with regard to geothermics.

Quantity	Symbol	Definition	Remarks	Name of unit	International unit
A. Commonly used in geothermics					
Thermodynamic temperature	T, θ	–	The thermodynamic temperature is defined according to the principles of thermodynamics.	kelvin[a]	K
Celsius temperature	t, θ[b]	$t = T - T_0$, where T_0 is fixed by convention to be $T_0 = 273.15$ K.	The thermodynamic temperature T_0 is by definition 0.01 K below the thermodynamic temperature of the triple point of water.	degree Celsius[c]	°C
Heat, quantity of heat	Q	–	The heat transferred in an isothermal phase transformation, formerly called 'latent heat' with symbol L, is better expressed as the change of the appropriate thermodynamic functions, e.g., $T \cdot \Delta S$ where ΔS is the change of entropy, or as ΔH.	joule	J
Heat flow rate, heat flow	φ	Rate of heat flow across a surface.	–	watt	W
Density of heat flow rate, heat-flow density	q	Heat flow rate divided by area.	–	watt per square metre	W m^{-2}
Thermal conductivity	$\lambda(k)$	Heat flow density divided by temperature gradient.	–	watt per metre kelvin	W m^{-1} K^{-1}
Thermal diffusivity	a, (α, ϰ)	$a = \lambda/(\rho \cdot c)$	–	square metre per second	m^2 s^{-1}
Heat capacity	C	When the temperature of a system is increased by dT as a result of the addition of a small quantity of heat dQ, the quantity dQ/dT is the heat capacity.	This quantity is not completely defined unless the type of change is specified.	joule per kelvin	J K^{-1}

Table 2.2 (Continued)

Quantity	Symbol	Definition	Remarks	Name of unit	International unit
Specific heat capacity	c	Heat capacity divided by mass.	–	joule per kilogramme kelvin	$J\ kg^{-1}\ K^{-1}$
Specific heat capacity at constant pressure	c_p	–	–	"	$J\ kg^{-1}\ K^{-1}$
Specific heat capacity at constant volume	c_v	–	–	"	$J\ kg^{-1}\ K^{-1}$
Specific heat capacity at saturation	c_{sat}	–	–	"	$J\ kg^{-1}\ K^{-1}$
Geothermal gradient	Γ	–	–	degree Celsius per metre	$°C\ m^{-1}$
Heat generation, heat production rate	H	Heat generated in 1 m^3 material during 1 second	–	watt per cubic metre	$W\ m^{-3}$
Density	ρ	Mass per unit volume.	–	kilogramme per cubic metre	$kg\ m^{-3}$
B. Rarely used in geothermics					
Linear expansion coefficient	α_l	–	The quantities α_l, γ, α_p, β, and $\varkappa$ are not completely defined unless the type of change is specified.	reciprocal kelvin, kelvin to the power minus one	K^{-1}
Cubic expansion coefficient	α_v, γ	$\alpha_v = \frac{1}{V}\frac{dV}{dT}$	–	"	K^{-1}
Relative pressure	α_p	$\alpha_p = \frac{1}{p}\frac{dp}{dT}$	–	"	K^{-1}
Pressure coefficient	β	$\beta = \frac{dp}{dT}$	–	pascal per kelvin	$Pa\ K^{-1}$
Compressibility	$\varkappa$	$\varkappa = -\frac{1}{V}\frac{dV}{dp}$	–	reciprocal pascal	Pa^{-1}

Table 2.2 (Continued)

Quantity	Symbol	Definition	Remarks	Name of unit	International unit
Coefficient of heat transfer	h, k, K, α	Heat-flow density divided by temperature difference.	The symbols h and α are used for surface coefficient of heat transfer; k and K for overall coefficient of heat transfer.	watt per square metre kelvin	$W\ m^{-2}\ K^{-1}$
Thermal insulance, coefficient of thermal insulation	M	Temperature difference divided by heat-flow density.	–	square metre kelvin per watt	$m^2\ K\ W^{-1}$
Thermal resistance	R	Temperature difference divided by heat flow rate.	The quantity $G = 1/R$ is called thermal conductance.	kelvin per watt	$K\ W^{-1}$
Ratio of the specific heat capacities	γ	$\gamma = c_r/c_v$	These quantities are dimensionless.	–	–
Isentropic exponent	$\varkappa$	$\varkappa = -\frac{V}{p}\left(\frac{\partial p}{\partial V}\right)_S$	For an ideal gas, γ is equal to $\varkappa$; dimensionless.	–	–
Entropy	S	When a small quantity of heat dQ is received by a system the thermodynamic temperature of which is T, the entropy of the system is increased by dQ/dT provided that no irreversible change takes place in the system.	–	joule per kelvin	$J\ K^{-1}$
Specific entropy	s	Entropy S divided by mass.	–	joule per kilogram kelvin	$J\ kg^{-1}\ K^{-1}$
Internal energy	$U, (E)$	–	–	joule	J
Enthalpy	$H, (I)$	$H = U + pV$	–	"	J
Helmholtz free energy, Helmholtz function	A, F	$A = U - TS$	–	"	J
Gibbs free energy, Gibbs function	G	$G = U + pV - TS$	$G = H - TS$	"	J
Specific internal energy	$u, (e)$	Internal energy divided by mass.	–	joule per kilogram	$J\ kg^{-1}$

Table 2.2 (Continued)

Quantity	Symbol	Definition	Remarks	Name of unit	International unit
Specific enthalpy	h, (i)	Enthalpy divided by mass.	–	joule per kilogram	$J\ kg^{-1}$
Specific Helmholtz energy	a, f	Helmholtz free energy divided by mass.	–	″	$J\ kg^{-1}$
Specific Gibbs free energy	g	Gibbs free energy G divided by mass.	–	″	$J\ kg^{-1}$
Massieu function	J	$J = -A/T$	–	″	$J\ K^{-1}$
Planck function	Y	$Y = -G/T$	–	″	$J\ K^{-1}$

[a] The kelvin, unit of thermodynamic temperature, is the fraction 1/273.16 of the thermodynamic temperature of the triple point of water.
[b] In geothermics also the symbol T occurs, because often both are used, thermodynamic temperature and Celsius temperature.
[c] Degree Celsius is a special name for the unit kelvin for use in stating values of Celsius temperature.

To avoid confusion with earlier nomenclature it is suggested that the use of HFU (heat flow units), HGU (heat generation units) and HPU (heat production rate units) be discarded. Where some confusion is likely to arise the older CGS units should be placed in parentheses.

To facilitate conversion from other systems to SI, Tables 2.3 to 2.6 are provided; the Tables give the various factors required to transfer from one system of units to another for temperature, specific heat capacity, thermal conductivity, and heat-flow density respectively.

It would be useful to have an agreed terminology for use by those involved in geothermal energy. Figure 2.1 illustrates a proposal by McKelvey taken from Muffler and Cataldi (1978); Table 2.7 defines some of the terms given in Figure 2.1. Geothermal reservoirs are often classified as being of either low enthalpy or high enthalpy, the latter term generally being applied to reservoirs that contain water or steam suitable for the generation of electric power, whereas low enthalpy reservoirs are usually those where the energy is principally of interest for heating purposes. However, neither term is really clearly defined since a proper classifica-

TABLE 2.3
Temperature scales, temperature conversions.

Unit		Kelvin (1824–1907) (K)	Celsius (1701–1744) (°C)	Fahrenheit[a] (1686–1736) (°F)	Reamur[a] (1683–1757) (°R)	Rankine[a] (1820–1872) (°Rank)
Absolute zero		0	−273.15	−459.67	−218.5	0
Water freezing point		273.15	0	32	0	491.67
Water boiling point		373.15	100	212	80	671.67
Kelvin	x K =	–	$x - 273.15$	$\frac{9}{5}\cdot x - 459.67$	$\frac{4}{5}\cdot x - 218.5$	$\frac{9}{5}\cdot x$
Celsius	x °C =	$x + 273.15$	–	$\frac{9}{5}\cdot x + 32$	$\frac{4}{5}\cdot x$	$\frac{9}{5}\cdot(x + 273.15)$
Fahrenheit	x °F =	$\frac{5}{9}\cdot(x + 459.67)$	$\frac{5}{9}\cdot(x - 32)$	–	$\frac{4}{9}\cdot(x - 32)$	$x + 459.67$
Reamur	x °R =	$\frac{5}{4}\cdot(x + 218.5)$	$\frac{5}{4}\cdot x$	$\frac{9}{4}\cdot x + 32$	–	$\frac{9}{4}\cdot(x + 218.5)$
Rankine	x °Rank =	$\frac{5}{9}\cdot x$	$\frac{5}{9}\cdot x - 273.15$	$x - 459.67$	$\frac{4}{9}\cdot x - 218.5$	–

[a] Unit is to be substituted by kelvin or degree Celsius.

TABLE 2.4
Specific heat capacity conversions.

	cal g^{-1} K^{-1}	J g^{-1} K^{-1}	W h g^{-1} K^{-1}	Btu g^{-1} K^{-1}	J kg^{-1} K^{-1}
1 cal g^{-1} K^{-1}	1	4.186	1.163×10^{-3}	3.967×10^{-3}	4.186×10^{3}
1 J g^{-1} K^{-1}	0.239	1	2.778×10^{-7}	0.948×10^{-3}	1×10^{3}
1 W h g^{-1} K^{-1}	859.8	3600	1	3.412	3600×10^{3}
1 Btu g^{-1} K^{-1}	252	1055	0.293	1	1055×10^{3}
1 J kg^{-1} K^{-1}	0.239×10^{-3}	1×10^{-3}	2.778×10^{-7}	0.948×10^{-6}	1

TABLE 2.5
Thermal conductivity conversions.

	kcal $m^{-1} h^{-1} K^{-1}$	cal $cm^{-1} s^{-1} K^{-1}$	W $m^{-1} K^{-1}$	Btu $ft^{-1} h^{-1}$ °F^{-1}	Btu $in^{-1} h^{-1}$ °F^{-1}	W $cm^{-1} K^{-1}$
1 kcal $m^{-1} h^{-1} K^{-1}$	1	2.778×10^{-3}	0.01163	0.6719	0.05599	1.163
1 cal $cm^{-1} s^{-1} K^{-1}$	360	1	4.186	241.9	20.16	418.6
1 W $m^{-1} K^{-1}$	86.0	0.239	1	57.79	4.815	100
1 Btu $ft^{-1} h^{-1}$ °F^{-1}	1.488	4.134×10^{-3}	1.731×10^{-2}	1	0.08333	1.731
1 Btu $in^{-1} h^{-1}$ °F^{-1}	17.858	4.964×10^{-2}	0.2077	12	1	
1 W $cm^{-1} K^{-1}$	0.86	0.239	1×10^{-2}	0.5779	4.815×10^{-2}	1

TABLE 2.6
Heat-flow density conversions.

	kcal $m^{-2} h^{-1}$	kW cm^{-2}	cal $cm^{-2} s^{-1}$	Btu $in^{-2} s^{-1}$	Btu $ft^{-2} s^{-1}$	mW m^{-2}
1 kcal $m^{-2} h^{-1}$	1	11.63×10^{-8}	27.78×10^{-6}	71.17×10^{-8}	1.024×10^{-4}	1.163×10^{3}
1 kW cm^{-2}	8.6×10^{6}	1	238.9	6.12	880.6	1×10^{10}
1 cal $cm^{-2} s^{-1}$	3.6×10^{4}	4.188×10^{-3}	1	2.562×10^{-2}	3.687	4.186×10^{7}
1 Btu $in^{-2} s^{-1}$	1.405×10^{6}	16.34×10^{-2}	39.05	1	144	16.34×10^{8}
1 Btu $ft^{-2} s^{-1}$	9.765×10^{3}	1.135×10^{-3}	0.2713	6.944×10^{-3}	1	1.136×10^{7}
1 mW m^{-2}	8.6×10^{-4}	1×10^{-10}	0.239×10^{-7}	6.12×10^{-10}	0.8806×10^{-7}	1

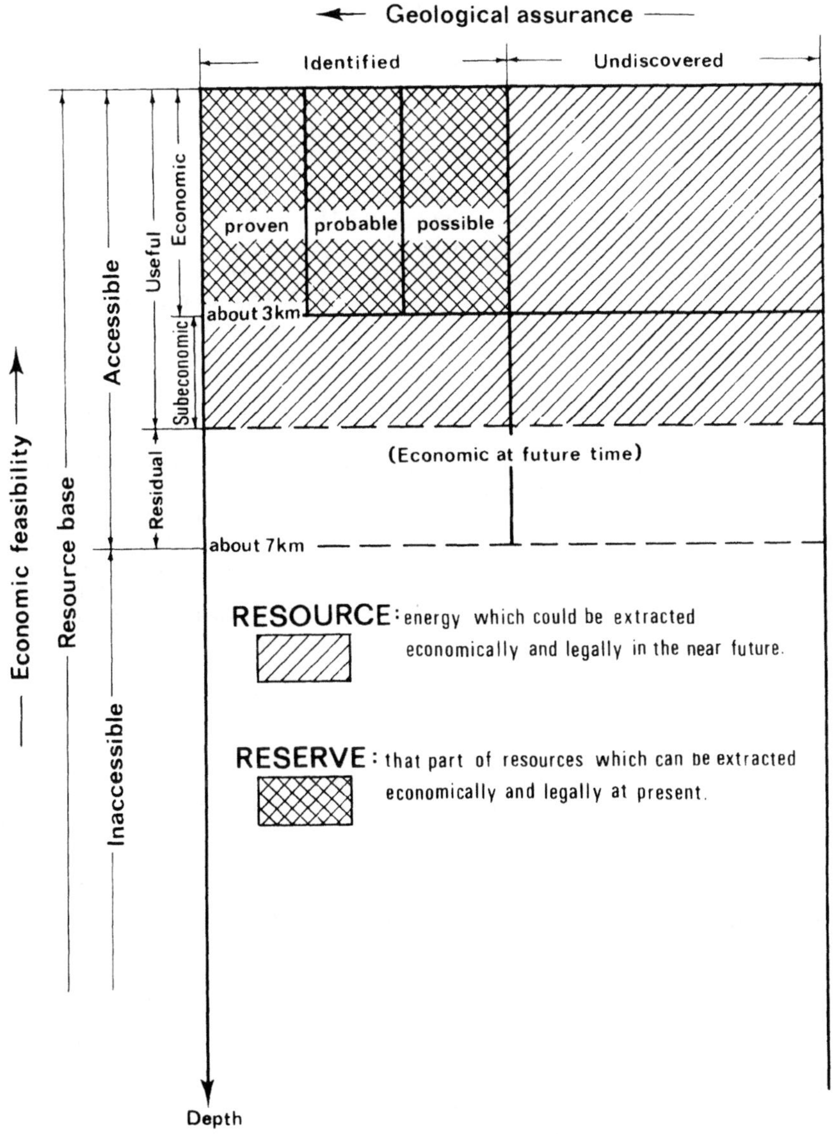

Fig. 2.1. McKelvey diagramme (Muffler and Cataldi, 1978), revised version for geothermal energy.

tion would also depend on reservoir pressure, rate of production, chemical content etc.

We adopt the generally accepted boundary between low enthalpy and high enthalpy reservoirs as beeing 150 °C, that ist:

$$T\,(\text{low enthalpy}) < 150\,^{\circ}\text{C} < T\,(\text{high enthalpy}).$$

TABLE 2.7
Definitions of geothermal energy (Muffler and Cataldi, 1978), see also Figure 2.1; T_b = transmissibility.

Name	Definition	Attributes, corollaries, and calculation
Resource base	All of the heat in the Earth beneath a specified area, measured from local mean annual temperature.	– Stored thermal energy at an instant time – Neglects transfer of heat from mantle – Takes no regard of whether or not it would ever be technically or economically feasible to recover the heat
Inaccessible resource base	All of the thermal energy stored between the base of the crust and a specified depth in the crust, beneath a specified area and measured from local mean annual temperature.	– Stored thermal energy at an instant in time – Neglects transfer of heat from mantle – Depth chosen for the upper limit is a matter of convenience, but must be specified in each case – Implies that thermal energy beneath the specified depth is unlikely to be tapped by production drilling at a reasonable time in the future.
Accessible resource base	All of the thermal energy stored between the Earth's surface and a specified depth in the crust, beneath a specified area and measured from local mean annual temperature, maximum drillable depth at present time is about 7 km.	– Stored thermal energy at an instant in time – Neglects transfer of heat from deeper levels – Depth chosen for the lower limit is a matter of convenience, but must be specified in each case
Residual accessible resource base	That part of the accessible resource base unlikely to be extracted economically and legally at some specified time in the future.	– Criterion for subdivision of accessible resource base is a subjective aggregate of predicted technology and economics at some reasonable and specified future time
Useful accessible resource base = Resource	That part of the accessible resource base that could be extracted economically and legally at some specified time in the future; $t \leqq 100$ a.	
Identified resource	The proportion of resource that is more likely to be available for exploitation now or in the future. It introduces a maximum recovery factor and it is a more conservative and realistic value.	

Table 2.7 (Continued)

Name	Definition	Attributes, corollaries, and calculation
Subeconomic resource	That part of the resource of a given area that cannot be extracted legally at a cost competitive with other commercial energy sources at the time of determination, but might be extracted economically and legally at some specified time in the future.	
Economic resource	That part of the resources of a given area that can be extracted legally at a cost competitive with other commercial energy sources at the time of determination.	
Undiscovered economical resource	That part of the economic resource in unexplored parts of regions known to contain geothermal resources, or in regions where geothermal resources are suspected but not yet discovered.	
Undiscovered economic resource, subeconomic	That part of the economic resource which can be extracted e.g. by the hot dry rock technology.	
Identified economic resource = Reserve	That part of the economic resource known and characterised by drilling or by geochemical, geophysical and geological evidence; $z \leqq 3$ km.	
Reserve, proven	That part of the identified economic resource supported by geological drilling evidence (e.g. direct measurements) which can produce at well head at competitive cost at the time of determination; $T_h \geqq 5$ Dm.	
Reserve, probable	That part of the identified economic resource based on geological, geophysical and/or geochemical (but not local drilling) evidence that could be produced at well head at competitive cost at the time of determination.	
Reserve, possible	That part of the identified economic resource based only on geological evidence which can be produced at well head at competitive cost at the time of determination.	

Turning to be standard of registration, documentation and publication of heat-flow density data, a very suitable form is worked out by Jessop *et al.* (1975). Using this form, it is possible to describe each measurement within the limit of one punch-card of 80 characters. The summary of format is given on Table 2.8.

TABLE 2.8
Summary of Format for documentation of heat-flow density data; Jessop *et al.* (1975).

Col.	Description	Units	Characters	Card Cols.
1	Data number		5 digits	1–5
2	Descriptive codes		7 letters	6–11, 75
3	Names		8 characters	13–20
4	Latitude	Deg., Min., Tenths	5 digits & 1 letter	21–26
5	Longitude	Deg., Min., Tenths	6 digits & 1 letter	27–33
6	Elevation of collar (land	m	Sign & 4 digits	34–38
	Elevation of sea floor	m		
	Elevation of lake floor	m		
7	Minimum depth used (land)	m	4 digits	39–42
	– blank – (sea)			
	Water depth (lake)			
8	Maximum depth used (land)	m	4 digits	43–46
	Penetration (sea)	m		
	Penetration (lake)	m		
9	No. of temperature meas.		3 digits	47–49
10	Temperature gradient	$mK\ m^{-1}$	3 digits	50–52
11	No. of conductivity meas.		3 digits	53–55
12	Average conductivity	$Wm^{-1}\ K^{-1}$	3 digits & decimal	56–59
13	No. of heat generation meas.		3 digits	61–63
14	Av. heat generation (land)	Wm^{-3}	3 digits & decimal	64–67
	Bottom water temp. (sea)	°C		
	Bottom water temp. (lake)	°C		
15	Heat-flow density	$mW\ m^{-2}$	3 digits & decimal	68–71
16	No. of sites		2 digits	73–74
17	Reference		3 digits	76–78
18	Year of publication		2 digits	79–80

2.2. EQUATIONS OF HEAT CONDUCTION

In accordance with the first law of thermodynamics, a temperature change $\triangle T$ taking place in a body, is accompained by a change of heat content $\triangle Q$, usually known as enthalpy. The increase of heat content must also be proportional to the mass m of the body considered:

$$\triangle Q \sim m \triangle T.$$

The equation is given by introducing a proportionality factor:

$$\triangle Q = cm \triangle T, \tag{2.1}$$

where c in J kg^{-1} K^{-1} is called specific heat capacity.

2.2.1. Fourier Equations

A fundamental hypothesis, which cannot be proved by theory, is Fourier's first equation of heat conduction. This equation follows from Fourier's experiment (Figure 2.2). The quantity of heat Q flowing through a planoparallel plate is

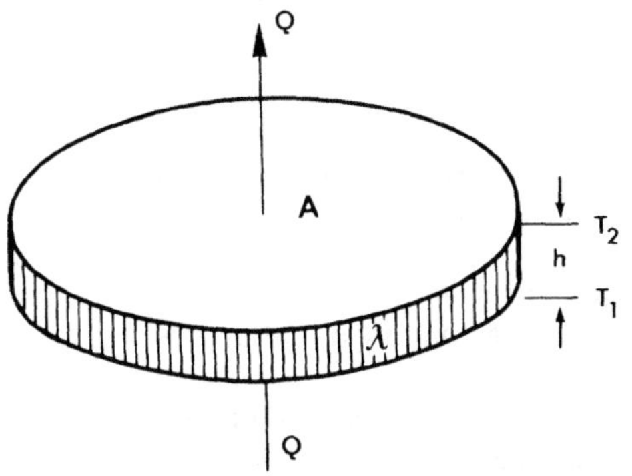

Fig. 2.2. Heat flow through a plate as demonstrated by Fourier.

proportional to the temperature difference $T_2 - T_1$ between the two boundary surfaces, to the area A, to the time t during which the heat flows, and is inversely proportional to the thickness h of the plate:

$$Q = -\lambda \frac{T_2 - T_1}{h} At = -\lambda \frac{\Delta T}{h} At.$$

For unit area and unit time, the heat-flow density q is given by:

$$q = -\lambda \frac{\mathrm{d}T}{\mathrm{d}h},$$

and in three dimensions:

$$\mathbf{q} = -\lambda \operatorname{grad} T = -\lambda \nabla T, \tag{2.2}$$

where q and ∇T are vectors:

$$\mathbf{q} = (q_x, q_y, q_z)$$

$$\nabla T = \left(\frac{\partial T}{\partial x}, \frac{\partial T}{\partial y}, \frac{\partial T}{\partial z}\right),$$

where ∇T is the temperature gradient, and λ, the thermal conductivity, – for anisotropic media – is a tensor:

$$\lambda = \begin{pmatrix} \lambda_{xx} & \lambda_{xy} & \lambda_{xz} \\ \lambda_{yx} & \lambda_{yy} & \lambda_{yz} \\ \lambda_{zx} & \lambda_{zy} & \lambda_{zz} \end{pmatrix}.$$

Because of the fact that heat generally flows perpendiculary to the Earth's surface geothermics uses mostly the one-dimensional equation:

$$q = \lambda_{zz} \frac{\mathrm{d}T}{\mathrm{d}z} = \lambda \Gamma. \tag{2.3}$$

However, it must be taken into consideration that:

- rocks, especially sedimentary rocks, are usually anisotropic, and one has to measure the thermal conductivity in the vertical direction
- the normally vertical heat-flow density is sometimes distorted by local heat sources and/or by the geometrical pattern of rocks with various thermal conductivity.

As practical units, geothermics preferentially uses 10^{-3} W m^{-2} (= 1 mW m^{-2}) for heat-flow density and sometimes 10^{-3} K m^{-1} = 1 mK m^{-1} for temperature gradient.

Another important equation is Fourier's second equation of heat conduction, which is a direct consequence of Fourier's first equation and the first law of thermodynamics.

Let us suppose that heat flows through the volume element $\mathrm{d}V = \mathrm{d}x\,\mathrm{d}x\,\mathrm{d}z$. Then, the difference of inflowing and outflowing heat has to be equal to the thermal energy remaining in $\mathrm{d}V$ (Figure 2.3). In the direction of the z-axis it

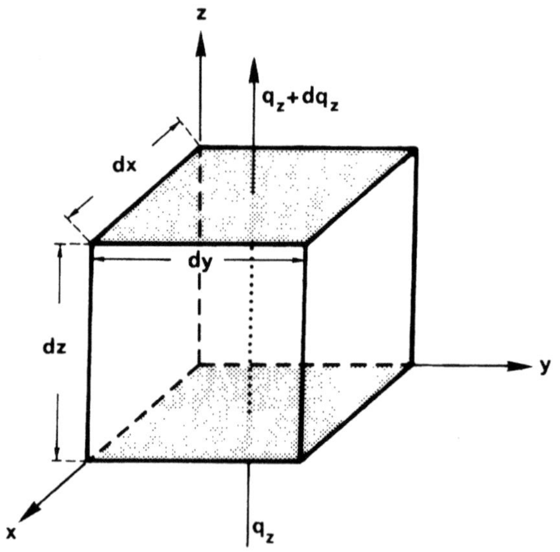

Fig. 2.3. Heat flowing through a volume element $\mathrm{d}V = \mathrm{d}x\ \mathrm{d}y\ \mathrm{d}z$.

amounts during unit time t to $\mathrm{d}q_z = \frac{\partial q_z}{\partial z}\mathrm{d}z$:

$$[(q_z) - (q_z + \mathrm{d}q_z)]\,\mathrm{d}x\,\mathrm{d}y = \left(q_z - q_z - \frac{\partial q_z}{\partial z}\mathrm{d}z\right)\ \mathrm{d}A = -\frac{\partial q_z}{\partial z}\mathrm{d}V$$

and similarly for all three components:

$$-\left(\frac{\partial q_x}{\partial x} + \frac{\partial q_y}{\partial y} + \frac{\partial q_z}{\partial z}\right)\mathrm{d}V.$$

The thermal energy stored in the volume element $\mathrm{d}V$ during unit time is equal to:

$$\rho c\,\mathrm{d}V\frac{\partial T}{\partial t},$$

where

c = specific heat capacity
ρ = density and
$\partial T/\partial t$ = temperature change during unit time.

By setting the two quantities equal we get Fourier's second equation of heat conduction:

$$-\left(\frac{\partial q_x}{\partial x} + \frac{\partial q_y}{\partial y} + \frac{\partial q_z}{\partial z}\right) = \rho c\frac{\partial T}{\partial t}, \tag{2.4a}$$

or

$$-\mathrm{div}\,(\mathbf{q}) = \rho c\frac{\partial T}{\partial t}. \tag{2.4b}$$

Substituting for $\mathbf{q}$ with the corresponding equations as given by (2.2):

$$\left(\frac{\partial\left(\lambda_x\frac{\partial T}{\partial x}\right)}{\partial x} + \frac{\partial\left(\lambda_y\frac{\partial T}{\partial y}\right)}{\partial y} + \frac{\partial\left(\lambda_z\frac{\partial T}{\partial z}\right)}{\partial z}\right) = \rho c\frac{\partial T}{\partial t}. \tag{2.4c}$$

In the case that the rock material is isotropic and homogeneous (i.e. $\lambda_x = \lambda_y = \lambda_z = \lambda$) one obtains:

$$\lambda\left(\frac{\partial^2 T}{\partial x^2} + \frac{\partial^2 T}{\partial y^2} + \frac{\partial^2 T}{\partial z^2}\right) = \rho c\frac{\partial T}{\partial t}, \tag{2.5a}$$

or

$$\frac{\partial T}{\partial t} = \frac{\lambda}{\rho c}\Delta T = a\,\Delta T = a\,\mathrm{div}\,(\mathrm{grad}\,T) = a\cdot\nabla^2 T, \tag{2.5b}$$

where

$$a = \lambda/\rho c \qquad = \text{thermal diffusivity}$$

$$\Delta = \nabla\cdot\nabla = \left(\frac{\partial^2}{\partial x^2} + \frac{\partial^2}{\partial y^2} + \frac{\partial^2}{\partial z^2}\right) = \text{LAPLACE operator}$$

$$\nabla = \left(\frac{\partial}{\partial x}, \frac{\partial}{\partial y}, \frac{\partial}{\partial z}\right) \qquad = \text{NABLA operator} = \text{DEL operator.}$$

2.2.2. Laplace and Poisson Equations

It is easy to extend Fourier's equation to the case in which a heat source (or heat sink) is present in dV. In this case, the warming of the material in dV is caused partly by heat flow and partly by the heat source:

$$\frac{\partial T}{\partial t} = a\,\Delta T + \frac{H}{\rho c}, \tag{2.6}$$

where H is the heat generated in the volume element of the medium during unit time(= heat production rate).

If no temperature change with time is considered (i.e. steady state, $\partial T/\partial t = 0$), Fourier's differential equation reduces to the Poisson equations:

$$\Delta T = -\frac{H}{\lambda}. \tag{2.7}$$

If heat production rate H is neglected, the equation reduces further to the well-known Laplace equation:

$$\Delta T = 0. \tag{2.8}$$

For Geotherms in layered media see p. 283.

2.2.3. Cylindrical and Spherical Coordinates

Sometimes, not only Cartesian, but also spherical and cylindrical coordinate systems are used in geothermics, too. Because NABLA is an invariant operator, any convenient system of coordinates may be chosen.

In cylindrical coordinates (Figure 2.4) Equation (2.6) becomes:

$$\frac{\partial^2 T}{\partial r^2} + \frac{1}{r}\frac{\partial T}{\partial r} + \frac{1}{r^2}\frac{\partial^2 T}{\partial \varphi^2} + \frac{\partial^2 T}{\partial z^2} = \frac{1}{a}\frac{\partial T}{\partial t} - \frac{H}{\lambda}, \tag{2.9}$$

and in the case of radial symmetry:

$$\frac{\partial^2 T}{\partial r^2} + \frac{1}{r}\frac{\partial T}{\partial r} + \frac{\partial^2 T}{\partial z^2} = \frac{1}{a}\frac{\partial T}{\partial t} - \frac{H}{\lambda}. \tag{2.10}$$

In spherical polar coordinates Equation (2.6) becomes:

$$\frac{\partial^2 T}{\partial r^2} + \frac{2}{r}\frac{\partial T}{\partial r} + \frac{\operatorname{ctg}\psi}{r^2}\frac{\partial T}{\partial \psi} + \frac{1}{r^2}\frac{\partial^2 T}{\partial \psi^2} + \frac{1}{r^2\sin^2\psi}\frac{\partial^2 T}{\partial \varphi^2} = \frac{1}{a}\frac{\partial T}{\partial t} - \frac{H}{\lambda}, \tag{2.11}$$

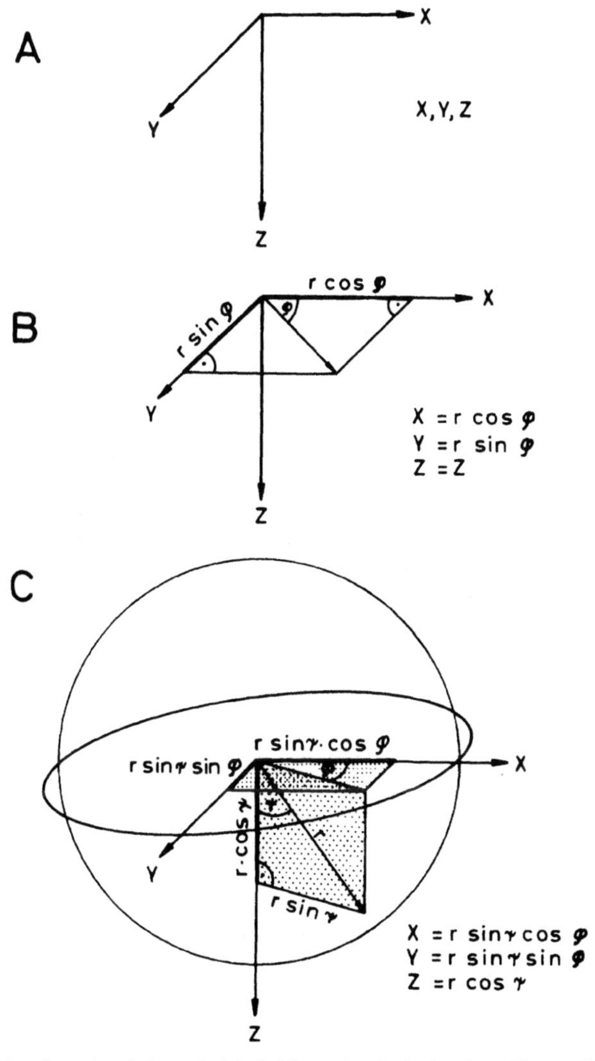

Fig. 2.4. The Cartesian (A), cylindrical (B), and spherical polar (C) coordinate systems.

and in the case of spherically symmetrical conditions:

$$\frac{\partial^2 T}{\partial r^2} + \frac{2}{r}\frac{\partial T}{\partial r} = \frac{1}{a}\frac{\partial T}{\partial t} - \frac{H}{\lambda}. \tag{2.12}$$

2.3. EQUATIONS OF HEAT CONVECTION

2.3.1. Introductory Remarks

The effect of water and gas movement can influence the heat-flow density tremendously. At present, only insufficient information is available for the purpose of the recognizing, separating and expressing this influence in terms of 'convective heat-flow density'. Therefore, more attention should be paid to this phenomenon in the future; see also bibliography of Clauser and Behrens (1987).

For greater depths (>5–10 km), the degassing effect (Polyak and Smirnov, 1966, 1968; Gold and Soter, 1980; O'Nions and Oxburgh, 1983) as well as the dehydration of the upper mantle and the lower crust can cause upward material transport and associated convective heat-flow contributions (Birch *et al.*, 1968; Lachenbruch and Marshall, 1968; Haenel and Zoth, 1982; Werner, 1982). At present, it is not well known which component of this heat transfer most effectively supports the purely conductive heat-flow density.

In the uppermost 5 to 10 km another effect occurs. Mountains, hills, tectonically affected areas, especially basins, are often areas of water migration which influences the geothermal field more or less strongly. As has been demonstrated by radiocarbon, tritium and other measurements, predominantly meteoric water is involved, whereas connate water, expelled by compaction of sediments, is of secondary importance. The descending water moves under the action of gravity. Consequently, water circulation is often controlled by the topography: elevated areas are the recharge areas with downward migration, and the areas of subsidence represent the discharge area.

A very generalised system for the subsurface water flow of younger and closed sedimentary basins is the following (Stegena, 1982); see also Figure 2.5.

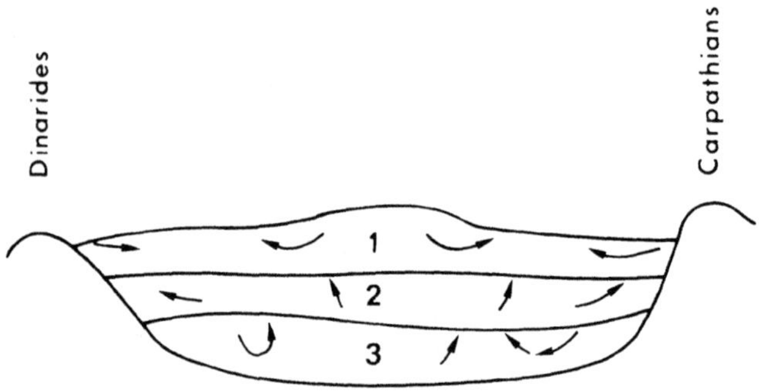

Fig. 2.5. A schematic representation of the subsurface hydraulic water provinces in the Pannonian basin. (1) zone of infiltration; (2) zone of expulsion; (3) zone of re-infiltration.

(1) The zone of infiltration which includes the uppermost layers of some hundred metres thickness: in this zone, there is a free communication between meteoric and subsurface water. As a consequence, fresh water and approximately hydrostatic pressures predominate in this zone. Slight hydraulic pressures are controlled by surface topography. Topographically elevated regions create areas of downward migration and vice versa.

(2) The zone of expulsion which extends from below the zone of infiltration the depth of the consolidated (compacted) sediments: in this zone the water migration is controlled by compaction (alteration by pressure) and consolidation (alteration in time) of sediments. The water of the layers (ancient connate sea water)

expelled from the sediments by compaction and consolidation moves upwards and sideways. Mineralised (connate) water and local or regional over-pressures characterise this zone.

(3) The zone of re-infiltration: at greater depths the process of compaction comes to an end, and the consolidated rocks may again allow a second infiltration where secondary permeability exists (karstic phenomena in carbonatic rocks, tectonic fractures etc.).

Water moving downward reduces and water moving upward increases the heat-flow density. Therefore, one generally has to distinquish between (Haenel, 1979):

(1) Heat-flow density values which give information about the heat-flow density from the deeper parts of the earth. These values are required for studies of the crustal temperature field as well as of the nature of deep-seated tectogenetic processes, and should be undisturbed by local factors such as water movement and others.

(2) Local heat-flow density values interpreted in the course of applied surveys in the search for, e.g., sources of natural steam, thermal water and/or in hydrogeological exploration. In this case, the distortion of the heat-flow density by the underground water movements provides important information.

2.3.2. Analytical Solutions

Subsurface convection can be of natural (geological) and of forced (man made) character. Natural convection is a consequence of hydraulic pressure differences. These pressure differences can be caused by niveau differences in the groundwater level, by the overburden or compaction of rocks, and by differences in the density of water, which may result from temperature differences.

Water more or less conserves its original (base) temperature during migration, because of its high specific heat capacity (relative to the surrounding rocks) and of the low thermal conductivity of rocks. Consequently, the temperature of moving water usually differs from that of the neighbouring rocks. The heat exchange across the boundaries between solids an fluids can be described by Newton's law of cooling:

$$q = \alpha(T_s - T_f), \tag{2.13}$$

where

q = heat-flow density, W m^{-2}
T_s = temperature at the boundary in the solid, K
T_f = temperature at the boundary in the fluid, K
α = coefficient of surface heat transfer which also includes the corresponding thermal radiative component, W m^{-2} K^{-1}.

The quantity q, expressed by Equation (2.13) may represent an additional heat flow, e.g., from the upper boundary of an aquifer.

Equations of Heat Transport

Let us assume water flow through porous rock. The heat transport by the water flow is then given by:

$$\operatorname{div}(\rho_w c_w \mathbf{v}_f T), \tag{2.14}$$

where

ρ_w = density of water, kg m^{-3}
c_w = specific heat capacity of water, J kg^{-1} K^{-1}
$\mathbf{v}_f$ = volume flow rate per unit area, Darcy velocity or filtration velocity, (m^3/s) · (m^{-2}) = m s^{-1}
T = temperature, K.

It is important to note that if the velocity denotes the (water) particle velocity, which is called also pore velocity, distance velocity or average velocity $\mathbf{v}_a$, the expression (2.14) is to be multiplied by effective porosity P; $\mathbf{v}_f = P\mathbf{v}_a$.

The expression conductive heat flow through the rock has already been derived, Equation (2.4). In the case of water filtration, this equation is to be modified by the term given by (2.14):

$$\operatorname{div}(\lambda_A \operatorname{grad} T) - \operatorname{div}(P\rho_w c_w \mathbf{v}_a T) = \rho_A c_A \frac{\partial T}{\partial t}, \tag{2.15a}$$

where

$\lambda_A = (1-P)\lambda_r + P\lambda_w$
$\rho_A c_A = (1-P)\rho_r c_r + P\rho_w c_w$
P = porosity
w refers to water or gas and r to rock.

If only vertical movement, either down- or upwards (parallel to the conductive heat flow) is considered with:

$$\operatorname{div}(P\rho_w c_w \mathbf{v}_a T) = P\rho_w c_w T \operatorname{div} \mathbf{v}_a + P\rho_w c_w \mathbf{v}_a \operatorname{grad} T,$$

and if div $\mathbf{v}_a = 0$, the following modified Fourier equation is obtained ($\mathbf{v}_a = v_z$):

$$\lambda_A \frac{\partial^2 T}{\partial z^2} - P\rho_w c_w v_z \frac{\partial T}{\partial z} = \rho_A c_A \frac{\partial T}{\partial t}. \tag{2.15b}$$

If are taken the heat generation in the rock into consideration, too, and istropie and homogeneous condition as well, the following relationship, similar to Equation (2.6), results:

$$\lambda_A \operatorname{div}(\operatorname{grad} T) + H - \operatorname{div}(P\rho_w c_w \mathbf{v}_a T) = \rho_A c_A \frac{\partial T}{\partial t}. \tag{2.16}$$

Equations of Water Transport

Darcy's law is of fundamental importance for describing the connection between filtration velocity, v, and hydraulic gradient:

$$\mathbf{v}_f = k_f \operatorname{grad} h, \tag{2.17}$$

where

$\mathbf{v}_f$ = filtration velocity;
k_f = hydraulic conductivity or k_f–value in m s^{-1}; $k_f = k\rho g/\mu$ with k = permeability in m^2, ρ = density in kg m^{-3}, g = gravitational acceleration in m s^{-2}, μ = dynamic viscosity in kg m^{-1} s^{-1};
grad h = hydraulic pressure gradient, expressed in water-level difference, as referred to a horizontal distance in m m^{-1} = dimensionless.

In accordance with Darcy's law, a variation of hydraulic pressure has to prevail in the hydraulic field to promote water movement, i.e. filtration.

Darcy's empirical law has a general validity for natural filtration processes, but not without limits. Figure 2.6 shows the limits of Darcy's law: At very high hydraulic pressure gradients and/or at very high values of k_f, the effect of turbulence can occur. Here, instead of Darcy's law (which is valid for laminar flow) the Helmholtz laws of turbulence are to be applied. At the other end of the curve, at very low hydraulic gradients (and/or at very low values of k_f), the velocity of filtrating water is less than that predicted by Darcy's law, and at a certain small hydraulic gradient, the filtration will cease ($v = 0$). Turbulence does not occur in the subsurface exept in special, mostly man-made cases: in boreholes, caves, etc. Microfiltration is a phenomenon which has not received much study; it is caused by molecular, Van der Waals and similar forces acting between rock matrix and pore fluid, and hinders the fluid movement. The importance of microfiltration, from the point of view of geothermics, is the following: In sedimentary basins,

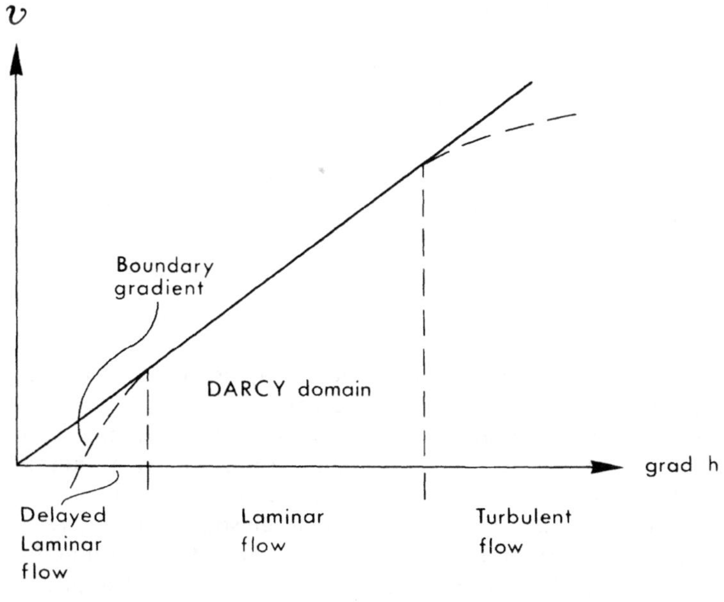

Fig. 2.6. Validity domain of Darcy's law.

surplus pressure (which exceeds the hydrostatic pressure) often occurs at greater depths. One can calculate the vertical water filtration in the upward direction with the knowledge of pressures and water permeabilities, on the basis of Darcy's law. The water movement resulting from this calculation and its influence on geothermics may, however, be overestimated, because of the fact that this geological water migration lies in the domain of microfiltration or even beneath the boundary value of the hydraulic gradient as shown in Figure 2.6.

Darcy's law is realised for many boundary and initial conditions, also in cylindrical and spherical coordinates (preferred for boreholes and point-like problems). These equations are presented in the textbook on subsurface hydraulics (Bear, 1979; Faust and Mercer, 1980; Langguth and Voigt, 1980) if are taken into consideration the conservation of mass and the capacity of the aquifer to store water, the Darcy equation yields the fundamental differential equation of groundwater movement (Jacob, 1940), in analogy with the second Fourier equation:

$$\mathrm{div}(\mathrm{grad}\ h) = \frac{S}{T_r}\frac{\partial h}{\partial t} \tag{2.18}$$

or, including water sources:

$$T_v\ \mathrm{div}(\mathrm{grad}\ h) + \bar{Q} = S\frac{\partial h}{\partial t}, \tag{2.19}$$

where

h = hydraulic head, which corresponds approximately to the unconfined water level of an aquifer, m

T_v = transmissivity, $m^2\ s^{-1}$

$T_v = \int k_f \mathrm{d}z$, and if k_f = const.: $T_v = k_f M$ with

M = thickness of aquifer in m

($T_b = kM$ = transmissibility, m^3 or Darcy metre)

$\bar{Q} = \dfrac{Q_v}{A}$ = source term (spring, injection borehole), $m\ s^{-1}$

Q_v = Extraction or injection in $m^3\ s^{-1}$, A = area in m^2

S = storativity, dimensionless

$S = S_S \cdot M$, S_S = fractional change in water volume in per cent, in the case that the water level is lowered by 1 m

t = time, s.

The three Equations (2.16), (2.17), and (2.19) describe a hydro-geothermal system, e.g., heat transport by water movement.

A simple analytical solution describing the vertical steady-state flow of water and heat through an isotropic, homogeneous, porous and fully saturated plano-parallel, horizontal layer is presented by Bredehoeft and Papadopulos (1965); see also Lubimova *et al.* (1965). The situation is illustrated in Figure 2.7A. The surface temperature at the top of the layer is T_0 and at the bottom T_L. Vertical water filtration occurs through the layer with the velocity v_z. The heat generation H is neglected; thus only the vertical, steady-state heat transport (one-dimensional model) is considered. Equation (2.15b) can be simplified to:

$$\frac{d^2T}{dz^2} - \frac{P\rho_w c_w}{\lambda_r} v_z \frac{dT}{dz} = 0. \tag{2.20}$$

This linear differential equation with the constant coefficient $\mu = (P\rho_w c_w v_z)/\lambda_r$ has a general solution given by:

$$T_z = C_1 e^{b_1 z} + C_2 e^{b_2 z}$$

with

$$b(b - \mu) = 0$$
$$b_1 = \mu \quad \text{and} \quad b_2 = 0,$$

and

$$T_z = C_1 e^{\mu z} + C_2.$$

The values C_1 and C_2 can be determined with the use of the boundary conditions:

$$T = T_0 \quad \text{at} \quad z = 0$$
$$T = T_L \quad \text{at} \quad z = L$$

with leads to:

$$\frac{T_z - T_0}{T_L - T_0} = \frac{e^{\beta z/L} - 1}{e^{\beta} - 1}$$

where

$$\beta = \mu L = \frac{P\rho_w c_w v_z L}{\lambda_r} \tag{2.21}$$

β, the Peclet number Pe (see Table 2.9), is positive or negative depending on whether v_z is directed downward or upward. β yields v_z:

$$v_z = \frac{\beta\lambda_r}{P\rho_w c_w L} \quad . \tag{2.22}$$

The parameter β is given in Figure 2.7B as type curves.

The integration of Equation (2.20) leads to the total heat-flow density q_t, which is the sum of conductive heat-flow density q_d and the convective heat-flow density q_v (Lubimova *et al.*, 1965; Mansure and Reiter 1979):

$$q_t(z) = q_d + q_v = -\lambda_r \frac{\Delta T}{\Delta z} - P\rho_w c_w v_z (T(z) - T_L)$$

$$q_t(z) = -\lambda_r \frac{T_L - T(z)}{L} - P\rho_w c_w v_z (T(z) - T_L), \tag{2.23}$$

and the maximum occurs at $z = 0$ for $T(z) = T_0$, with v_z directed upward:

$$q_{t,\max} = \left| \lambda_r \frac{T_L - T_0}{L} \right| + P\rho_w c_w v_z (T_L - T_0). \tag{2.24}$$

Figure 2.7 demonstrates the possible distortions of temperatures in a layer (or borehole), in the case of vertical water migration. The undisturbed temperature-versus-depth curve is located at $\beta = 0$ ($v_z = 0$). The figure permits the calculation of temperature distortions if β is known, the calculation of v by means of the measured $T(z)$ curve, or the calculation of v_z, by using Darcy's law (with the knowledge of the hydraulic gradient and water permeability) and comparison with v_z calculated from the temperature distortion.

However, areal (areally distributed) vertical diffusion seldom occurs because

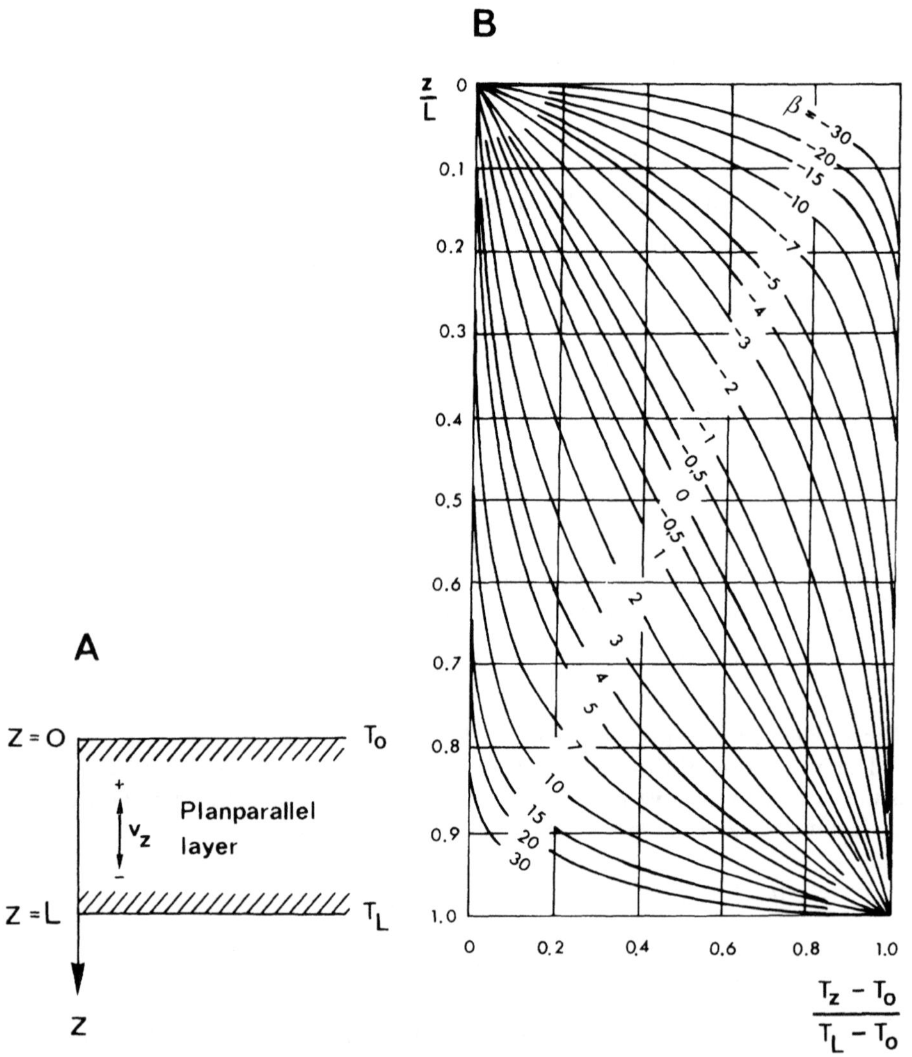

Fig. 2.7. Temperature distortion in a layer penetrated by vertical water filtration, Bredehoeft and Papadopulos (1965). (A) Schematic crust model; (B) Type curves of function β, which is identical with the Peclet number Pe.

the vertical hydraulic gradients are usually lesser than the boundary gradient as shown in Figure 2.6. Furthermore, there always exist fissures and fracture zones, which are preferred because of the low hydraulic resistance. On the other hand, if a large area is considered, the numerous fissures result in rock behaviour similar to that of intergranular rock; this is also valid for karstic rock and basement rocks.

The derived equations as well as Figure 2.7 are based on a constant thermal conductivity λ_r for the vertical section L considered. Under natural conditions this is not always the case, and the curved temperature log (or the β-values) can be explained by variation in thermal conductivity, too. Therefore, it is very important also to know, the variation of thermal conductivity with depth, and thus the heat-flow density variation with depth (see Section 2.3.3.).

For describing the natural convection caused by temperature differences, the application of the so-called dimensionless numbers is useful (Table 2.9). These numbers can be derived from the laws of similarity (Gröber *et al.*, 1955). The advantage of this approach is that the differential equation, which otherwise can be solved in only a few cases, may thus be solved by experiment. The procedure of similarity laws allows a generalisation of the experimental results by means of the rules of similarity.

TABLE 2.9

Dimensionless numbers applied in geothermics. The dimensionless numbers Pr, Re, and Gr originally related to momentum transport, and Fo, Pe, Ra, Nu, and St to heat transport.

Name	Number	Comments
Characteristic length, aspect ratio	$A = \frac{l}{z}$	Length and depth of model considered.
Nusselt number, $\lambda \simeq \lambda_{\text{fluid}}$ (= Biot number, $\lambda \simeq \lambda_{\text{matrix}}$)	$\text{Nu} = \frac{\alpha z}{\lambda} = \frac{qz}{\Delta T \lambda}$	Ratio of the total heat transmitted to that transmitted by conduction alone, and therefore an index of the magnitude of the convective heat transfer.
Peclet number (= Re Pr)	$\text{Pe} = \frac{\upsilon z}{a} = \frac{\upsilon z}{\frac{\lambda}{\rho c}} = \frac{\rho c \upsilon \Delta T}{\lambda \frac{\Delta T}{z}}$	Ratio of heat transferred convectively to heat transferred conductively in the bulk flow far from any boundaries.
Prandtl number	$\text{Pr} = \frac{\nu}{a}$	Characteristic of the ratio of momentum diffusivity to thermal diffusivity of a fluid.
Fourier number	$\text{Fo} = \frac{at}{z^2}$	Ratio of transported heat to stored thermal energy.
Stanton number (= Margoulis number)	$\text{St} = \frac{\alpha}{c\rho\upsilon}$	Heat transfer from solid into flowing fluid.
Ratio $2\frac{z}{r}\frac{\text{St}}{\text{Nu}}$	$\text{Sn} = \frac{2\lambda z}{c\rho\upsilon r^2}$	Related to cylindrical problems as borehole.

Table 2.9 (Continued)

Name	Number	Comments
Reynolds number	$\mathrm{Re} = \dfrac{\upsilon z}{\nu}$	Transition from laminar to turbulent fluid movements.
Grashof number as in velocity	$\mathrm{Gr} = \dfrac{\alpha_v g \Delta T z^3}{\upsilon^2}$	Characteristic of the ratio of buoyancy to viscosity.
Rayleigh number (= Gr Pr)	$\mathrm{Ra} = \dfrac{\alpha_v g \Gamma z^4}{a\nu}$	Transition from conduction to convection. Convection due to mass flow between parallel horizontal layers of thickness z.
	$\mathrm{Ra} = \dfrac{\alpha_v g z^4 q}{\nu a \lambda}$	As before, but expressed in terms of given heat-flow density q.
	$\mathrm{Ra} = \dfrac{g \alpha_v H z^5}{a^2 \nu \rho c}$	Convection caused by internal heating.
	$\mathrm{Ra} = \dfrac{\rho_w c_w \alpha_v k g \Delta T z}{\nu \lambda_r}$	Convection of fluid in porous medium if Ra > 40.
	$\mathrm{Ra} = \dfrac{l^2 g \alpha_v \Delta T}{a\nu}$	Convection with an infinite medium starts above Ra ≈ 1700, and above Ra ≈ 2200 for flow in pipes.
Rayleigh number, extended	$\mathrm{Ra}' = \dfrac{l^2 g \alpha_v \Delta T}{a^2}$	

where
l = length
z = height, depth
α = coefficient of surface heat transfer
λ = thermal conductivity
v = velocity
a = thermal diffusivity
ν = kinematic viscosity
t = time
c = specific heat capacity
ρ = density
r = radius
α_v = coefficient of volumetric thermal expansion
ΔT = temperature difference
Γ = temperature gradient
q = heat-flow density
k = permeability
g = gravitational acceleration
H = heat production
r, w = r refers to rock and w to water

Two particularly important dimensionless numbers are the Rayleigh number and the Nusselt number: By means of these numbers it is possible to examine whether or not water convection occurs through porous rocks in the subsurface, thus promoting heat transport.

The two dimensionless numbers are given by:

$$\mathrm{Ra} = \frac{\rho_w c_w \alpha_v k g \Delta T z}{\nu \lambda_r} \tag{2.25}$$

$$\mathrm{Nu} = \frac{\alpha z}{\lambda_r}, \tag{2.26}$$

where

ρ_w = density of water, kg m^{-3}
c_w = specific heat capacity of water, J kg^{-1} K^{-1}
α_v = coefficient of volumetric thermal expansion of water, K^{-1}
k = water permeability of rocks, m^2
g = gravitational acceleration, m^2 s^{-1}
$\triangle T$ = temperature difference at vertical distance z, K
z = vertical distance considered, m
ν = kinematic viscosity of water m^2 s^{-1}
λ_r = thermal conductivity of water-saturated rock, W m^{-1} K^{-1}
α = coefficient of surface heat transfer, W m^{-2} K^{-1}.

Elder (1958, 1965, 1981) found from experimental results with a Hele-Shaw cell, which is filled with glass spheres and water, that convection starts at the critical Rayleigh number Ra_c with:

$$\mathrm{Ra} = \mathrm{Ra}_c \geq 40, \tag{2.27}$$

and that the relation between the Rayleigh number Ra and the Nusselt number Nu is given by (Figure 2.8):

$$\mathrm{Nu} = \frac{\mathrm{Ra}}{40}. \tag{2.28}$$

Furthermore, he found the total heat-flow density q_t transmitted from the bottom to the top of the Hele-Shaw cell to be ($z = L$):

$$q_t = \mathrm{Nu}\,\lambda_r \frac{\triangle T}{L}. \tag{2.29}$$

This equation corresponds to Equation (2.24). Both equations give ($\triangle T = T_L - T_0$):

$$q_t = -\lambda_r \frac{\triangle T}{L} + P\rho_w c_w v_z \triangle T$$

$$v_z = \frac{\lambda_r\left(1 + \dfrac{\mathrm{Ra}}{40}\right)}{P\rho_w c_w L}. \tag{2.30}$$

It should be remarked that a quadratic relation between q_t and ΔT results, if Equations (2.28) and (2.25) are substituted into Equation (2.29). Furthermore, the right-hand side is independent of λ_r and the aspect ratio l/L.

For practical purposes, Ra first, has to be calculated from Equation (2.25). Of course, the parameters of Equation (2.25) must thereby be known or estimated. In the case where Ra < 40, the Nusselt number has to be set at Nu = 1 and Equation (2.29) reduces to Fourier's first equation of heat conduction; Equation (2.3). Thus no water convection in the subsurface can be expected. In the case where Ra ≥ 40, water convection occurs. Witherspoon (1977) believes that twelve

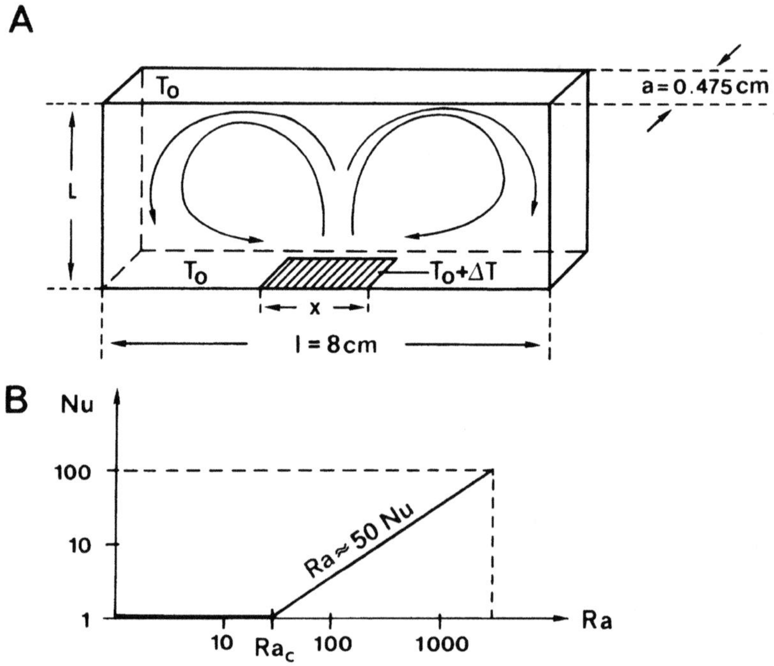

Fig. 2.8. (A) Hele-Shaw cell filled with glass spheres and water or suspended aluminium particles for modelling heat conduction and heat convection caused by temperature difference, Elder (1958, 1965, 1981). (B) Result from model given in A, Nu = Nusselt nubmer, Ra = Rayleigh number, Ra_c = critical Rayleigh number, T = temperature.

contiguous convection cells in a linear array are present at Wairakei (New Zealand); each has a roughly square aspect ratio $l = L = 5$ km and Ra ranges up to 5000.

2.3.3. Recognition of Water Movements

It is not easy to recognize the influence of water movements on heat-flow density values. Besides indications from hydrogeological studies, temperature measurements performed in a borehole can constitute a useful tool.

The effects of water movements on temperature are shown schematically in Figure 2.9. On graph I reference temperatures are assumed (solid straight line) and the relevant temperature gradients are noted. By means of dashed curves, graphs II to V show the temperature distributions that may possibly occur in the case of the water movement.

As shown in Figure 2.9 (Haenel, 1980), it is possible to recognize water movement with the knowledge of undisturbed reference temperature and the actual mean annual temperature T_0 at the Earth's surface, at depth $z = 0$ m. It is usually difficult to get an undisturbance as illustrated in Figure 2.9, II, can be easily recognized, because of the temperature difference between T_0 and T_0'. This again shows the importance of knowing T_0 as exactly as possible.

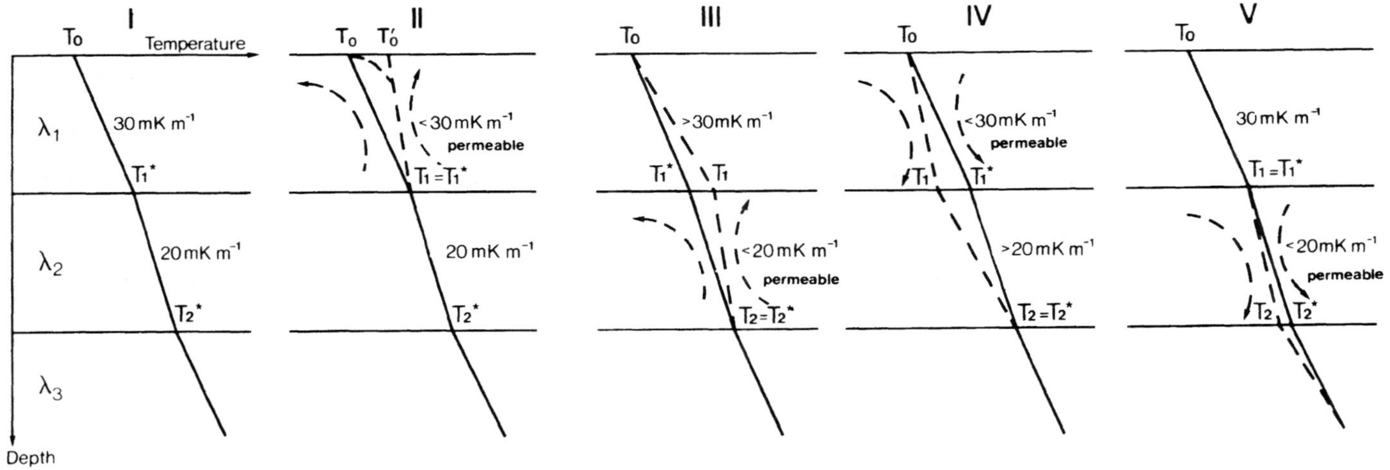

Fig. 2.9. Influence of water movement on temperature.

I – Reference model
II, III Ascending water
IV, V – Descending water

T_0 = Mean annual temperature on the Earth's surface
T'_0 = Extrapolated surface temperature
T_1, T_2 = Measured temperature
T_1^*, T_2^* = Undisturbed temperature
λ = Thermal conductivity.

Furthermore, where the thermal conductivity of rocks is unknown, the results are not unequivocal, since the plotted temperature distributions can, in principle, be attributed also to corresponding thermal conductivity distributions.

Conclusions concerning any water movements can be reached if heat-flow density values for various depth intervals have been determined independently and the heat generation in the rocks in the zone under consideration is known.

In Figure 2.10 examples which have been derived from Figure 2.9 are given.

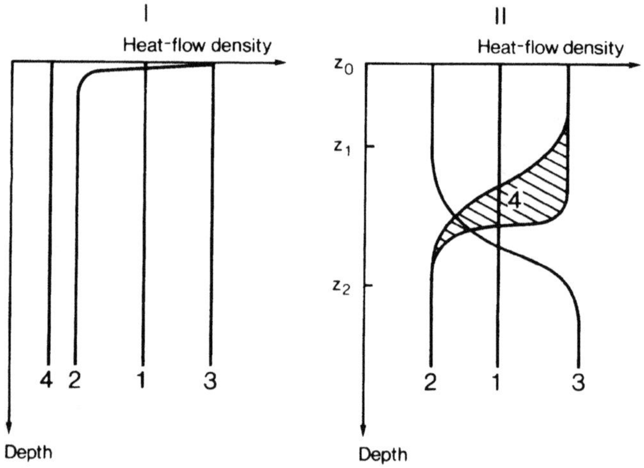

Fig. 2.10. Influence of water movement on heat-flow density, in mW m^{-2} with reference to Figure 2.9.

I – Influenced heat-flow density in the total vertical section.
- 1 = Undisturbed heat-flow density.
- 2 = Decreased heat-flow density, corresponding to II in Figure 2.9.
- 3 = Increased heat-flow density, corresponding to III (upper part) in Figure 2.9.
- 4 = Decreased heat-flow density, corresponding to IV (upper part) in Figure 2.9.

II – Influenced heat-flow density in a part of the vertical section.
- 1 = Undisturbed heat-flow density.
- 2 = Heat-flow density caused by ascending water in the region deeper than z_2; corresponding to III in Figure 2.9.
- 3 = Heat-flow density caused by descending water in the region between z_0 and z_1; corresponding to IV in Figure 2.9.
- 4 = Curve in this region is influenced by thickness of transition zone in which the movement is stopped.

Normally, of course, the heat-flow density-versus-depth plot is only a straight line if the temperature (or temperature gradient) is also a straight line (No. 1 of graphs I and II). However, if the temperature gradient is higher or lower than the reference gradient (see Figure 2.9), the heat-flow density is also lower or higher (graph I, No. 2, 3, and 4) than the reference heat-flow density (graph I, No. 1). The deviation of the parallel-running heat-flow density from the undisturbed heat-flow density is a measure of the water velocity. It can be seen that even small water velocities are sufficient to cause thermal anomalies such as that known, for example, in the Upper Rhine Graben (Werner and Parini, 1980).

The width of the transition zone between low and high heat-flow density (graph II) is determined primarily by the thickness of the zone in which the water movement decreases to zero vertical velocity.

In the case of only slight water convection, and as a result of the normal scattering of the heat-flow density values, it usually cannot be recognized. For that purpose a more sensitive procedure is necessary, e.g. plotting $\Delta T/\Delta z$ versus T between each measuring point at vertical distance L.

Rearranging Equation (2.24) (Mansure and Reiter, 1979):

$$\frac{\Delta T}{\Delta z} = \left(\frac{\beta}{L}\right)(T_0 - T_L) - \frac{q_t}{\lambda_r}, \quad \beta = -\frac{P\rho_w c_w v_z L}{\lambda_r},$$

which yields a straight line when β = constant, and horizontal lines when $\beta = 0$ (no water movement); see Figure 2.11. From this figure, β/L can also be determined from the slope, and q_t/λ_r form the intercept.

Because P, ρ_w, L and λ_r are assumed to be known, the vertical component of water velocity v_z and the total heat-flow density q_t can be calculated.

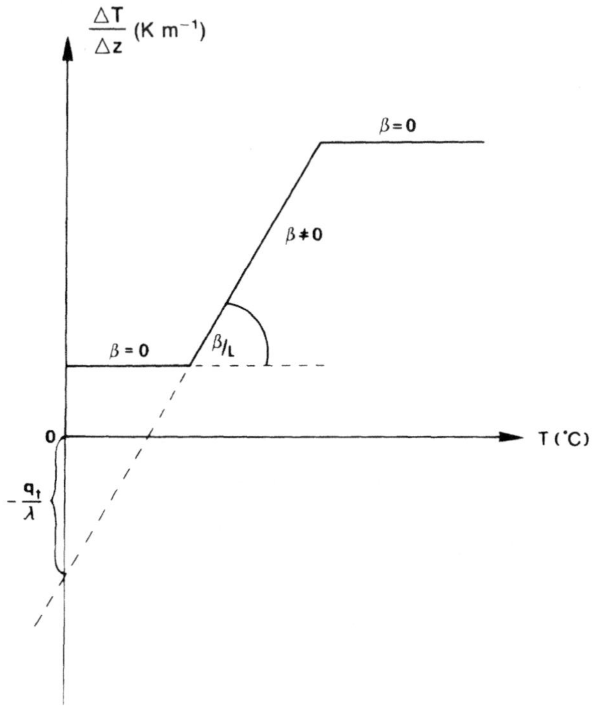

Fig. 2.11. Plot for recognition of water movement; see text.

2.4. NUMERICAL METHODS

2.4.1. Finite-Difference Method (FDM)

For many problems it is very difficult or impossible to find a proper solution from the above ordinary and partial differential equations. Therefore, the use of numerical methods is customary.

Two methods are used for the purpose: the finite-difference method and the finite-element method. Numerous procedures and iteration techniques exist for both (Collatz, 1966; Marsal, 1976). In the following, only the simplest procedure for the FDM will be explained.

The FDM involves the substitution of finite differences for the differential quotients.

Let us consider Equation (2.8) in two dimensions in the (x, z) plane:

$$\Delta T = 0$$

i.e.

$$\frac{\partial^2 T}{\partial x^2} + \frac{\partial^2 T}{\partial z^2} = 0. \tag{2.31}$$

A graph of temperature versus x-coordinate is shown in Figure 2.12A. For small steps of x, the differential quotient of the first partial derivative can be approximated by a quotient of differences as follows:

$$\frac{\partial T}{\partial x} = \frac{T(x + \Delta x) - T(x)}{\Delta x} = \frac{T_{x+1} - T_x}{\Delta x} = \left(\frac{\partial T}{\partial x}\right)_f \quad \text{forward step}$$

or

$$\frac{\partial T}{\partial x} = \frac{T(x) - T(x - \Delta x)}{\Delta x} = \frac{T_x - T_{x-1}}{\Delta x} = \left(\frac{\partial T}{\partial x}\right)_b \quad \text{backward step}$$

or

$$\frac{\partial T}{\partial x} = \frac{T_{x+1} - T_{x-1}}{2\Delta x} = \left(\frac{\partial T}{\partial x}\right)_c . \quad \text{central difference.}$$

The second partial derivative can be approximated by:

$$\frac{\partial^2 T}{\partial x^2} = \frac{\partial}{\partial x}\left(\frac{\partial T}{\partial x}\right) = \frac{\Delta\left(\frac{\partial T}{\partial x}\right)}{\Delta x} = \frac{\left(\frac{\partial T}{\partial x}\right)_f - \left(\frac{\partial T}{\partial x}\right)_b}{\Delta x}$$

$$= \frac{\frac{T_{x+1} - T_x}{\Delta x} - \frac{T_x - T_{x-1}}{\Delta x}}{\Delta x}$$

$$= \frac{T_{x+1} + T_{x-1} - 2T_x}{\Delta x^2},$$

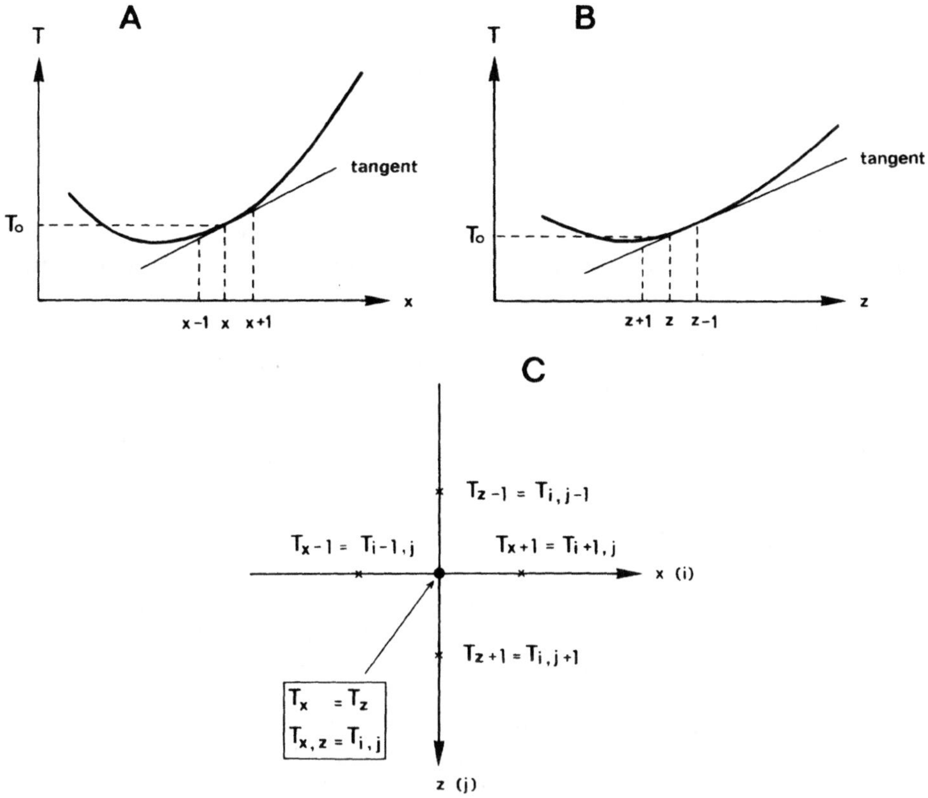

Fig. 2.12. Principles of the finite difference method.

and similarly for direction z, Figure 2.12B:

$$\frac{\partial^2 T}{\partial z^2} = \frac{T_{z+1} + T_{z-1} - 2T_z}{\Delta z^2}.$$

A top view of both graphs centred at the tangent point (x_0, z_0) is given in Figure 2.12C. By superposition of partial derivatives:

$$\nabla^2 T = \frac{T_{i+1,j} + T_{i-1,j} - 2T_{i,j}}{\Delta x^2} + \frac{T_{i,j+1} + T_{i,j-1} - 2T_{i,j}}{\Delta z^2} = 0,$$

which gives

$$T_{i,j} = \frac{1}{2\left(\frac{1}{\Delta x^2} + \frac{1}{\Delta z^2}\right)}\left(\frac{T_{i+1,j} + T_{i-1,j}}{\Delta x^2} + \frac{T_{i,j+1} + T_{i,j-1}}{\Delta z^2}\right), \tag{2.32}$$

and in the case $\Delta x = \Delta z$:

$$T_{i,j} = (T_{i+1,j} + T_{i-1,j} + T_{i,j+1} + T_{i,j-1})/4, \tag{2.33}$$

with $i = 0, 1, 2 \ldots, n$ and $j = 0, 1, 2 \ldots, m$.

The corresponding solutions of the most common equations of heat conduction are given in Table 2.10. In the case of filtration the velocity term is to be solved and added similarly, e.g. for the (x, z) plane:

$$\rho_w c_w \left(v_x \frac{T_{i+1,j} - T_{i-1,j}}{2\Delta x} + v_z \frac{T_{i,j+1} - T_{i,j-1}}{2\Delta z} \right). \tag{2.34}$$

For related hydrogeothermal problems as given by Equation (2.19), see Freeze and Cherry (1979). To avoid instability, the calculation must be carried out in the direction 'against the flowing water', called upwind scheme; Thomasset (1981).

The above-mentioned procedure is also called the explicit finite difference method, because the new calculated temperature $T_{i,j}$ is explicitly represented. In contrast, there also exist implicit methods, in which the new $T_{i,j}$ are implicitly expressed within the difference equation (Minear and Toksöz, 1970).

For most of the explicit methods the stability condition for non-steady state models is given by:

$$\Delta t \leqslant \frac{\Delta x \, \Delta z}{2a}, \tag{2.35}$$

where Δx, Δz (or Δz, Δr) are the step lengths, Δt the time interval of used calculations, and a = thermal diffusivity of the medium under consideration.

The advantage of the implicit methods is that the convergence condition can be neglected. In practice this is not always the case.

To apply the difference method described, the area under consideration has to be covered by a grid system with the distances Δx, Δz between the nodal points.

With the use of the equations compiled in Table 2.10, the following boundary conditions must be taken into account:

- The temperature at the Earth's surface has to be given

$$T = T_0 \quad \text{at} \quad z = 0.$$

- At great distance from the disturbance, the temperature field can be assumed to be undisturbed

$$T = T_0 + z\left(\frac{\partial T}{\partial z}\right)_{\text{undisturbed}} \qquad \text{for} \quad x \to \infty.$$

- In the case of a symmetrical model (e.g. to avoid having too many nodal points), with the x-axis as the axis of symmetry:

$$\frac{\partial T}{\partial x} = 0 \quad \text{at} \quad x = 0.$$

- At the boundary two domains 1 and 2 with different thermal conductivities λ_1 and λ_2 the temperatures T_1 and T_2 have to be the same; the same applies to heat-flow density component perpendicular to the boundary of both domains:

$$\begin{aligned} T_1 &= T_2 \\ \lambda_1(\text{grad } T_1)_n &= \lambda_2(\text{grad } T_2)_n, \end{aligned}$$

TABLE 2.10

Transformation of differential equations for heat conduction into finite difference equations for explicite determination of $T_{i,j}$ (two dimensional solutions); $r_0 = i$; for practical application H, a, λ, ρ, and c are to be supplied with the indices i, j; H = heat production rate.

1. Steady-state condition, Equation (2.7), λ = const.
   1.1. Longitudinal

$$\lambda\left(\frac{\partial^2 T}{\partial x^2} + \frac{\partial^2 T}{\partial z^2}\right) + H = 0$$

$$T_{i,j} = \frac{1}{2\left(\frac{1}{\Delta x^2} + \frac{1}{\Delta z^2}\right)}\left(\frac{T_{i+1,j} + T_{i-1,j}}{\Delta x^2} + \frac{T_{i,j+1} + T_{i,j-1}}{\Delta z^2} + \left(\frac{H}{\lambda}\right)\right)$$

$\Delta x = \Delta z$:

$$T_{i,j} = \frac{1}{4}\left(T_{i+1,j} + T_{i-1,j} + T_{i,j+1} + T_{i,j-1} + \Delta x^2\left(\frac{H}{\lambda}\right)\right).$$

   1.2. Cylindrical

$$\lambda\left(\frac{\partial^2 T}{\partial r^2} + \frac{1}{r}\frac{\partial T}{\partial r} + \frac{\partial^2 T}{\partial z^2}\right) + H = 0$$

$$T_{i,j} = \frac{1}{2\left(\frac{1}{\Delta r^2} + \frac{1}{\Delta z^2}\right)}\left(\frac{T_{i+1,j} + T_{i-1,j}}{\Delta r^2} + \frac{T_{i+1,j} + T_{i-1,j}}{2r_0\Delta r} + \frac{T_{i,j+1} + T_{i,j-1}}{\Delta z^2} + \left(\frac{H}{\lambda}\right)\right)$$

$\Delta r = \Delta z$:

$$T_{i,j} = \frac{1}{4}\left(T_{i+1,j} + T_{i-1,j} + T_{i,j+1} + T_{i,j-1} + \frac{\Delta r}{2r_0}\left(T_{i+1,j} - T_{i-1,j}\right) + \Delta r^2\left(\frac{H}{\lambda}\right)\right).$$

Table 2.10 (Continued)

1.3. Spherical

$$\lambda\left(\frac{\partial^2 T}{\partial r^2} + \frac{2}{r}\frac{\partial T}{\partial r} + \frac{1}{r^2}\frac{\partial^2 T}{\partial \psi^2} + \frac{1}{r^2 \tan\psi}\frac{\partial T}{\partial \psi}\right) + H = 0$$

$$T_{i,j} = \frac{1}{2\left(\frac{1}{\Delta r^2} + \frac{1}{r_0 \Delta \psi^2}\right)}\left[\frac{T_{i+1,j} + T_{i-1,j}}{\Delta r^2} + \frac{T_{i+1,j} - T_{i-1,j}}{r_0 \Delta r} + \frac{T_{i,j+1} + T_{i,j-1}}{r_0^2 \Delta \psi^2} + \frac{T_{i,j+1} - T_{i,j-1}}{2r_0^2 \Delta\psi \tan\psi} - \left(\frac{H}{\lambda}\right)\right]$$

$\Delta r = \Delta \psi$:

$$T_{i,j} = \frac{1}{2 + \frac{2}{r_0^2}}\left[T_{i+1,j} + T_{i-1,j} + \frac{1}{r_0^2}(r_0 \Delta r(T_{i+1,j} - T_{i-1,j}) + T_{i,j+1} + T_{i,j-1}) + \frac{\Delta r}{2r_0^2 \tan\psi}(T_{i,j+1} - T_{i,j-1}) + \Delta r^2\left(\frac{H}{\lambda}\right)\right].$$

2. Steady-state condition, $-\mathrm{div}(\mathbf{q}) + H = 0$, $\lambda = f(x, z)$, determination for $\Delta x = \Delta z$, $\Delta r = \Delta z$, $\Delta r = \Delta \psi$ similar to that under 1.

2.1. Longitudinal

$$\lambda\left(\frac{\partial^2 T}{\partial x^2} + \frac{\partial^2 T}{\partial z^2}\right) + \left(\frac{\partial \lambda}{\partial x}\frac{\partial T}{\partial x} + \frac{\partial \lambda}{\partial z}\frac{\partial T}{\partial z}\right) + H = 0$$

$$T_{i,j} = \frac{1}{2\left(\frac{1}{\Delta x^2} + \frac{1}{\Delta z^2}\right)}\left[\left(\frac{T_{i+1,j} + T_{i-1,j}}{\Delta x^2} + \frac{T_{i,j+1} + T_{i,j-1}}{\Delta z^2}\right) + \frac{1}{4\lambda_{i,j}\Delta x^2}(\lambda_{i+1,j} - \lambda_{i-1,j})(T_{i+1,j} - T_{i-1,j})\right.$$

$$\left. + \frac{1}{4\lambda_{i,j}\Delta z}(\lambda_{i,j+1} - \lambda_{i,j-1})(T_{i,j+1} - T_{i,j-1}) + \left(\frac{H}{\lambda_{i,j}}\right)\right].$$

Table 2.10 (Continued)

2.2. Cylindrical

$$\lambda\left(\frac{\partial^2 T}{\partial r^2} + \frac{1}{r}\frac{\partial T}{\partial r} + \frac{\partial^2 T}{\partial z^2}\right) + \left(\frac{\partial \lambda}{\partial r}\frac{\partial T}{\partial r} + \frac{\partial \lambda}{\partial z}\frac{\partial T}{\partial z}\right) + H = 0$$

$$T_{i,j} = \frac{1}{2\lambda_{i,j}\left(\frac{1}{\Delta r^2} + \frac{1}{\Delta z^2}\right)}\left[\lambda_{i,j}\left(\frac{T_{i+1,j} + T_{i-1,j}}{\Delta r^2} + \frac{T_{i,j+1} + T_{i,j-1}}{\Delta z^2} + \frac{T_{i+1,j} - T_{i-1,j}}{2r_0 \cdot \Delta r}\right)\right.$$

$$\left. + \frac{1}{4\Delta r^2}(\lambda_{i+1,j} - \lambda_{i-1,j})(T_{i+1,j} - T_{i-1,j})(T_{i+1,j} - T_{i-1,j}) + \frac{1}{4\Delta z^2}(\lambda_{i,j+1} - \lambda_{i,j-1})(T_{i,j+1} - T_{i,j+1}) + H\right].$$

2.3. Spherical

$$\lambda\left(\frac{\partial^2 T}{\partial r^2} + \frac{2}{r}\frac{\partial T}{\partial r} + \frac{1}{r^2}\frac{\partial^2 T}{\partial \psi^2} + \frac{1}{r^2 \tan \psi}\frac{\partial T}{\partial \psi}\right) + \left(\frac{\partial \lambda}{\partial r}\frac{\partial T}{\partial r} + \frac{1}{r^2}\frac{\partial \lambda}{\partial \psi}\frac{\partial T}{\partial \psi}\right) + H = 0$$

$$T_{i,j} = \frac{1}{2\lambda_{i,j}\left(\frac{1}{\Delta r^2} + \frac{1}{r_0 \Delta \psi^2}\right)}\left[\frac{T_{i+1,j} + T_{i-1,j}}{\Delta r^2} + \frac{T_{i+1,j} - T_{i-1,j}}{r_0 \Delta r} + \frac{T_{i,j+1} + T_{i,j-1}}{r_0^2 \Delta \psi^2} + \frac{T_{i,j+1} - T_{i,j-1}}{2r_0^2 \Delta \psi \tan \psi}\right.$$

$$\left. + \frac{(\lambda_{i+1,j} - \lambda_{i-1,j})(T_{i+1,j} - T'_{i-1,j})}{4\Delta r^2} + \frac{(\lambda_{i,j+1} - \lambda_{i,j-1})(T_{i,j+1} - T_{i,j-1})}{4r_0^2 \Delta \psi^2} + H\right].$$

3. Nonsteady-state condition, Equation (2.6), λ = const., the fundamental equation below 1. is extended on the right-hand side by $\rho c \partial T/\partial t$, determination for $\Delta x = \Delta z$, $\Delta r = \Delta z$, $\Delta r = \Delta \psi$ similar to that under 1.

3.1. Longitudinal, Equation (2.6)

$$T_{i,j}^{n+1} = a\Delta t\left(\frac{T_{i+1,j} + T_{i-1,j} - 2T_{i,j}}{\Delta x^2} + \frac{T_{i,j+1} + T_{i,j-1} - 2T_{i,j}}{\Delta z^2}\right) + \Delta t\left(\frac{H}{\rho c}\right) + T_{i,j}.$$

3.2. Cylindrical

$$T_{i,j}^{n+1} = a\Delta t\left(\frac{T_{i+1,j} + T_{i-1,j} - 2T_{i,j}}{\Delta r^2} + \frac{T_{i+1,j} - T_{i-1,j}}{2r_0 \Delta r} + \frac{T_{i,j+1} + T_{i,j-1} - 2T_{i,j}}{\Delta z^2}\right) + \Delta t\left(\frac{H}{\rho c}\right) + T_{i,j}.$$

Table 2.10 (Continued)

3.3. Spherical

$$T_{i,j}^{n+1} = a\Delta t\left(\frac{T_{i+1,j} + T_{i-1,j} - 2T_{i,j}}{\Delta r^2} + \frac{T_{i+1,j} - T_{i-1,j}}{r_0\Delta r} + \frac{T_{i,j+1} + T_{i,j-1} - 2T_{i,j}}{r_0^2\Delta\psi^2} + \frac{T_{i,j+1} - T_{i,j-1}}{2r_0^2\Delta\psi \tan\psi}\right) + \Delta t\left(\frac{H}{\rho c}\right) + T_{i,j}.$$

4. Nonsteady-state condition, $(-\mathrm{div}(\mathbf{q}) + H = \rho c\partial T/\partial t$, $\lambda = f(x, z)$, the fundamental equation below 2. is extended on the right-hand side by $\rho c\partial T/\partial t$, determination for $\Delta x = \Delta z$, $\Delta r = \Delta z$, $\Delta r = \Delta\psi$ similar to that under 1.

4.1. Longitudinal

$$T_{i,j}^{n+1} = a\Delta t\left(\frac{T_{i+1,j} + T_{i-1,j} - 2T_{i,j}}{\Delta x^2} + \frac{T_{i,j+1} + T_{i,j-1} - 2T_{i,j}}{\Delta z^2}\right) + \frac{\Delta t}{4\rho c}\left(\frac{(\lambda_{i+1,j} - \lambda_{i-1,j})(T_{i+1,j} - T_{i-1,j})}{\Delta x^2} + \frac{(\lambda_{i,j+1} - \lambda_{i,j-1})(T_{i,j+1} - T_{i,j-1})}{\Delta z^2}\right) + \Delta t\left(\frac{H}{\rho c}\right) + T_{i,j}.$$

4.2. Cylindrical

$$T_{i,j}^{n+1} = a\Delta t\left(\frac{T_{i+1,j} + T_{i-1,j} - 2T_{i,j}}{\Delta r^2} + \frac{T_{i+1,j} - T_{i-1,j}}{2r_0\Delta r} + \frac{T_{i,j+1} + T_{i,j-1} - 2T_{i,j}}{\Delta z^2}\right) + \frac{\Delta t}{4\rho c}\left(\frac{(\lambda_{i+1,j} - \lambda_{i-1,j})(T_{i+1,j} - T_{i-1,j})}{\Delta r^2} + \frac{(\lambda_{i,j+1} - \lambda_{i,j-1})(T_{i,j+1} - T_{i,j-1})}{\Delta z^2}\right) + \Delta t\left(\frac{H}{\rho c}\right) + T_{i,j}.$$

4.3. Spherical

$$T_{i,j}^{n+1} = a\Delta t\left(\frac{T_{i+1,j} + T_{i-1,j} - 2T_{i,j}}{\Delta r^2} + \frac{T_{i+1,j} - T_{i-1,j}}{r_0\Delta r} + \frac{T_{i,j+1} + T_{i,j-1} - 2T_{i,j}}{r_0^2\Delta\psi^2} + \frac{T_{i,j+1} - T_{i,j-1}}{2r_0^2\Delta\psi \tan\psi}\right) + \frac{\Delta t}{4\rho c}\left(\frac{(\lambda_{i+1,j} - \lambda_{i-1,j})(T_{i+1,j} - T_{i-1,j})}{\Delta r^2} + \frac{(\lambda_{i,j+1} - \lambda_{i,j-1})(T_{i,j+1} - T_{i,j-1})}{r_0^2\Delta\psi^2}\right) + \Delta t\left(\frac{H}{\rho c}\right) + T_{i,j}.$$

where $(\text{grad } T)_n$ is the component of grad T perpendicular to the surface of separation.

– The heat-flow density at the base of the grid system is normally used to be constant:

$$q_b = \lambda \frac{\partial T}{\partial z} = \text{const.} \quad \text{at} \quad z = mz,$$

which is realised for:

$$T_{i,m-1} = \text{const.} = b_1 \quad \text{and} \quad T_{i,m} = \text{const.} = b_2.$$

The numerical computation starts with estimated values for each point $T_{i,k}$ (and given values $\lambda_{i,j}$ or estimated values $\lambda_{i,j}(T)$). These values are improved by iteration, starting at point ($i = 0, j = 1$) and running along the line $j = 1$, then $j =$ 2, 3, etc. until $j = m - 1$ to the last point ($i = n, j = m - 1$), depending on the boundary conditions. Fore more details see Balling (1976), Kappelmeyer and Haenel (1974), Mundry (1966, 1968).

The convergence can be checked by means of the test number:

$$p(n) = \sum_{i,j} |T_{i,j}^{(n)} - T_{i,j}^{(n-1)}|, \tag{2.36}$$

which can be set equal to 1°C, 0.1°C, etc. depending on the problem.

In many cases the stability condition Equation (2.35) is a serious problem if small steps Δx, Δz and large time steps Δt are necessary (e.g. for a borehole or a fracture).

For those conditions the nodal distance can be taken with continuously increasing distance as shown in Figure 2.13. Other configurations are also possible, of course; but for more details see Beck and Shen (1985). If the model is complicated, the calculation can be split as demonstrated in Figure 2.13. In this case, a borehole is considered (radial symmetry about the borehole axis) and a fracture perpendicular to the borehole axis (and also radial symmetry about the borehole axis). Cold water is injected through the borehole into the fracture, and after a certain time, the injection is stopped and the heated water is recovered (called huff- and -puff heat extraction). In both cases the temperature near the borehole wall and near the fracture well, for which small nodel distances are necessary, are of interest. Because of the radial symmetry it is impossible to have small nodal distances simultaneously for both borehole and fracture. Therefore, the network is split into two parts: one partial net for the borehole and one for the fracture. When the water arrives at the entry to the fracture after a time step within the borehole at a temperature T_x (point 1), the temperature T_x is set at point 2 in the fracture and injection starts there; and vica versa. The points 1 and 2 correspond to a source and sink, respectively.

2.4.2. Finite-Element Method (FEM)

The FEM is a numerical method employed for finding an approximate solution to ordinary as well as partial differential equations. Where as FDM approximates

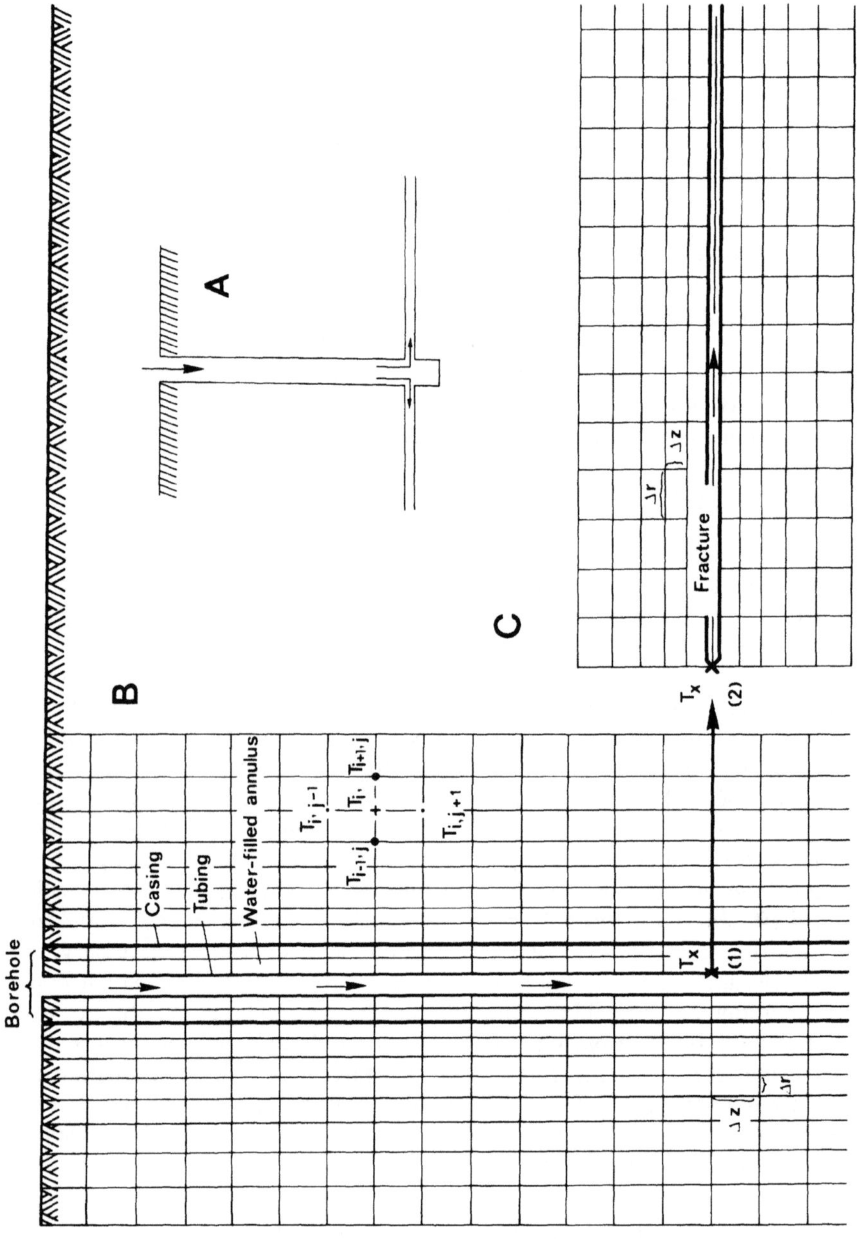

Fig. 2.13. Net for finite-difference method in the case of water injection through a borehole into a fracture. (*A*) Principle of the radially symmetric arrangement of borehole and fracture, (*B*) net for borehole; (*C*) net for fracture, (T_x) 'connection point' for net B and net C.

differential quotients by means of differences, FEM approximates differential equations by an integral approach.

Again, different procedures and iteration techniques exist, and in the following only one of the most common procedures will be discussed. This is based on Galerkin's method (Cavendish *et al.*, 1969; Zienkiewicz, 1971; Marsal, 1976; Wang and Anderson, 1982) due to the parabolic equation.

Galerkin applies a particular weighted residual principle which turns out to be equivalent to a variational principle. The philosophy behind a variational principle is that a physical quantity, such as the rate of energy dissipation, be minimized over the domain of the problem (Neuman and Witherspoon, 1971).

Let us consider again the Laplace equation (2.8) in two dimensions in the (x, z) plane:

$$\frac{\partial^2 T}{\partial x^2} + \frac{\partial^2 T}{\partial z^2} = 0.$$

(1) With reference to Figure 2.14, it is first necessary to transform the partial differential equation into an integral equation.

For that purpose the area A under consideration has to be subdivided into elements. The unknown temperature T_p, which is being sought, prevails at the nodal points p of the elements. Furthermore in the area A an approximate temperature solution $t(x, z)$ will be assumed to be a polynomial:

$$t(x, z) = c_1 + c_2 x + c_3 y + c_4 x^2 + c_5 y^2 + c_6 xy + \dots .$$

For demonstration only the first-order polynominal is used:

$$t(x, z) = c_1 + c_2 x + c_3 y. \tag{2.37}$$

The correlation between $t(x, z)$ and T_p is defined as:

$$t(x, z) = \sum_{p=1}^{N} T_p f_p(x, z), \tag{2.38}$$

where f_p is called a basis function (see Equation (2.42)), and the subscript p indicates the nodal number of the area A which has N nodes.

The next step is to specify a total of N conditions to determine the T_p values (all together N numbers for the total area A). In Galerkin's method, the N conditions are that the residuals of the governing equations weighted by each of the N basis functions yields the value zero when integrated over the entire domain of A:

$$\iint_A \left(\frac{\partial^2 t}{\partial x^2} + \frac{\partial^2 t}{\partial z^2} \right) f_p(x, z)\, \mathrm{d}x\, \mathrm{d}z = 0, \tag{2.39}$$

where the term in parenthesis is the residual. The application of the first form of Green's theorem for two-dimensional consideration:

$$\iint_A f_p \nabla^2 t\, \mathrm{d}x\, \mathrm{d}z = -\iint_A \nabla f_p \nabla t\, \mathrm{d}x\, \mathrm{d}z + \int_B f_p \frac{\partial t}{\partial n}\, \mathrm{d}s$$

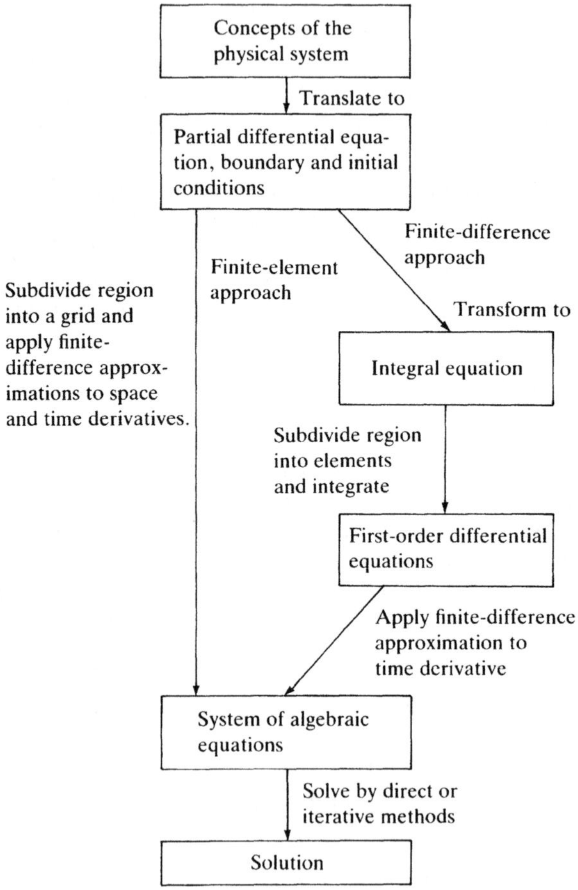

Fig. 2.14. Generalised model development by finite-difference (FDM) and finite-element (FEM) methods (Faust and Mercer, 1980).

with

$$\frac{\partial t}{\partial n} = \frac{\partial t}{\partial x} n_x + \frac{\partial t}{\partial z} n_z,$$

which leads to:

$$\iint_A \left(\frac{\partial^2 t}{\partial x^2} + \frac{\partial^2 t}{\partial z^2} \right) f_p(x, z) \, dx \, dz = - \iint_A \left(\frac{\partial t}{\partial x} \frac{\partial f_p}{\partial x} + \frac{\partial t}{\partial z} \frac{\partial f_p}{\partial z} \right) dx \, dz \; +$$

$$\int_B \left(\frac{\partial t}{\partial x} n_x + \frac{\partial t}{\partial z} n_z \right) f_p(x, z) \, ds, \tag{2.40}$$

where B is the boundary of A, s is a generalised variable representing distance along the boundary in a counterclockwise sense, and n_x and n_z are the components of a unit vector outwardly perpendicular to B. Note: if the flow is zero on the boundary, the boundary integral term is also zero, and the integrands on the right side contain only first derivatives which simplifies the problem considerably.

(2) With reference to Figure 2.14 the region is now subdivided into elements and the integration is performed.

The common procedure of to subdividing the are A into triangular elements which are defined by three nodes will be used. Because each finite element e is treated separately, each corner point can be designated by nodal coordinates (x_p, z_p) as shown in Figure 2.15. In Equation (2.38) i, j, k now have to be substituted

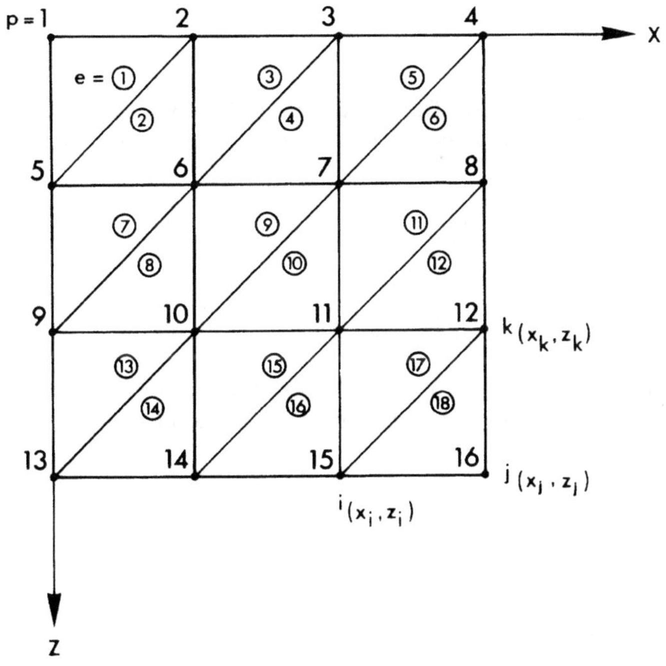

Fig. 2.15. Triangular finite element mesh including node numbers for the node points p and the element number e (circled). Each element e can be represented by its node point coordinates i, j, and k.

for p; this yields three equations. If these three equations are solved for c_1, c_2, c_3, and if the expressions are substituted back into Equation (2.38), Equation (2.38) can be rewritten for an element as:

$$t^e(x, z) = f_i^e(x, z)\,T_i + f_j^e(x, z)\,T_j + f_k^e(x, z)\,T_k, \tag{2.41}$$

where $e = 1, 2, \ldots, M$ and:

$$
\begin{aligned}
f_i^e(x, z) &= [(x_j z_k - x_k z_j) + (z_j - z_k)x + (x_k - x_j)z]/S^e \\
f_j^e(x, z) &= [(x_k z_i - x_i z_k) + (z_k - z_i)x + (x_i - x_k)z]/S^e \\
f_k^e(x, z) &= [(x_i z_j - x_j z_i) + (z_i - z_j)x + (x_j - x_i)z]/S^e \qquad (2.42) \\
S^e &= (x_i z_j - x_j z_i) + (x_k z_i - x_i z_k) + (x_i z_k + x_k z_j). \qquad (2.43)
\end{aligned}
$$

Now, the first-order differential equation (see Figure 2.14) can be found by means of Equation (2.40); if it is borne in mind that the elements e are to be summed, whereas the boundary integral does not change:

$$
\sum_{e=1}^{M}\left[\iint_e \left(\frac{\partial t^e}{\partial x}\frac{\partial f_p^e}{\partial x} + \frac{\partial t^e}{\partial z}\frac{\partial f_p^e}{\partial z}\right)\mathrm{d}x\,\mathrm{d}z\right] =
$$

$$
\int_B \left(\frac{\partial t}{\partial x}n_x + \frac{\partial t}{\partial z}n_z\right)f_p\,\mathrm{d}s. \qquad (2.44)
$$

In this equation f_p^e are known from Equations (2.42) and (2.43), which are valid for each element e, and the values t^e can be found by differentiation of Equation (2.41):

$$
\begin{aligned}
\frac{\partial t^e}{\partial x} &= \frac{\partial f_i^e}{\partial x}T_i + \frac{\partial f_j^e}{\partial x}T_j + \frac{\partial f_k^e}{\partial x}T_k \\
\frac{\partial t^e}{\partial z} &= \frac{\partial f_i^e}{\partial z}T_i + \frac{\partial f_j^e}{\partial z}T_j + \frac{\partial f_k^e}{\partial z}T_k. \qquad (2.45)
\end{aligned}
$$

The differentiation of Equation (2.42), substitution into Equation (2.45), and this in turn into the left-hand side of Equation (2.44), lead to:

$$
M_{p,1}T_1 + M_{p,2}T_{p,2} + \ldots + M_{p,N}T_N = b_p, \qquad (2.46)
$$

or rewriting:

$$
[M]\{T\} = \{b\}, \qquad (2.47)
$$

where M is called the conductance matrix. These are $N \cdot N$ equations with the general boundary conditions b_p for determining the required temperature T_p at all nodal points p. The matrix $M_{p,s}$ ($s = 1, 2, \ldots, N$) itself consists of submatrices. As already mentioned, the FEM solves the problem successively, element by element by summing over all elements e:

$$
M_{p,s} = \sum_{e=1}^{M} M_{p,s}^e. \qquad (2.48)
$$

$M_{p,s}^e$ is called the element conductance matrix, $M_{p,s}$ the global conductance matrix, and s is now a function of the chosen elements; in our case simple triangular elements with the coordinates i, j, k. Hence, i, j, k are to be substituted for s:

$$
\begin{aligned}
M^e_{p,i} &= S^e\left(\frac{\partial f_i^e}{\partial x}\frac{\partial f_p^e}{\partial x} + \frac{\partial f_i^e}{\partial z}\frac{\partial f_p^e}{\partial z}\right)\\
M^e_{p,j} &= S^e\left(\frac{\partial f_j^e}{\partial x}\frac{\partial f_p^e}{\partial x} + \frac{\partial f_j^e}{\partial z}\frac{\partial f_p^e}{\partial z}\right)\\
M^e_{p,k} &= S^e\left(\frac{\partial f_k^e}{\partial x}\frac{\partial f_p^e}{\partial x} + \frac{\partial f_k^e}{\partial z}\frac{\partial f_p^e}{\partial z}\right),
\end{aligned}
\tag{2.49}
$$

where $p = i$, j or k. The S^e are known from Equation (2.43) and the derivative f from the corresponding differentiation of Equation (2.42).

If a boundary across which the normal heat-flow density is specified, e.g. by Fourier's first equation (2.3) is considered, the boundary integral in Equation (2.44) gives:

$$
\int_B \left(\frac{\partial t}{\partial x}n_x + \frac{\partial t}{\partial z}n_z\right) f_p \,\mathrm{d}s = \int_i^p \frac{q_1}{\lambda} f_p(s)\,\mathrm{d}s + \int_p^k \frac{q_2}{\lambda} f_p(s)\,\mathrm{d}s,
$$

q is positive for inflow into the element along the two line segments ip and pk. This leads to (Wang and Anderson, 1982):

$$
b_p = \frac{q_1}{\lambda}\frac{ip}{2} + \frac{q_2}{\lambda}\frac{pk}{2}. \tag{2.50}
$$

(3) Finally, the system of algebraic equations has to be solved; refer to Figure 2.14.

The Equation (2.47) includes the desired temperature T_p at each node point; it can be solved by using the well-known procedures as given by the Gauss algorithm, Gauss-Seidel iteration, the Thomas algorithm, or other techniques.

In Figure 2.16, after Cheng (1978), the element-by-element treatment is re-explained in a graphical manner. An area A is covered by triangular elements $e = 1$ to 6 with nodal points 1 to 8. Because the FEM successively calculates the required temperature element by element with the coordinates i, j, k, a small appropriate table is given. All nodal points are now placed in a field $p \times p = 8 \times 8$ (or 8×8 equations to solve the problem).

Within this field, all elements (or element matrices) are placed in a configuration which correspond to Equation (2.48). The element matrices (right-hand side of Equation (2.48)) furnish the global matrix (left-hand side of Equation (2.48)). Only a 'part' of the conductance matrix of Equation (2.46) is necessary. The conductance matrix is represented by $p \times p = 8 \times 8$. The 'part' in question is given by the global conductance matrix which does not fill all fields, as can be seen in the 'global matrix' in Figure 2.16.

A systematic study for comparison of FDM and FEM (Lühr, 1973; Myers, 1971, p. 387) has shown that both numerical techniques have nearly the same accuracy and the same weaknesses (poor accuracy for large grid system, alternating solutions for low flow rates); see Table 2.11. If the Equation (2.37) is a quadratic equation, the accuracy in FEM is better than in FDM. Furthermore, the advantage of FEM is given in the case of complicated grid systems as necessary for bridges, tunnels etc.

Element No. e	Nodes i	Nodes j	Nodes k
1	1	2	3
2	2	5	3
3	2	4	5
4	4	6	5
5	5	6	7
6	6	8	7

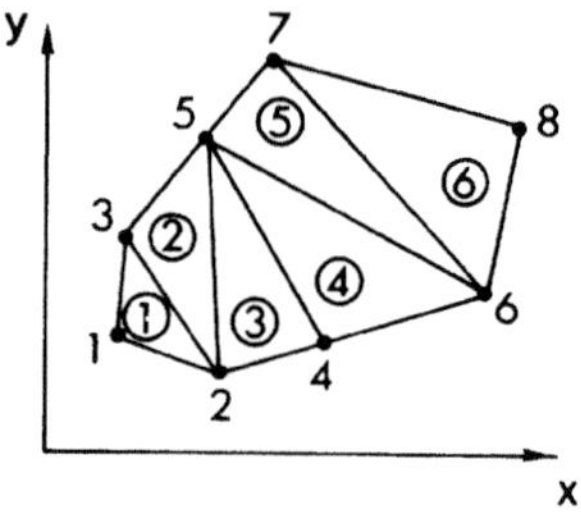

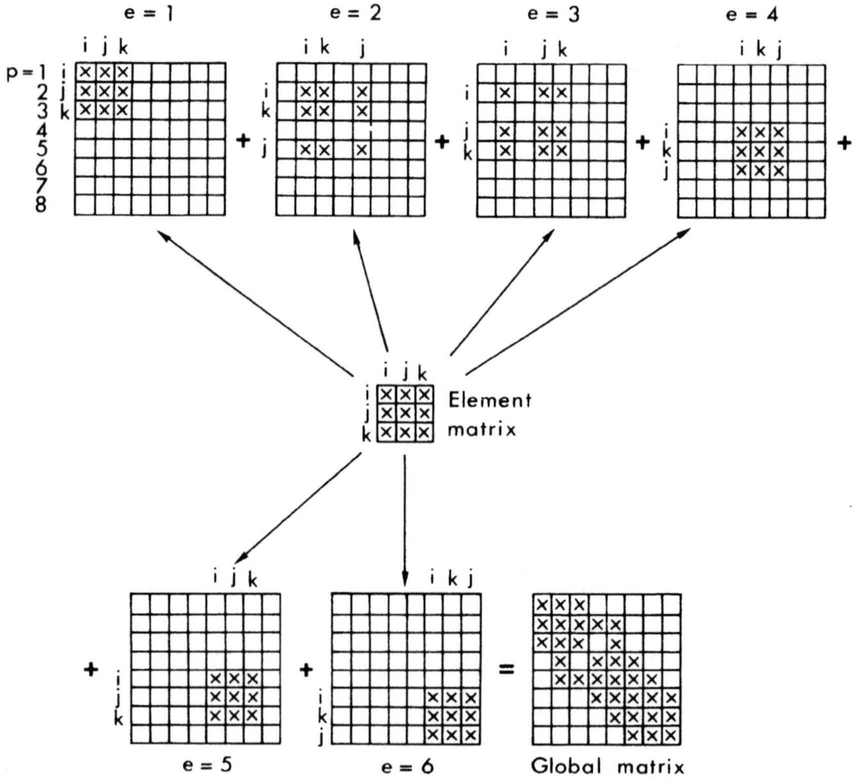

Fig. 2.16. The process of FDM for forming the global matrix from the elements after Cheng (1978); $N = 8$.

TABLE 2.11
Advantages and disadvantages of FEM and FDM.

Advantages	Disadvantages
FDM	
Easily understandable mathematical basis	Grid geometry not very flexible
Easy data input	Poor stability for certain problems
Efficient matrix techniques	
Easy programming	
FEM	
Flexible grid geometry	Advanced mathematical basis
Relatively high stability	Difficult data input
Many parameters can be easily introduced	Difficult programming
High accuracy if using a quadratic equation (2.37)	

2.5. SUMMARY OF RECOMMENDATIONS

1. The geothermal parameters are associated by convention with symbols and units as given in Table 2.2. These should be used.
2. As practical units, the following are used in geothermics: 10^{-3} W m^{-2} = 1 mW m^{-2} for heat-flow density and 10^{-3} K m^{-1} = 1 mK m^{-1} for the temperature gradient.
3. The abbreviation for geothermal temperature gradient grad $T = \Gamma$ and for heat-flow density q = HFD can be introduced.
4. For applications in the field of geothermal energy, the terms and definitions should be adapted as given in Tables 2.2 and 2.3.
5. The possible influence of convective heat transport on the conductive heat-flow density should be checked carefully. In the event that an influence is recognized, the result should be rejected as a convective heat-flow density. Otherwise, the convective influence must be separated, and if possible, represented in mW m^{-2}.
6. The heat-flow density values should be documented as given in Table 2.8.
7. In practice, problems often cannot be solved analytically. Therefore, a numerical treatment should be applied.
8. The finite-difference method is recommended, because this method can be adapted to suit most of the common problems encountered in geothermics and/or hydrogeothermics.

Acknowledgements

We are indebted to many colleagues who read the manuscript but particularly to Alan Beck who made many useful suggestions.

2.6. REFERENCES

Balling, N. P.: 1976, 'Geothermal Models of the Crust and Upper-most Mantle of the Fennoscandian Shield in South Norway and the Danish Embayment, *J. Geophys.* **42**, 237–256.

Bear, J.: 1972, *Dynamics of Fluids in Porous Media*, 2nd ed., New York, 76 p.

Beck, A. E. and Shen, P. Y.: 1985, 'Temperature Distribution in Flowing Liquid Wells', *Geophysics* **50**, 1113–1118.

Birch, F., Roy, R. F., and Decker, E. R.: 1968, 'Heat Flow and Thermal History in New England and New York', in E-an Zen, W. S. White, J. B. Hadley, and J. B. Thomson (eds.), *Studies of Appalachian Geology: Northern and Maritime*, Interscience, New York, 437–451.

Bredehoeft, J. D. and Papadopulos, I. S.: 1965, 'Rates of Vertical Groundwater Movements Estimated from the Earth's Thermal Profile', *Water Resources Research* **1**, 325–328.

Cavendish, J. C., Price, H. S., and Varga, R. S.: 1969, 'Galerkin Method for the Numerical Solution of Boundary Value Problems', *Soc. of Petr. Eng. J. Transaction*, **246**, 204–220.

Cheng, R. T. S.: 1978, 'Modeling of Hydraulic Systems by Finite-Element methods', in *Advances in Hydroscience*, **11**, New York, Academic Press, 207–284.

Clauser, C. and Behrens, J.: 1987, *Bibliography on Heat and Mass Transfer in Porous Media*, TU Berlin (Fed. Rep. of Germany), ISBN 3 7983 1183 8, 328 p.

Collatz, L.: 1966, *The Numerical Treatment of Differential Equations*, Berlin (Springer), 568 p.

Elder, J. W.: 1958, Ph.D. Thesis, Cambridge, England, 215 p.

Elder, J. W.: 1965, 'Physical Processes in Geothermal Areas', in H. K. Lee, *Terrestrial Heat Flow*, Geophys. Monogr. 8, Baltimore, Maryland (Port City Press), 211–239.

Elder, J. W.: 1981, *Geothermal Systems*. London (Academic Press), 508 p.

Faust, C. R. and Mercer, J.W.: 1980, 'Ground-Water Modeling: Numerical Models', *Groundwater* **18**, 395–409.

Freeze, R.A. and Cherry, J.A.: 1979, *Groundwater*, Englewood Cliffs, N.J., Prentice Hall, 604 p.

Gold, I. and Soter, St.: 1980, *Enthält das Erdinnere große Mengen Naturgas?*, Spektrum der Wissenschaft, Weinheim, 41–48.

Gröber, Erk and Grigull, U.: 1955, *Wärmeübertragung*, 3rd ed., Springer, Berlin, 428 p.

Haenel, R.: 1979, 'A Critical Review of Heat Flow Measurements in Sea and Lake Bottom Sediments', in V. Čermák and L. Rybach (eds.), *Terrestrial Heat Flow in Europe*, Springer, Berlin, 49–73.

Haenel, R (ed.): 1980, *Atlas of Subsurface Temperature in the European Community*, Th. Schäfer Druckerei GmbH, Hannover, 36 p. + 43 maps.

Haenel, R. and Zoth, G.: 1982, 'Temperature Measurements and Determination of Heat-Flow Density', in R. Haenel (ed.), *The Urach Geothermal Project, Swabian Alb, Germany*, Schweizerbart'sche Verlagsbuchhandlung, Stuttgart, 81–88.

ISO Standards: 1982, *Units of Measurements: ISO Standards Handbook 2*, ISO Central Secretariat, CH-1211 Genève 20 Switzerland, 258 p.

Jacob, C. E.: 1940, 'On the Flow of Water in an Elastic Artesian Aquifer', *Trans. Am. Geophys. Union* **21**, 574–586.

Jessop, A. M., Hobart, M. A., and Sclater, J. G.: 1975, 'The World Heat Flow Data Collection 1975', Geothermal Service of Canada, Geoth. Ser. No. 5, Ottawa, Canada, 125 p.

Kappelmeyer, O. and Haenel, R.: 1974, 'Geothermics with Special Reference to Practical Application, *Geoexplor. Monogr.* **1**, No. 4, Borntraeger, Berlin, 238 p.

Lachenbruch, A. H. and Marshall, B. V.: 1968; 'Heat Flow and Water Temperature Fluctuations in the Denmark Strait', *J. Geophys. Res.* **73**, 5829–5842.

Langguth, H. R. and Voigt, R.: 1980, *Hydrogeologische Methoden*, Springer-Verlag, Berlin, 486 p.

Lubimova, E. A., Von Herzen, R. P., and Udintsev, G. B.: 1965, 'On Heat Transfer Through the Ocean Floor', in H. K. Lee (ed.), *Terrestrial Heat Flow* Geophys. Monogr. 8, Baltimore, Maryland (Port City Press), 78–86.

Lühr, H. P.: 1973, *Vergleich der Lösungsmethoden für stationäre Grundwasserströmungen mit Hilfe einer Kostensparenden-Effektivitäts-Analyse*, Wasser und Abwasser in Forschung und Praxis, Bielefeld, 8, 115 p.

Mansure, A. J. and Reiter, M.: 1979, 'A Vertical Groundwater Movement Correction for Heat Flow, *J. Geophys. Res.* **84**, 3490–3496.

Marsal, D.: 1976, *Die numerische Lösung partieller Differentialgleichungen*, Wissenschaftsverlag, Zürich, 574 p.

Mercer, J. W. and Faust, Ch. R.: 1980, 'Ground-Water Modeling: Mathematical Models', *Groundwater* **18**, 212–227.

Myers, G. E.: 1971, *Analytical Methods in Conduction Heat Transfer*, McGraw-Hill Book Comp., London, 508 p.

Minear, J. W. and Toksöz, M. N.: 1970, 'Thermal Regime of a Downgoing Slab and a New Global Tectonics', *J. Geophys. Res.* **75**, 1397–1419.

Muffler, L. J. P. and Cataldi, R.: 1978, 'Methods for Regional Assessment of Geothermal Resources', *Geothermics* **7**, 53–89.

Mundry, E.: 1966, 'Berechnung des gestörten geothermischen Feldes mit Hilfe eines Relaxationsverfahrens', *Zt. f. Geophys.* **32**, 158–162.

Mundry, E.: 1968, *Über die Abkühlung magmatischer Körper*, Geol. Jb. 85, Hannover, 755–766.

Neumann, S. P. and Witherspoon, P. A.: 1971, 'Analysis of Nonsteady Flow with a Free Surface Using the Finite Element Method', *Water Resources Research* **7**, 611–623.

O'Nions, R. K. and Oxburgh, E. R.: 1983, *Heat and Helium in the Earth*, Report No. 381, Dept. of Earth'Sciences, University of Cambridge, UK.

Polyak, B. G. and Smirnov, Ya. B.: 1966, 'Heat Flow on Continents', in *Doklady of the Academy of Sciences of the USSR*, Earth science section, Vol. 168, Washington, 26–29.

Polyak, B. G. and Smirnov, Ya. B.: 1968, *Relationship between Terrestrial Heat Flow and the Tectonics of Continents*, Geotectonics, Acad. Sci. USSR, 4, Washington 1969, 205–213.

Stegena, L.: 1982, 'Water Migration Influences on the Geothermics of Basins', *Tectonophysics* **83**, 91–99.

Thomasset, F.: 1981, *Implementation of Finite Element Methods for Navier-Stokes Equations*, Springer Verlag, New York, 165 p.

Wang, H. F. and Anderson, M. P.: 1982, *Introduction to Groundwater Modeling, Finite Difference and Finite Element Methods*, W.H. Freeman and Company, San Francisco, 238 p.

Werner, D. and Parini, M.; 1980; 'The Geothermal Anomaly Landau/Pfalz: An Attempt of Interpretation', *J. Geophys.* **48**, 28–33.

Werner, D.: 1982, 'The Relationship Between the Volcanism and the Geothermal Anomaly in the Urach Region', in R. Haenel (ed.), *The Urach Geothermal Project, Swabian Alb, Germany*, Schweizerbart'sche Verlagsbuchhandlung, Stuttgart, 223–228.

Witherspoon, P. A.: 1977, *Geothermal Reservoir Engineering*, Report CATMEC/3, Division of Engineering. Brown University, Providence, R. I.

Zienkiewicz, O. C.: 1971, *The Finite Element Method in Engineering Science*, London, McGraw-Hill Book Co., 521 p.

3. DETERMINATION OF VIRGIN ROCK TEMPERATURES

A. E. BECK
Department of Geophysics, The University of Western Ontario, London, Canada

and

N. BALLING
Laboratoriet for Geofysik, Geologisk Institut, Aarhus Universitet, Denmark

3.1. INTRODUCTION

For a valid terrestrial heat-flow density value, the equilibrium temperature gradient must be obtained in combination with a value for the thermal properties of the materials through which a temperature gradient has been measured.

In this chapter, we will describe techniques for determining virgin rock temperatures (VRT) – that is, the temperature we would observe if we could insert a temperature sensor into the crust without using procedures such as drilling which perturb the natural regime. From these VRT's we can obtain a temperature-depth ($T-z$) profile which, in the absence of natural perturbing sources, such as climatic change or water flow, will give the radial component of the equilibrium temperature gradient. This gradient Γ is then combined, Equation (3.1), with the thermal conductivity λ of the uniform section over which Γ has been determined, to give q, the normal component of the equilibrium heat-flow density:

$$q = -\lambda\Gamma, \tag{3.1}$$

where $\Gamma = \mathrm{d}T(z)/\mathrm{d}z$, $T(z)$ is the temperature measured at depth z and z is positive downwards. If the thermal conductivity varies with depth then q is found from (see also Equations 5.2 and 7.30a):

$$T(z) = T_0 + q\sum_{n=1}^{N}(\Delta z_n/\lambda_n), \tag{3.2}$$

where Δz_n is the depth range over which the conductivity λ_n is reasonably uniform; $\sum_{n=1}^{N}(\Delta z_n/\lambda_n)$ is often referred to as the thermal depth.

The detection of, and corrections for, natural perturbing sources will be dealt with in Chapter 5.

3.2. SENSORS FOR MEASUREMENTS

3.2.1. Types of Sensor

Although platinum resistance and quartz crystal thermometers have been used to measure temperatures, and differential thermocouples to measure temperature gradients, in boreholes and oceanic sediments, by far the most used temperature

R. Haenel, L. Rybach, and L. Stegena (eds.)
Handbook of Terrestrial Heat-Flow Density Determination, 59–85.

sensor is the thermistor. The advantages of the thermistor over a Pt resistance element are that: (1) it has a very low thermal mass and therefore a very small time constant (the time required for temperature to reach $1/e$ of the initial temperature), (2) the minute sensitive element can be made with very high resistances so that even though the long cable wire resistances may still need to be taken into account to obtain absolute temperatures, variations of these wire resistances with temperature are negligible, and (3) the temperature coefficient of resistance of approximately −4 % at 20 °C is about 10 times that of Pt, – that is, an element with a nominal resistance of 10000 Ω at 20 °C will have a resistance of approximately 9600 Ω at 21 °C. Since it is easy to measure resistances very accurately it is not difficult to measure temperature differences of 1–10 mK and even 0.1 mK. A Pt–resistance thermometer calibrated against a standard (or using the well defined characteristic phase change points of standard substances) is, however, essential for calibration purposes if absolute temperatures of high accuracy are required; the quartz crystal thermometer is a useful alternative standard.

The normal resistance-temperature relation for thermistors is usually given as $R_T = a e^{-b/T}$ where T is the absolute temperature and a and b are positive 'constants'. However, both a and b are slightly temperature dependent and the standard equation does not fit to better than 10 mK over more than about 10 °C. Better fitting equations have been used (Steinhart and Hart 1968) but Lewis (1975) finds that:

$$R_T = \exp(a + b/T + c/T^2) \tag{3.3}$$

is the best fitting equation to 10 mK over a range of about 50 K but the range is narrower for an inaccuracy of <1 mK.

3.2.2. Calibration of Field Sensors

A thermostatic bath with accurate temperature control is essential for the calibration of field thermistors; this usually presents no problem since most geothermics laboratories use a technique of conductivity measurement which requires two such thermostatic baths. Although thermostatic baths capable of controlling temperatures to within 1 mK are available they are expensive. Baths with temperature control to 10 or 20 mK are very much cheaper and with the addition of a large thermal mass into the bath, the water fluctuations in the bath at the surface of the thermal mass are readily damped to the 1 or 2 mK level at its centre.

A typical mass for damping the bath temperature fluctuations consists of an aluminum cylinder 10–15 cm diameter, 10–20 cm long, standing on relatively thin legs so that the bath water may flow all around it; the price paid for this simple addition is a longer time constant when changing from one bath temperature to another. This usually is not too inconvenient since it takes several minutes to change the bath temperature by 2 or 3 K even without the thermal mass; a more sophisticated approach of using layers of thermal insulation and high specific thermal capacity material will reduce the temperature fluctuations still further but results in a much longer time constant.

Holes parallel to the cylinder axis are bored to the central plane of the thermal mass, the diameters being appropriate to the diameters of the calibrated Pt resistance (or other) thermometer and the field (or laboratory) sensors; the holes must be sufficiently long that axial heat flow along the thermometers does not cause errors that exceed the level of inaccuracy chosen. Therefore, in undertaking a calibration the bulb of the calibrated thermometer is placed in the central region of the thermal mass and the field sensors to be calibrated are grouped around it.

In a typical calibration a minimum of 10 and as many as 20 temperature points at no more than 2 or 3 K intervals over the temperature range of interest should be used in the calibration. The constants in Equation (3.3) are then found from a least squares fit and, most importantly, the measured temperatures, the calculated temperatures (using the constants derived as above) and the residuals (the difference between the observed temperatures and the calculated temperatures) are printed out. This is necessary because occasionally a thermistor will be found where Equation (3.3) does not fit over the whole range and this is detected when there is systematic curvature of the residuals, particularly noticeable at each end of the range. In such a case either another thermistor must be chosen or two sets of constants should be calculated for the upper and lower part of the range, steps being taken to ensure that the overlap is smooth – i.e. that using either equation in the region of overlap will give the same temperature to within the chosen level of inaccuracy. Table 3.1 shows the elements of a typical calibration.

Full calibration should always be made just before a field trip or oceanographic cruise is undertaken to give reassurance about the long term stability of the temperature sensors by comparison with previous calibration data. This is parti-

TABLE 3.1

Typical results of calibration against a Pt-resistance thermometer.

Pt (observed) temp (°C)	Thermistor resistance (Ω)	Calculated temp (°C)	Residual (mK) (Obs−calc)
4.6141	19699.2	4.6159	−1.8
7.1986	17578.7	7.1987	−0.1
9.7429	15742.3	9.7404	2.4
12.7646	13836.2	12.7643	0.3
14.7599	12722.4	14.7607	−0.8
17.2679	11466.9	17.2657	2.2
19.8711	10308.4	19.8730	−1.9
22.3320	9338.2	22.3293	2.7
24.9148	8427.8	24.9163	−1.5
27.5424	7605.6	27.5448	−2.4
32.6143	6266.1	32.6170	−2.7
35.1057	5710.2	35.1026	3.1
37.6420	5200.0	37.6427	−0.7
40.1231	4752.1	40.1220	1.0

Constants: $a = -4.2256836$, $b = 4411.5227$, $c = -136419.31$.

Maximum deviation of temperature = 3.1 mK.
Sums of residuals = −0.1 mK.
RMS deviation = 0.5 mK.

cularly important for an oceanographic type probe where multiple thermistors are used so that the important parameter is the absolute level of inaccuracy since temperature differences have to be taken between different sensors spaced only 50–100 cm apart. A further necessary step in oceanographic work is the final check just above the sea bottom where there is normally an isothermal layer. Before the probe is released to bury itself in the sediment it is suspended for several minutes so that one final set of thermistor readings can be obtained while all are at a given temperature.

It may be possible to carry out full calibrations at sea if the calibration equipment can be made available on board ship. However, this is not practical for field trips in many areas on land and short term stability can be checked on the return journey to the surface. A borehole is usually logged going down but the return is made without taking temperature measurements; a borehole depth is chosen where a measurement was taken on the way down and where there is no indication of water movement or other transient effects and the temperature is monitored for several minutes; if the thermistor has remained stable the reading gives the same temperature as was obtained on the way down. Usually a depth of 100 m is suitable. Medium term stability is checked on return to the laboratory simply by a short recalibration at the upper and lower temperatures of the range found in the field.

3.3. MEASUREMENTS OF SUBSURFACE TEMPERATURES IN BOREHOLES

There are two basic ways of determining temperature using a thermistor; by measuring its resistance directly, the most popular method, or by making it the resistive element in an R-C circuit in the probe and measuring a frequency at the surface. The quartz crystal technique is similar to the latter technique in that a frequency which depends on temperature is measured at the surface.

3.3.1. Resistance Measuring Technique

The thermistor is usually placed in one arm of a Wheatstone bridge with 1:1 ratio arms and a variable resistance box in the fourth arm, the resistance being determined by balancing the bridge for zero current through the central detector, Figure 3.1A. For deep borehole work the thermistor is at the end of a very long cable which introduces a cable resistance in series with the thermistor. This can be measured beforehand and an equivalent resistance put in series with the variable resistor, or, because the lead resistances may change as the cable is stressed and changes in temperature occur a 3 lead system, as shown in Figure 3.1A, is preferred; the principal assumption here is that at least two wires in the three wire cable have equal physical properties so that these can be placed in opposite arms of the bridge, an assumption which usually holds good. In any case variations with temperature are second order effects and differences in behaviour should not cause significant errors in temperature differences over small distances (of the order of a few metres); however, if there are slight differences in wire resistance that cannot be balanced with suitable series resistors, there might be significant

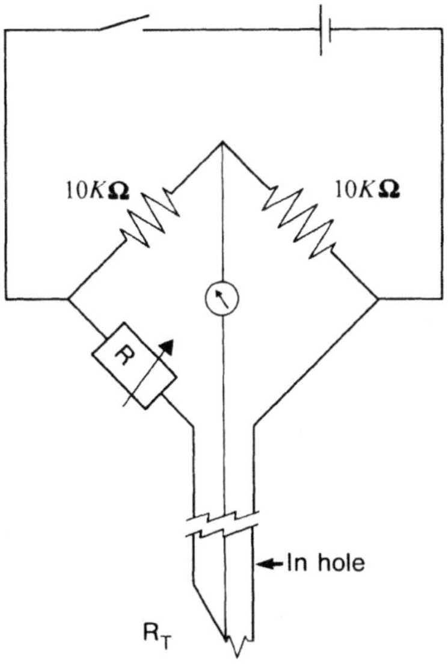

Fig. 3.1A. Schematic of 1:1 ratio arm wheatstone bridge for measuring thermistor resistances.

errors in temperature (because of errors in the wire resistance) and in temperature differences over large distances (because of cumulative effects of differences in wire behaviour). A four wire system which also compensates for the insulation shunt resistance has been described by Beck (1963).

More recently, the resistance may be measured directly with a digital multimeter; in modern instruments it is possible to use a four wire technique or to 'buck out' wire resistances and, with a small microchip suitably programmed, either resistance or temperature may be read and recorded directly. Once again the principal problem here is the assumption of equality of the physical properties of each wire in the cable.

In all cases, it is necessary to make sure that the voltage across the thermistor is sufficiently small that self heating effects are below the level of imprecision required, or else that the self heating effects are accurately known and allowed for. These effects vary according to the thermistor material used but the data are always given by the supplier in the form of a dissipation constant which is usually expressed as the number of watts needed to raise the temperature of the thermistor bead by 1 K, at 20 or 25 °C, with the bead suspended in still air; clearly the dissipation constant will be higher if the bead is suspended, or encased, in a more highly conducting medium. It is not difficult to obtain the requisite accuracy and precision without creating significant self heating errors; typical currents in a thermistor of 10 kΩ nominal resistance, usually given at 20 or 25 °C, to keep self heating effects to less than 1 mK are 10–100 μA. In other words the voltage across such a thermistor is of the order of 1 V and to measure changes of temperature of 1 mK we must measure changes in current of about 5 nA or in voltage of about 50 μV. Higher voltages can be tolerated across higher resistances, the usual law of power dissipation ($W = V^2/R = I^2R$) being observed. For work in cold regions, such as the Arctic, lower nominal resistances, e.g. 5 kΩ, might be chosen, while in hot regions nominal resistances as high as 25 kΩ might be used.

3.3.2. Frequency Measuring Technique

At the cost of increased complexity of the probe, frequency transmission systems avoid the problem of series and shunt effects of the cable leads, and voltages induced by natural phenomena such as magnetotelluric currents can be handled more easily by appropriate filtering.

For borehole use, efforts were made to use temperature sensitive elements in an R-C circuit as early as 1949 (Guelke *et al.*, 1949; Campanella, 1962, Doig *et al.*, 1961; Judge, 1972), while Hammond *et al.* (1964) appear to be the first to attempt to use a quartz thermometer. However, the early sensors, and in particular the thermistors, did not have the necessary stability and proper development or suitably sensitive instruments had to wait until the necessary technology had been developed to make the important steps of providing very stable voltages in the probe and of converting an analog signal to a frequency. The next important technical step will be to digitize the signal in the sensor head and transmit this data to the surface, thus eliminating the need to consider cable length at all.

A typical system (Bristow and Conaway, 1984) uses an integrated circuit voltage to frequency converter, in the probe head, with a sensitivity of 2 kHz K^{-1}; the usable temperature range is 0–70 °C with a chip stability of 50 ppm K^{-1} and a required power supply stability of ±0.03 % of any voltage chosen from 4.5 to 36 V. The frequency, ν_T, is measured to the necessary precision by determining the length of time necessary to accumulate a given number, N, of pulses of frequency ν_T against a crystal clock of frequency f on the surface; typically, N and f are on the orders of 10^4 and 10^7 respectively.

If m is the number of clock pulses of frequency f accumulated in N cycles of frequency ν_T, then:

$$\nu_T = (N/m)f. \tag{3.4}$$

Since ν_T is inversely proportional to R_T, the thermistor resistance at temperature T, and if D is the constant of proportionality, the relation between m and R_T is given by:

$$m = (Nf/D)\,R_T \tag{3.5}$$

or, from equation (3.3):

$$\ln(m) = A' - b/T + c/T^2, \tag{3.6}$$

where $A' = a + \ln(Nf/D)$ and the temperature may be found by solving the quadratic in T.

Apparent resolutions as high as 10 μK can be attained by choosing a clock with a high enough frequency, but the limiting factor on reliability is the stability of the chip and in practice the resolution of gradient is similar to that of the direct resistance measuring techniques.

A precise quartz thermometer for borehole use has been described by Balling *et al.* (1981). In brief the sensitive element is a quartz crystal with a nominal frequency of 9 MHz; because such a high frequency leads to transmission problems over long borehole cables the frequency is divided by 2^{12}, using a CMOS divider, to yield a convenient transmission frequency of 2 kHz. At the surface the signal is filtered, amplified and finally multiplied back again to yield a frequency of about 28 MHz, where the temperature sensitivity is about 1 $kHz\,K^{-1}$ and almost linear over the typical temperature range of 0–100 °C. The current of less than 10 mA is supplied to the probe from the surface via the one or two conductors also used to carry the signal.

A prototype system was first used in 1979 and has now been developed into a self-contained trailer mounted temperature logging unit. At present the system measures temperatures up to 125 °C with an imprecision of <1 mK and an inaccuracy of <10 mK in holes up to 6 km deep using a 3.2 mm outer diameter single conductor armoured cable.

3.3.3. Precautions for Maintaining Accuracy of Depth Measurement

For boreholes up to 1 km deep steel strengthening armour is not necessary and the cable can be made quite light, with a consequence being that light weight reels may be used, and it is possible to have a complete field equipment, including borehole pulleys and counter and measuring instruments, that weighs less than 25 kg; for holes deeper than about 1 km, armoured cable becomes necessary with a consequent very great increase in overall equipment weight and cost.

In addition to having a means of accurately measuring the length of cable payed out, either a good mechanical revolution counter or an electronic device can be used, it is necessary to mark the cable at regular intervals, say every 5 or 10 m, so that any stress lengthening of the cable can be observed; it is also a useful back-up should the counter fail for any reason. Usually, the 'stretch' that is seen in a new cable on the first two or three runs as measurements are made by the 'stop and go' or 'point by point' method on the way down the borehole, can be attributed far more to unwinding of the twisted wires than to pure elastic strain; since rewinding is usually a continuous operation there is less time for the cable to recover on the way up than there is for it to unwind on the way down. It is therefore adviseable to re-mark the cable after the first two or three runs. In practice, subsequent untwisting and stretching is negligible.

In most cases, measurements are made in water or liquid filled boreholes and, because of the way the cables are constructed, the cable may be almost buoyant. In fact, it would not be difficult to design a cable which has neutral buoyancy in water. However, if measurements are made in air, something which is not particularly recommended because of the long times required for the temperature sensor to reach equilibrium with the borehole at any given depth, then the stress on the cable is considerably greater and more stretch might occur if the cable is used in a dry hole after many uses in liquid filled holes.

3.3.4. Protection of the Sensing Head from a Hostile Environment

Nearly all types of sensor are pressure dependent; the dependency over the range of pressures encountered in borehole work is negligible for most types of thermistor material but is significant for the quartz crystal sensor. The problem may be avoided if precautions are taken to protect the sensor from pressure effects. Water, particularly if it is highly conductive because it contains salts in solution, would tend to act as a low resistance shunt of uncertain value if allowed to come in contact with the bare leads to the sensor. By placing the sensor in a pressure tight container, but making good thermal contact by means of a high conductivity cream, protection is obtained against errors arising from both pressure and shunt resistance effects. The container surrounding the sensor should be metal and of small thermal mass to allow rapid response times.

A typical sensing head is shown in Figure 3.1B. The sensor head itself is very

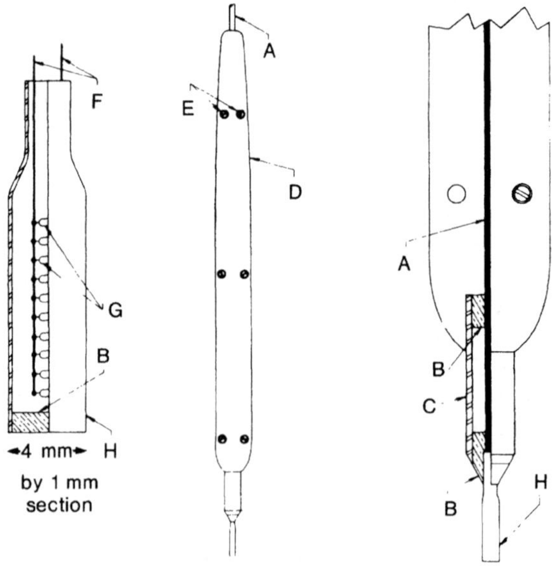

Fig. 3.1B. Schematic of probe and protective container with small time constant. A = cable, B = flexible epoxy, C = thermal barrier, D = split brassweight, E = assembly screws, F = electrical leads, G = glass bead thermistor assembly, H = low mass, high conductivity thermistor guard (after Conaway and Beck, 1977).

small and very light; because the cable may have close to neutral buoyancy in water filled boreholes a weight may be required to move it down the hole with ease. These weights usually consist of metal tubes, rigidly attached to the sensor head via a thermal insulator, to prevent conduction of heat down the metal weights to the sensor, and need weigh only a few kg, at most.

3.3.5. Continuous Logging of Temperature

More recently, there have been equipments developed to measure temperatures on a 'continuous' basis. Detailed descriptions are given by Conaway and Beck (1977), Balling *et al.* (1981) and by Beck (1982) but in brief, instead of using the 'stop and go' technique, with intervals of 5 or 10 m, the cable is driven down the borehole at a constant velocity; the constant cable velocity is obtained either by driving the cable directly (Conaway and Beck, 1977) or to drive the reel and measure the cable velocity, using this information in a feed back circuit to the reel drive electronics to vary the angular velocity of the reel in such a way as to give a constant cable velocity (Bristow and Conaway, 1984). Low noise brushes are used on a commutator on the reel hub, and resistances or frequencies are measured at regular intervals which depend upon the rate at which the cable is driven, the sample rate of the shaft encoder and the maximum level of imprecision that is required. For a typical velocity of lowering of 5–10 m min^{-1} and sample rates of 2 or 3 s^{-1}, measurements are obtained every 3–9 cm. The thermal time constant, typically about 1 s, presented by the thermal mass of the small sensor assembly is treated as a filter and its effect removed by a deconvolution process (Conaway, 1977); the constant velocity of descent simplifies the treatment of the filter effects. In practice, the thermal mass of the assembly may be so small that it is unnecessary to deconvolve to obtain accurate temperature differences over short distances (Reiter *et al.*, 1980) although this procedure may result in a small depth offset. A general discussion of suitable compromises between lithologic resolution and speed and convenience of logging is given by Nielsen and Balling (1984).

The data can be presented either as temperatures or processed to give temperature gradients directly. The inaccuracy of temperatures obtained depends principally upon the inaccuracy with which a calibration against a standard platinum resistance thermometer has been carried out, but <10 mK is easily obtained and <1 mK inaccuracy can be obtained with care; the imprecision is an order of magnitude better and changes in gradient of 0.1 mK m^{-1} can be detected over distances of 0.5 to 1 m in a thermally stable hole; an example is given in Figure 3.2.

3.4. PERTURBATIONS DUE TO CONVECTION

There are two types of convection that have to be considered: that which is induced within a borehole by the act of introducing a long cylinder of free standing fluid into the host rock, and free convection where the regional temperature gradient exceeds the critical temperature gradient in a permeable medium. The latter type is considered in Chapter 5.6.; we consider here only induced convection within a borehole.

This was first considered by Hales (1937) and Krige (1939). For a long cylinder filled with fluid the critical gradient (Γ_c), i.e. that above which convection will commence, is given by:

$$\Gamma_c = (g\alpha_v(T + 273))/c_p + (Bav)/g\alpha_v r^4 \qquad (3.7)$$

where

a = thermal diffusivity
α_v = volume coefficient of thermal expansion
c_p = the specific heat capacity at constant pressure
ν = kinematic viscosity of the fluid at T °C
g = gravitational acceleration
r = radius of the borehole
B = constant which depends on the aspect ratio of the hole; B = 216 for a large aspect ratio.

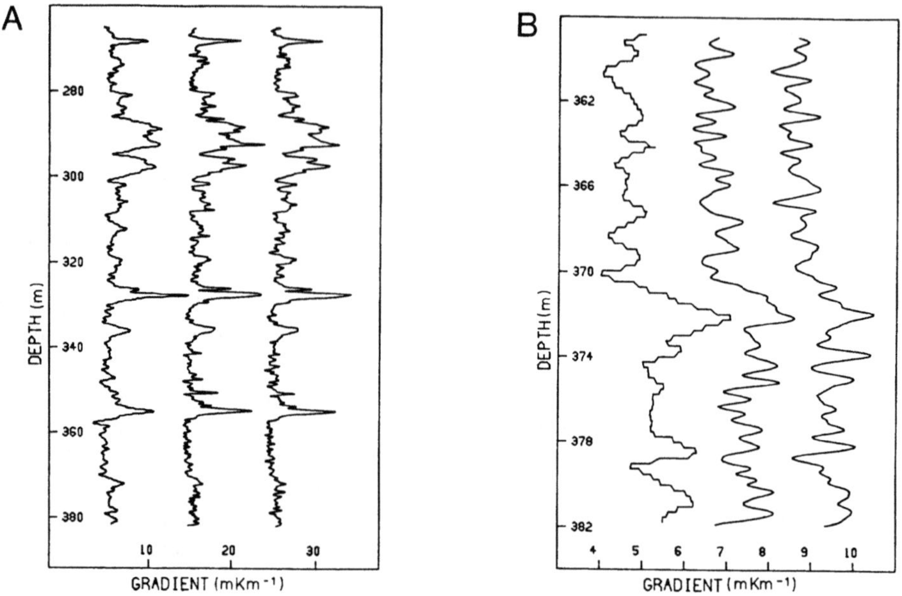

Fig. 3.2. Temperature gradient – depth plots taken several months apart using a continuous logging device through a limestone – dolostone – shale sequence of formations. The left hand log is on scale but for clarity the others are displaced by 10 mK m^{-1} and 20 mK m^{-1} respectively in (A) and by 2 mK m^{-1} and 4 mK m^{-1} respectively in (B). The left hand log was obtained with equipment that measures depth only at 25 cm intervals; the other two logs were obtained with equipment that measures depth at about 0.5 cm intervals. Note the small but definite change of 1 mK m^{-1} gradient at a depth of 371 m.

The first term is the adiabatic gradient which is usually negligible, being about 0.2 mK m^{-1} for water at 20 °C. Since borehole radius appears to the fourth power in the denominator of the second term, it can be seen that for large diameter boreholes there is a high probability that even a normal temperature gradient, in the range of 20 to 70 mK m^{-1}, will exceed the critical gradient; Figure 3.3 shows how Γ_c varies with a few given temperatures in water filled boreholes. Fortunately the value of Γ_c found from Equation (3.7) is an upper limit; in practice somewhat higher critical gradients are likely to be found (Krige, 1939).

Even if the critical gradient is exceeded whole hole convection is not possible. Cellular convection occurs, breaking down into cells which are approximately three times the hole diameter (Diment, 1967; Gretener, 1967). It therefore can be seen that in some types of stratigraphic columns, for a constant heat-flow density with depth, some sections of the hole with very low thermal conductivity, and therefore high gradients, may convect while others are stable. This means that although true regional gradients can still be obtained, the inaccuracy of temperature, and therefore gradient and heat-flow density, in any section where convection cells are formed will be greater than that for the stable sections; the convecting sections can be detected by observing periodic oscillations of temperature when the probe is stationary. In a section containing coal bearing formations, where the

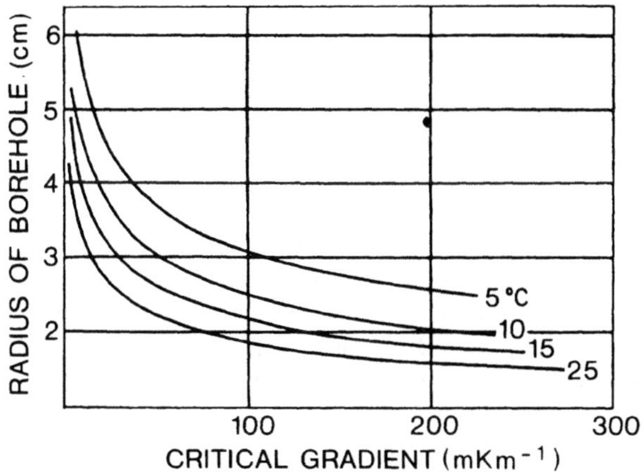

Fig. 3.3. Critical gradient, the temperature gradient above which thermal convection commences, at various temperatures (shown on the curves in °C) in a water filled borehole of radius shown (after Misener and Beck, 1960).

gradient was approximately five times that of adjacent sections, temperature oscillations of only a few hundredths of a degree were observed (Beck, 1976). If a long section of borehole exhibits a very low, or even negative, gradient, then hydraulically controlled water movement or free convection, or forced convection in the formations must be suspected.

3.5. DATA FROM HOLES THAT ARE THERMALLY PERTURBED BY DRILLING FLUIDS

Up to now we have implicitly assumed that all measurements have been made in holes that have returned to equilibrium after the formations have been thermally disturbed by the act of drilling.

Two difficulties which prevent a great increase in HFD measurement activities are the cost of drilling boreholes specifically for HFD work, and the regulatory requirements in most countries which require that the vast number of non-producing holes drilled for oil and gas, and other economic deposits, be closed off and sealed when the site is abandoned. This has led to the search for methods of making use of whatever data can be obtained from special logging procedures or from temperature data collected as an ancillary to other types of physical parameter logging.

Because of the inaccuracies involved, neither of the techniques described below is particularly recommended but there may be times when there is no alternative if some HFD data are to be obtained from a specific region.

3.5.1. Use of Bottom Hole Temperatures

In the oil and gas industry it is usual to measure the bottom hole temperature (BHT) at the end of each logging run for such parameters as resistivity and SP;

there may be as many as three or four logging runs before a well is completed so that up to four bottom hole temperatures may be available from a single hole. These temperatures are auxiliary measurements made for such purposes as applying corrections to temperature dependent physical properties, locating a zone of setting cement etc. where inaccuracies of a few degrees can be tolerated. Each individual temperature can be as much as 10 K in error, partly because of relatively crude temperature measuring devices and partly because when the logging device stops moving it may be several meters off the true bottom because the mud solids have settled out. This latter problem has two consequences. First, the temperature sensor, which is usually below the logging probe, is not at the true bottom which is the part of the hole which has been least thermally disturbed by the circulating drilling mud of very different temperature; second, the point at which the probe stops is often a region which is particularly sensitive to induced convection as the borehole fluid starts to redistribute itself in the approach back to thermal equilibrium.

In an early attempt to use this type of data, from several thousand points, Anglin and Beck (1965) found very poor correlation between depth and temperature except for the relatively few observation wells which could legally be left open for reservoir management studies and which were periodically logged for various parameters.

Over the past two decades, there has been a marked improvement in the accuracy with which the auxiliary temperature measurements are made; the level of inaccuracy is still one or two orders of magnitude larger than is typical for HFD apparatus but coupled with an increase by about an order of magnitude in the amount of data available, statistical techniques can be used to gain some useful information (Lam *et al.*, 1982; Majorowicz *et al.*, 1986). Even though it may be dangerous to try to extract too much detail from such an approach, there is one big advantage in that the results, provided they have reasonably low levels of inaccuracy, give a reliable average for the region; results from a single borehole, no matter how accurate they may be for the hole, may not be valid for regions that are more than a few borehole depths away. Considerable improvements in the quality of the thermal data are obtained when a series of temperature measurements versus time are available from the bottom of the hole, or from within 20 m of the bottom, and a suitable model used for extrapolating the time series to infinite time.

The most widely used technique is the 'Horner plot' (Horner, 1951) which was an empirical approach based upon the similarity between the diffusion and conduction equations. The equation found is:

$$T(t_s) = VRT + (\bar{Q}/4\pi\lambda)\ln(1 + t_c/t_s), \tag{3.8}$$

where

T = temperature
$\bar{Q}$ = heat supplied per unit length per unit time
λ = thermal conductivity
t_c = time before circulation ceased
t_s = shut-in time.

The technique, and its deficiencies, are illustrated in Figure 3.4. Here it is assumed that the equilibrium temperature is 80 °C and that the temperature of 40 °C of the circulating drilling fluid has been maintained for 5 h (= t_c) before circulation ceased. If readings of temperature are taken after 1, 2, 3, and 4 h shut-in time (t_s) and plotted as shown, a least squares straight line through the points would give an estimated temperature, at $\log[t_s/(t_c + t_s)] = 0$, of 66 °C, i.e. 14 °C in error. On the other hand, if readings are taken after 5, 10, 15, and 20 h shut-in time, the error is less but is still 4 °C. The problem with the Horner method is that there is an implicit assumption of no physical property constrast between the borehole mud and rock, and that the borehole is infinitesimally thin – that is, acts as a line source. Attempts have been made to improve the model by taking these factors into account, the approaches being classified into two main classes – those that assume a constant mud temperature, (Dowdle and Cobb, 1975; Middleton, 1979; Luheshi, 1983), who all use numerical solutions, and those that assume a constant heat supply from a cylindrical source. In the latter model for the special case of a homogeneous medium and line source supply, if $r = a$, the solution reduces to that of Bullard (1947); as implied earlier, if $r = 0$ the solution is that of Horner (1951). The effect of using the improved models is also shown in Figure 3.4.

A further complication is that drillers frequently report loss or gain of fluid for reasons that are sometimes, but not always, known; for example, to prevent possible blow out of the hole the mud density and pressure are usually kept at pressures higher than that of the formation fluids. This means that in permeable formations, drill fluid may be forced into the formations creating an invaded zone; not only might this affect the thermal properties, but heat is exchanged by means of the mass transfer. Model solutions must therefore be sought which include heat transfer by mechanisms other than pure conduction. Luheshi (1983) has obtained numerical solutions for the case of symmetric radial fluid flow while Shen and Beck (1986) have obtained analytical solutions for lateral fluid flow in a homogeneous medium, constant mud temperature being assumed in both cases. Ribeiro and Hamza (1986) have obtained solutions for the special case of zero circulation time and lateral fluid flow. The effect of radial loss or gain of fluid is shown in Figure 3.5 for various Peclet numbers Pe (a measure of the ratio of heat transferred by mass transfer to that transferred by conduction; see also Table 2.9); with the particular parameters used for Figure 3.5, a Pe = + 1 corresponds to a fluid loss of 0.014 $m^2\ h^{-1}$ per unit length or a movement of approximately 0.1 $m\ h^{-1}$, an amount which is not uncommon in drilling operations. As can be seen, Figure 3.5, the borehole temperature will reach to within 1% of its equilibrium value after only 5 h shut-in time provided the Pe = −5. On the other hand, without fluid gain (Pe = 0) almost 50% of the anomaly is still present, with the situation becoming worse as Pe increases.

For lateral fluid flow, because of the loss of cylindrical symmetry, the model once again has to be that of a homogeneous medium. As can be seen from Figure 3.5 the fluid flow carries the temperature anomaly away from the borehole while the amplitude of the anomaly decays slowly by conduction.

We have dealt with this problem at some length because it is important to recognize that the area of proper use of BHT's is one that requires significant

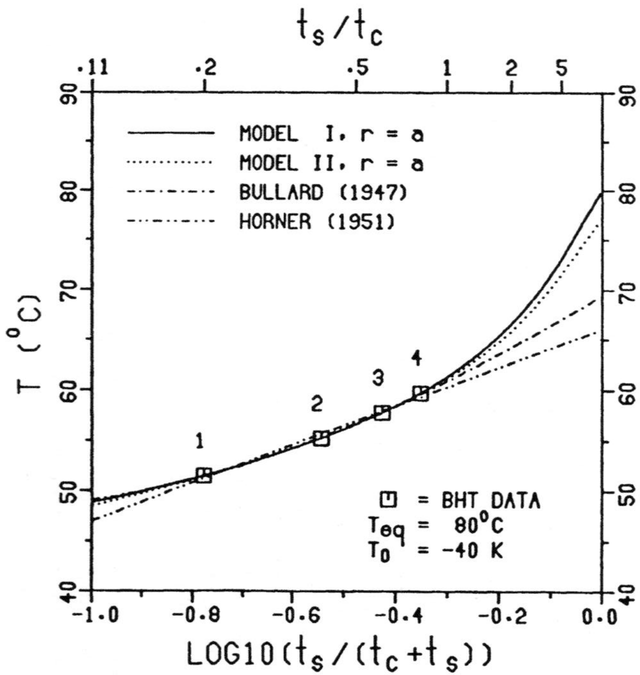

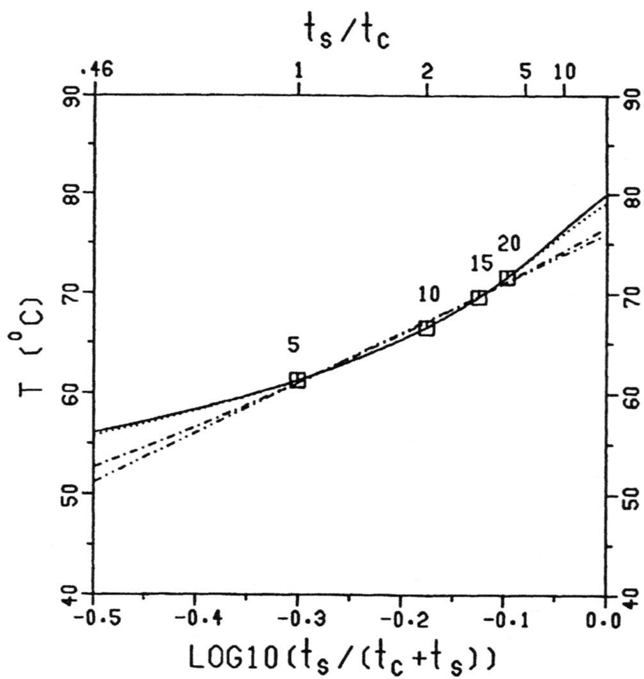

Fig. 3.4. Illustration of errors arising in use of 'Horner Plot'. VRT (T_{eq}) = 80°C, temperature of flushing mud = 40°C (i.e., temperature difference between VRT and mud is −40 K), circulation time t_c is 5 h, conductivity contrast between formation and mud = 2. Model I assumes the inner surface of the borehole at r = a is maintained at a constant temperature during circulation. Model II assumes that circulation maintains a constant heat flux across $r = a$. Temperature measurements are made at shut-in times (t_s) shown on points.

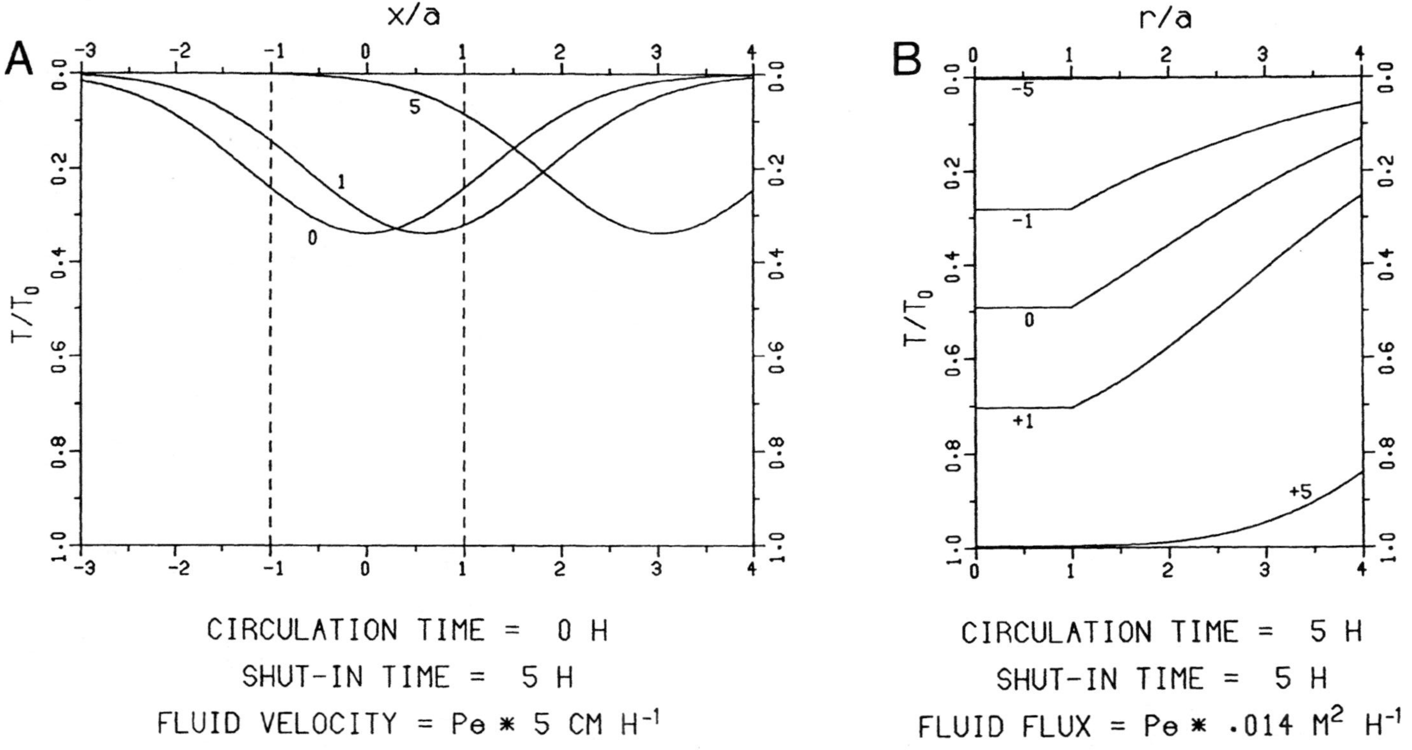

Fig. 3.5. Effect of radial loss or gain of fluid on time for borehole, of radius a = 0.11 m, and surrounding regions, porosity = 0.2, to return to thermal equilibrium. T_0 is temperature anomaly in the mud introduced by mud circulation at time circulation ceases (shut in time = 0) and all temperature anomaly distributions are relative to T_0. (A) Temperature distribution along the direction of a lateral fluid flow five hours after mud circulation ceased (shut-in time = 5 hours) and for Peclet numbers 0 (no flow), 1 and 5 with fluid velocity as shown. Other parameters used are thermal conductivities of the formation and mud = 1.23 W m^{-1} K^{-1}, specific thermal capacities of mud and formation = 3.14 J cm^{-3} K^{-1}, specific thermal capacity of formation fluid = 4.19 J cm^{-3} K^{-1}. (B) Radial temperature distributions after circulation time of 5 h, with borehole mud at T_0, and a shut in time of 5 h for a BHT stabilization model in which heat transfer is by radial fluid flow and conduction. The borehole mud is assumed to have the properties of water and behaves as a perfect conductor during both the circulation time and shut-in time. The rate of fluid flow is given by the Peclet numbers which are shown on the curves. Other parameters used are thermal conductivity of the formation = 2.5 W m^{-1} K^{-1}, specific thermal capacity of formation = 2.1 J cm^{-3} K^{-1}, specific thermal capacities of borehole mud and formation fluid = 4.19 J cm^{-3} K^{-1}.

improvement. The Horner plot always underestimates the VRT, the error increasing with decreased shut-in time and increasing initial temperature anomaly; furthermore, because the initial temperature anomaly is usually a function of depth, it can be seen that the consistent use of the Horner plot will also lead to an underestimate of the temperature gradient.

Some of the factors that need to be taken into account to produce further improvement in the quality of BHT data are variations in circulation rate, mud composition, mud inlet temperature and lithological properties, together with convection in the borehole, proximity of the hole bottom, the geothermal gradient, shut-in time, borehole radius and non-zero drilling time.

In the rare circumstance that the borehole is penetrating unconsolidated, or only lightly consolidated, sediments it may be possible to avoid some of the problems associated with BHT measurements by driving a probe into the sediments ahead of the drillhole during a drilling break when the rods are withdrawn a few meters from the bottom; the gradient can be measured over a short section below the bottom of the hole where the thermal perturbation may be expected to be minimal (Sass *et al.*, 1981).

3.5.2. Boreholes that are not in Equilibrium

For detailed investigations some workers try to make use of temperature measurements in all sections of boreholes, not just BHT's, before they have returned to equilibrium. To avoid the worst effects of the drilling procedures, it is usual to wait several hours, or perhaps a few days, before making the first of a series of temperature logs to be obtained over a period of weeks to years.

If valid corrections are to be applied it is essential to know the flushing rate of the fluid, the thermal properties of the fluid and formations, and the time over which each section of the hole has been disturbed.

Theoretically, similar techniques to those used in correcting BHT's to VRT's can be applied to any section of the hole (Bullard, 1947) but the theory is best used to estimate how long it will be before the hole, or a given section of it, has returned to equilibrium because there are additional sources of error which are very difficult to incorporate into a theory, some of which were mentioned in the previous section.

A more reliable way to use a borehole where it is not possible to wait for it to return to thermal equilibrium, is to measure the temperature at selected points, or even log continuously, at various times after drilling has ceased. Lachenbruch and Brewer (1959) find that the same equation as was used for estimating the VRT at the bottom of the hole can also be used for any depth provided t_c is now regarded as the time duration of the drilling disturbance (usually assumed to be continuous) at the chosen depth, and t_s is the time since drilling stopped. This is not surprising because Equation (3.8) is the exact solution for the cases when $\mathrm{Fo} = at/r^2 \gg 1$ (Bullard, 1947) where r is the radius of the borehole.

Therefore, at a given depth, z, a plot of temperature, $T(z, +t)$, versus the logarithmic time function will yield a straight line from which the VRT(z) at that depth can be estimated; it is, of course, not necessary to know source strength of the thermal disturbance (provided it is reasonably constant) nor the conductivity of the formation before applying the equation. Several days, and preferably weeks (which satisfies the requirement of Fo ≫ 1), are needed to define the line adequately. The technique has been used with three or four points covering a period of about one month after drilling to determine the line and VRT estimates (Čermák, 1968).

Yet another possibility may be to take measurements only in a 'transition' zone of small disturbance. Glaser and Hurtig (1984) found that in a 6 km deep hole the gradient measured over the 2–3 km range was 17 mK m^{-1} after 48 h and 16.7 mK m^{-1} after 4.5 a whereas the sections above and below this were seriously perturbed. The problem here is that even if 'zones of transition' turn out to be general, their location and length probably depend on a number of factors such as hole diameter and length, flow rate of the drilling fluid, thermal properties of the formations, temperature etc., and the location of such a zone would be difficult to predict without much more theoretical and experimental work.

3.6. DATA FROM BOREHOLES WITH HYDRAULICALLY DRIVEN WATER FLOWS

There are two types of water flow to be considered. (1) Regional flow, or advection (sometimes called forced convection), where the water is forced through permeable formations by the hydraulic regime; this type will be dealt with in detail in Chapters 2.3 and 9.4.3. (2) Flows induced by the act of drilling; these can be divided into two types. (a) Drilling fluid is forced into a permeable formation and later flows back into the hole disturbing the thermal regime; this will not be discussed further since it is a transient phenomenon of relatively short duration and the effects have usually dissipated during the time required for the hole to return to conductive thermal equilibrium. (b) Flows which are initiated when the hole connects two previously unconnected aquifers; these are discussed below.

Figure 3.6A shows typical temperature – depth plots for holes in which water was known to be flowing only along the hole. The flows are sufficiently rapid that the thermal effects are obvious and the results would not be used in HFD work. If only temperature – depth plots are used, ambiguities in interpretation of the flow regime may arise, as can be seen from Figure 3.6B. However, with high resolution logging, the use of gradient – depth plots may give much more information; Figure 3.6C shows schematically the change in shape with time for both temperature and gradient plots for two cases where water enters the hole at a cooler temperature than ambient temperature. Although the general shapes of the long time curves, particularly for temperature, are similar, detailed analyses, especially of the gradient plot, should show differences; with the development of suitable inversion techniques (Beck and Shen, 1985) it should be possible to extract information on flow rate and time since onset in those cases where flow rates are too small to be detected by traditional flow rate logging techniques. For comparison. Figure 3.6D shows a typical plot for a hole drilled in a region through which water diffuses vertically in the formations. Unfortunately, although the regional diffusion flow rate is about six orders of magnitude less than for the very local induced flow, it can be seen that the two temperature – depth plots are similar. Methods developed by Mansure and Reiter (1979) to test for diffusive flows are discussed in Section 2.3.3.

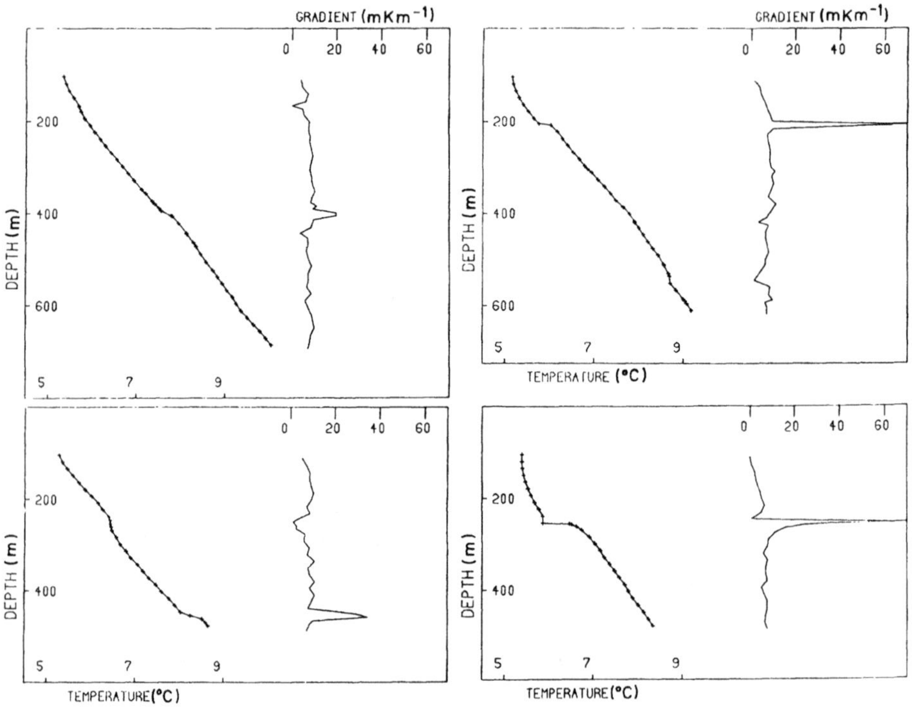

Fig. 3.6A. Temperature and gradient versus depth plots for holes with water flow induced by drilling.

3.7. MEASUREMENTS OF TEMPERATURES IN LAKE AND OCEANIC SEDIMENTS

For work in the sediments at the bottom of a lake a special probe, originally designed for use in oceanic sediments, must be used (Bullard, 1954, Langseth, 1965, Haenel, 1970); the probe penetrates the top few meters, and sometimes as much as 10 or 20 m. This means that to obtain temperature gradients, a number of thermistors have to be mounted along the probe. The probe might be of the Ewing type, Figure 3.7A, where the thermistors are mounted on outriggers or of the multiple penetration type, Figure 3.7B, which is basically a modified Bullard type probe (Hyndman *et al.*, 1979), where a thermistor string and heater are mounted in a thin tube, held in tension via a strength member, and allowing in situ conductivity measurements to be made. The former system is usually used in conjunction with a piston corer which also recovers sediment samples (for on board conductivity measurements) in the same drop as temperatures are obtained. Probes vary in length from 2 m (Davis *et al.*, 1980) to 10 m (Finckh, 1981) with the longest probes being of the Ewing type. The sensor separation is usually 1 m but more recently shorter separations are becoming favoured (Lindqvist, 1984; Francheteau *et al.*, 1984) particularly with shorter length probes. There is at

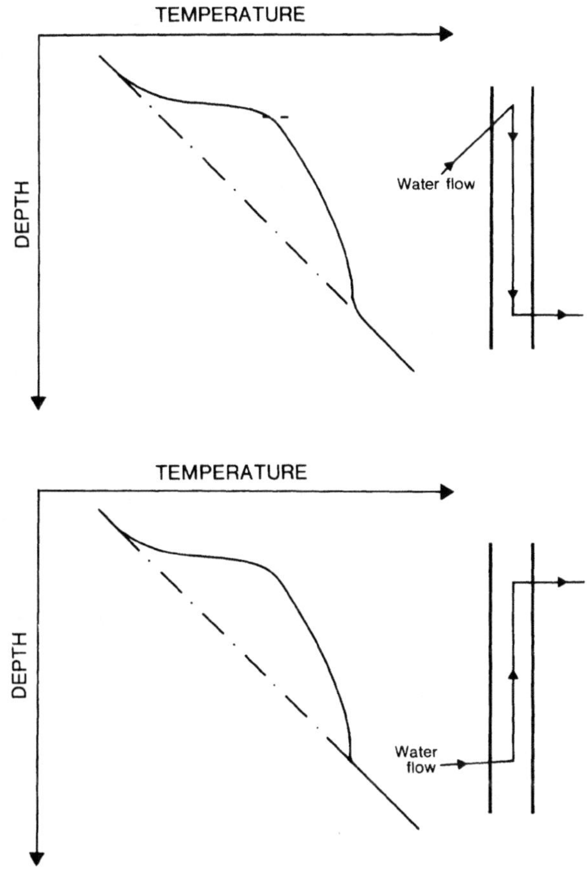

Fig. 3.6B. Schematic illustrating similarity of temperature-depth plots in thermally perturbed holes with different perturbation sources.

least 1 hollow probe where the thermistor may be moved internally for detailed measurements (Matsubara *et al.*, 1982). Because a value of the gradient, typically 70 mK m^{-1}, has to be obtained over only a few meters using different sensors, much more care has to be taken in the calibration procedures if gradients are to be obtained with inaccuracies of less than a few per cent; the inaccuracy of an individual temperature measurement should be <1 mK and the imprecision <0.1 mK.

The other principal difference between borehole and oceanographic techniques arises because the kinetic energy of the falling probe is converted mostly into heat. This heating raises the probe temperature by various amounts, which are difficult to predict, at different lengths along the probe. For lake bottom measurements there is more opportunity than in oceanic measurements to allow the thermal-kinetic energy of the probe to dissipate since there is usually no concern about ship drift and ship-board time allocation. However, if it is not possible to

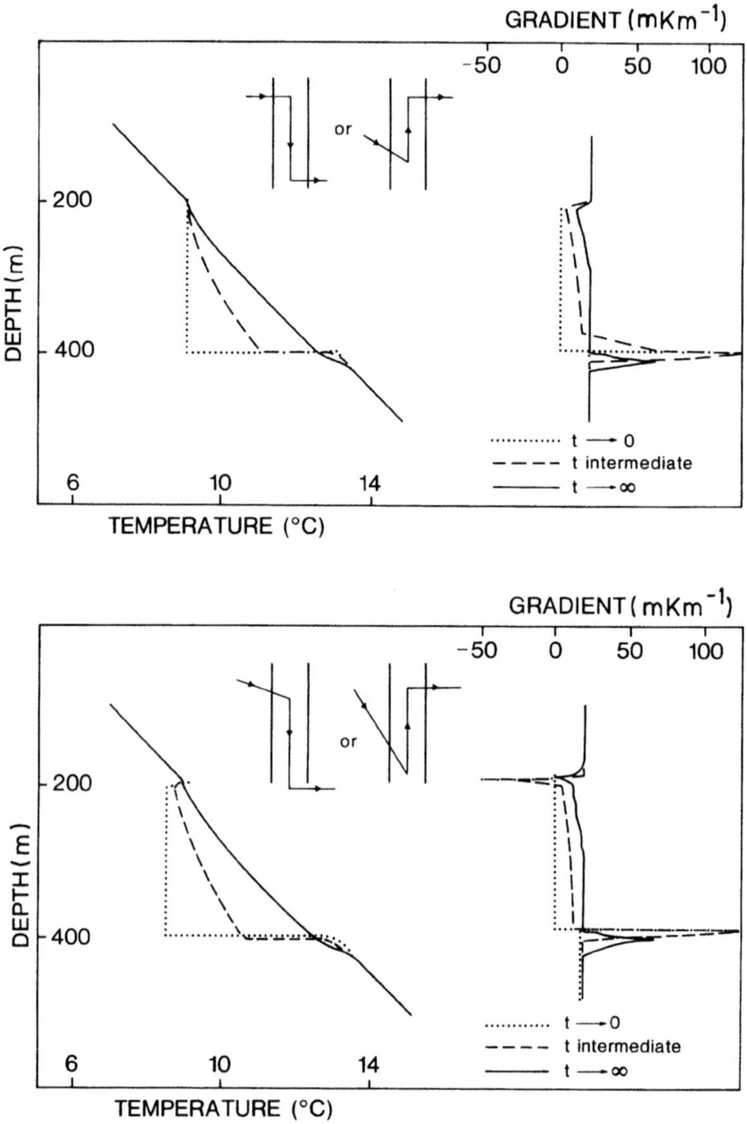

Fig. 3.6C. Illustration of how temperature and gradient profiles change with time for two types of perturbing sources.

allow time for the heat to dissipate completely, the temperature of each sensor is recorded for several minutes and the undisturbed temperature found by extrapolation of the temperature-time plot. This extrapolation is usually carried out analytically; there are several techniques available but that of Hyndman *et al.* (1979) is as good as any; see also Sections 6.4.1. and 6.4.5.

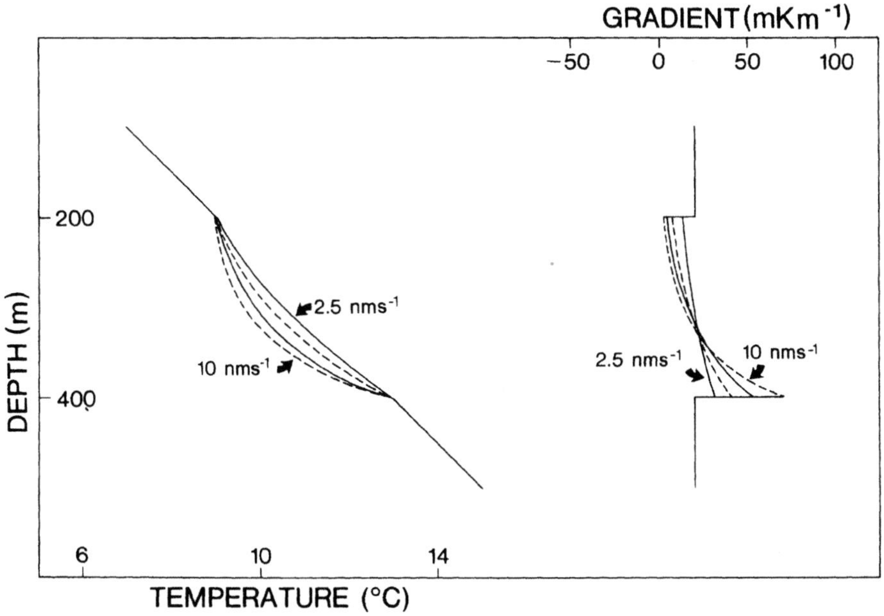

Fig. 3.6D. Illustration of temperature depth profile in a hole drilled through an area regionally perturbed by diffusive flows.

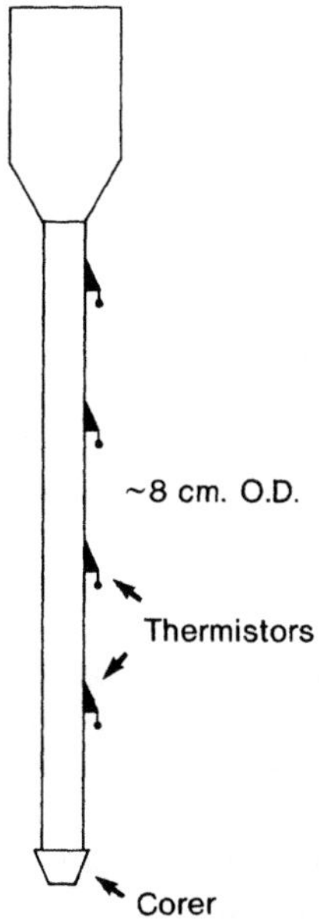

Fig. 3.7A. Ewing type probe and piston corer for lake and oceanic temperature measurements and sediment core collection; O.D. = outer diameter.

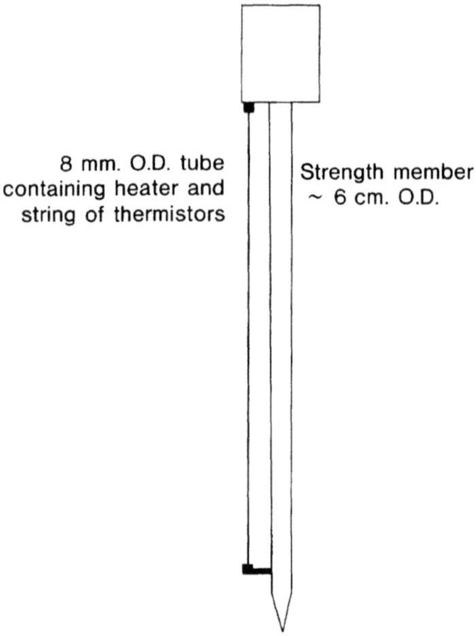

Fig. 3.7B. Multiple penetration type probe for lake and oceanic temperature measurements and in situ sediment conductivity measurements; O.D. = outer diameter.

The recording usually takes place in a pressurized container attached to the probe or corer (Langseth, 1965) but, more recently, techniques have been developed for transmitting the information in digital form to recorders at the surface (Hyndman *et al.*, 1979). The techniques do not differ significantly although there may be considerable differences in details of the electronics. The problem of correcting measured temperatures and gradients to take account of possible variations in bottom water temperature is dealt with in Chapter 5.

3.8. MEASUREMENTS IN MINES AND TUNNELS

Only occasional use has been made of holes drilled into the sides of tunnels and mine drifts (Birch, 1950; Misener *et al.*, 1951; Rybach *et al.*, 1982). As a mine is developed, exploratory holes are often drilled horizontally or at some angle to the horizontal, and occasionally vertically, to delineate the economic deposit as accurately as possible. Sometime relatively short holes may be drilled for other purposes such as to monitor the stress regime within the mine or tunnel. If the holes remain open and are accessible from within the tunnel, they may be used to determine VRT with only minor changes in technique.

In general, the holes are air-filled and not water-filled; it is therefore important to make the thermal mass of the sensor, and hence its time constant, as small as possible. The probe is pushed into the hole by means of light rods, usually broken down into sections about 2 m long which can quickly be screwed together – rather like a miniature drill rod assembly. Because the ventilation systems in mines and tunnels require the circulation of air which is at a different temperature from the VRT at various levels, the natural thermal regime beyond the tunnel walls may have been perturbed; the degree of perturbation depends on the time span of the surface disturbance as well as the magnitude of the temperature difference imposed by the ventilation. The theoretical treatment of this is discussed in Jaeger (1956) and in Section 9.4.5. In a horizontal hole the perturbation can be detected by obtaining temperature measurements at a number of distances along the hole and noting whether or not the temperature changes with distance; if the hole has dip, due allowance should be made for the natural temperature gradient. The VRT can be obtained by extrapolation of the temperature-distance plot to large distances or, since the theory predicts a logarithmic fall-off with distance, by extrapolating a log(distance) versus temperature plot back to the zero axis.

If there are a number of levels in a mine and several holes are available at each level, the advantage of using such opportunities is that a good regional value of HFD may be obtained. However, in general, it may be possible to obtain better values by making use of the numerous exploratory holes that are drilled from the surface when the mine is being proved (Lewis and Beck, 1977).

There are two major problems in using holes in and around mines. First, the nature of a mineral deposit is such that it is usually located in a structurally complex area and even extensive sampling for thermal conductivity measurements does not necessarily guarantee a reliable HFD value; considerable knowledge of the structural details of the region is needed. Second, the act of putting in shafts and tunnels may seriously disturb the hydrogeological regime (Beck and Neophytou, 1968) and lead to erroneous HFD values.

If mining regions are to be used it would be best if the surface spudded holes are used for temperature measurements before the first shaft is put down, and therefore before the hydrogeological regime has reacted to the intrusion of large diameter shafts and tunnels.

3.9. SUMMARY OF RECOMMENDATIONS

EQUIPMENT AND PRESENTATION OF DATA

1. Temperatures should be measured with equipment capable of giving imprecisions of less than 1 mK and inaccuracies of less than 10 mK; generally only electrically based sensors can achieve these figures.
2. The temperature sensing element should be protected against pressure and fluid invasion, and be separated from the main mass of the probe by a thermal resistance.
3. The time constant of the probe should be less than 5 s.
4. Temperature data should be processed to account for any significant temperature offset associated with the time constant of the element.
5. Vertical data spacing should be such that interval gradients have an inaccuracy of less than 5 %; interval gradients should not be given without also giving the depth intervals.

6. Temperatures should be given in °C and temperature gradients in mK m^{-1}. Depths should be given in meters and be referred to ground level or to water-sediment interface.
7. Each site at which a temperature-depth log has been obtained should be defined by Greenwich coordinates, and the altitude of the zero depth measurement given in m relative to mean sea level.
8. Where industrially measured temperatures, particularly bottom hole temperatures, are to be used, the inaccuracy of a single measurement should be stated and the sample number over an area should be very large. Such data sets should be used only for regional studies and not for drawing conclusions about a site.
9. Temperature data presented in maps for international use should be prepared as isotherm maps and contoured at depths of 250, 500, 3000, 4000, 5000 m, etc.; where possible the mean annual surface temperature at 0 m should be estimated; see also 5. in Chapter 7.5.

MEASUREMENTS IN BOREHOLES

10. Boreholes should be cased to minimize intrahole and interhole water movements, and to preserve them for further work – including measurements at thermal equilibrium.
11. Every effort should be made to ensure that boreholes to be used for detailed measurements are water or mud filled and in thermal equilibrium.
12. For holes not in thermal equilibrium in addition to quoting the inferred virgin rock temperature at a given depth either a time series of temperatures (preferably at logarithmic intervals of time) at that depth should be given, or details given of the corrections applied in the light of fluid circulation time, temperature and rate.
13. Holes open to depths greater than 500 m are preferred. It is important that near surface (<200 m deep) measurements be supported by evidence that there have no significant perturbing influences (climatic changes, erosion, sedimentation and topographic – see also Chapter 5.).
14. Holes preferably should be logged continuously for temperature but in any case at intervals that are not more than 10 m apart in apparently uniform material, and closer if the formations have highly variable thermal properties. Continuous logging should not be carried out in air filled holes unless the logging speed can be significantly reduced.
15. If it is necessary to compress data for publication, temperatures should be given at a minimum of 50 m depth intervals, preferably beginning at 50 m.
16. A gradient profile should be obtained before a decision is made about sampling procedures for conductivity, and other physical property, measurements.
17. Sections of holes where induced water flows are obviously perturbing the thermal regime should not be used for heat-flow density work although the data in the affected sections may be useful for other purposes.
18. Measurements in mines and tunnels should, where possible, be made at distances from the tunnel wall which are beyond the depth of penetration of the thermal disturbance caused by mine conditioning.

MEASUREMENTS IN SEDIMENTS OF LAKES AND SHALLOW SEAS

19. The minimum penetration length of the probe should be 5 m and preferably 10 m.
20. The minimum number of temperature points for a single profile should be 5 and the spacing of the sensors preferably at 1/2 m intervals.
21. Data should be processed to allow for the effects of the frictional heating which occurs when probes are forced into sediments.
22. An accurate record of λ and a variations with depth, preferably a minimum of 5 values with special emphasis on the first 2 m, should be taken at the same locations as the temperature data.
23. The precise depth of penetration and the angle of penetration should be measured independently and not be deduced from an assumption that non-linearity in the temperature-depth plot is due to miscalculation of these two parameters.
24. An accurate record of lake bottom temperature variations should be obtained, preferably with weekly readings over at least 3 a at the precise site where probe measurements will be made, and preferably three depths of 0, 1, and 2 m (see also Chapter 5.4.3.).
25. Temperature-depth measurements should be made at least twice at the same site and separated in time by at least 3 and preferably up to 6 months (see also Chapter 5.4.3.)

GENERAL

26. The depth range over which a gradient applies should be clearly stated with error limits, or α_{95} – the 95 % confidence limits, given; if the gradient changes significantly with depth the above procedure should be applied to each section for which different gradients have been obtained.
27. In tabulating the data set, not only should location and elevation be given but all stages of gradient correction should be noted down from the raw data through data corrected for topography, structure, climate to the final completely corrected gradient.

3.10. REFERENCES

Anglin, F. A. and Beck, A. E.: 1965, 'Regional Heat Flow Pattern in Western Canada', *Can. J. Earth Sci.* **2**, 176–182.

Balling, N., Kristiansen, J. I., Breiner, N., Poulsen, K. D., Rasmussen, R., and Saxov, S.: 1981, *Geothermal Measurements and Subsurface Temperature Modelling in Denmark*, GeoSkrifter 16, Aarhus University, 172 p.

Beck, A. E.: 1963, 'Light Weight Borehole Temperature Measuring Equipment for Resistance Thermometers', *J. Sci. Instr.* **40**, 452–454.

Beck, A. E.: 1976, 'The Use of Thermal Resistivity Logs in Stratigraphic Correlation', *Geophysics* **41**, 300–309.

Beck, A. E.: 1982, 'Precision Logging of Temperature Gradients and the Extraction of Past Climate', *Tectonophysics* **83**, 1–11.

Beck, A. E. and Neophytou, J. P.: 1968, 'Heat Flow and Underground Water Flow in the Coronation Mine Area', in A. R. Byers (ed.), *Symposium on the Geology of Coronation Mine, Saskatchewan*, Paper No. 68–5, Geological Survey of Canada, Ottawa, 229–239.

Beck, A. E. and Shen, P. Y.: 1985, 'Temperature Distribution in Flowing Liquid Wells', *Geophysics* **50**, 1113–1118.

Bristow, Q. and Conaway, J. G.: 1984, 'Temperature Gradient Measurements in Boreholes Using Low Noise High Resolution Digital Techniques', *Current Research, Part B, Geol. Surv. Can.* **84–1B**, 101–108.

Birch, F.: 1950, 'Flow of Heat in the Front Range, Colorado', *Bull. Geol. Soc. Am.* **61**, 567–630.

Bullard, E. C.: 1947, 'The Time Necessary for a Borehole to Attain Temperature Equilibrium', *Monthly Notices Roy. Astron. Soc. Geophys. Supp.* **5**, 127–130.

Bullard, E. C.: 1954, 'The Flow of Heat Through the Floor of the Atlantic Ocean', *Proc. Roy. Soc. A.* **222**, 408–429.

Čermák, V: 1968, 'Heat Flow in the Upper Silesian Coal Basin' *Pure and Appl. Geophys.* **69**, 119–130.

Campanella, A. J.: 1962, 'A Telemetering Thermometer', *Marine Sciences Instrum.* **1**, 39–49.

Conaway, J. G.: 1977, 'Deconvolution of Temperature Gradient Logs', *Geophysics*, **442**, 823–837.

Conaway, J. G. and Beck, A. E.: 1977, 'Continuous Logging of Temperature Gradients', *Tectonophysics* **41**, 1–7.

Davis, E. E., Lister, C. R. B., Wade, U. S., and Hyndman, R. D.: 1980, 'Detailed Heat Flow Measurements over the Juan de Fuca Ridge System', *J. Geophys. Res.* **85**, 299–310.

Diment, W. H.: 1967, 'Thermal Regime of a Large Diameter Borehole; Instability of the Water Column and Comparison of Air and Water Filled Conditions', *Geophysics* **32**, 720–726.

Doig, R., Saull, V. A., and Butler, R. A.: 1961, 'A New Borehole Thermometer', *J. Geophys. Res.* **66**, 4263–4264.

Dowdle, W. L. and Cobb, W. M.: 1975, 'Static Formation Temperature from Well Logs – an Empirical Method', *J. Petrol. Technol.* **27**, 1326–1330.

Finckh, P.: 1981, 'Heat Flow Measurements in 17 Perialpine Lakes', *Geol. Soc. Am. Bull. Part II*, **92**, 452–514.

Francheteau, J., Jaupart, C., Shen, X., Kang, W., Lee, D., Bai, J., Wei, H., and Deng, H.: 1984, 'High Heat Flow in Southern Tibet', *Nature* **307**, 32–36.

Glaser, S. K. and Hurtig, E.: 1984, 'The Establishment of the Thermal Equilibrium in a Deep Borehole in the Northern Part of the German Democratic Republic', *Tectonophysics* **103**, 47–54.

Gretener, P. E.: 1967, 'On the Thermal Instability of Large Diameter Wells – an Observational Report', *Geophysics* **32**, 727–738.

Guelke, R., Heydenrych, J. C. R., and Anserson, F.: 1949, 'Measurement of Radioactivity and Temperature in Narrow Holes and Development of Instruments for this Purpose' *J. Sci. Instrum.* **26**, 150.

Haenel, R.: 1970, 'A New Method for the Determination of the Heat Flow in Lakes', *Z.f. Geophysik* **36**, 725–742.

Hales, A. L.: 1937, 'Convection Currents in Geysers', *Monthly Notices Roy. Astron. Soc. Geophys. Supp.* **4**, 122–131.

Hammond, D. L., Adams, C. A., and Schmidt, P.: 1964, *Linear Quartz Crystal Temperature Sensing Element*, 19th Annual I.S.A. Conf.

Horner, D. R.: 1951, 'Pressure Build-up in Wells', *Proc. Third World Petroleum Congress* **34**, 316, The Hague.

Hyndman, R. D., Davis, E. E., and Wright, J. A.: 1979, 'The Measurement of Marine Geothermal Heat Flow by a Multipenetration Probe with Digital Acoustic Telemetry and *in situ* Conductivity', *Marine Geophys Res.* **4**, 181–205.

Jaeger, J. C.: 1956, 'Numerical Values for the Temperature in Radial Heat Flow', *J. Math. Phys.* **34**, 316–321.

Judge, A. S.: 1972, 'Geothermal Measurements in a Sedimentary Basin', Ph.D. Thesis, U. of Western Ontario, London, Canada, 377 p.

Krige, L. J.: 1939, 'Borehole Temperatures in the Transvaal and Orange Free State', *Proc. Roy. Soc. A.* **173**, 450–474.

Lachenbruch, A. H. and Brewer, M. C.: 1959, 'Dissipation of the Temperature Effect of Drilling a Well in Arctic Alaska', *U.S.G.S. Bull.* **1083–C**, 73–109.

Lam, H. L., Jones, F. W., and Lambert, C.: 1982, 'Geothermal Gradients in the Hinton Area of West-Central Alberta', *Can. J. Earth Sci.* **19**, 755–766.

Langseth, M. G.: 1965, 'Techniques of Measuring Heat Flow Through the Ocean Floor', in W. H. K. Lee (ed.), *Terrestrial Heat Flow* Monograph No. 8, A.G.U., Washington, 58–77.

Lewis, T. J.: 1975, 'A Geothermal Survey at Lake Dufault, Quebec', Ph.D. Thesis, U. of Western Ontario, London, Canada, 340 p.

Lewis, T. J. and Beck, A. E.: 1977, 'Analysis of Heat Flow Data: Detailed Observations in many Holes in a small Area', *Tectonophysics* **41**, 41–59.

Lindqvist, J. G.: 1984, 'Heat Flow Density Measurements in the Sediments of Three Lakes in Northern Sweden', *Tectonophysics* **103**, 121–140.

Luheshi, M. N.: 1983, 'Estimation of Formation Temperature from Borehole Measurements', *Geophys. J. Roy. Astr. Soc.* **74**, 747–776.

Majorowicz, J. A., Jones, F. W., and Jessop, A. M.: 1986, 'Geothermics of the Williston Basin in Canada in Relation to Hydrodynamics and Hydrocarbon occurrences', *Geophysics* **51**, 767–779.

Mansure, A. J. and Reiter, M.: 1979, 'A Vertical Groundwater Movement Correction for Heat Flow', *J. Geophys. Res.* **84**(B7), 3490–3496.

Matsubara, Y., Kinoshita, H., Uyeda, S., and Thienprasert, A.: 1982, 'Development of a New System for Shallow Sea Heat Flow Measurements and its Test Application in the Gulf of Thailand', *Tectonophysics* **83**, 13–31.

Middleton, M. F.: 1979, 'A Model for Bottom-Hole Temperature Stabilization', *Geophysics* **44**, 1458–1462.

Misener, A. D. and Beck, A. E.: 1960, 'The Measurement of Heat Flow over Land', in S. K. Runcorn (ed.) *Methods and Techniques in Geophysics* Interscience, New York, 10–61.

Misener, A. D., Thompson, L. G. D., and Uffen, R. J.: 1951, 'Terrestrial Heat Flow in Ontario and Quebec', *Trans. Am. Geophys. Un.* **32**, 729–738.

Nielsen, S. B. and Balling, N.: 1984, 'Accuracy and Resolution in Continuous Temperature Logging', *Tectonophysics* **103**, 1–10.

Ribeiro, F. B. and Hamza, V. M.: 1986, 'Stabilization of Bottom Hole Temperature in the Presence of Formation Fluid Flows', *Geophysics* **51**, 410–413.

Reiter, M., Mansure, A. J., and Peterson, B. K.: 1980, 'Precision Continuous Temperature Logging and Comparison with Other Types of Logs', *Geophysics* **45**, 1857–1868.

Rybach, L., Bodmer, Ph., Weber, R., and England, P. C.: 1982, 'Heat Flow and Heat Generation in the New Gotthard Tunnel, Swiss Alps', in V. Cermak and R. Haenel (eds.), *Geothermics and Geothermal Energy*, Schweizerbartsche Verlagsbuchhandlung, Stuttgart, 63–69.

Sass, J. H., Kennelly, J. P. Jr., Wendt, W. E., Moses, T. H. Jr., and Ziagos, J. P.: 1981, '*In situ* Determination of Heat Flow in Unconsolidated Sediments', *Geophysics* **46**, 76–83.

Shen, P. Y. and Beck, A. E.: 1986, 'Stabilization of Bottom Hole Temperature with Finite Circulation Time and Fluid Flow', *Geophys. J. Roy. Astr. Soc.* **86**, 63–90.

Steinhart, J.S. and Hart, S. R.: 1968, 'Calibration Curves for Thermistors', *Deep-sea Res.* **15**, 497–503.

4. THERMAL PROPERTIES

4.1. METHODS FOR DETERMINING THERMAL CONDUCTIVITY AND THERMAL DIFFUSIVITY

A. E. BECK
Department of Geophysics, The University of Western Ontario, London, Canada

4.1.1. Introduction

Heat may be transported in a number of different ways but the mechanisms fall into two classes – those that are of purely thermal origin and those that are not. In the latter class we would place advective processes – the movement of fluids driven by, for example, a hydraulic gradient; although of considerable importance in terrestrial heat-flow density work, these processes are not discussed in this chapter (but see Chapters 2.3., 5.5., 5.6., 9.4.3.). In the former class the three principal mechanisms are convection, radiation, and lattice (or phonon) conduction. The relative contributions from each of these three mechanisms depends upon the prevailing local conditions. For example, radiation transfer is not very significant below 1000 K (see also Chapter 4.3.) but above this temperature can be represented by a coefficient λ_R where:

$$\lambda_R = 16\, n^2 \bar{\sigma} T^3 / 3\bar{\varepsilon}, \tag{4.1}$$

where

n = mean refraction index
$\bar{\varepsilon}$ = mean extinction coefficient
$\bar{\sigma}$ = Stefan-Boltzmann constant
T = absolute temperature.

The term 'convection' is used here sensu strictu in that the transfer of heat by convection occurs only when the temperature gradient exceeds the critical gradient, the latter being a function of the temperature of the region, the permeability and diffusivity of the containing matrix, and the expansion coefficient, the kinematic viscosity and the diffusivity of the fluid in the containing medium.

The heat transferred by conduction depends only upon the thermal diffusivity, a, and the temperature gradient, where a is the ratio of the thermal conductivity to the thermal capacity of the medium; in the steady state case, that is when the system does not change temperature with time, the amount of heat transferred per unit area per unit time is proportional only to the thermal conductivity and the temperature gradient.

For simple isotropic single crystals at temperatures well above the Debye temperature, the lattice thermal conductivity λ_L can be defined in terms of lattice

R. Haenel, L. Rybach, and L. Stegena (eds.)
Handbook of Terrestrial Heat-Flow Density Determination, 87—124.

parameters by:

$$\lambda_L = c_v v A_0 / 3\alpha_v \gamma T^n, \tag{4.2}$$

where

c_v = specific heat capacity at constant volume
v = mean sound velocity
A_0 = lattice constant
α_v = volume coefficient of thermal expansion
T = absolute temperature
γ = Grüneisen parameter.

γ is essentially a measure of the fractional change of pressure that occurs when the material is heated at constant volume; and the exponent range is: $1.25 > n > 1$.

Theoretical values for conductivity obtained using Equation (4.2), and equivalent derivations, agree well with measured values obtained for single crystals of cubic structure. However, even for complex materials the conductivities are found to follow the inverse temperature relation quite well over a wide range of temperatures, even if it is impossible to obtain a reliable calculated absolute value.

Specific thermal capacity c_t of a medium is defined as the amount of heat per unit volume required to raise the temperature 1 K and is the product of the specific heat capacity c_p at constant pressure and of the density ρ of the material. It is preferable to accept c_t as a thermal property, rather then just the specific heat capacity, since the c_t has a built in self-compensating factor in that for the great majority of minerals and impervious rocks it lies within 20 % of 2.3×10^{-6} J m^{-3} K^{-1}.

For porous rocks, c_t can be calculated from the c_t of the matrix and fluid, weighted according to their volume proportions. Because of the relative constancy of c_t it is often much simpler to obtain the thermal diffusivity a by measuring the thermal conductivity λ with an established technique and determining the diffusivity from the equation:

$$a = \lambda_L / \rho c. \tag{4.3}$$

In this chapter, discussion is limited mainly to methods and techniques of measuring thermal conductivity and thermal diffusivity at normal temperatures (from zero to a few tens of °C) and pressures (up to a few kPa); only brief reference will be made to techniques required for measurements at high pressures and/or temperatures since the pressure and temperature coefficients of conductivity have opposite signs and the situations where the effects become important in heat-flow density work are rare.

The most commonly used methods for measuring thermal conductivity may be divided broadly into two classes, (a) one dimensional (linear) steady state and (b) two dimensional (cylindrical) transient types.

In principle, both classes may be used to give absolute values or relative values, compared to some standard, but in practice the great majority of people use the steady-state method in the comparative mode and the transient method in the absolute mode. The steady-state techniques give only the thermal conductivity while the transient techniques give either or both of thermal conductivity and thermal diffusivity. For this reason the discussion of transient techniques in Section 4.1.3. will cover the determination of both λ and a.

At this stage it must be pointed out that in every technique, without exception, the introduction of a specimen into the system (or of a probe into the specimen) creates a thermal property discontinuity which disturbs what would otherwise be the natural flow of heat in a continuous specimen. Since the contact between two dissimilar materials is never perfect, a thermal contact resistance is created which can never be eliminated and depends on many things, including the structure of the specimen, the size of the grains, the shape of the grains, the radii of curvature of the contacting surfaces, and the conductivities of the materials involved; however, in many cases steps can be taken to make it negligible or else susceptible to determination and the application of appropriate corrections. Figure 4.1A and 4.1B illustrates how the problem arises.

4.1.2. Steady State Methods for Measuring Thermal Conductivity

Some investigators take elaborate precautions to eliminate errors due to small transverse heat losses and thermal contact resistance, whatever the technique being used, to obtain high accuracy (low inaccuracy – see Section 4.1.6(2)) for the measurement of the thermal conductivity of a given specimen. In some special circumstances (Ratcliffe, 1959) such precautions may be necessary; however, in most cases low inaccuracy (e.g. <1 %) is quite unnecessary for two reasons. First, in many problems, e.g. calculating the effects of a sinusoidal or step change of temperature at one boundary of the medium, the diffusivity appears as a square root and relatively large errors can be tolerated. Second, the measurements are usually made on a small specimen of rock which is not necessarily representative of the rock in bulk, and conductivities of coarse grained materials can differ by as much as 50 % for contiguous specimens: therefore it is much better to establish a method which will give, on reasonably sized specimens (about 10 cm^3), results moderately quickly and then take the average of a number of measurements from different specimens of the same rock, rather than spend the same amount of time, or even more time, striving for low inaccuracy on one specimen. For this reason discussion is limited to methods requiring only relatively simple and inexpensive equipment designed to give results fairly quickly (in about 5–10 min) with inaccuracies <5 %, although with a little extra care inaccuracies of <1 or 2 % can be obtained.

(1) LINEAR HEAT FLOW TECHNIQUES

Simple Apparatus for most Situations

For solid rock specimens the most commonly used technique is the steady-state comparison method. The basic idea is to compare the conductivity of an unknown specimen with that of a reference material in the system, the heat flowing through the system having to pass through both the reference material and the unknown specimen (as well as any contact resistances). Figure 4.1C shows the elements used in simple modelling of the plane contact case for typical steady state

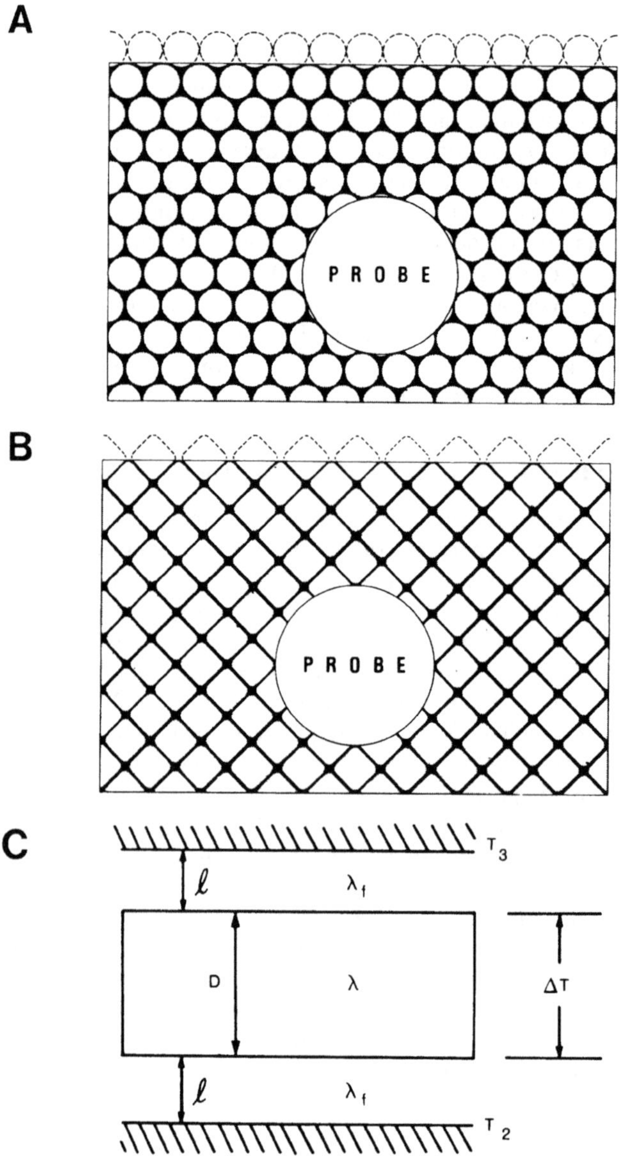

Fig. 4.1. Illustration of how granular thermal contact resistance for both divided bar and probe type measurements is a function of macroscopic structure, conductivity of the components, and size and shape of grains. Solid black portions represent pores, light shading represents grains, clear areas represent extra voids created, usually by grain plucking, when a surface of larger radius of curvature than the grains is introduced into the medium. (A) Spherical grains. (B) 'Round cube' grains; a side length of the rounded cube grain equals the diameter of spherical grains in (A). (C) Simplified equivalent film representation of thermal contact resistance between a specimen and divided bar surface; if contact resistance is negligible, $\Delta T = T_3 - T_2$.

techniques. Using the symbols in Figure 4.1 and summing the thermal resistances of the segments the apparent conductivity, λ_{app}, in the presence of a contact resistance is given by:

$$(D + 2l)/\lambda_{app} = (2l/\lambda_f) + D/\lambda \qquad (4.4)$$

for apparatus where the temperatures T_3 and T_2 of the contacting apparatus surfaces are determined. D is usually very much larger than $2l$; if $2l/\lambda_f$ is negligible with respect to D/λ then the experiment will give the correct value of conductivity with small error. Further discussion of the contact resistance problem will appear later.

The reference material may be of relatively high conductivity, in which case it is used in the form of a bar several centimeters long (Misener and Beck, 1960), or it can be a relatively low conductivity material, perhaps of the same order of magnitude as the unknown specimens, in which case it would be used as a disk of approximately the same thickness as the unknown specimen (Birch, 1950). In both cases the separation between the temperature sensors is known and if, as is usual, the transverse heat losses are negligible the amount of heat flowing along the system can be determined by finding the temperature gradient in, or across, the reference material and then using this value of heat-flow density to determine the conductivity of the unknown specimen, having first found the temperature gradient in the unknown specimen. In a few instances, the unknown specimen is itself a bar (e.g. a drill core) in which temperature sensors are placed at known separations (Zierfuss and van der Vliet, 1956).

A number of people believe that it takes a long time to establish thermal equilibrium in a steady state equipment. This is only true in poorly designed equipment. In early designs (Benfield, 1939; Bullard, 1939) an electrical heater was used in an asymmetrical apparatus resulting in long times for equilibrium to be established; however, with proper design, and usually a symmetrical system will satisfy this criterion, equilibrium for specimens approximately 1 cm thick can be established within 5 to 10 min (Beck, 1957). There have been occasional reports of continual drift during measurements on saturated porous materials; since it is not possible for heat to be transferred in the vapour phase in a completely saturated material (Hutchinson *et al.*, 1948; Kuzmak and Sereda, 1958a, b), and for moist (incompletely saturated) material movement in the vapour phase is arrested after a short time if there is no loss of moisture from the system (Beck and Beck, 1958), the cause of the drift is almost certainly due to the loss of moisture from the sides of the specimen and can be prevented by coating the cylindrical outside surface of the specimen with a waterproof paint or grease.

Figure 4.2 illustrates three basic types of comparator apparatus; there are many mixed variations on these themes. In all cases pressure of a few hundred kPa is applied to stabilize the system and make the thermal contact resistance from specimen to specimen more uniform. There is a common misconception that the thermal contact resistance decreases as the pressure is increased and that it is therefore good practice to apply very high axial pressures. To a certain extent this is true, up to the stabilizing pressures of a few kPa (or a few tens of bars); there is little advantage in going beyond this sort of pressure and there may be some

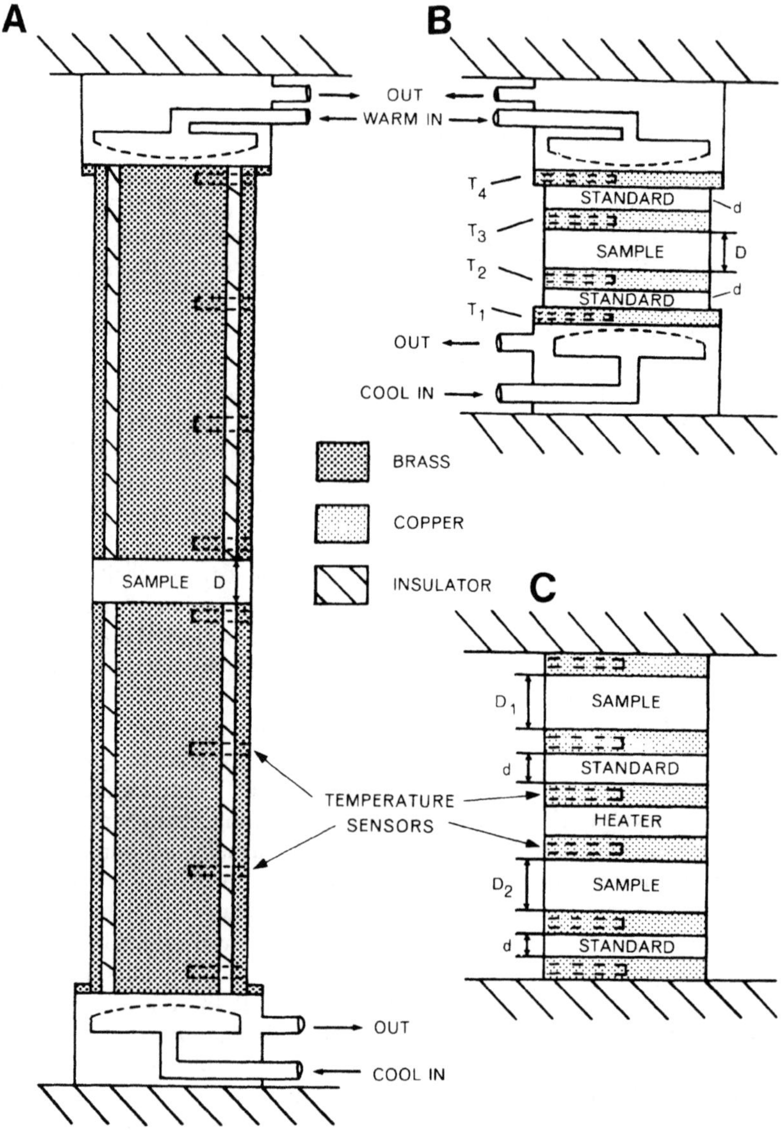

Fig. 4.2. (A) Divided long bar with simple guard ring. (B) Stacked disks or divided short bar apparatus with constant temperature difference across the stack. (C) Stacked disks with heat supplied electrically. Cross-sections of apparatus are normally, but not necessarily, circular. Some workers use insulation around the bars, others use simple draught screens.

significant disadvantages. With hard specimens, especially if poorly prepared, there is a risk of damaging the contact surfaces and even of cracking the standards; with soft, porous specimens there is a risk of damaging the natural structure and therefore of obtaining erroneous results.

Figure 4.2A is a divided bar with simple guard ring and a constant temperature difference maintained across the ends. The conductivity of the bar material is found using standard materials (e.g. fused quartz and crystalline quartz) and using disks of different thicknesses to determine the thermal contact resistance effects. The principal advantages of the simple guard ring are that the errors due to (a) minor mismatch of diameters between bar and specimen, (b) slight asymmetry of specimen placement, and (c) small chips in the edges of the specimen are minimized; the bath temperatures can be chosen so that the mean specimen temperature is the same as the ambient temperature. In a typical experiment with the material of unknown conductivity in place, the heat flowing down the bar is found from the product of the bar conductivity and the temperature gradient along the bar. In principle a correction should be applied to allow for the distortion in what should be parallel heat flow lines in the specimen, due to the fact that the specimen is placed across the central bar and guard ring. However, provided the thicknesses of the specimen disks are within the range of the thicknesses of the standard disks used for calibration, the errors in the value obtained are less than 2 %. The principal disadvantage of the method is the long length of bar over which lateral heat losses might occur, although the effects are considerably reduced by the guard ring.

Using Equation (4.4) and combining the nomenclatures in Figures 4.1 and 4.2, the thermal conductivity is found from:

$$D/\lambda = (\Delta T/\lambda_b \Gamma_b) - (2l/\lambda_f), \tag{4.5}$$

where λ_b is the conductivity of, and Γ_b is the temperature gradient in, the reference bar. If $2l/\lambda_f$ is not small enough to be ignored, then by making the reasonable assumption that for disks of the same material which have been prepared in the same manner the contact resistance for disks of different thicknesses will not vary significantly, the contact resistance can be found and eliminated since a plot of total thermal resistance against the thickness D should give a straight line the slope of which will give λ and the intercept will give the contact resistance, Figures 4.2D and 4.2E.

Figure 4.2B shows a variation of Figure 4.2A with the 'bars' now being disks of standard material. In general, the procedures are similar to those for the apparatus of Figure 4.2A, but it is usually assumed that the thermal contact resistance can be ignored compared to the total thermal resistance of the standard and/or the unknown disk.

Clearly, we now have contact resistances between the standards and the temperature measuring disks. From measurement to measurement these contact resistances remain constant but since the product of the temperature gradient across the standard (plus contact) and the conductivity of the standard are used to calculate the heat flowing along the system, there will be an error in this value if the contact resistance in not negligible. Relative values of specimen conductivity will be reasonably accurate but the absolute values could be in serious error.

Fortunately, the contact resistance between the standard and the temperature measuring disks can be made negligible, by good surface preparation and using either water or, if a more permanent union is required, a highly conducting silver

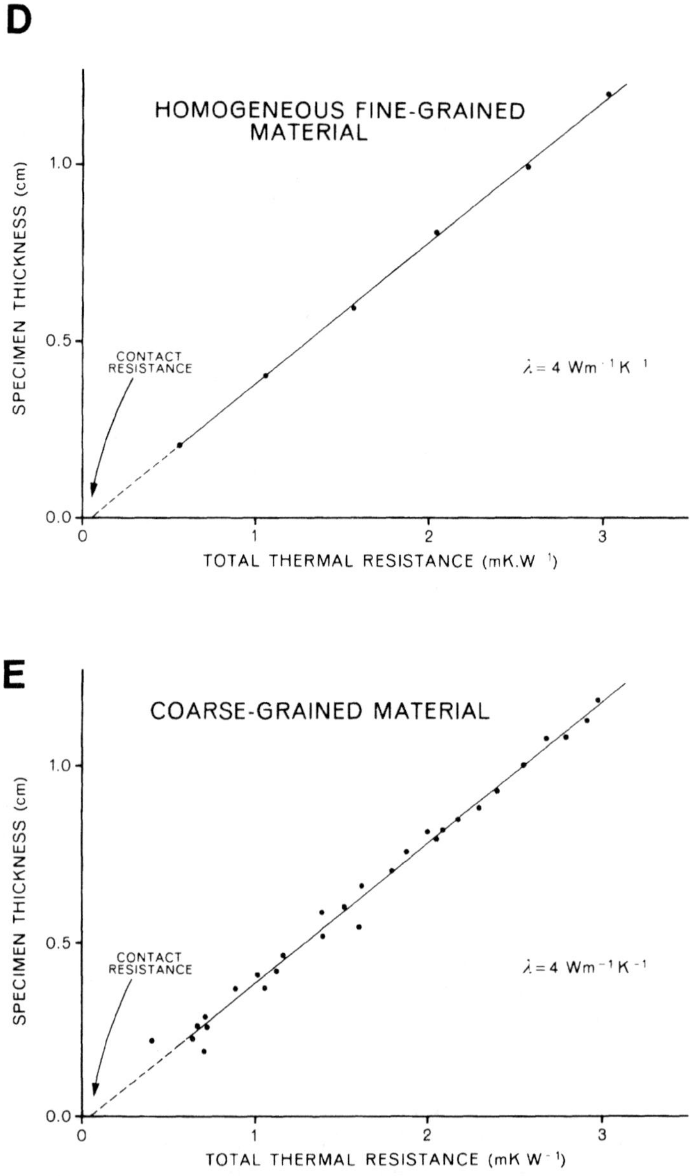

Fig. 4.2. (D) Plot of total thermal resistance (specimen + contact film) versus thickness for a homogeneous fine grained rock; in this example a standard contact resistance could be subtracted from any one total value and a reliable value of the bulk conductivity of the rock obtained. (E) Similar to (D) but for a coarse grained rock; in this case the standard contact resistance subtracted from the total thermal resistance of some of the single specimens would give an accurate value for that specimen but an inaccurate value for the conductivity of the rock in bulk.

loaded epoxy resin; if we carry the argument further and assume that the contact resistances between the specimen and the temperature measuring disks are also negligible, appropriate choice of standard thicknesses can lead to considerable simplification of procedures.

If contact resistances are negligible, and considering only the lower part of Figure 4.2B, the relevant equation for the heat-flow density q is:

$$q = \lambda_s(T_2 - T_1)/d = \lambda(T_3 - T_2)/D, \tag{4.6}$$

where λ_s is the thermal conductivity of the standard being used. A suitable standard is fused silica, which is cheap, well characterized and easily obtained with optically finished surfaces. The conductivity is 1.36 $Wm^{-1} K^{-1}$ at 22°C, a typical room temperature; if we now choose $d = 0.34$ cm and $D = 1$ cm, which for a typical rock means that the thermal resistances of standard and specimen are about the same, then:

$$\lambda = 4(T_2 - T_1)/(T_3 - T_2) = 4(\Delta T_1/\Delta T_2) = 4(\Delta V_1/\Delta V_2) \tag{4.7}$$

if, further, differential thermocouples are used to measure ΔT_1 and ΔT_2 and if, as is usual, the temperature differences are in the 2 to 5 K range, the voltage output (ΔV) from the differential thermocouple is a linear function of the temperature difference and the conductivity can be obtained from a direct readout on a two coil meter with a simple in the head multiplication of 4. Of course, one can achieve the same effect these days with even inconvenient parameters by means of appropriate on-line micro-computers; if a typical contact resistance for a given apparatus and disk parameters is known, this too may be taken into account for single disks by appropriate programming and use of Equation (4.5).

Figure 4.2C shows an arrangement where the heat is supplied to the stack at its centre and two specimens of the same material, but of different thicknesses, are measured simultaneously. The stack can be arranged asymmetrically as shown in the figure, or symmetrically by interchanging the specimen and standard on one side. It is sometimes used with only one specimen in the stack. In the mode shown, the equipment is used as a comparator but it could also be used as an absolute apparatus if the heat is supplied by an electrical heater and the heat input is measured accurately. Of the three types shown in Figure 4.2, this arrangement takes the longest time to reach equilibrium; this time to equilibrium could be reduced by placing each end in a thermostatically controlled bath. Yet another variation is to supply heat at the centre by means of thermostatically controlled water, giving what is essentially a double bar type of Figure 4.2A. The appropriate procedures and equations are essentially the same as for the technique shown in Figure 4.2B.

Modifications for Special Situations

Any of these methods can readily be modified for use when unusual situations arise. For example, Sass *et al.* (1971), have described a technique used for unconsolidated sediments or rock cuttings which required only a minor modification to apparatus of the type shown in Figure 4.2B, although the measurements must be combined with computational methods (Section 4.1.4.).

On the relatively rare occasions when data are required far outside the normal range of temperatures (say greater than several tens of °C) then a separately controlled guard heater may be required to reduce transverse heat losses to negligible proportions. This is fairly simple in principle but the complication introduces long time delays into the process, for example see McElroy and Moore (1969), and the advantage of rapid measurements is lost.

A variant of the long divided bar method, Figure 4.2A, was used by Zierfuss and van der Vliet (1956). The unknown material was used as one bar into which holes were drilled for temperature sensors to give the gradient in the specimen; the other bar was a material of known but similar conductivity to the unknown and used as the standard for comparison. This technique removes the problem of contact resistances associated with the specimen and standard but introduces two new problems – that of precise location of the temperature sensors, and the possibility of erroneous temperature measurement because the sensor leads of high conductivity are now buried in a medium of low conductivity, and extra care must be taken to ensure that there are no heat losses or gains along the leads. The advantage of sampling a large volume of the material is countered by the disadvantage of lengthy and expensive preparation for each specimen.

(2) RADIAL HEAT FLOW TECHNIQUES

A radial steady-state technique could also be used. Once again the advantage of sampling a reasonably large volume of material is countered by the disadvantage of having to drill an axial hole for the heat source and two parallel holes along generators at radii r_1 and r_2 for the measurement of temperatures T_1 and T_2 once the steady state has been reached. When the steady state has been reached, the conductivity is given by:

$$\lambda = \bar{Q} \ln(r_2/r_1)/2\pi(T_1 - T_2), \tag{4.8}$$

where $\bar{Q}$ is the amount of heat supplied by an axial heater per unit length per unit of time.

Because of the complexity of specimen preparation and the fact that further complications arise if a specimen is strongly anisotropic, the method has rarely been used and further discussion is not warranted and the application is not recommended.

4.1.3. Transient Techniques for Determining Conductivity and/or Diffusivity

(1) LINEAR HEAT FLOW TECHNIQUES

Instantaneous Source

The 'flash' technique gives a direct measurement of diffusivity only. A typical technique is that of Hanley *et al.* (1978). A laser 'flash' is used as an instantaneous source of energy applied to one surface of a specimen, usually disk shaped and about 2 cm diameter and 2–5 cm thick. The temperature is measured as a function of time at the back face and the thermal diffusivity found from:

$$a = 0.139D^2/t, \tag{4.9}$$

where D is the thickness of the disk and t is the time required for the rear face temperature to reach half its maximum value.

Two particular advantages of this method are: (a) absolute temperatures are not required thus eliminating the need for accurate calibration of the temperature sensors, especially if they are essentially linear over the several degrees of temperature rise that normally occurs, and (b) the relative ease with which the specimen can be used in a furnace if high temperature measurements are required.

Some disadvantages are: (a) the narrow range of thicknesses that can be used; if too thin (<2 mm) individual grains may control the heat transfer mechanism (i.e. the usual sampling problem) and if too thick (say >4 or 5 mm) the temperature rise at the rear face may be too small to give sufficient sensitivity; (b) because of the previous problem some specimens may be too transparent to laser light and have to be coated with an opaque thin metallic film; (c) the heat transfer may not be one dimensional if the laser beam diameter is significantly smaller than the specimen diameter, or if the specimen has significant anisotropy. The thermal conductivity cannot be determined unless a method is found for accurate measurement of the energy density of the 'flash' at the sample surface.

Constant Plane Heat Source

Mongelli (1968) has used a technique that requires minimal specimen preparation. A core is simply cut at right angles to its axis; a flat heater is sandwiched between the two pieces, Figure 4.3, and a temperature sensor placed in a small hole drilled at a known distance, x, from the cut surface. Heat is then supplied to the system at a known rate $\bar{Q}$ per unit area and the temperature rise, $T(t)$, versus time, t, is recorded; this follows the form given by Equation (4.10), (Carslaw and Jaeger, 2.9(8), 1959):

$$T(t) = [\bar{Q}x/\lambda][(\sqrt{\mathrm{Fo}/\pi})\exp(-1/4\mathrm{Fo}) - 0.5\,\mathrm{erfc}(1/2\sqrt{\mathrm{Fo}})], \tag{4.10}$$

where Fo is the dimensionless Fourier number at/x^2 (see Table 2.9). Equation (4.10) may be generalized to read:

$$T(t) = (\bar{Q}x/\lambda)f(\mathrm{Fo}). \tag{4.11}$$

Following a method first proposed by Jaeger (1959a) for radial heat flow methods and later generalized (Jaeger, 1959b), the following relation can be formed:

$$R(\mathrm{Fo}) = T(2t)/T(t) = f(2\mathrm{Fo})/f(\mathrm{Fo}), \tag{4.12}$$

where Fo is found from an experimental value of $T(2t)/T(t)$ using a theoretical curve of $f(2\mathrm{Fo})/f(\mathrm{Fo})$. Since t and x are known, a is determined. Constant values of a at various units of time serve as a check on whether or not there is systematic drift during the experiment.

The value of a may then be used with Equation (4.10) to give λ. From a and λ the value of the specific thermal capacity c_t can be found and hence that of the specific heat capacity c if the density ρ is known.

In this method the length of the two pieces of core should be sufficiently long to be regarded as of infinite length, with x being small with respect to the length; Loddo (1970) found that in practice this means that, for a typical 2 to 4 cm

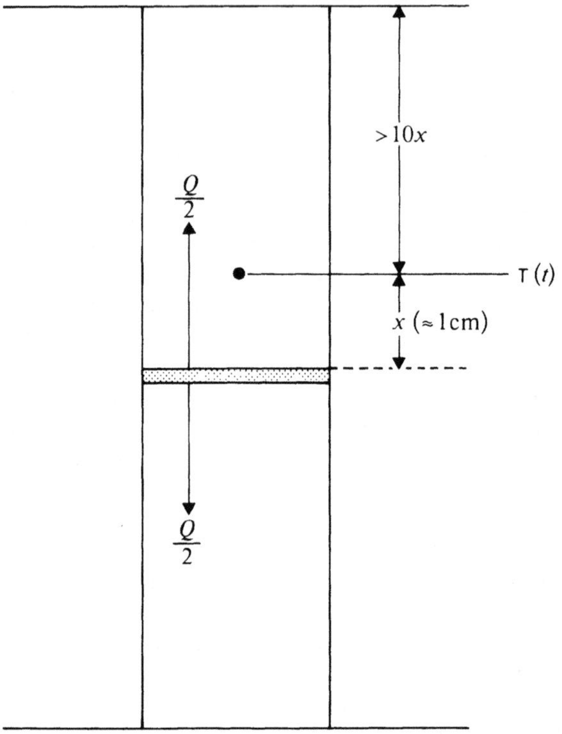

Fig. 4.3. (A) Principles of the 'cut core' method. The heater may give a constant heat output (Mongelli method) or a sinusoidal output (Ångström method).

diameter core, the specimen length should be $>5x$ and x should be between 1 and 2 cm.

Periodic Heating Technique

The method originated with Ångström (1863) and makes use of phase and intensity relationships as a disturbance, due to a sinusoidally varying surface temperature, propagates along a semi-infinite specimen. If the disturbance at one face of the core is $T(t) = T_0 \cos(2\pi t/\tau)$, where τ is the period, the temperature at some point distance x from the surface is given by:

$$T(x) = TT + T_0 \exp(-2\pi x/\Lambda) \cos[2\pi(t/\tau - x/\Lambda)], \tag{4.13}$$

where TT is a transient term which becomes negligible a few periods after switching on, and Λ $(= 2\sqrt{\pi a \tau})$ is the wavelength.

The point at which measurements are made in the material should be sufficiently far from the undisturbed end to satisfy the semi-infinite length boundary condition. If this is not possible, e.g. if only relatively thin disks are available, then corrections have to be applied for end effects or the far more complex two solid theory must be used (Carslaw and Jaeger, 3.7, 1959).

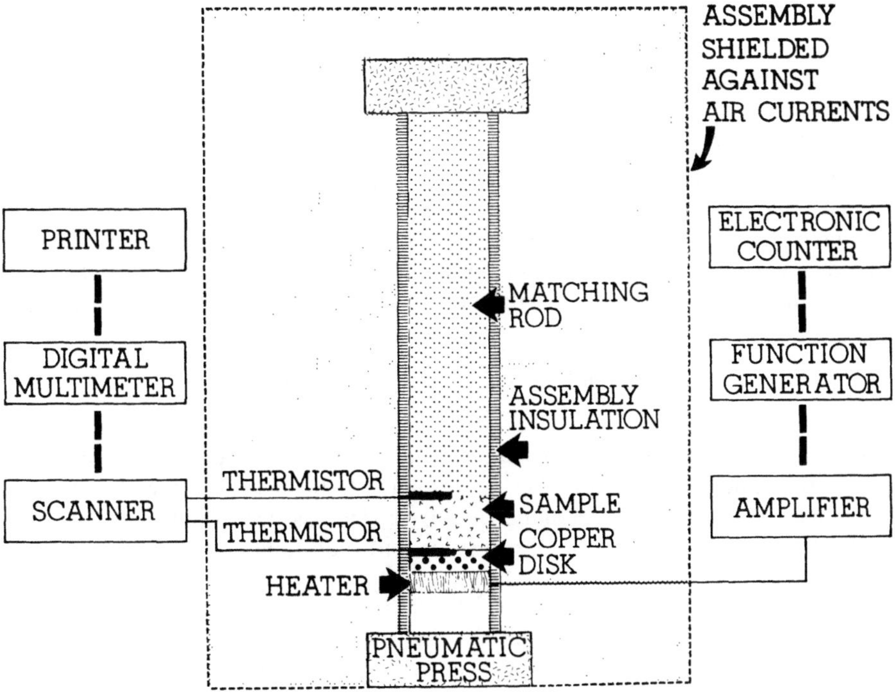

Fig. 4.3. (B) A modern version of the Ångström method (Drury *et al.*, 1984).

It can be seen from Equation (4.13) that at any point x the amplitude has decreased to $T_0 \exp(-x\sqrt{\pi/a\tau})$ of its original value, that there is a phase shift of $\alpha = x\sqrt{\pi/a\tau}$, and that the disturbance propagates at a velocity of $v = 2\sqrt{\pi a/\tau}$. It is clear that either amplitude or phase measurements relative to the initial disturbance will yield a; the same result may be obtained by differential measurements at two points x_1 and x_2. In principle, a could also be found from measurement of the velocity of propagation of a particular phase – say a maximum; however, at a distance of one wavelength ($x = \Lambda$) from the disturbed surface the amplitude of the disturbance has fallen to $\exp(-2\pi)$ of the surface amplitude and this makes it difficult to find a from observations of the velocity of propagation of a particular phase.

For a typical rock of diffusivity 1 mm^2 s^{-1} (= 0.01 cm^2 s^{-1}) and a period $\tau = 10$ s, the wavelength is 1.12 cm; therefore a few centimeters of core would be sufficient to satisfy the boundary condition. However, this means that x should be known accurately. If $\tau = 100$ s then $\Lambda = 3.5$ cm.

The equipment illustrated in Figure 4.3 could be used for an Ångström type experiment if the heater is programmed for sinusoidal heating. More modern usage of the Ångström method has been described by Bosanquet and Aris (1954) and Drury *et al.* (1984).

A periodic heating technique using cylindrical geometry has been used by Stiller *et al.* (1976).

(2) TECHNIQUES USING RADIAL (TWO DIMENSIONAL) HEAT FLOW

Instantaneous Line Source

This technique, which gives both a and λ, has been used by Lubimova *et al.* (1961). Two holes, separated by a distance D ($\cong$ 3 cm), are drilled into the specimen; heat is supplied impulsively in one hole and the temperature, $T(t)$, is recorded versus time t in the other hole. If the dimensions are such that only radial heat flow occurs between the heater and sensor, then a and λ are found from:

$$a = D^2/4t_{\max} \quad \text{and} \quad \lambda = \bar{Q}/4\pi T_{\max} t_{\max} \exp(1) \\ = \bar{Q}/34.16 T_{\max} t_{\max}, \tag{4.14}$$

where $\bar{Q}$ is the total amount of heat supplied per unit length of the probe, and $t_{\max}$ is the time required for $T(t)$, at distance D from the source, to reach its maximum value $T_{\max}$. A technique similar to this has been used by Stiller *et al.* (1978) for measurements at high pressures and high temperatures.

Constant Linear or Cylindrical Heat Sources

The instantaneous line source technique and the three linear transient techniques discussed above have not gained widespread use. In general, the transient technique most commonly used makes use of a central probe to provide heat at a constant rate; the system is of sufficient axial extent that in the central section orthogonal to the axial heater only radial heat flow takes place, i.e. the system is regarded as being infinitely long for the period of the experiment, and the temperature rise at the probe is recorded versus time. A typical laboratory probe arrangement is shown in Figure 4.4.

For a probe of radius r with a contact resistance at the boundary and supplying heat at the constant rate $\bar{Q}$ per unit length per unit time, the temperature rise at time t is given by:

$$T(t) = (\bar{Q}/4\pi\lambda)[2h^* - (2h^* - B^*)/2B^*\text{Fo} \\ + [1 - ((B^* - 1)/2B^*\text{Fo})] \ln Dt + \ldots], \tag{4.15}$$

where

$D = 4a/e^g r^2$
$g = 0.5772$ = Euler's constant
r = radius of probe
$h^* = \lambda/rc_y$ = thermal contact resistance, dimensionless
c_y = thermal conductance per unit length of the contact layer
$\text{Fo} = at/r^2$, dimensionless
$B^* = (\rho C)_m/(\rho C)_p$, dimensionless
m refers to medium and p refers to probe material.

B^* is valid for all values of r and large values of Fo. If Fo is very large, which

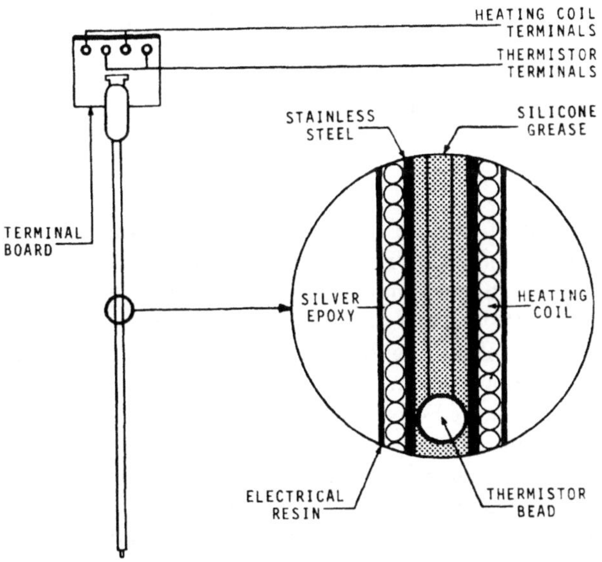

Fig. 4.4. Typical 'needle' probe arrangement. Probe diameter is 1–3 mm, and the aspect (probe length to diameter) ratio A* should be greater than 20 and preferably 50, depending on the time required for an experiment.

usually means using probes of small radius, then Equation (4.15) reduces to the more familiar line source form:

$$T(t) = (\bar{Q}/4\pi\lambda) \ln t + S, \tag{4.16}$$

where $S = (\bar{Q}/4\pi\lambda)(2h^* + \ln D)$.

Transient line source methods are usually chosen because for typical values of r ($\cong 0.5 - 1.5$ mm) the thermal conductivity and/or thermal diffusivity can be obtained in experimental times of 2 or 3 min. For this reason, all 'Quick' methods suffer from the disadvantage that the thermal properties of the specimen are averaged over only a small volume compared to the volumes used in typical steady-state techniques. This may not be a serious problem for fine grained materials, such as oceanic sediments, but may cause significant errors for coarse grained material with further problems arising because of the contact resistance between the probe and sample.

There is a commonly accepted myth that for radial heat flow methods, there are no contact resistance problems; that there are problems can be seen from Figure 4.1. If results are obtained too quickly seriously erroneous values may be obtained because it is no longer possible to ignore the thermal mass of the probe and the contribution of the contact film. The numerical value of λ (or a) obtained in an experiment is some weighted average of the value of the probe material, the contact film, and a small volume of the unknown material being sampled, with the greatest weight in the averaging process being on the material in the region of highest temperature gradient – that closest to the heating probe.

To avoid significant errors due to axial heat flow the aspect (or length to diameter) ratio, A^*, of the probe must be carefully chosen to match the expected duration of an experiment (Blackwell, 1956); for most 'Quick' experiment A^* should be about 50, although this can often be halved without significant errors creeping in.

In-situ Techniques

Relative to the radii of 'needle' probes used to determine the conductivity of oceanic sediments by means of Equation (4.16), borehole radii are large – not less than 1 cm and frequently much larger than this. Therefore an experiment utilizing Equation (4.15) must be designed. Since advantages and disadvantages of this technique have been thoroughly discussed elsewhere (Beck *et al.*, 1971; Beck, 1977) only brief mention of some difficulties will be made here. It can be seen from Equation (4.15) that if Fo is large enough and h^* small enough, the slope of a plot of $T(t)$ versus ln t will approach an asymptotic value. In general, the time required will increase as r increases, although there are some special cases where a combination of particular values of the dimensionless parameters will result in considerably shortened times; for example, if $B^* = 1$ and $h^* = 0.5$ the terms in Equation (4.15) converge very rapidly and the asymptotic region is reached quite quickly. Table 4.1 indicates how closely the slope of the $T(t)$ – ln t curve approaches the asymptotic value at given experimental times for probe parameters as shown in the table; the figures after the decimal point are therefore the percentage errors at the times shown. Table 4.2 shows the length of experimental times required to reach within 25 % and 10 % of the asymptotic values for probes of the geometries indicated. It can be seen that even in holes of quite modest radius, experimental times of several hours will be required to obtain conductivities (and/or diffusivities) to within 5 % by the logarithmic asymptote method. This leads to further serious design problems. To avoid significant contributions to heat loss by axial flow, A^* must be very large (Blackwell, 1956); this in turn requires a very large power source in the hole if typical powers of 5 W for each cm^3 of hole material displaced are used to give a few K rise in temperature – i.e. about

TABLE 4.1

Value of apparent conductivity obtained from slope of $T(t)$ – ln t curve at time shown, relative to correct conductivity obtained from true logarithmic asymptote; assorted probe parameters as shown in headings. h^*, D and B^* are defined in connection with Equation (4.15).

Time (min)	$r = 12.5$ mm $h^* = 0.19$, $D = 0.0010$		$r = 8$ mm $h^* = 0.30$, $D = 0.0027$
	$B^* = 2.5$	$B^* = 3.5$	$B^* = 3.0$
11	1.75	1.65	1.26
20	1.40	1.47	1.17
30	1.27	1.35	1.12
55	1.15	1.22	1.08
100	1.09	1.14	1.05
332	1.03	1.05	1.02

TABLE 4.2

Time to reach within 25 % and 10 % of slope of true asymptote for various probe parameters; diffusivity of host material is 1.3 $mm^2 s^{-1}$. h^* and B^* are defined in connection with Equation (4.15).

r (cm)	h^*	B^*	25 %	10 %
0.1	0–7	1.5–3.0	less than 2 min	
1.0	0–0.7	1.5–3.0	5 min	15 min
2.0	0.35	1.5	3	5
	0.35	3.0	3	30
	0	1.5	11	20
	0	3.0	16	55
10.0	0.2	1.5	2.5 h	4.5 h
	0.2	3.0	5.5	17
	0.05	1.1	3.5	4.5
	0.05	70.0	15	45

100 W m^{-1} for a borehole of 3 cm radius or a total power of the order of kW for a probe with $A^* \cong 200$. Alternative approaches to avoiding errors due to axial flow are to use guard end pieces, leading to consequent complexity of probe design with appropriate in-hole controls, or to use the complete three dimensional equation, which has no exact solution, or to use a curve fitting technique on the early part of the data set before the asymptotic region is approached. This last technique is that preferred and discussed by Beck *et al.* (1971); they state hat inaccuracy is unlikely to be less than 10 %.

The in-situ measurement of thermal properties in boreholes therefore suffers from many disadvantages peculiar to the method; in addition, it suffers from the disadvantage of all radial heat flow methods in that the result gives the average over two dimensions which in many cases are orthogonal to the direction over which terrstrial heat-flow density is being determined. However, this last is not likely to present a serious problem very often since at the 10 % level of inaccuracy, anisoptropy in thermal properties is significant in only a few special cases, e.g. in regions of poorly consolidated shales.

Only if there is an uncored borehole where it is crucial that thermal properties be known better than can be obtained by estimation methods is the in-situ technique likely to be the one of choice. The technique does have the advantage of sampling large volumes, because of the length of the probe and of the experimental times; furthermore, it may be useful in uncored deep holes where thermal properties may be significantly temperature and pressure dependant.

An interesting technique which is a bridge between the above methods and those mentioned next, has been described by Sass *et al.* (1981); at convenient times during drilling in soft sedimentary formations, the drill rods are pulled back and the conductivity determined from measurements on a thin probe driven into the bottom hole sediments.

The in-situ measurement of conductivity of oceanic sediments is much easier than for boreholes. The heating probe is integral with the gradient measuring probe which is plunged into the sediment; after allowing time for the temperatures to settle down and making measurements of the 'natural' temperature at various depths (up to a few meters) below the sediment surface the heater is switched on and the temperature sensors are used to observe the temperature rise versus time over a few minutes (Christoffel and Calhaem, 1969). Alternatively, an instantaneous heat source approach may be used (Lister, 1979) and the temperature rise followed at the probe rather than at some other radius as in Lubimova *et al.* (1961). Because the contact resistance is so small and the probe is only a few millimeters in diameter, Equation (4.16) may be used to give reasonable values of λ for the steady heat source case; the solution for the pulsed heat source case follows from Carslaw and Jaeger (10.31(3), 1959). Some of the advantages and disadvantages of the instantaneous source type of probe, and a description of extensive testing, are given by Hyndman *et al.* (1979); one of the principal advantages is that all other things being equal, the pulse technique is more than twice as efficient as the constant source technique – a very important consideration in oceanic work. Some probes are designed so that an average of the data may be taken from a few closely spaced sensors.

Laboratory Techniques

In the laboratory it is much easier to control many design parameters. In particular, since the probe does not have to be robust enough to stand up to rough field use, small values of probe radius can be used; values of $r = 1$ mm are quite common. The asymptotic region of the $T(t) - \ln t$ curve is therefore reached quite quickly, see Table 4.2. One advantage of using the logarithmic asymptote is that not only can the conductivity be obtained from the value of the slope but, as can be seen from Eq. (4.16), the intercept of the asymptote on the $\ln t$ axis can be found. Since the intercept at $T(t) = 0$ contains the diffusivity, it is possible in principle to find a and hence the specific thermal capacity c_t and specific heat capacity c. If the value of c_t is unreasonable, then the value of λ may not be valid. This may arise if, particularly in the presence of unknown significant contact resistance effects (i.e. $h^* \neq 0$), a 'false' asymptote has been used to obtain λ. In fact, if poor agreement is found, it may be wise to iterate (Jaeger, 1958) until reasonable agreement is obtained.

A 'false' or poorly determined asymptote may arise from a number of causes including: significant but unsuspected contact resistance, effectively giving a two component, concentric cylinders arrangement; convection induced by too high a power input; unsuspected axial heat losses. In this last case it is worth cautioning against running an experiment for too long a time, perhaps in an effort to 'improve' the statistics for the logarithmic asymptote, since the plot of $T(t) - \ln t$ may start to depart from linearity for large times because of the contributions from axial heat flow; this can usually be seen quite readily but use, at the equipment design stage, of the Blackwell (1956) criteria for choosing A^* would avoid the problem. A simple technique for observing when axial heat losses become important will be pointed out in Section 4.1.5. If there is no contact resistance, a useful result for small values of F_0 is, from Carslaw and Jaeger (13.5(1–8), 1959):

$$T(t) = (2r\bar{Q}/\lambda)(0.5642\sqrt{\mathrm{Fo}} - 0.25\mathrm{Fo}); \tag{4.17}$$

since the expression will not hold if there is a contact resistance, and values for all parameters have been found from the experiment, it may be used to check if a contact resistance is present.

Nearly all the radial heat flow techniques are absolute in the sense that the thermal properties are calculated from a knowledge of heat input without reference to a standard material. A recent commercial development has been a comparator technique which allows the simplicity of laying a heater wire on a flat surface of the specimen, e.g. a split core. The system has a microprocessor which allows for the automatic subtraction of a typical contact resistance and is calibrated against a number of standards from which the temperature expected at given times can be determined for materials covering a range of conductivities and diffusivities. The value for an unknown material is then, basically, picked off a set of calibration curves. The disadvantages of this technique are that great care has to be exercised with the preparation of the contact surfaces, that some unknown specimens may have properties outside the range of calibration, and the system depends upon sophisticated and expensive micro-computer read-out. In all other respects it offers the same advantages (rapid measurement) and disadvantages (small volume sampled) as other 'Quick' methods. In addition, for layered rocks corrections are necessary as shown by Grubbe *et al.* (1983).

Sass *et al.* (1984) have compared the system with a typical divided bar apparatus and find that for results of similar reliability the experimental times are of the same order, but larger sample sizes are required for the transient system.

4.1.4. Methods of Estimating Thermal Properties from Mineral Content

Sometimes it is either impossible or inconvenient to determine thermal properties by direct experiment. In these cases it may be necessary to estimate the thermal properties of the whole rock from a knowledge of its mineralogical composition. In other instances it may be necessary to know the thermal properties of a porous rock filled with a fluid of known thermal properties (e.g. water) which causes the rock to disintegrate when saturated with the required fluid but which retains its integrity when filled with another fluid (e.g. oil) of very different thermal properties. If suitable models can be found, the measurements can be made with the specimen saturated with the undesired fluid, the thermal properties of the matrix determined and then the required value computed for the specimen if it could have been saturated with the desired fluid.

Numerous models have been proposed but all have certain deficiencies. The general picture is shown in Figure 4.5 which illustrates the typical dispersive and limiting cases.

The principal problem lies in finding a suitable physical model and associated equation for the general case of a randomly dispersed multi-component system which is usable over a wide range of structure and component thermal properties. One of the first models was proposed by Maxwell (1891), originally for the electrical conductivity case but applicable to thermal conductivity cases as well.

The relevant equation giving the rock conductivity for a two component dispersive system is:

$$\lambda_D = \lambda_c[(X^* - 2V^*Y^*)/(X^* + V^*Y^*)], \qquad (4.18)$$

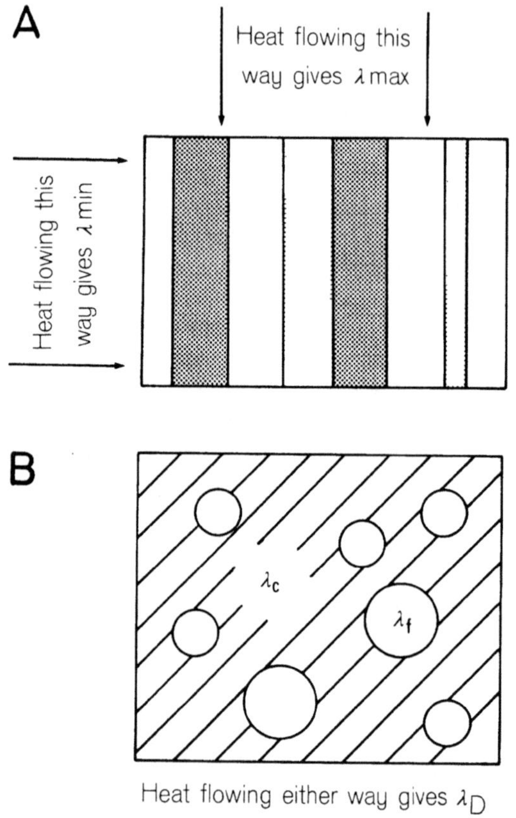

Fig. 4.5. (A) Models for computing maximum and minimum conductivities with components arranged in parallel and series respectively. (B) Model for computing conductivity when a discontinuous phase (λ_f) is randomly dispersed throughout the matrix (λ_c) as spheres.

where

V^* = ratio of volume of the discontinuous phase (the porosity if the discontinuous phase consists of voids) to total volume
$X^* = (2r^* + 1)$
$Y^* = (r^* - 1)$
$r^* = \lambda_c/\lambda_f$
λ_c = conductivity of the continuous phase (the matrix)
λ_f = conductivity of the discontinuous phase (the fluid filling the pores in a porous rock).

For a multi-component system with the components arranged in parallel and in series respectively:

$$\lambda_{max} = (\Sigma V_n^* \lambda_n)/\Sigma V_n^* \quad (4.19)$$

$$\lambda_{min} = \Sigma V_n^*/(\Sigma V_n^*/\lambda_n), \quad (4.20)$$

where V_n^* is the proportion by volume of the n-th component of conductivity λ_n. If V_n^* is expressed as the fractional volume proportion then $\Sigma V_n^* = 1$; if expressed as percentages, as is frequently done, $\Sigma V_n^* = 100$.

It can be shown that for a two component system λ_{max} and λ_{min} are respectively the upper and lower limits of the Maxwell model. λ_{max} and λ_{min} are sometimes called the Hashin and Shtrickman (1962) limits but, once again, it can be shown that they are identical with the Maxwell limits. The principal difference in the approaches was that Maxwell assumed that the discontinuous phase consisted of spheres and solved the problem on this basis; on the other hand, Hashin and Shtrickman did not assume that the discontinuous phase consisted of spheres and obtained a more general expression which was intractable unless spherical shapes were later assumed.

Brailsford and Major (1964) modified Maxwell's model for use with three component systems to obtain:

$$\lambda_{B3} = \lambda_c[V_1^* + 3(Z_2^* + Z_3^*)]/[V_1^* + 3(r_2^*Z_2^* + r_3^*Z_3^*)], \tag{4.21}$$

where $Z_n^* = r_n^*V_n^*/(2r_n^* + 1)$, V_n^*, λ_n and λ_c $(= \lambda_1)$ have the same meanings as before, and $r_n^* = \lambda_c/\lambda_n$. For a two phase random mixture this equation may be modified, by interchanging phase 2 and phase 3 and solving, to give:

$$\lambda_{B2} = (\lambda_c/4r^*)[U^* - (U^{*2} + 8r^*)^{1/2}], \tag{4.22}$$

where $U^* = [(2r^* - 1) - 3V^*(r^* - 1)]$ and the other symbols have the same meaning as for the Maxwell model.

A further equation which essentially takes a modest account of tortuosity was derived by Adler *et al.* (1973):

$$(\lambda_A)^{1/2} = \Sigma V_n^*(\lambda_n)^{1/2}/\Sigma V_n^*. \tag{4.23}$$

Another much used 'model', but empirically and not physically based, is that given by the equation for the geometric mean value of conductivity:

$$\lambda_G = (\Pi\lambda_n^{V_n^*})^{1/\Sigma V_n^*}. \tag{4.24}$$

The behaviour of these various models has been discussed by Roy *et al.* (1981) and curves for two phase systems are shown in Figure 4.6 for typical values of r^* which might be encountered if porous rocks are saturated with water $(2 < r^* < 6)$, oil $(10 < r^* < 30)$ or air $(50 < r^* < 150)$ (dry), the approximate range of r^* being shown in brackets for each case. It can be seen that if $r^* < 5$, there is little to choose between any of the models represented by Equations (4.18), (4.22), or (4.23); even Equation (4.24) gives values within about 15 % of the others over most of the range. However, as r^* increases it becomes increasingly important to choose the 'right' model, although there is still little difference between the results from Equations (4.22) and (4.23) over the range of porosities most likely to be encountered in nature. In brief, Roy *et al.*, believe that Equations (4.22) or (4.23) cover a much wider range of r^* and typical conditions than other models and that the estimate for bulk conductivity is probably best found using them. As mentioned earlier, if it is necessary to estimate the specific thermal capacity as well, this

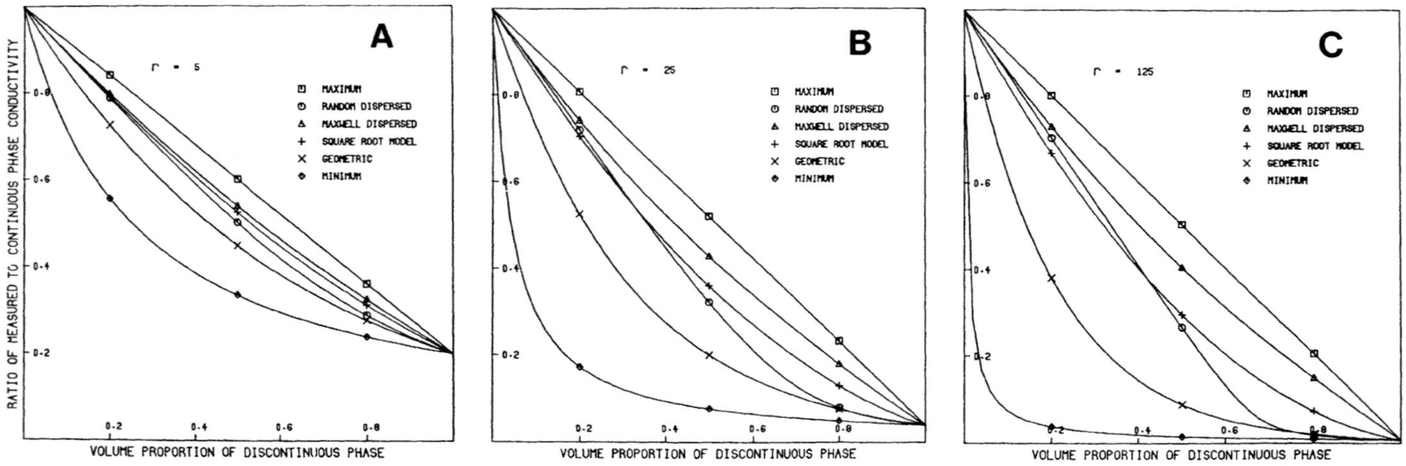

Fig. 4.6. Curves of normalized specimen conductivity versus volume proportion of the discontinuous phase for two phase systems and selected values of conductivity ratio r^* ($= \lambda_c/\lambda_f$); r corresponds to r^*, and (A) to $r^* = 5$, (B) to $r^* = 25$, (C) to $r^* = 125$. Specimen conductivity is normalised to continuous phase conductivity. For porous materials the continuous phase is the matrix and the discontinuous phase is the fluid filling the pores.

may be done by taking the mean of the various components weighted according to the proportions present.

Although the discussion in this section pertains to estimating the conductivity of a small specimen of rock based on its mineralogical composition and structure, the models could be used to estimate gross properties of a large volume of rock consisting of formations of different conductivities. In this context we have to consider two problems. First, how to estimate the conductivity of a formation knowing the proportions of constituent rock types, e.g. clay and limestone fractions and, if the rock is not impervious, the porosity; for this estimate Equation (4.21) is probably the most satisfactory. Second, having estimated the conductivity of a number of formations by this, or an alternative technique, if we now wish to estimate the mean conductivity of a layered sequence of formations, the most useful model would be that represented by Equation (4.20) with the V_n^* now representing formation thicknesses as a fraction of the total thickness of the sequence being considered. In both cases it must be emphasized that use of a particular model equation implies that we accept the structure implied by that model; that is, a random dispersion of the components for Equation (4.21) and a sequence of layers perpendicular to the direction of heat flow for Equation (4.20). This latter assumption is likely to be easier to confirm than the former one.

4.1.5. Sources of Experimental Error

(1) HEAT LOSSES FROM THE SYSTEM

In the linear heat flow apparatus there is always a danger of errors arising because of lateral heat losses from the system, thus violating the conditions under which the relevant equations are applicable. With only moderate care this source of error can be made negligible at the 5 % level for most measurements. If the design parameters are chosen so that the mean specimen temperature is close to ambient, and the temperature drop across any of the symmetrical systems shown in Figure 4.2 is kept to less than 10 K, then by taking the average value for q, in Equation (4.6) for example, from the upper and lower parts of the system, heat lost from the upper part is compensated by the heat gained in the lower part. There is little need for elaborate guard arrangements or a thermostatically controlled environment; the only real need is for the provision of a simple draught screen around the apparatus.

(2) ERRORS IN TEMPERATURE MEASUREMENT

Another source of error arises from incorrect measurement of temperatures, largely because of loss or gain of heat along the wires to the temperature sensors, usually thermocouples or thermistors. Once again, reasonable care in design can make this a negligible problem; the principal requirement is to ensure that a sufficient length of the lead wires (about 20 lead diameters) is close to the same temperature environment as the sensor.

(3) CONTACT RESISTANCE

As indicated in Figure 4.1, the complex pattern of inter-surface contact points can be represented by an equivalent thin film of the contact fluid in series with the rest of the system. For this film to be negligible, its thermal resistance must be small compared with that of the specimen and standards. The contact film equivalent thickness will vary according to the surface roughness which in turn depends very much on the type of material in the specimen; in a typical experiment with reasonably well prepared specimens, that is the macroscopic surface is flat, the equivalent film thickness may vary from about 10 μm for an igneous or metamorphic rock of low porosity, to 20 or 30 μm for a porous sandstone. If the contact fluid is water ($\lambda = 0.59$ W m^{-1} K^{-1}) the thermal resistance of each film may range from 1.7 to 5×10^{-5} K W^{-1}; this may be compared with the thermal resistance of a 1 cm thick specimen of fresh quartzite (a high conductivity material of $\lambda \cong 7.1$ W m^{-1} K^{-1}) which is approximately 141×10^{-5} K W^{-1}. Clearly the contribution of the contact film can be ignored at the 5 % level of inaccuracy. On the other hand, if air ($\lambda \cong 0.025$ W m^{-1} K^{-1}) is the contact fluid, then the contact resistance for each film ranges from 40 to 120×10^{-5} K W^{-1} and clearly cannot be ignored.

(4) HEAT TRANSFER BY CONVECTION

One of the least considered sources of error is the possibility of heat transfer by convection during experiments on fluid saturated permeable media. In the simplest terms, the onset of convection is determined by the relative effects of the buoyancy forces and the gravitional forces.

For the case of a permeable layer of finite thickness and infinite areal extent, theoretical treatments have been given by Lapwood (1948), Wooding (1957), Elder (1967), and Straus (1974) for various boundary conditions and with experimental confirmation from laboratory model experiments. For this case it is convenient to define a dimensionless parameter, known as the Rayleigh number (Ra^*), which contains the temperature gradient, and to find a critical value (Ra_1^*) above which convection occurs; for a given medium the value of the temperature gradient at which this occurs is the critical gradient. Straus finds a second critical number ($Ra_2^* > 10Ra_1^*$) above which convection is turbulent.

Nothing has been published for the more complex case of the infinite permeable medium heated internally by a cylinder, although Genceli (1980) has discussed a case for pure fluids. However, it is clear that there will be general similarities: the first critical Rayleigh number which defines the onset of convection must depend on the intrinsic permeability and thermal diffusivity of the medium, on the kinematic viscosity and volume coefficient of expansion of the pore fluid, on the magnitude of the component of the temperature gradient parallel to the gravitational acceleration and, for incompletely saturated media (which are here defined as moist media), surface tension; some of these parameters are strongly temperature dependant. Whatever the geometry, convection will occur when the temperature gradient in the specimen exceeds the critical gradient. For this reason, convection is most likely to occur in the transient radial heat flow experiments since such experiments have, under typically used conditions, gradients at small radii that are at least an order of magnitude greater than is found in the steady state linear heat flow type of apparatus. Of the two limiting cases for the cylindrical probe the onset of convection is more likely to occur when the probe is horizontally oriented, because there are significant components of temperature gradients parallel to the gravitational field; for vertically oriented probes the onset of convection is inhibited because the gravitional field, which controls the buoyancy forces, is orthogonal to the temperature gradient.

The errors which can arise are indicated in Figures 4.7, 4.8, and 4.9. The data were obtained by continuously recording temperatures versus time during a 'needle' probe experiment. Equation (4.16) is used to give the value of λ versus time over small increments of time.

Figure 4.8 shows how the contact resistance and thermal mass of the probe affect the early part of an experimental curve.

Figure 4.9 shows the effect of orientation of the probe axis and different power inputs (which affect the time at which the critical gradient is exceeded) on media with the same porosity and thermal properties but having different permeabilities.

Figure 4.9 shows how the contact resistance and thermal mass of the probe affect the early part of an experimental curve.

Detailed descriptions of these effects are given by Fodemisi (1980).

It is possible to allow for all these effects by proper design of an experiment. The design process is helped by an empirically derived graph based on the experiments of Fodemisi and shown in Figure 4.10 (Fodemisi and Beck, 1983). Fodemisi's approach was to use the fact that for sufficiently long times, and sufficiently small radial separations, the steady state linear heat flow equation could also be applied to the radial transient case, which indicates that the work to date on convection in permeable layers could be easily modified. The equations for the critical Rayleigh number in the linear case were therefore manipulated to produce a 'Convection Parameter' *Ra* for the radial case (see also Table 2.9):

$$Ra = k\alpha_v g q A^*/(\rho c)_f a_e^2 \nu, \tag{4.25}$$

where

k = permeability
$a_e = \lambda_s/(\rho c)_f$ = effective thermal diffusivity (Elder, 1965)
λ_s = thermal conductivity of the fluid saturated medium
α_v = volume coefficient of thermal expansion
$(\rho c)_f$ = heat capacity of the saturating fluid
ν = kinematic viscosity of the saturating fluid
g = gravitational acceleration
q = heat-flow density
A^* = aspect ratio of the system.

All temperature dependant variables are to be evaluated at the initial temperature of the medium before any heating takes place.

From Figure 4.10 the critical convection parameters, that is the values of *Ra*

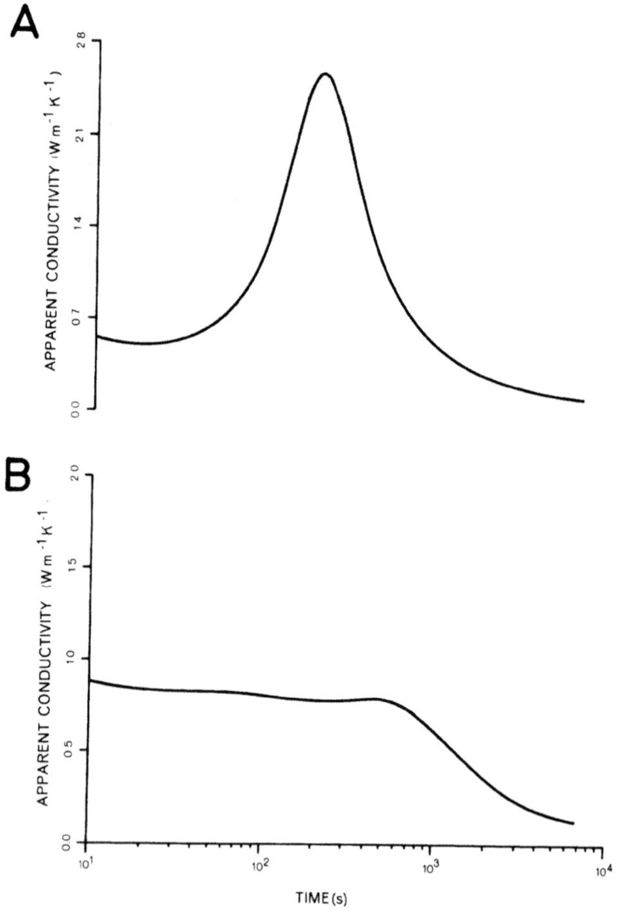

Fig. 4.7. Illustrating significant and insignificant convection contributions to heat transfer and, at large times, the effects of finite size of the model; the medium is contained in a lucite cylinder, 5 cm in diameter and 13 cm long, surrounded by foamed thermal insulating material, see Figure 4.8(A). (A) Medium consists of matrix of glass beads of 3 mm diameter giving an intrinsic permeability of 17.6 × 10^{-5} cm^2. Pore space is filled with ethyl alcahol, heat input is 0.2 W cm^{-1} to a 3 mm diameter probe and the probe axis is horizontal. Convection is dominant in the 40–700 s interval; the effects of the low conductivity insulation surrounding the container are becoming dominant after 700 s, possibly modified by axial losses. (B) Medium consists of a matrix of glass beads of 1 mm diameter giving an intrinsic permeability of 6.3 × 10^{-5} cm^2. Pore space is filled with water, heat input is 0.13 W cm^{-1} to a 3 mm diameter probe and the probe axis is vertical. There is little convection but the effects of the low conductivity insulation dominate after 500 s, but are possibly modified by axial losses.

above which errors due to convection become increasingly important (see also Figure 2.8), are:

$$Ra_{hc} = 930 \text{ for the case where the probe axis is horizontal,}$$

$$Ra_{vc} = 1490 \text{ for the case where the probe axis is vertical.}$$

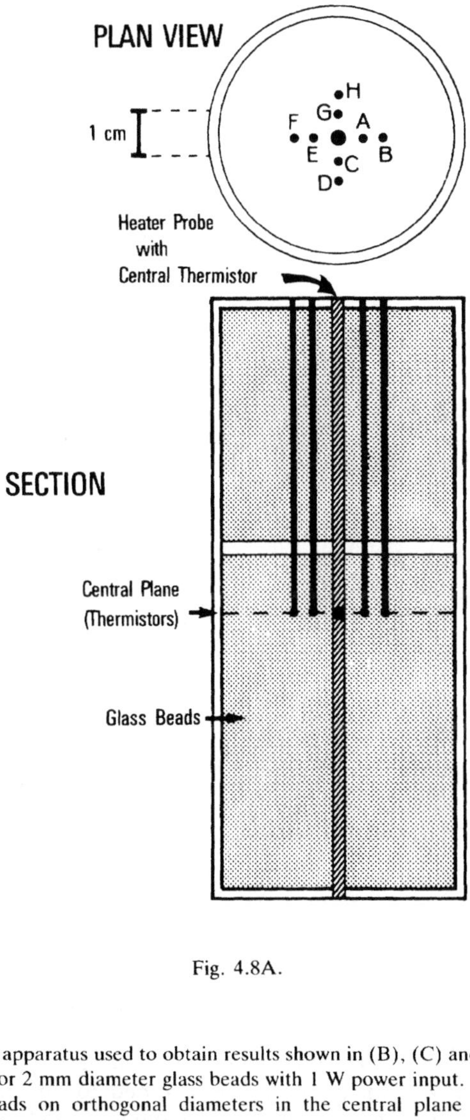

Fig. 4.8A.

Fig. 4.8. Schematic of apparatus used to obtain results shown in (B), (C) and Figure 4.9. (B) and (C) show complete results for 2 mm diameter glass beads with 1 W power input. The letters on the curves refer to thermistor beads on orthogonal diameters in the central plane and at radii of 0.5 cm (thermistors A, C, E, G) and 1 cm (thermistors B, D, F, H), one pair of thermistors per quadrant. For each labelled set of curves, e.g. in (a) upper curve is the central probe ($r = 0$) temperature and the lowest curve is the temperature at $r = 1$ cm (thermistors B, D, F, H); in (c) each curve labelled is the temperature difference between the thermistor pair on one radial. (B) Water saturated. Note that constancy of the conductivity versus time from 10–1000 s; the large excursion when the axis is horizontal is due to a single electronic noise pulse. (C) Air saturated. The thermal mass of the probe, of high conductivity, together with the large thermal contact resistance, cause a longer time to be required before the volume of specimen being penetrated by heat is sufficiently large to mask the probe effect; true conductivity values are not obtained until after 60 s.

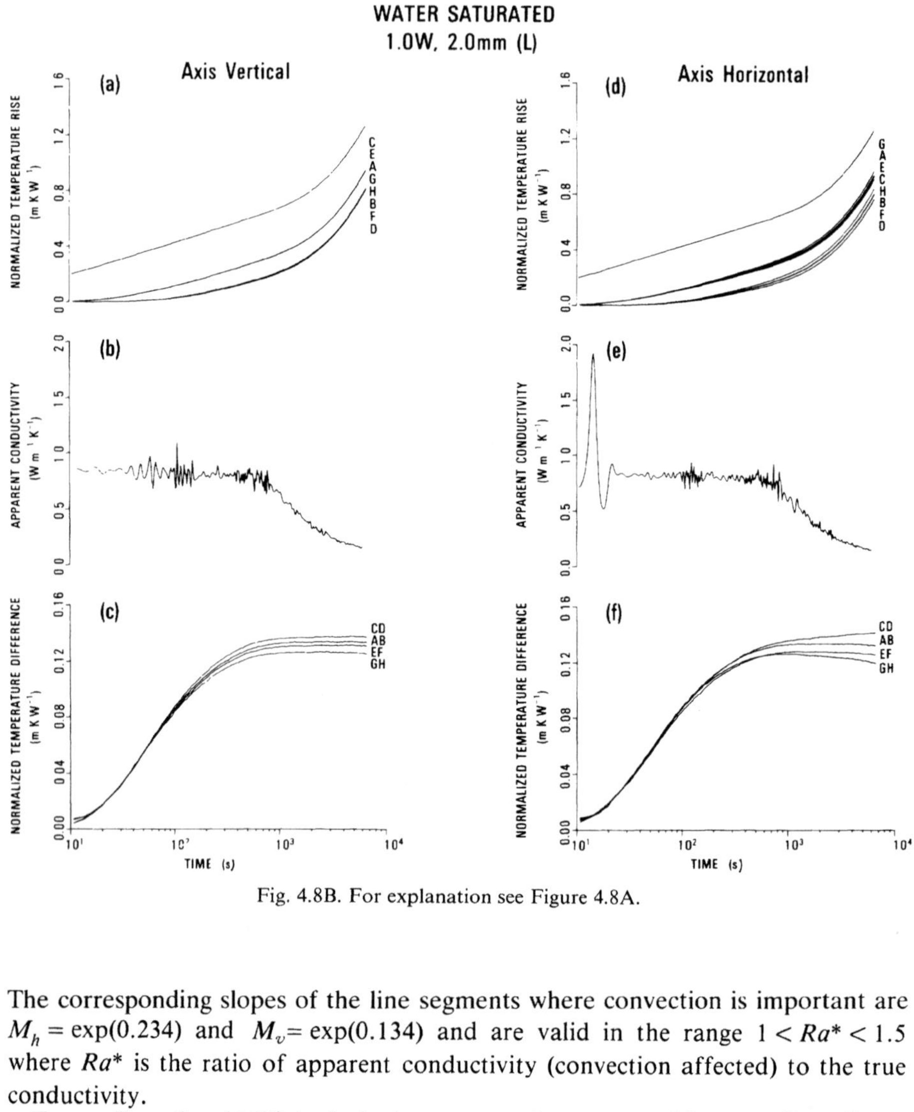

Fig. 4.8B. For explanation see Figure 4.8A.

The corresponding slopes of the line segments where convection is important are $M_h = \exp(0.234)$ and $M_v = \exp(0.134)$ and are valid in the range $1 < Ra^* < 1.5$ where Ra^* is the ratio of apparent conductivity (convection affected) to the true conductivity.

To use Equation (4.25) in designing an experiment to avoid convection effects, a_e and λ_s must be known. Since these are assumed to be unknown, a suitable estimate can be made by using one of the component models given by Equations (4.22) or (4.23) to find λ_s.

If, after the apparatus has been designed, it is found that on some occasions convection is significant, corrections can be applied using:

$$C_h = \exp(-0.234)\ \ln(Ra_h/930) \text{ and} \tag{4.26}$$

$$C_v = \exp(-0.134)\ \ln(Ra_v/1490). \tag{4.27}$$

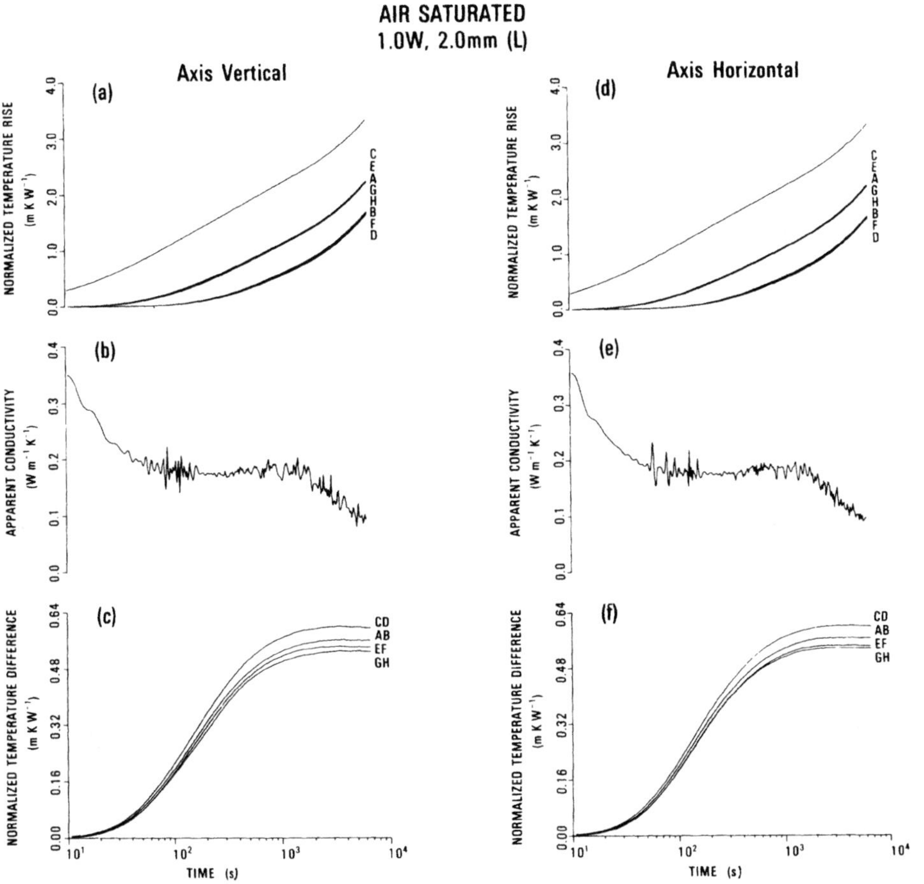

Fig. 4.8C. For explanation see Figure 4.8A.

for the horizontal and vertical axis cases respectively, both correction factors being based upon the data in Figure 4.10.

4.1.6. Standards

One of the weaknesses in terrestrial heat-flow density work has been the lack of formally accepted standard materials or even of methods of reporting and presenting data.

(1) CALIBRATION MATERIALS

Part of the problem with regard to standard materials is a misunderstanding of internationally accepted nomenclature in other branches of science. A standard reference material (SRM) is one where a particular batch has had the physical properties, not necessarily just thermophysical properties, carefully measured at a

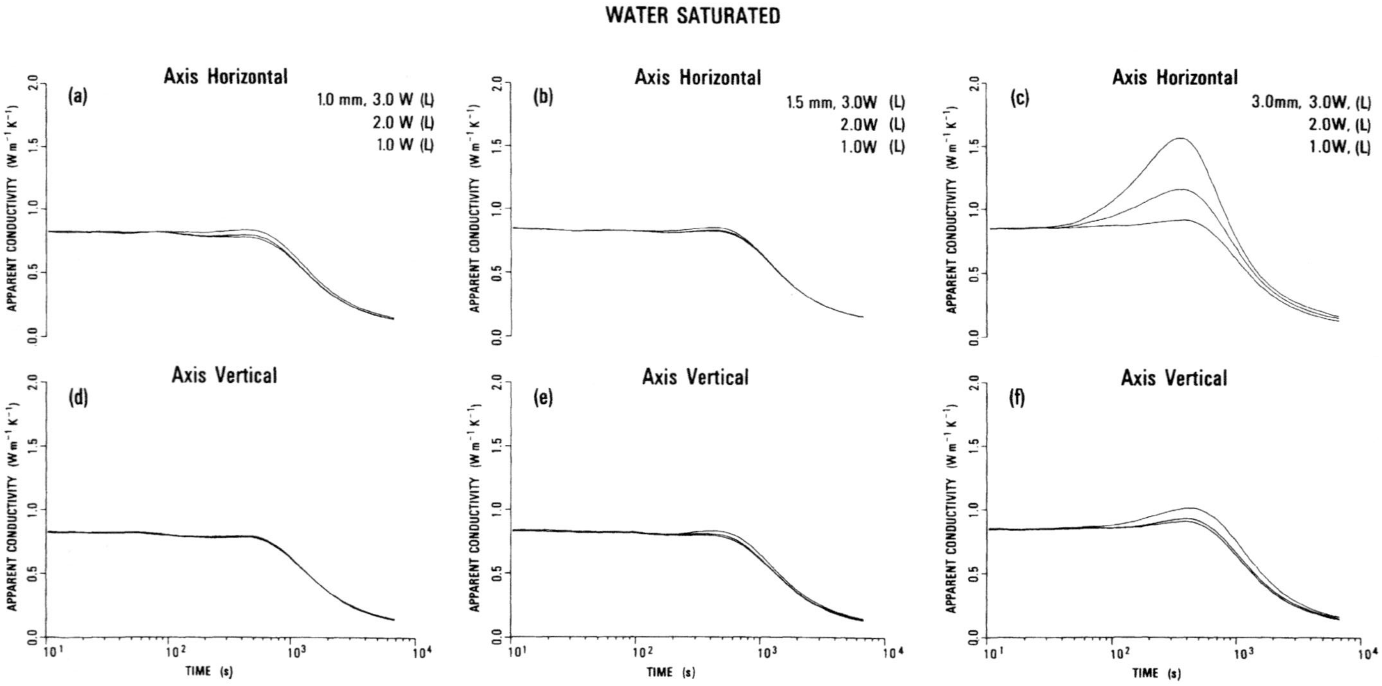

Fig. 4.9. Results for a typical series of experiments using three media of intrinsic permeabilities (6.3, 7.3 and 17.6 × 10^{-5} cm^2 for (a) and (d), (b) and (e), (c) and (f) respectively), three power inputs (1, 2, and 3 W) and two orientations (horizontal and vertical). Where curves are separable, the upper curve corresponds to the highest power input. Convection effects are clearly negligible at the lower power inputs whatever the permeability of the medium, but are becoming quite significant for the highest permeability medium at the higher power inputs, especially when the axis is horizontal; even with the axis vertical, convection is becoming noticeable. 4.9(d) contains the curve shown in Figure 4.7(B).

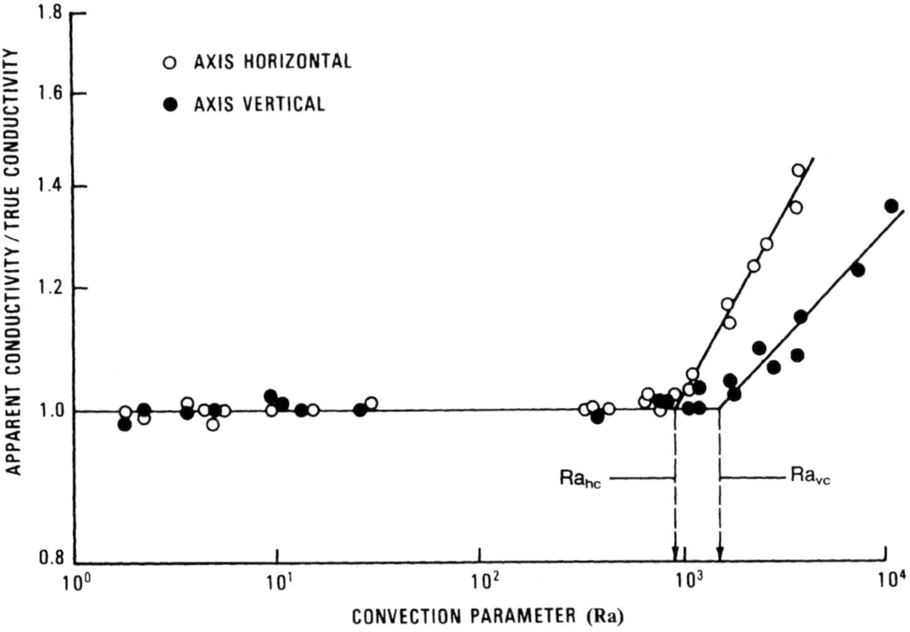

Fig. 4.10. Experimentally determined heat transfer characteristics for free convection in a permeable medium heated axially, showing critical values (Ra_{hc} – axis horizontal, and Ra_{vc} – axis vertical) above which convection contributes increasingly large errors.

number of laboratories, perhaps using very different techniques. The laboratories where the measurements are made are either National Standards Laboratories or universally regarded as such. Several specimens are used and the material is well characterized as to structure, composition, impurity content (which is especially important at very low temperatures) etc. After interlaboratory comparison (and numerous committee meetings) a value for a particular property, together with a statistical statement of inaccuracy, is agreed upon. At least some of the specimens are kept by the laboratories making the original measurements and remeasurements are occasionally made to check the stability of the apparatus (and of the material). Other groups may request specimens of that SRM and be confident that if their equipment is operating properly they should obtain values within the stated level of inaccuracy. Once all the SRM has been consumed, that is the end of that particular SRM. If more of that material is required for SRM use a new batch is manufactured and the whole process of measurement is repeated creating a new SRM.

There are some materials which are so easy to reproduce to high levels of purity, and used over temperature and pressure ranges where small differences in impurity levels are not important, that they are for all practical purposes the equivalent of SRM's and can be used as calibration standards. Two such materials in terrestrial heat-flow density work have been informally recognized – Fused Quartz (FQ) and Crystalline Quartz (CQ). Two particular advantages of FQ and

CQ are that they are not expensive to obtain and that if CQ is cut so that heat flows perpendicular to the optic axis (CQ1), their thermal conductivities span the range of about 90 % of the consolidated rocks, i.e. 1.2–6.2 $Wm^{-1} K^{-1}$. Since the thermal conductivities of both materials are temperature dependant, CQ particularly so, the value of λ should not be quoted without giving the relevant temperature. In this context the Thermophysical Properties Research Centre recommended values, which result from an exhaustive survey and quality analysis of all data in the literature, should be used (see Section 4.1.7., calibration materials).

If an apparatus is calibrated with, for example, FQ and if CQ1 is used as an 'unknown' specimen and found to give correct values, it can confidentally be assumed that the results for materials within the range of FQ and CQ1 will be reliable (barring occasional operator error). For calibration of a divided bar apparatus, it is advisable to have three different thicknesses of both FQ and CQ, say 2, 4, and 6 mm or 3, 6 and 9 mm; this will permit the calculation and elimination of contact resistances; Equations (4.4) and (4.5). A number of laboratories have sets of disks which they are willing to lend to others who are just setting up.

For some special purposes it may be desirable to have an intermediate or, especially for unconsolidated or oil saturated sediments, a lower conductivity standard.

For the intermediate conductivities it is convenient to stack FQ and CQ1 disks, a 2 mm disk of each being equivalent to a 4 mm disk of conductivity given by Equation (4.20); the present author has found that a water contact film between two such disks is negligible at the 0.5 % level. However, the CODATA Task Group on Thermophysical Properties is giving consideration to Pyroceram 9606, an accepted SRM, as a possible intermediate standard; it is a completely impervious ceramic which has the advantages of already being well characterized and is available in quantity, but suffers from the disadvantage of being difficult and expensive to machine accurately; the Task Group is also investigating candidate materials with known porosities and permeabilities.

For a lower conductivity interim standard, Hyndman *et al.* (1979) have found that less than 3 % gelatine inhibits convection in distilled water and does not significantly alter the thermal conductivity; this is a particularly useful procedure for checking 'needle' probe apparatus which hitherto have not usually been calibrated against an accepted standard material.

If a higher conductivity standard is desired, say if a suite of ore material is to be measured, then CQ cut so that heat flows parallel to the optic axis (CQ2) may be used.

(2) REPORTING APPARATUS PERFORMANCE

In general there is little to be added to the recommendations of the CODATA Task Group on Thermophysical Properties (Minges *et al.*, 1980); they make a number of general and detailed proposals and it is perhaps worth pointing out two important points here.

First, information in publications should be given in sufficient detail that other workers could reproduce them independently if they wish, especially if the publication deals with new techniques or approaches. This may involve a description of the specimens. For example, there is little point in quoting the results for a sandstone that is not closely specified; to understand the meaning of the value it is necessary to know the porosity and permeability, whether the measurements were made with the specimens saturated with water or some other fluid of known conductivity, whether the sandstone is weathered or fresh, the location and name of the formation, etc.

Second, the Task Group follow the current international trend by suggesting that apparatus performance should be stated in terms of imprecision (a statement of the probability that an observer can reproduce the result for a specific material on that apparatus within the stated limits of imprecision), and inaccuracy (a statement of the probability that other observers on different apparatus would obtain the same results within the stated limits of inaccuracy). An observer may be able to quote, quite correctly, an imprecision of 1 % but have an inaccuracy of several per cent because, for instance, of poorly measured separation of temperature sensors in a long divided bar equipment. Probably the best way of determining inaccuracy is to undertake inter-laboratory comparison on a number specimens, preferably calibration standards covering a wide range of conductivity.

(3) SAMPLING

In laboratory experiments on controlled man made materials a statement of imprecision and inaccuracy gives all the information necessary for readers to make a qualitative judgement of the results. However, the earth scientist is faced with a further problem. Usually the interest lies in trying to determine a representative properly value for a large mass of material, say a few cubic meters or even cubic kilometers, from measurements on small 'representative' specimens. Several such specimens, making up a statistical sample, may be necessary to obtain reasonable statistics.

There is little one can do here except rely on the experience and common sense of the observer. Clearly, to obtain a formation value, fewer specimens would be required of a fine grained siltstone than of a coarse grained granite. Enough specimens of any formation material should be used to give a mean value of the formation properties with an inaccuracy of less than 10–15 %; in this context, formation inaccuracy is a statement of the probability that another observer sampling the same formation in similar numbers would obtain a mean value within the stated limits.

(4) REPORTING OF DATA

Most observers report the results of their sample of n specimens in terms of a mean value $\bar{\lambda}$ and a standard deviation s or coefficient of variation (the standard deviation expressed as a percentage of the mean value). In making the calculations it is usually implicitly assumed that an infinite population with a standard deviation σ and a normal distribution about the mean μ has been sampled. In this context $\bar{\lambda}$ is an estimate of μ and s, an estimate of σ, may be regarded as a statement that about one third of the values in the sample set fall outside the stated limits indicated by s.

If the n specimens are chosen at random (i.e. each individual specimen of the infinite population has the same chance of being selected) it is obvious that in general the sample mean (of the n specimens) will not be the same as the mean of the whole population. It follows that if two sets of specimens, n_1 and n_2, are used, the means will in general be different; if n_1 and n_2 are large the differences may not be great but if one or both are small the differences may be large. Therefore, a simple reporting of the mean and standard deviation does not give the complete picture; the number of specimens used, or another parameter which allows for the incompleteness of sampling, should also be quoted.

A useful way of combining everything is to use tables of Student's t distribution (Beyer, 1968) to produce confidence limits. A simple rule of thumb is that if there are n specimens in a sample and $n > 5$, then multiplying the standard deviation of the sampling mean (= standard error $\alpha = s/\sqrt{n-1}$) by 2 will give the 95 % confidence limits without sensible error; this is usually indicated by the symbol α_{95}. The 95 % confidence limit is essentially a statement that there is only 1 chance in 20 that the mean of a different sample will fall outside the stated limits; see also Section 5.2.9.

The use of α alone is basically another way of stating the probability that the mean of two samples will lie within a certain range. For example, if $n = 10$, $\bar{\lambda} = 11$ and $s = 0.3$, then the result may be quoted as 11 ± 0.3 (standard deviation, $n = 10$) or 11 ± 0.1 (standard error, $n = 10$). This implies that if we select another and much larger sample ($n \gg 10$), there is 1 chance in 3 that the mean will lie outside the range 10.9–11.1.

4.1.7. Summary of Recommendations

The International Heat Flow Commission has for some years been concerned about the lack of coherent procedures for data collection, measurement, and interpretation. Varied approaches are necessary for the continued health of any lively and productive group, but if they are too varied it becomes impossible, from the published literature alone, to make accurate qualitative judgements and comparisons of values quoted by different authors; this is particularly important when two quite different values are quoted for apparently similar materials.

Therefore, working groups have been set up to look at various aspects and the following are recommended guidelines pertaining to the measurement of thermophysical properties of rocks for use in geothermal work. Many of them may be regarded as a summary of the discussions in some parts of this chapter.

(1) THERMAL CONDUCTIVITY

Units

1. The units recommended for reporting conductivity are $Wm^{-1} K^{-1}$.

Apparatus

2. A divided bar method should be used wherever possible for competent cores.

3. A radial geometry line source method should be used on fine grained, homogeneous, soft sediments.
4. A rapid half-space line source transient method may be used but needs replicate measurements and careful assessment.
5. The in situ method may be used in cased and uncased holes when the inaccuracy of a single measurement does not exceed 10 %.
6. Each laboratory apparatus should have an imprecision of less than 2 % and an inaccuracy of less than 5 %.

Calibration Materials

7. For apparatus used for competent rocks, crystalline quartz cut so that heat flows perpendicular to optic axis (CQ1) and fused quartz (FQ) should be the basic calibration materials. If a higher conductivity calibration material is required, crystalline quartz cut so that heat flows parallel to the optic axis (CQ2) may be used. The values to be used at given temperatures over the range $0 < T < 300\,°C$ may be obtained from the following relations, where λ is in W m^{-1} K^{-1} and T is in °C; the FQ relation is from Ratcliffe (1959), while the CQ relations are from a least squares fit to TPRC data and show the coefficients of variation.

$$\text{FQ: } \lambda \times 10^3 = 1323 + 1.93T - 0.0067T^2$$
$$\text{CQ1: } 1/\lambda_{CQ1} = (443 \times 10^{-6} \pm 4\,\%)\,T + (0.1544 \pm 2\,\%)$$
$$\text{CQ2: } 1/\lambda_{CQ2} = (347 \times 10^{-6} \pm 1\,\%)\,T + (0.0872 \pm 1\%).$$

A useful interim intermediate conductivity calibration material is to place a disk of FQ and a disk of equal thickness of CQ1 in series and use the series computed conductivity value for the compound specimen.
8. FQ may be used as a standard for line source experiments provided it is properly prepared. CQ1 may also be used but, because of its anisotropy, the axial hole to contain the heater must be drilled parallel to the optic axis. If a lower conductivity calibration material is required for needle probe methods, distilled water with not more than 3 % silica gel added to inhibit convection under normal power usage may be used as an interim standard.
9. The interim, and other, materials will be fully recommended if they pass appropriate assessment by an international group.

Procedures

10. Wherever possible, measurements should be made on rock specimens which are representative of rock formations in their natural state, and under natural conditions, preferably on cores from the hole in which temperature gradient measurements have been made.
11. If measurements are made on simple apparatus and the specimens come from considerable depth, corrections should be applied to allow for the temperature and pressure dependence of thermal conductivity.
12. The porosity of every specimen should be determined; it is also desirable to attempt to determine the permeability.

13. If measurements cannot be made shortly after porous specimens have been collected, the specimens should be stored in a closely fitting water tight container to preserve the natural fluid content.
14. If evaporation takes place between the times of collection and measurement, every effort should be made to restore the natural fluid distribution before measurement.
15. Moisture movement and evaporation during measurement should be prevented by covering all external surfaces of the specimen with a waterproof material.
16. Replicate specimens should be measured in sufficient number to give a formation (or bulk) conductivity inaccuracy of less than 10–15 % (coefficient of variation); in addition, the number of specimens used and the 95 % confidence limits of the mean (α_{95}) should be given.
17. If representative specimens are not available, some attempt should be made to estimate conductivity only when a reliable correlation or computation technique is available.
18. In all cases, details of the conditions of measurements, problems encountered and error estimates should be clearly stated.

(2) THERMAL DIFFUSIVITY

The history and extent of determining thermal diffusivity of rocks and minerals is less than that for thermal conductivity. Furthermore, no standard materials have been identified and universally accepted. In the interim, the following recommendations are made but it must be realised that much more work and experience is required before great confidence can be placed in the validity of the recommendations.

19. The recommended units for reporting thermal diffusivity are $mm^2\ s^{-1}$.
20. For fine grained homogeneous specimens, a plane (one dimensional) instantaneous source technique may be used.
21. For less homogeneous materials requiring larger specimens to give valid values for bulk thermal diffusivity, an Ångström (sinusoidally varying source) technique should be used.
22. The same materials as are used for thermal conductivity standards (Recommendation 7) should also be regarded as thermal diffusivity standards unless alternative and better materials are later identified; in this context, if the diffusivity is to be found indirectly from the conductivity, the specific thermal capacity c_t of crystalline quartz and fused silica may be taken as 2.00 J $cm^{-3}\ K^{-1}$ within 5 % at 300 K.
23. In all other respects the same general procedures (10–18) should be followed, and attempts be made not to exceed the same levels of inaccuracy (5–6).

Acknowledgements

I am indebted to many colleagues who read the manuscript but particularly to Christoph Clauser, Ralph Haenel, Francesco Mongelli, John Sass, and Herman Villinger who made many useful suggestions. This work was carried out with the aid of an operating grant from the Natural Sciences and Engineering Research Council of Canada.

4.1.8 References

Adler, D., Flora, L. P., and Sentura, S. D.: 1973, 'Electrical Conductivity in Disordered Systems', *Solid State Comm.* **12**, 9–12.

Ångström, A. J.: 1863, 'New Method of Determining the Thermal Conductivity of Bodies', *Phil. Mag.* **25**, 130–142.

Beck, A. E.: 1957, 'A Steady State Method for the Rapid Measurement of the Thermal Conductivity of Rocks', *J. Sci. Instr.* **34**, 185–189.

Beck, A. E.: 1977, 'Geothermal Measurements in Five Small Lakes of Northwest Ontario: Discussion', *Can. J. Earth Sci.* **14**, 332–334.

Beck, A. E. and Beck, J. M.: 1958, 'On the Measurement of the Thermal Conductivity of Rocks by Observations on a Divided Bar Apparatus', *Trans A.G.U.* **30**, 1111–1123.

Beck, A. E., Anglin, F. M., and Sass, J. H.: 1971, 'Analysis of Heat Flow data – *in situ* Thermal Conductivity Measurements', *Can. J. Earth Sci.* **8**, 1–19.

Benfield, A. E.: 1939, 'Terrestrial Heat in Great Britain', *Proc. Roy. Soc. A* **173**, 428–450.

Beyer, W. H.: 1968, *Handbook of Tables for Probability and Statistics*, Chemical Rubber Co. Cleveland.

Birch, F.: 1950, 'Flow of Heat in the Front Range, Colorado', *Bull. G.S.A.* **61**, 567–630.

Blackwell, J. H.: 1956, 'The Axial Flow Error in the Thermal Conductivity Probe', *Can. J. Phys.* **34**, 412–417.

Bosanquet, C. H. and Aris, R.: 1954, 'On the Application of Ångström's Method of Measuring Thermal Conductivity', *Brit. J. Appl. Phys.* **5**, 252–255.

Brailsford, A. D. and Major, K. C.: 1964, 'The Thermal Conductivity of Aggregates of Several Phases, Including Porous Materials', *Brit. J. Appl. Phys.* **15**, 313–319.

Bullard, E. C.: 1939, 'Heat Flow in South Africa', *Proc. Roy. Soc. A* **173**, 474–502.

Carslaw, H. S. and Jaeger, J. C., 1959: *Conduction of Heat in Solids*, 2nd. edition, O.U.P., London, 510 p.

Christoffel, D. A. and Calhaem, I. M.: 1969, 'A Geothermal Heat Flow Probe for *in situ* Measurement of both Temperature Gradient and Thermal Conductivity', *J. Sci. Instr.* **2** (2nd. series), 457–465.

Drury, M. J., Allen, V. S., and Jessop, A. M.: 1984, 'The Measurement of Thermal Diffusivity of Rock Cores', *Tectonophysics* **103**, 321–333.

Elder, J. W.: 1965, 'Physical Processes in Geothermal Areas', in W. H. K. Lee (ed.) *Terrestrial Heat Flow* Monograph No. 8, A.G.U., Washington, 211–240.

Elder, J. W.: 1967, 'Steady Free Convection in a Porous Medium Heated from Below', *J. Fluid. Mech.* **27**, 29–48.

Fodemisi, S. P.: 1980, 'The Removal of Convective Effects Induced During Needle Probe Thermal Conductivity Measurements', M.Sc. Thesis, University of Western Ontario, London, Canada, 129 p.

Fodemisi, S. P. and Beck, A. E.: 1983, 'Induced Convection During Cylindrical Probe Conductivity Measurements on Permeable Media', in J. D. Hust (ed.) *Proceedings of the Seventeenth International Thermal Conductivity Conference* Plenum, N.Y. 619–634.

Grubbe, K., Haenel, R., and Zoth, G.: 1983, 'Determination of the Vertical Component of Thermal Conductivity by Line Source Methods', *Zbl. Geol. Paläont.*, Teil, I. H. 1/2, Stuttgart, 49–56.

Genceli, O. F.: 1980, 'The Onset of Manifest Convection from Suddenly Heated Horizontal Cylinders', *Wärme u. Stoffübertragung* **13**, 163–169.

Hanley, E. J., DeWitt, D. P., and Roy, R. F.: 1978, 'The Thermal Diffusivity of Eight Well Characterized Rocks for the Temperature Range 300–1000 K', *Engineering Geol.* **12**, 31–47.

Hashin, Z. and Shtrickman, S.: 1962, 'A Variational Approach to the Theory of the Effective Magnetic Permeability of Multiphase Materials', *J. Appl. Phys.* **33**, 3125–3133.

Hutchinson, H. P., Dixon, I. S., and Denbigh, K. G.: 1948, 'The Thermo-Osmosis of Liquids Through Porous Materials', *Discussions Faraday Soc.* **3**, 86–94.

Hyndman, R. D., Davis, E. E., and Wright, J. A.: 1979, 'The Measurement of Marine Geothermal Heat Flow by a Multipenetration Probe with Digital Acoustic Telemetry and *in situ* Thermal Conductivity', *Marine Geophys. Res.* **4**, 181–205.

Jaeger, J. C.: 1958, 'The Measurement of Thermal Conductivity and Diffusivity with Cylindrical Probes', *Trans. A.G.U.* **39**, 708–710.

Jaeger, J. C.: 1959a, 'The Analysis of Aquifer Test Data or Thermal Conductivity Measurements Which Use a Linear Source', *J. Geophys. Res.* **64**, 561–564.

Jaeger, J. C.: 1959b, 'The Use of Complete Temperature-time Curves for Determination of Thermal Conductivity with Particular Reference to Rocks', *Austral. J. Phys.* **12**, 203–217.

Kuzmak, J. M. and Sereda, P. J.: 1958a, 'The Mechanism by Which Water Moves Through a Porous Material Subjected to a Temperature Gradient: 1. Introduction of a Vapour Gap into a Saturated System, *Soil Sci.* **84**, 291–299.

Kuzmak, J. M. and Sereda, P. J.: 1958b, 'The Mechanism by Which Water Moves Through a Porous Material Subjected to a Temperature Gradient: 2. Salt Tracer and Streaming Potential to Detect Flow in the Liquid Phase', *Soil Sci.* **84**, 419–422.

Lapwood, E. R.: 1948, 'Convection of a Fluid in a Porous Medium', *Proc. Cam. Phil. Soc.* **44**, 508–521.

Lister, C. R. B.: 1979, 'The Pulse Probe Method of Conductivity measurement', *Geophys. J. Roy. Astr. Soc.* **57**, 451–461.

Loddo, M.: 1970, 'Temperature in an Insulated Slab Heated by a Plane Source: Application to Thermal Conductivity Measurements of Rocks', *Geothermics*, Special Issue, No. 2, 437–442.

Lubimova, E. A., Lusova, L. M., Firsov, F. V., Starikova, G. N., and Shushpanov, A. P.: 1961, 'Determination of Surface Heat Flow in Mazesta (U.S.S.R.), *Annali di Geophys.* **XIV**, 157–167.

Maxwell, J. C.: 1954, *A Treatise on Electricity and Magnetism 1891*, New York, (Dover Publications).

McElroy, D. L. and Moore, J. P.: 1969; 'Radial Heat Flow Methods for the Measurement of the Thermal Conductivity of Solids', in R. P. Tye (ed.), Thermal Conductivity, Vol. 1, Academic Press, New York.

Minges, M. L., Beck, A. E., Berman, R., Cabannes, R., Shpilrain, E. E., Touloukian, Y. S., and White, G. K.: 1980, 'Criteria for the Presentation in the Primary Literature of Scientific and Technical Information on Thermophysical Properties of Solids', *Int. J. Thermophys.* **1**, 135–140.

Misener, A. D. and Beck, A. E.: 1960, 'The Measurement of Heat Flow over Land', in S. K. Runcorn (ed.), *Methods and Techniques in Geophysics*, Vol. 1, Interscience, New York, 10-61.

Mongelli, F.: 1968, 'Un metodo per la determinazione in laboratorio della conducibilità termica delle rocce', *Boll. Geof. Teor. Appl.* **X**, 51–58.

Ratcliffe, E. H.: 1959, 'Thermal Conductivities of Fused and Crystalline Quartz', *Brit. J. Appl. Phys.* **10**, 22–25.

Roy, R. F., Beck, A. E., and Touloukian, Y. S.: 1981, 'Thermophysical Properties of Rocks, in Y. S. Touloukian, W. R. Judd and R. F. Roy (eds.), *Physical Properties of Rocks and Minerals*, McGraw-Hill, New York, 409-502.

Sass, J. H., Lachenbruch, A. H., and Munroe, R. J.: 1971, 'Thermal Conductivities of Rocks from Measurements on Fragments and its Application to Heat Flow Determinations', *J. Geophys. Res.* **76**, 3391–3401.

Sass, J. H., Kennelly, J. P. Jr., Wendt, W. E., Moses, T. H. Jr., and Ziagos, J. P.: 1981, '*In situ* Determination of Heat Flow in Unconsolidated Sediments', *Geophysics* **48**, 76–83.

Sass, J. H., Stone, C., and Munroe, R. J.: 1984, 'Thermal Conductivity Determinations on Solid Rock: A Comparison Between a Steady-State Divided Bar Apparatus and a Commercial Transient Line-Source device', *J. Volcan. Geoth. Res.* **20**, 145–153.

Stiller, H., Vollstadt, H., and Seipold, U.: 1978, 'Investigations of Thermal and Electric Properties of Rocks by Means of a Cubic Press', *Phys. Earth Planet. Int.* **17**, 31–34.

Stiller, H., Vollstädt, H., Seipold, U., and Wasch, R.: 1976, 'Electrical and Thermal Investigations of Rocks and Minerals Under Extreme p,T – Conditions', *Pure and Appl. Geophys.* **114**, 263–272.

Straus, J. M.: 1974, 'Large Amplitude Convection in Porous Media', *J. Fluid. Mech.* **64**, 51–63.

Wooding, R. A.: 1957, 'Steady State Free Thermal Convection of Liquid in a Saturated Permeable Medium', *J. Fluid. Mech.* **2**, 273–285.

Zierfuss, H. and Van der Vliet, G.: 1956, 'Laboratory Measurements of Heat Conductivity of Sedimentary Rocks', *Bull. G.S.A.* **40**, 2475–2488.

4.2 DETERMINATION OF HEAT PRODUCTION RATE

L. RYBACH
Institute of Geophysics, ETH Zürich, Switzerland

The heat produced by naturally radioactive elements in rocks is one of the key factors in geothermics, especially in the interpretation of continental heat-flow density data. Combined determinations of heat-flow density and of radioactive heat production rate yield basic information about the temperature field and the structure of the Earth's crust and are indispensable to understand the interrelations between heat-flow density and geology.

4.2.1. Definitions

During radioactive decay mass is converted to energy. This energy is set free as the kinetic energy of the involved particles and nuclei (α- and β-particles, recoil nuclei) and as the energy of the accompanying γ-radiation. Except the amount carried away by the neutrino the whole decay energy is converted to heat (Hurley and Fairbairn, 1953). All natural radioactive isotopes generate heat to a certain extent but only the contributions of the decay series of ^{238}U, ^{235}U, ^{232}Th and of the isotope ^{40}K are geologically significant.

Radioactive heat production rate is a scalar and isotropic petrophysical property independent of in situ temperature and pressure. The dimension of heat production rate is W m^{-3}; in geothermics it is customary to use the unit μW m^{-3}. Heat production rate of rocks depends on the abundances of the natural radioelements U, Th, and K in a given sample and on the heat production constants.

The heat production rate of a given radioactive isotope depends upon its radiation energy and half life. For uranium and thorium the decay energy $E_{\Delta m}$ can be readily determined from the mass differences between starting isotopes and stabile end products of the decay series, taking the number of the emitted α-particles (^4He) into account. The $E_{\Delta m}$-values (Table 4.3) include the contributions of the emitted β-particles, γ-radiation and of the recoil nuclei. For β-emission the amount of energy carried away by the neutrino (2/3 of the maximum β

TABLE 4.3
Decay energies which are converted to heat for uranium and thorium, based on data from Brunée and Voshage (1964) and Hyde *et al.* (1964).

Isotope	Mass balance	$E_{\Delta m}$ (MeV)	$\Sigma E_{\beta max}$ (MeV)	E_{abs} (MeV)
^{238}U	$m(^{238}U) - 8m(^{4}He) - m(^{206}Pb)$	51.667	7.99	46.34
^{235}U	$m(^{235}U) - 7m(^{4}He) - m(^{207}Pb)$	46.392	1.71	45.26
^{232}Th	$m(^{232}Th) - 6m(^{4}He) - m(^{208}Pb)$	42.793	5.71	38.99

R. Haenel, L. Rybach, and L. Stegena (eds.)
Handbook of Terrestrial Heat-Flow Density Determination, 125—142.

energy) must be accounted for. By summing up the individual maximum β energies according to the decay schemes ($\Sigma E_{\beta\max}$) the total absorbed energy per atom is thus:

$$E_{abs} = E_{\Delta m} - 0.667\,\Sigma E_{\beta\max}. \tag{4.28}$$

The heat production constants H are usually given in watts per unit mass of radioactive isotope. Therefore:

$$H(\mathrm{W\ kg^{-1}}) = \frac{6.025 \times 10^{23}}{M} \cdot \frac{0.693}{T_{1/2}} u E_{abs}\ (\mathrm{MeV}), \tag{4.29}$$

where M is the atomic mass number, $T_{1/2}$ the half life of the isotope i in seconds and u the conversion factor 1 MeV = 1.60×10^{-13} J.

For potassium the heat production constant can be calculated from the β and γ decay rates: (28 260 ± 50) β $\mathrm{kg^{-1}\ K^{-1} s^{-1}}$ and (3250 ± 70) γ $\mathrm{kg^{-1}\ K^{-1}\ s^{-1}}$; the mean β and γ energies are (0.60 ± 0.01) MeV and (1.460 ± 0.001) MeV, respectively (Leutz *et al.*, 1965). From these, $H_K = 3.48 \cdot 10^{-9}$ W $\mathrm{kg^{-1}}$. The revised heat production constants for natural uranium (99.28 % ^{238}U 0.711 % ^{235}U), ^{238}U, ^{235}U, thorium and potassium (with 0.0117 % ^{40}K) are given in Table 4.4. They are, in general, several percent lower than the corresponding values of Birch (1954).

The radioactive heat production rate in a given rock can be calculated, using the constants of Table 4.4, if its uranium, thorium and potassium contents (c_U, c_{Th}, and c_K) are known. Practical concentration units are weight ppm for uranium and thorium and weight percent for potassium. Using these units, heat production rate is usually given in μW $\mathrm{m^{-3}}$:

TABLE 4.4

Revised heat production constants (from Rybach, 1976a).

Element/decay series in equilibrium	$H(\mathrm{W\ kg^{-1}})$
^{238}U	9.17×10^{-5}
^{235}U	5.75×10^{-4}
U (natural)	9.52×10^{-5}
^{232}Th	2.56×10^{-5}
K (natural)	3.48×10^{-9}

$$H = 10^{-5}\rho(9.52c_U + 2.56c_{Th} + 3.48c_K),\ \text{in}\ \mu\mathrm{W\ m^{-3}}. \tag{4.30}$$

The density ρ (in kg $\mathrm{m^{-3}}$) of the rock in question must be known.

4.2.2. Measurement Techniques

The main requirements for a suitable analytical method to determine the abundances c_U, c_{Th}, and c_K are high sensitivity and simplicity to perform measurements on large series of samples. In common rocks uranium and thorium are present in trace amounts only. Numerous analytical techniques exist which are sensitive enough to determine such low concentrations (i.e. in the ppm range, 1 ppm = 10^{-6} kg kg^{-1}): fluorimetry, x-ray fluorescence, spectrophotometry, activation analysis, delayed neutrons, fission track counting, mass spectrometry, γ-ray spectrometry. Potassium can be determined by flame photometry, atomic absorption, x-ray fluorescence, activation analysis, mass spectrometry and γ-ray spectrometry. For advantages and disadvantages of the individual methods see Stuckless *et al.* (1977), Hart *et al.* (1980), and Bajo (1980). A comparison of results, obtained by different methods on the same samples, is given in Table 4.5.

TABLE 4.5
Interlaboratory comparison on different standards.

Standard	K (%)	U (ppm)	Th (ppm)
Potassium feldspar IAEA			
Tugsavul *et al.* (1970)[a]	12.6 ±0.5		
Rybach (1973)	12.1 ±0.1		
Soil sample IAEA			
Tugsavul *et al.* (1970)[b]		1.64±0.93	7.0±2.1
Rybach (1973)		1.3 ±0.3	6.5±0.9
Potassium feldspar BRZ[c]			
Gamma spectrometry			
Berkeley	9.15	5.6	2.0
Risø	10.3 ±0.2	5.7 ±0.3	2.2±0.2
Zürich	9.9 ±0.1	5.2 ±0.3	2.3±0.2
X-ray fluorescence			
Geol. Survey Denmark	9.9		
Atomic absorption			
A/S Norfloat, Norway	10.0		
Concrete sample BRZ[c]			
Gamma spectrometry			
Berkeley	0.97	0.82	2.3
Risø	1.06±0.03	0.7 ±0.1	2.0±0.2
Zürich	0.99±0.03	0.9 ±0.2	2.0±0.4
X-ray fluorescence			
Inst. of Petrology Univ. Copenhagen	1.08		

[a] Average of results from 32 laboratories.
[b] Average of results from 11 laboratories.
[c] BRZ: Intercomparison Berkeley – Risø – Zürich.

From these methods, γ-ray spectrometry is the only one which enables simultaneous determination of U, Th, and K. A further simplicity is the easy sample preparation. The assumption of secular radioactive equilibrium within the ^{238}U and ^{232}Th series and of a constant $^{40}K/K$ ratio is fulfilled in most rock types (cf. Table 4.6). However, due to the low γ-radiation intensity of common rocks, sen-

TABLE 4.6
Comparison of U and Th determinations by different analytical methods. Agreement of U results is indicative of radioactive equilibrium.

	Uranium (ppm)			Thorium (ppm)	
Sample	γ-spectrometry	fission track[a]	spectrophot.[b]	γ-spectrometry	spectrophot.[b]
Bergell granite	7.7	7	7.0	25.5	27.5
Rotondo granite	13.6	13	14.3	27.6	28.5
Mittagsfluh granite	11.7	11	10.8	32.7	30.7
Giuv syenite	20.0	18	19.7	67.0	65.6
Böttstein granite drillcores					
– 443.3 m	3.3		3.27		
– 471.6 m	2.6		2.33		
– 566.8 m	11.4		10.6		
– 812.8 m	3.7		3.94		
North Sea sediments					
– shale	4.1		4.34		
– limestone	0.9		0.84		
– sandstone	2.7		2.05		

[a] Fission track etching (cf. Bajo, 1980).
[b] Spectrophotometry (cf. Keil, 1979, 1981).

sitive detectors like large thallium-activated sodium iodide crystals [NaI(Tl)] must be used in combination with appropriate shielding against background radiation.

A γ-ray spectrometer consists of a NaI(Tl) detector, optically coupled to a photomultiplier (PM) tube and a multi-channel pulse-height analyzer for collection and storage of the energy spectra. Figure 4.11 shows an arrangement with two opposing scintillation detectors to enhance sensitivity. In order to ensure high instrumental stability of the spectrometer system, it is indispensable to install the detector as well as the multichannel analyzer in a climatized room (temperature constant to ±0.5 °C).

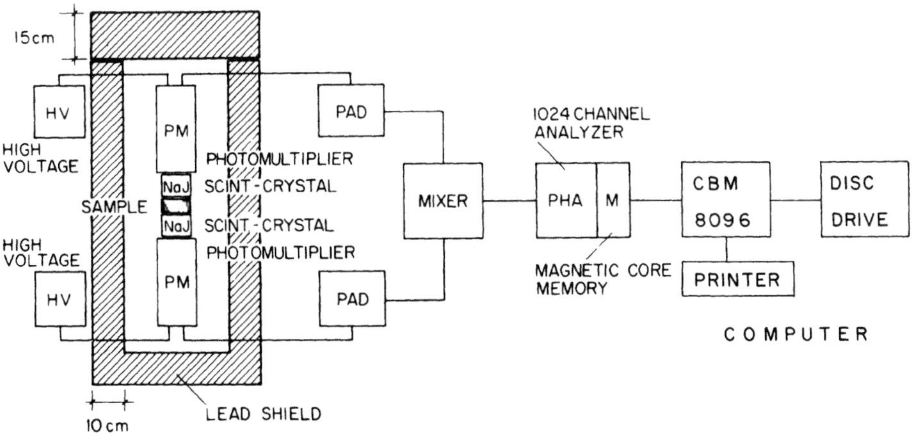

Fig. 4.11. Schematics of γ-spectrometric equipment, including digital data acquisition system, for heat production determinations. Two opposing scintillation detectors (7.5 cm × 7.5 cm) are used to enhance sensitivity. Their outputs are matched and summed up before pulse height analysis (PHA). PAD: pre-amplifier + amplifier + discriminator.

The electronics of the spectrometer system is usually adjusted to encompass the energy range between 0.2 and 3.5 MeV. Figure 4.12 shows a typical spectrum with characteristic peaks of the ^{238}U and ^{232}Th desay series and of the 1.46 MeV peak of ^{40}K. The 1.76 MeV ^{214}Bi peak of the ^{238}U series is commonly used for uranium analysis, whereas thorium is determined from the 2.62 MeV ^{208}Tl peak of the ^{232}Th series.

Besides NaI(Tl) detectors, solid-state detectors like lithium-drifted germanium detectors find increasing application in modern low-level γ-spectrometry. The great advantage of Ge(Li) detectors is their high peak resolution (Smith and Wollenberg, 1972). On the other hand, their sensitivity is much lower than with NaI(Tl) crystals. Therefore, only larger-volume Ge(Li) detectors (several 10 cm^3) can be utilized. The solid-state detector must be cooled by an accompanying cryostat as well as the attached field-effect transistor preamplifier. With these detectors, low-energy peaks like the 48 keV ^{238}U peak or the 143 keV ^{235}U peak, of primary importance for equilibrium/disequilibrium studies, can be detected over sufficiently long counting times.

Sample preparation consists of crushing the rocks to be analyzed and encasing

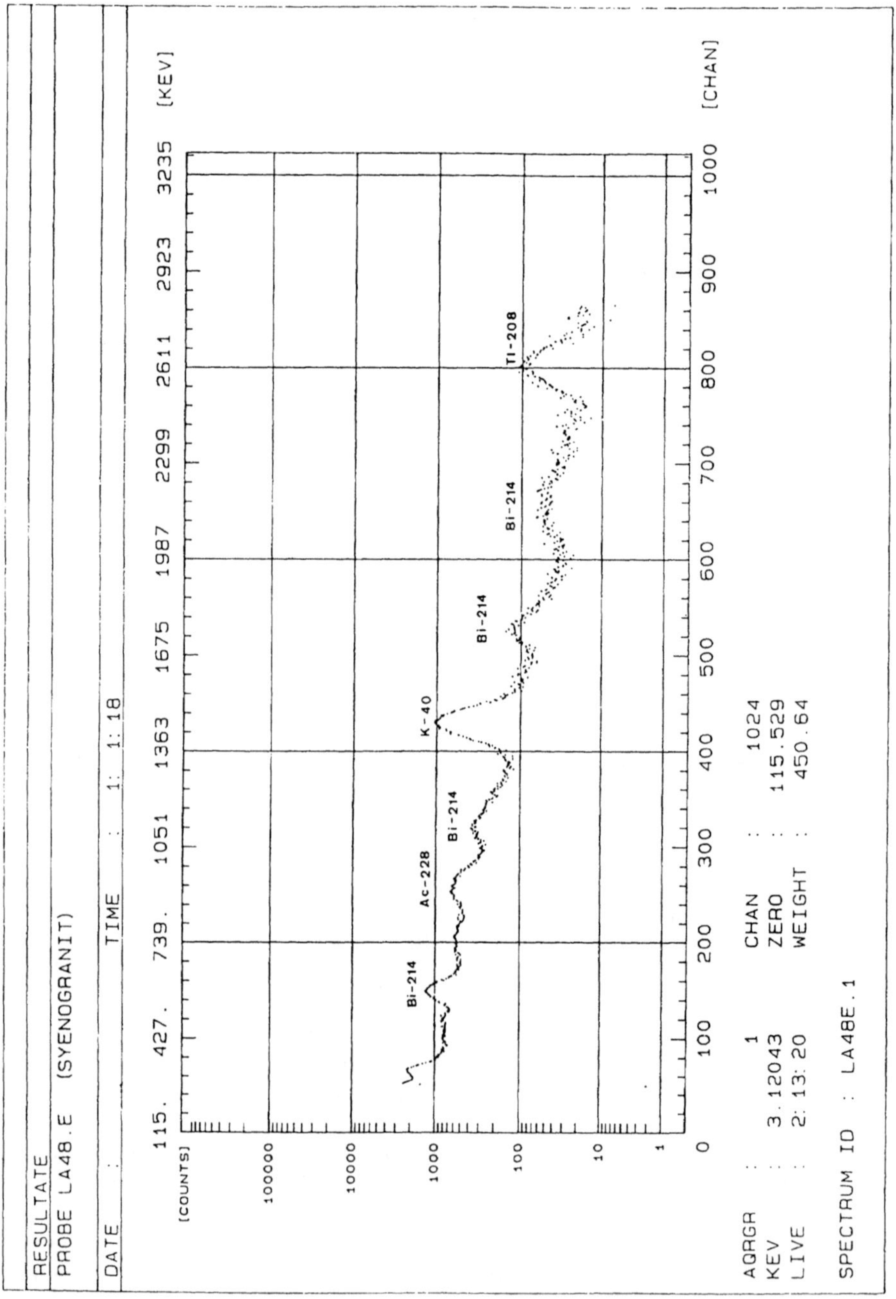

Fig. 4.12. Typical γ-ray spectrum (1024 channels) of a rock sample with characteristic photopeaks.

in plastic or aluminium boxes taking up several 0.1 to 1 kg material. The size and form of the sample containers must be matched to the detector dimensions to ensure optimum sample-detector configuration. Figure 4.13 shows a perspex sample holder for a solid-state detector. Complete filling of the sample container is necessary to avoid corrections. Variations in the filling density should be kept to minimum (e.g. by constant sample weight) or must be accounted for by a corresponding correction for self-absorption.

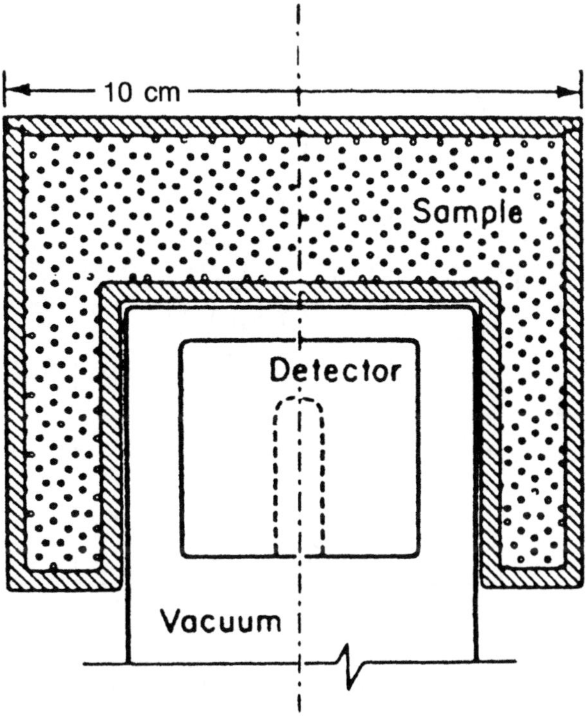

Fig. 4.13. Perspex sample holder surrounding the solid-state Ge(Li) detector to enhance sensitivity (Lewis, 1974).

4.2.3. Calibration, Data Processing

The determination of uranium, thorium and potassium is based on standard spectra of samples with known U, Th, and K concentrations. Usually three standard spectra (of pure U, Th, and K standards) are used. The spectrum of a sample with unknown U, Th, and K concentrations will consist of a linear combination of these standard spectra and of the background spectrum. Figure 4.14 displays three standard spectra.

Several methods can be used to determine the proportions of the component spectra; these will be proportional to the amounts of U, Th, and K present in the sample.

One method involves the three characteristic photo-peaks in the γ spectrum:

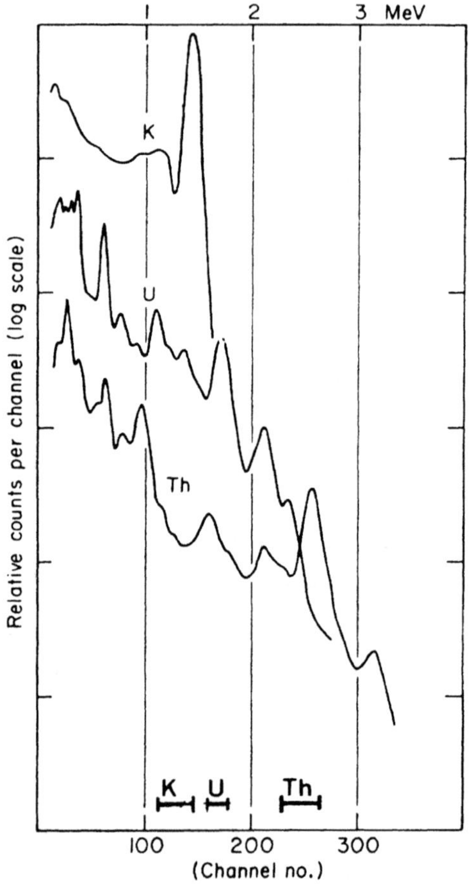

Fig. 4.14. Standard γ-spectra of uranium, thorium and potassium (after Wollenberg, 1977, modified). The energy windows utilized for data processing are indicated at the bottom of the figure (K, U, Th).

1.46 MeV ^{40}K, 1.76 MeV ^{214}Bi and 2.62 MeV ^{208}Tl (cf. Figure 4.12). The peak areas are integrated by summing up their counts over an energy 'window', centered at the middle of the peak. The window width corresponds to the detector resolution and is usually $2\Delta E$, where ΔE is about 0.05 E; e.g. for potassium $\Delta E = 70$ keV and the window extends from $E = 1.39$ MeV to 1.53 MeV.

The further calculation is based on the quantity 'specific activity':

$$\alpha_i = \frac{\text{peak area in window } i}{\text{sample weight}}$$

$$= \frac{\dfrac{\text{sum of counts}}{\text{counting time}} - \text{background count rate}}{\text{sample weight}}$$

$$= \frac{(N_i/t) - b_i}{C_T}, \text{ in cpm g}^{-1}, \tag{4.31}$$

where the index i denotes window i: $i = 1$ for the 2.62 MeV ^{208}Tl peak, $i = 2$ for the 1.76 MeV ^{214}Bi peak and $i = 3$ for the 1.46 MeV ^{40}K peak. With $i = 1$, 2 and 3 the solution of three simultaneous equations yields the three unknown concentrations (c_U, c_{Th}, and c_K):

$$c_{Th} = k_1(\alpha_1 - k_2\alpha_2), \text{ in ppm} \tag{4.32}$$

$$c_U = k_3(\alpha_2 - k_4 c_{Th}), \text{ in ppm} \tag{4.33}$$

$$c_K = k_5(\alpha_3 - k_6 c_U - k_7 c_{Th}), \text{ in \%.} \tag{4.34}$$

The constants $k_1 \ldots k_7$ are the calibration constants. The calibration procedure (using standard spectra) is described in detail in Rybach (1971).

Another method utilizes the full spectrum rather than selected windows. The counts in each channel of a spectrum measured by a multichannel analyzer are the sum of U, Th, and K contributions and of the background. Thus as many equations as channel (usually 526 or 1024) can be written up and the three unknowns c_U, c_{Th}, and c_K are calculated by a least squares algorithm (e.g. matrix inversion) which also calculates a 'best fit' spectrum. The method is further described in Rybach (1971). A 'best fit' spectrum is shown in Figure 4.15. High electronic system stability and sufficient computer memory available are prerequisites of this method. According to Höhndorf (1973) the results obtained by

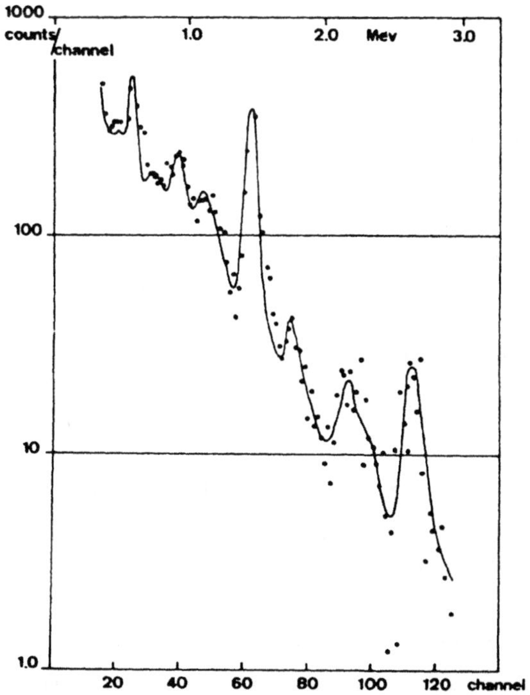

Fig. 4.15. Measured γ-ray spectrum and computed least-squares fit for a Mont Blanc granite sample, Switzerland (3.5 ppm U, 13 ppm Th, 2.5 % K). Dots: counts in 1000 s, solid line: best fit. Sample weight: 595 g. From Rybach *et al.* (1966).

the least squares method are generally not superior of those yielded by the simple three-equation method described above.

For the processing of the γ spectra (calculation of the α's, of c_U, c_{Th}, and c_K, of the errors (see next Section) and of the heat production) computer programs are employed.

4.2.4. Accuracy, Precision

It is customary to base the evaluation of the accuracy of γ spectrometric determinations on counting statistics. It can be shown that the standard error of the U, Th, and K contents can be written as:

$$m(c_U) = \pm(k_3/Gt)(N_1 + tb_1)^{1/2}, \text{ in ppm} \quad (4.35)$$

$$m(c_{Th}) = \pm(k_1/Gt)(N_2 + tb_2)^{1/2}, \text{ in ppm} \quad (4.36)$$

$$m(c_K) = \pm(k_5/Gt)(N_3 + tb_3)^{1/2}, \text{ in \%} \quad (4.37)$$

for definitions of G, t, N_i, and b_i see Equation (4.31).

Correspondingly, the standard error of the heat production H is:

$$m(H) = 10^{-5} \cdot \rho(90.82\ \mathrm{m}^2(c_U) + 6.55\ \mathrm{m}^2(c_{Th}) + 12.11\ \mathrm{m}^2(c_K))^{1/2}. \quad (4.38)$$

It can be seen that the accuracy of uranium determination is decisive.

The applicability of counting statistics can be tested by repeated measurements of the same sample. A satisfactory test result was obtained by many authors (see e.g. Höhndorf, 1973). The counting statistics can be improved by extending the counting time but measurements going on longer than for a day per sample yield no further improvement in accuracy.

The precision of γ-spectrometric U, Th, and K determinations can be checked by interlaboratory comparison of determinations on the same sample and/or by applying different analytical techniques (see Tables 4.5 and 4.6). Figure 4.16 shows a comparison of γ-spectrometric versus x-ray fluorescence results (potassium determinations). Further γ-ray spectrometry/x-ray fluorescence comparisons can be found in Galson *et al.* (1983).

In general, the analytical error of γ-spectrometric U, Th, and K determinations in common rocks (except for ultrabasites which have exceptionally low U, Th, and K contents) is generally ±(5−10) %.

4.2.5. Sensitivity, Detection Limits

The sensitivity of γ-ray spectrometry is generally sufficient to determine U, Th, and K in most ordinary rocks. The detection limit for 95.5 % probability (2σ) e.g. for uranium can be given (Watt and Ramsden, 1964) as:

$$L_U = 2\sqrt{2}(k_3/G)(b_2/t)^{1/2}, \text{ in ppm.} \quad (4.39)$$

Analogous equations can be written up for Th and K. The sensitivity for heat production determinations can then be calculated using Equation (4.30).

In principle, the detection limit can be lowered by applying large-size samples

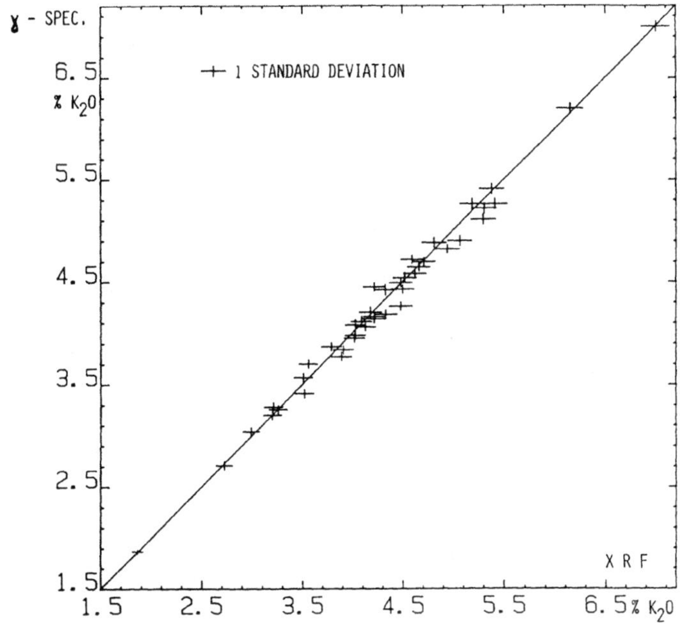

Fig. 4.16. Comparison of γ-spectrometric results with x-ray fluorescence (XRF) determinations.

and long counting times. However, sampies over 1 kg and measurement times in excess of a day yield no significant improvement of the detection limit; with these experimental conditions, γ-ray spectrometry with NaI(Tl) detectors of reasonable size (e.g. 20 cm in diameter and 10 cm high) can reach detection limits of 0.1 ppm for U and Th and 0.05 % for K. These limits correspond, for an average rock density of 2700 kg m^{-3} to a sensitivity of about 0.04 μW m^{-3} in the heat production rate determination.

Ultrabasic rocks like peridotites which are rare and atypical representatives of crustal rocks but are most probably the main constituents of the upper mantle, exhibit heat production figures which are clearly below γ-spectrometric detection limits. Therefore, more sensitive analytical techniques like fission track counting (for uranium) or mass spectrometric isotope dilution must be applied. By all means, the determinations will be tedious and time-consuming because separate treatments will be necessary for the three radioelements U, Th, and K to be determined.

A further problem arises if secular radioactive equilibrium is absent in the ^{238}U and ^{232}Th decay series e.g. due to selective leaching or by other geochemical effects. While this is hardly ever a problem for the ^{232}Th series, uranium series disequilibrium can be present in some samples, especially if they are weathered. No equilibrium problems are posed in γ-spectrometric potassium determinations.

If disequilibrium is suspected it is adviseable to apply to selected samples analytical techniques for uranium like spectrophotometry which enable direct determination. Table 4.6 gives results of such checks on different rock types. Also the use of the low-energy portion of the γ-ray spectrum with Ge(Li) detectors can help to solve disequilibrium problems (see e.g. Lewis, 1974 or Rieppo, 1978).

4.2.6. Heat Production Rate of Various Rock Types

It is evident from Equation (4.30) that heat production in a given rock is governed by the amounts of U, Th, and K present, which vary greatly with rock type but exhibit certain regularities due to the similar geochemical behaviour of these elements during the processes which determine the distribution of uranium, thorium and potassium in rocks (magmatic differentiation, sedimentation, metamorphism). Tables 4.7, 4.8, and 4.9 list typical heat production figures for magmatic, sedimentary and metamorphic rocks.

TABLE 4.7
Uranium, thorium and potassium abundances and radioactive heat production rates in igneous rocks (from Rybach, 1976b)

Rock	U ppm	Th ppm	K %	Th/U	K/U $\times 10^4$	H $\mu W\ m^{-3}$
Granite/Rhyolite	3.9	16.0	3.6	4.1	0.9	2.45
Granodiorite/Dacite	2.3	9.0	2.6	3.9	1.1	1.48
Diorite, Quartzdiorite/Andesite	1.7	7.0	1.1	4.1	0.7	1.08
Gabbro/Basalt	0.5	1.6	0.4	3.2	0.8	0.309
Peridotite	0.02	0.06	0.006	3.0	0.3	0.0117
Ultramafic						
Dunite	0.003	0.01	0.0009	3.3	0.3	0.00188

TABLE 4.8
Uranium, thorium and potassium abundances and radioactive heat production rates in sedimentary rocks (from Rybach, 1986)

Rock type	U ppm	Th ppm	K %	Th/U	K/U $\times 10^4$	H $\mu W\ m^{-3}$
Carbonates						
Limestone	2.0	1.5	0.3	0.75	0.2	0.62
Dolomite	1.0	0.8	0.7	0.80	0.7	0.36
Evaporites						
Salt	0.02	0.01	0.1	0.50	5.0	0.012
Anhydrite	0.1	0.3	0.4	3.0	4.0	0.090
Shales and Siltstones	3.7	12.0	2.7	3.2	0.7	1.8
Black shales	20.2	10.9	2.6	0.54	0.1	5.5
Sandstones						
Quartzite	0.6	1.8	0.9	3.0	1.5	0.32
Arkose	1.5	5.0	2.3	3.3	1.5	0.84
Graywacke	2.0	7.0	1.3	3.5	0.7	0.99
Deep sea sediments	2.1	11.0		5.2	1.2	0.74

TABLE 4.9
Uranium, thorium and potassium abundances and radioactive heat production rates in metamorphic rocks with granodioritic bulk chemistry (from Rybach, 1976b)

Metamorphic grade	U ppm	Th ppm	K %	Th/U	K/U $\times 10^4$	H $\mu W\ m^{-3}$
Greenschist/low	3.5	26.5	3.4	7.57	0.97	3.15
Amphibolite facies						
High amphibolite facies	1.2	9.4	2.0	7.83	1.67	1.18
Low granulite facies	0.9	4.1	2.1	4.65	2.33	0.732
High granulite facies	0.4	0.9	2.9	2.25	7.25	0.447

4.2.7. Indirect Methods for Determining Heat Production Rate

For combined heat-flow density/heat production rate studies, performed to define continental heat-flow density provinces (see Chapters 8.1. and 9.1.4.), surface sampling of representative rocks is carried out. It is adviseable to average the heat production rate values of surface samples collected within a radius of about 1 km around the heat-flow density site (Wollenberg and Smith, 1968) and, is possible, to supplement the surface information by heat production rate determinations on drillcores.

In some cases, heat production rate mapping has been carried out like in the French Massif Central (Royer and Cuney, 1980; Lucazeau, 1982). Airborne radiometric surveys by γ-ray spectrometry yield uranium, thorium and potassium contents of surface rocks; thus heat production rate can be calculated and mapped (Granar 1982). The limited depth range, about which information can be obtained by airborne γ-spectrometric measurements (less than 1 m for hard rock) must be, however, kept in mind.

Deeper-going information about heat production rate is supplied by modern spectral γ-ray logging in boreholes. The NGS (= Natural Gamma Spectrometry) log gives a continuous record of the uranium, thorium and potassium contents of the rock units traversed by the borehole (Schlumberger 1982). An example is given in Figure 4.17. Since the method used in NGS logging assumes secular equilibrium to calculate the uranium content, it is adviseable to perform control analyses by some method which determines uranium directly.

Before the introduction of NGS logging, ten thousands of boreholds have been logged by 'total count' detectors, i.e. by measuring the integrated γ-ray spectrum. Such GR (= Gamma Ray) logs are usually recorded in API (= American Petroleum Institute) units. An empirical relationship between heat production rate and the deviations of the GR log was established recently by Rybach (1986). Figure 4.18 shows the empirical relationship found. Radioactivity measurements in a tunnel by a portable scintillation counter led to a similar empirical relationship for crystalline rocks of the Gotthard Road Tunnel, Swiss Alps (Figure 4.19).

In many cases no direct sampling or access by drilling is possible to the objects of heat production studies. This is especially the case for deeper crustal layers. Especially for constructing and calculating geothermal models the vertical and

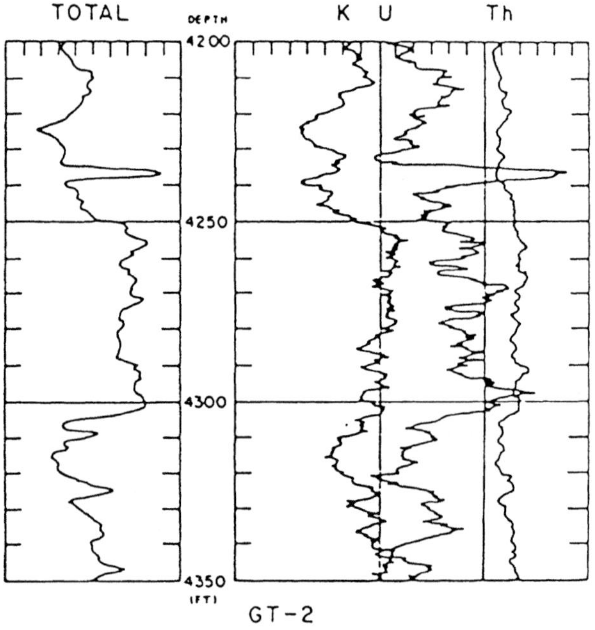

Fig. 4.17. Example of a natural gamma spectrometry (NGS) log: spectral gamma logging signature of a leucocratic monzogranite dyke extending from 4,250 to 4,305 ft in drillhole GT-2 in Los Alamos/USA (from West and Laughlin, 1976).

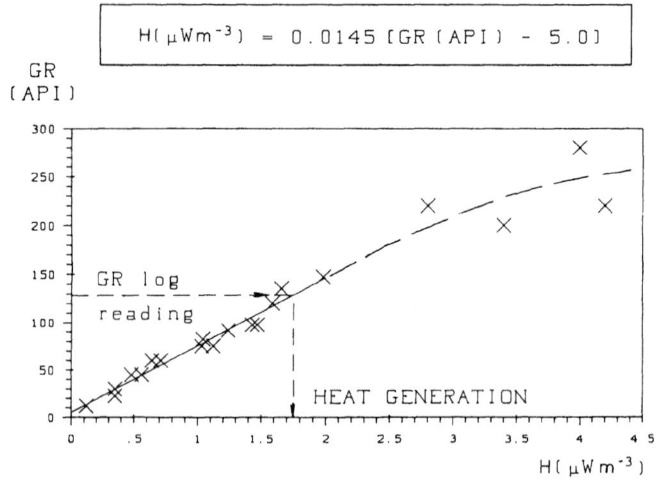

Fig. 4.18. Empirical relationship to obtain heat production rate (H) estimates from gamma ray (GR) logs (readings in API units). The range 0–150 API encompasses practically all common sediments. Above 150 API the scatter of data points increases (from Rybach, 1986).

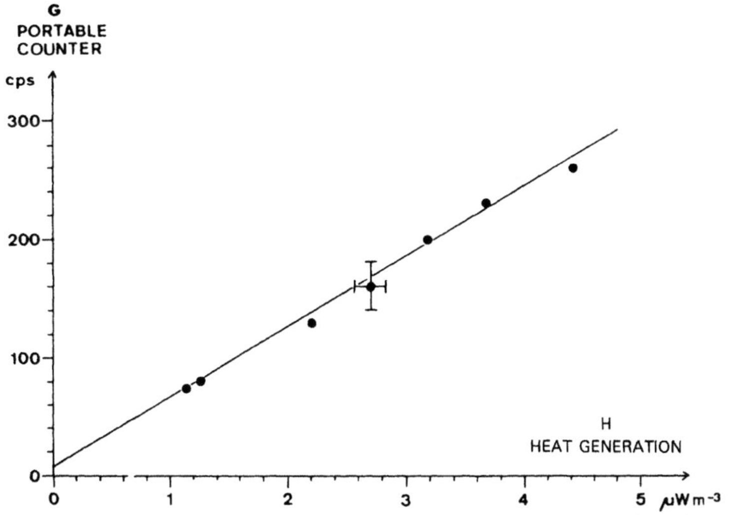

Fig. 4.19. Empirical relationship between radioactivity measurements on crystalline rocks of the Gotthard Road Tunnel, Switzerland (scintillation counter readings in cps, 100 cps = 25 μr h^{-1}) and heat production rate of corresponding rock samples. Regression analysis yields G (cps) = 14.3 + 57.4 H (μW m^{-3}) with r = 0.977 (from Rybach *et al.*, 1982).

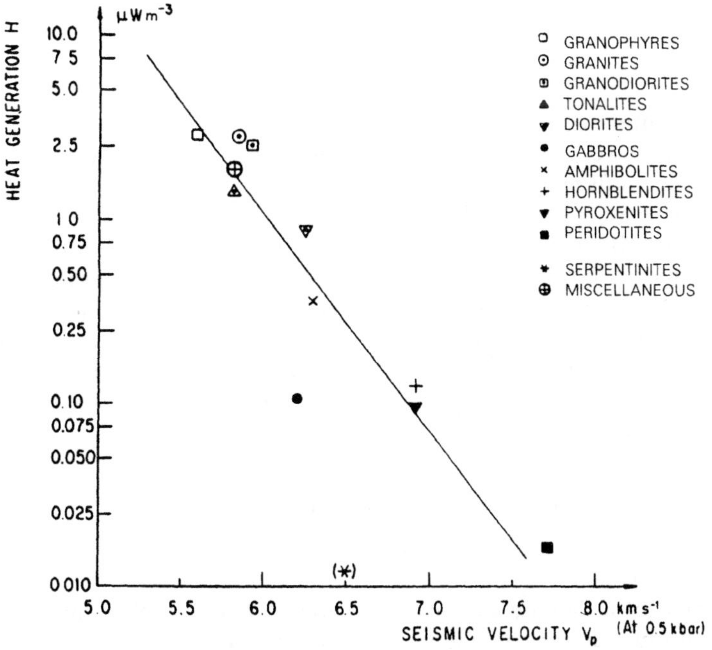

Fig. 4.20. Relationship between seismic velocity v_p and heat production rate H: ln H = 16.5 − 2.74 v_p. v_p measured at 50 MPa.

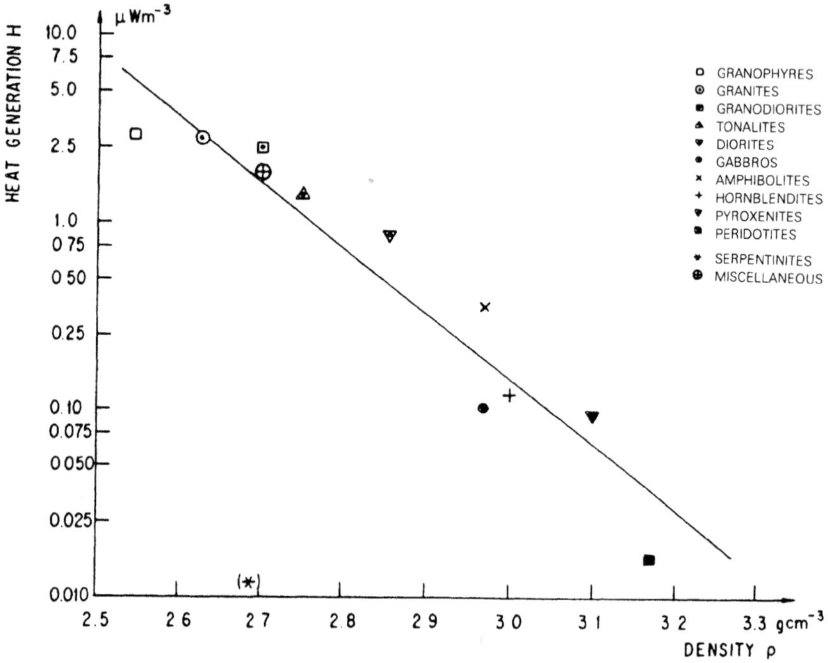

Figure 4.21. Relationship between bulk density ρ and heat production rate (H): $\ln H = 22.5 - 8.15\,\rho$.

lateral distribution of heat production rate is needed. The velocity of compressional seismic waves (v_p) and its variation in the Earth's crust can be determined, however, with sufficient accuracy from the Earth's surface. Figure 4.20 depicts a relationship between heat production rate H and seismic velocity v_p as established by Rybach and Buntebarth (1982). The relationship was derived from laboratory measurements at 50 MPa pressure and at room temperature; before applying the relationship to v_p values measured *in situ*, these must be recalculated to match the above-mentioned laboratory conditions, using the temperature and pressure derivatives of v_p. A similar relationship exists between the bulk density ρ of rocks and H (Figure 4.21). The dependence of H, v_p and ρ on rock type is also evident from Figures 4.20 and 4.21.

4.2.8. Summary of Recommendations

1. Heat production rate determinations should be carried out on representative, unweathered samples.
2. An analytical technique should be employed which enables simultaneous uranium, thorium and potassium determinations, preferably γ-ray spectrometry.
3. Large, sensitive γ detectors should be used: NaI(Tl) scintillation detectors or Ge(Li) solid-state detectors.

4. Sample size should be at least several hundred grams; the form of sample containers should ascertain optimum sample-detector configuration.
5. The spectrometer system should be kept under temperature-stabilized conditions.
6. Cross-checks with other analytical techniques like x-ray fluorescence should be made, especially if radioactive disequilibrium might be present in the samples investigated.
7. The bulk density of the samples studied should be determined.
8. If no direct sampling is possible, indirect methods for heat production rate determination should be used.

4.2.9 References

Bajo, C.: 1980, 'Extraction du Th et de l'U de granites Suisses', Ph.D. Thesis ETH Zürich, No. 6738, 167 p.

Birch, F.: 1954, 'Heat from Radioactivity', in H. Faul (ed.), *Nuclear Geology*, Wiley & Sons, New York, 148–174.

Brunée, C. and Voshage, H.: 1964, *Massenspektrometrie*, Thiemig, München, 316 p.

Galson, D. A., Atkin, B. P., and Harvey, P. K.: 1983, 'The Determination of low Concentrations of U, Th, and K by XRF Spectrometry', *Chem. Geol.* **38**, 225–237.

Granar, L.: 1982, *The Impression of Computer-Made Coloured Geophysical Maps*, Abstract, 8th European Geophysical Society Meeting, Uppsala, 119 p.

Hart, R. J., Reid, D. L., and Welke, H. J.: 1980, 'Comparison of Three Techniques for the Determinations of Uranium and Thorium in Rocks', *Chem. Geol.* **29**, 345–350.

Höhndorf, A.: 1973, 'Auswertung von γ-Spektren zur quantitativen Analyse von Uran, Thorium und Kalium in Gesteinen', *Geol. Jb.* **E 1**, 81–92.

Hurley, P. M. and Fairbairn, H.: 1953, 'Radiation Damage in Zircon: A Possible Age Method', *Bull. Geol. Soc. Amer.* **64**, 659–673.

Hyde, E. K., Perlman, I., and Seaborg, G. T.: 1964, *The Nuclear Properties of the Heavy Elements. II. Detailed Radioactive Properties*, Prentice Hall, Englewood Cliffs, N.J., 1107 p.

Keil, R.: 1979, 'Hochselektive spektralphotometrische Spurenbestimmung von Uran(VI) mit Arsenazo III nach Extraktionstrennung', *Fresenius Z. Anal. Chem.* **297**, 384–387.

Keil, R.: 1981, 'Selektive spektralphotometrische Spurenbestimmung von Thorium mit Arsenazo III nach Extraktionstrennung', *Fresenius Z. Anal. Chem.* **305**, 374–378.

Leutz, H., Schultz, G., and Wenninger, H.: 1965, 'Decay of Potassium 40', *Z. Physik* **187**, 151–172.

Lewis, T. J.: 1974, 'Heat Production Measurement in Rocks Using a Gamma Ray Spectrometer with a Solid State Detector', *Can. J. Earth Sci.* **11**, 526–532.

Lucazeau, F.: 1982, 'Flux de chaleur, production de chaleur et evolution geodynamique récente du Massif Central', Ph.D. Thesis, Université des Sciences et Techniques du Languedoc, Montpellier, 181 p.

Rieppo, R.: 1978, 'A Method to Investigate the Secular Equilibrium of Natural Radioactive Series', *Geoexploration* **16**, 177–183.

Royer, J. J. and Cuney, M.: 1980, *Le flux thermique par les granites de la Montagne Bourbonnais*, 8e Reunion Annuelle des Sciences de la Terre, Marseille, 315. Soc. Géol. Fr. édit. Paris, 315–319.

Rybach, L., von Raumer, J., and Adams, J. A. S.: 1966, 'A Gamma Spectrometric Study of Mont-Blanc Granite Samples', *Pure Appl. Geophys.* **63**, 153–160.

Rybach, L.: 1971, 'Radiometric Techniques', in R. E. Wainerdi and E. A. Uken (eds.), *Modern Methods of Geochemical Analysis*, Plenum Press, New York, 271–318.

Rybach, L.: 1973, *Wärmeproduktionsbestimmungen an Gesteinen der Schweizer Alpen*, Habilitationsschrift ETH Zürich, Dep. Hauptbibl. ETH Zürich, 95 p.

Rybach, L.: 1976a, 'Radioactive Heat Production in Rocks and Its Relation to Other Petrophysical Parameters', *Pure Appl. Geophys.* **114**, 309–318.

Rybach, L.: 1976b, 'Radioactive Heat Production: A Physical Property Determined by the Chemistry of Rocks', in R. G. J. Strens (ed.), *The Physics and Chemistry of Minerals and Rocks*, Wiley & Sons, London, 309–318.

Rybach, L., Bodmer, Ph., Weber, R., and England, Ph. C.: 1982, 'Heat Flow and Heat Generation in the New Gotthard Tunnel, Swiss Alps', in V. Čermàk and R. Haenel (eds.) *Geothermics and Geothermal Energy*, Schweizerbart'sche Verlagsbuchhandlung, Stuttgart, 63–70.

Rybach, L. and Buntebarth, G.: 1982, 'Relationships Between the Petrophysical Properties Density, Seismic, Velocity, Heat Generation and Mineralogical Constitution', *Earth Planet. Sci. Lett.* **57**, 367–376.

Rybach, L.: 1986, 'Amount and Significance of Radioactive Heat Sources in Sediments', in J. Burrus (ed.), *Thermal Modeling in Sedimentary Basins*, Editions Technip, Paris, 311–322.

Schlumberger; 1982, *Well Evaluation Developments – Continental Europe*, 296 p.

Smith, A. R. and Wollenberg, H. A.: 1972, 'High-Resolution Gamma-Ray Spectrometry for Laboratory Analysis of the Uranium and Thorium Decay Series', in J. A. S. Adams, W. M. Lowder, and T.F. Gesell (eds.), *Proc. Second Int. Symp. on the Natural Radiation Environment*, U.S. Dept. of Commerce, Springfield, Virginia, 181–231.

Stuckless, J. H., Millard, H. T., Bunker, C. M., Nkomo, I. T., Rosholt, J. N., Bush, C. A., Hiffman, C., and Keil, R. J.: 1977, 'A comparison of Some Analytical Techniques for Determining Uranium, Thorium, and Potassium in Granitic Rocks', *Research US Geol. Survey* **5**, 83–91.

Tugsavul, A., Merten, D., and Suschny, O.: 1970, *The Reliability of Low-Level Radiochemical Analysis*, International Atomic Energy Agency, Vienna, 32 p.

Watt, D. E. and Ramsden, D.: 1964, *High Sensitivity Counting Techniques*, Pergamon Press, London.

West, F. G. and Laughlin, A. W.: 1976, 'Spectral Gamma Logging in Crystalline Basement Rocks', *Geology* **4**, 616–617.

Wollenberg, H. A.: 1977, 'Radiometric Methods', in J. G. Morse (ed.), *Nuclear Methods in Mineral Exploration*, Elsevier, Amsterdam, 5–36.

Wollenberg, H. A. and Smith, A. R.: 1968, 'Radiogeologic Studies in the Central Part of the Sierra Nevada Batholith', *J. Geophys. Res.* **73**, 1481–1495.

4.3 OPACITY – THE CONCEPT OF RADIATIVE THERMAL CONDUCTIVITY

C. CLAUSER
Institute of Applied Geophysics, TU Berlin, FRG

4.3.1. Introduction

Besides conduction and convection radiation is the third possible mechanism for heat transfer. Knowledge of this process as a function of temperature and pressure is therefore necessary when extrapolating temperatures down to great depths such as the mantle. Also, cooling histories of a liquid Earth or of magmatic intrusions are sensitive to the amount of heat transferred by radiation.

Birch and Clark (1940) were the first to suspect a mechanism other than phonon conduction from their measurements of thermal conductivities of rocks at elevated temperatures, but quantitative work on radiative heat transfer was not started until 1956, when Clark published a series of papers on this subject (Clark, 1956, 1957a, b). He introduced the concept of a radiative thermal conductivity into geophysics, together with the material property 'opacity', that describes how 'transparent' the material in question is to heat (i.e. to electro-magnetic radiation in the wavelength-range of about 0.1–10 μm). Opacity, which is defined as the inverse of the mean free path of radiation, is sometimes also refered to as 'extinction coefficient' or, neglecting contributions from scattering, simply as 'absorption coefficient'.

It seems that the credit for first investigating heat transmission by radiation in 'partially transparent bodies' goes to Schuster (1903), who derived a formula similar to Clark's (1957b). This is quite remarkable because his theoretical work preceeded the development of spectrophotometry and hence the availability of experimental data. After Schuster, but before Clark, Rosseland (1930) had derived an expression for opacity investigating heat transfer in stellar interiors. He called the new parameter 'stellar opacity' or 'stellar absorption coefficient'. Again in quite a different field of science the same concept had been applied to the thermal properties of glass (see Gardon (1961) for a review) where the temperature distribution in tanks of molten glass could not be adequately explained by pure phonon conduction alone.

4.3.2. Some Theory

The basis of all thermal radiation measurements is Planck's distribution law that defines the spectral radiance L_Λ or L_ν of a blackbody as a function of temperature T and wavelength Λ or frequency ν respectively:

$$L_\Lambda = 2hc^2n^2/(\Lambda^5[\exp(hc/\Lambda kT) - 1]), \text{ in W sr}^{-1}\text{ m}^{-3} \quad (4.40a)$$

$$L_\nu = 2h\nu^3n^2/(c^2[\exp(h\nu/kT) - 1]), \text{ in W sr}^{-1}\text{ m}^{-2}\text{ s}, \quad (4.40b)$$

R. Haenel, L. Rybach, and L. Stegena (eds.)
Handbook of Terrestrial Heat-Flow Density Determination, 143—165.

where

n = real part of the refractive index (1.5–2.0 for many materials; cf. Shankland *et al.*, 1979),
Λ = wavelength of radiation in vacuum,
ν = frequency of radiation in vacuum,
T = absolute temperature,
c = velocity of light in vacuum ($2.997925 \cdot 10^8$ m s^{-1}),
k = Boltzmann constant ($1.380662 \cdot 10^{-23}$ J K^{-1}),
h = Planck's constant ($6.626176 \cdot 10^{-34}$ J s)
sr = steradian.

In radiation terminology radiance defines the energy per time that is propagated in a given direction per unit angle about that direction, and per unit area projected normal to that direction. Figure 4.22 illustrates the blackbody spectral radiance as a function of temperature and wavelength.

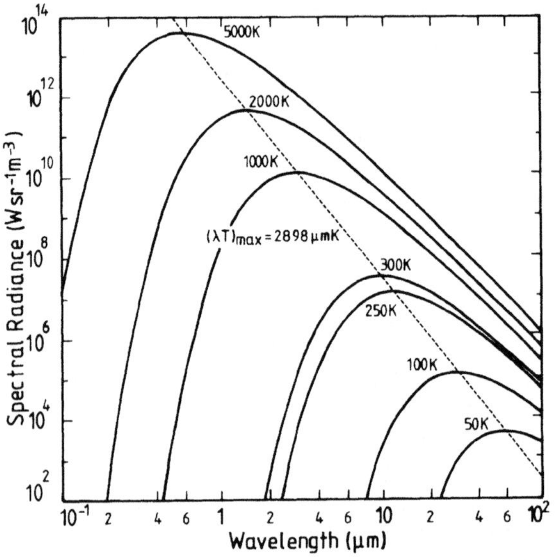

Fig. 4.22. The Planck distribution law: spectral radiance of blackbody radiation as a function of temperature and wavelength. The broken line indicates Wien's displacement law.

In an absorbing and scattering medium such as the Earth radiation can still contribute to the net heat transfer by repeated absorption and re-emission of energy within that medium. If the mean free path of radiation is small compared to the distance to a material boundary, and if the temperature gradients within the medium are small enough not to cause large anisotropies in the intensity of radiation (i.e. to permit a linear series expansion), the equilibrium diffusion approximation applies (for a derivation see e.g. Schärmeli, 1977), and the total heat-flow density q propagated through the medium is given by:

$$q = -(\lambda_p + \lambda_r)\,\Gamma. \tag{4.41}$$

λ_p is the lattice or phonon conductivity and λ_r an equivalent 'radiative conductivity'.

In this context a third component is sometimes added, the 'excition conductivity'. Earlier work by Clark (1957b), Lawson and Jamieson (1958), and Lubimova (1967) suggested that heat transfer by exciton migration becomes the dominant non-convective mechanism in heat transfer for the mantle. However, estimates of the exciton contribution are extremely dependent on exact knowledge and correct interpretation of high-temperature absorption spectra of the minerals in question. More recent work by Schärmeli (1977, 1979) based on high-temperature spectroscopic investigations of mantle silicates by Nitsan and Shankland (1976) reveals that earlier workers had misinterpreted an absorption edge in the UV (around 0.5 μm) as the edge of the electronic conduction band (which Nitsan and Shankland (1976) found to be close to 0.1 μm). This led them to a drastic overestimation: Schärmeli's (1977) calculations clearly show that the contributions by exciton migration (as well as other potential electronic heat transfer mechanisms such as polar or ambipolar charge transfer processes) are several orders of magnitude smaller than the radiative component and can safely be neglected.

The 'radiative conductivity' was given by Rosseland (1930) and Clark (1957b) as:

$$\lambda_r = (4\pi/3)\int_0^\infty \varepsilon_\nu^{-1}\ \partial L_\nu(T)\ \mathrm{d}\nu/\partial T. \tag{4.42}$$

Depending on which parameters are assumed to be independent of temperature and wavelength, the temperature derivative of L_ν leads to different practical expressions for λ_r:

– The so-called gray body approximation, introduced by researchers concerned with the thermal properties of hot glass (Kellett, 1952; Van der Held, 1952; Genzel, 1953) requires the opacity ε to be finite, constant and independent of wavelength (or frequency). If the refractive index n and the spectral radiance L_ν are also wavelength and temperature independent Equation (4.42) simplifies into:

$$\lambda_r = 16\sigma n^2 T^3/(3\varepsilon), \tag{4.43}$$

if one observes that the integral over L_ν equals $4\sigma n^2 T^3/\pi$, according to the Stefan-Boltzmann law, where σ is the Stefan-Boltzmann constant ($\sigma = 5.67032 \cdot 10^{-8}$ W m^{-2} K^{-4}).

– Rosseland (1924, 1930) removed the restriction to gray materials by defining a blackbody averaged opacity ε:

$$\varepsilon^{-1} = \left(\int_0^\infty \varepsilon_\nu^{-1}\frac{\partial L_\nu}{\partial T}\mathrm{d}\nu\right)\Big/\left(\int_0^\infty \frac{\partial L_\nu}{\partial T}\mathrm{d}\nu\right). \tag{4.44}$$

Observing again that the denominator in Equation (4.44) equals $4\sigma n^2 T^3/\pi$ this also leads to Equation (4.43) when inserted into Equation (4.42).

– Clark (1957b) proceeded slightly differently: differentiation of L_ν in the numerator of Equation (4.44) leads to:

$$\varepsilon^{-1} = (15/4n^2\pi^4)\int_0^\infty \varepsilon_x^{-1}x^4e^xn_x^2[e^x - 1]^{-2}\,\mathrm{d}x, \tag{4.45}$$

where x was substituted for $(h\nu/kT)$, $(2n^2k^4\pi^5/15c^2h^3)$ was used for σ, and n is again assumed to be independent of temperature. Inserted into Equation (4.42) this gives:

$$\lambda_r = (20\sigma T^3/\pi^4)\int_0^\infty \varepsilon_x^{-1}x^4e^xn_x^2[e^x - 1]^{-2}\,\mathrm{d}x. \tag{4.46}$$

– Shankland *et al.* (1979) allowed n to vary with temperature and wavelength and, starting with Equation (4.42), arrived at:

$$\lambda_r = (8\pi k^4T^3/3c^2h^3)\cdot$$

$$\int_0^\infty \varepsilon_x^{-1}(x^4e^xn_x^2[e^x - 1]^{-2} + 2Tx^3n_x[e^x - 1]^{-1}\,\partial n_x/\partial T)\,\mathrm{d}x. \tag{4.47}$$

– Schärmeli (1977, 1979) extended the theory to account for the temperature and wavelength dependence of both the mean free path and the scattering coefficient. This introduces a second term into λ_r:

$$\lambda_r = \lambda_{r1} + \lambda_{r2}, \tag{4.48}$$

where λ_{r1} is given by (4.42), (4.43), (4.46) or (4.47) and:

$$\lambda_{r2} = (4\pi/3)\int_0^\infty \varepsilon_\nu^{-1}\, L_\nu\, \partial(\ln(\alpha_\nu/\varepsilon_\nu))\,\mathrm{d}\nu/\partial T. \tag{4.49a}$$

Opacity ε is defined by the absorption coefficient α_ν and the scattering coefficient β_ν or rather by their weighted sum (Pitt and Tozer, 1970b; Nitsan, 1976; Shankland *et al.*, 1979) as:

$$\varepsilon_\nu^{-1} = (\alpha_\nu + \beta_\nu[1 - \langle\cos\vartheta\rangle])^{-1}, \tag{4.50}$$

where $\langle\cos\vartheta\rangle$ is the mean taken over all scattering angles ϑ. As can be seen from Equation (4.50) and Equation (4.49a), the term λ_{r2} in Equation (4.48) becomes negligible if $\beta_\nu[1 - \langle\cos\vartheta\rangle]$ is small compared to α. However, this requires knowledge of both β and $\langle\cos\vartheta\rangle$. If n and β are independent of temperature Equation (4.49a) simplifies into:

$$\lambda_{r2} = (4\pi n^2/3)\int_0^\infty \varepsilon_\nu^{-2}(\partial\varepsilon_\nu/\partial T)(\beta_\nu/\alpha_\nu)\, L_\nu\,\mathrm{d}\nu. \tag{4.49b}$$

For isotropic scattering ($\langle\cos\vartheta\rangle = 0$) Schärmeli (1979) computed $\lambda_{r1}/\lambda_{r2}$ in olivine as a function of β (Figure 4.24), using absorption data as a function of temperature (Fukao *et al.*, 1968) and pressure (Shankland, 1970). Even for scattering coefficients as small as 3 cm^{-1} λ_{r2} is already as big as 20 % of λ_{r1}.

Finally, since all expressions (4.42)–(4.49) for λ_r were derived using a linear series expansion, Schärmeli (1977, 1979) gives a detailed discussion on tolerable temperature gradients for this approximation to hold: for laboratory measurements of the effective thermal conductivity $\lambda = \lambda_p + \lambda_r$ the maximum temperature gradient within a sample should not exceed:

$$(\text{grad } T)_{max} = \min(0.1 \cdot \varepsilon_\Lambda^2/(\varepsilon_\Lambda/T)) \tag{4.51}$$

for all wavelengths Λ. High temperature measurements of the effective thermal conductivity, however, implicitely assume the radiative component to be linear in grad T. Figure 4.23 illustrates for olivine this linearization error as a function the

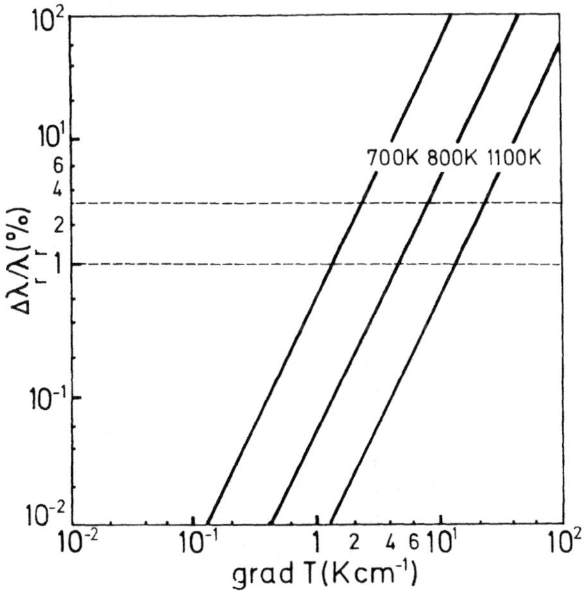

Fig. 4.23. Linearization-error for the radiative thermal conductivity as a function of the ambient temperature gradient in an olivine sample under the most restrictive conditions: at 2.5 μm with no scattering. There was no great pressure change in the range from 0–3 GPa (redrawn after Schärmeli, 1977).

temperature gradient across the sample (for a wavelength of 2.5 μm and no scattering, Schärmeli's (1979) 'most restrictive case'). One can see that the permissible gradient increases with increasing temperature; since λ_r is important mostly for elevated temperatures (T > 200–600 °C, as will be seen later) this poses no serious experimental constraint.

4.3.3. Experimental Work on Opacity

Absorption received much more attention in the literature than scattering, probably due to the fact that its contribution to the total opacity is generally greater. Often the scattering contribution to opacity is simply neglected. In the following, scattering is therefore treated first, and papers explicitely discussing it are presented. Then the remaining papers are presented that either concentrate on the absorption coefficient α or deal with opacity ε as the combined effect of α and β.

(1) THE SCATTERING COEFFICIENT β

A detailed discussion on the size and importance of β and $\langle\cos\vartheta\rangle$ is given by Schärmeli (1979), Shankland *et al.* (1979), Nitsan (1976), Pitt and Tozer (1970b), and Aronson *et al.* (1967, 1970). There are two major scattering mechanisms that are relevant in this context: grain boundary scattering and multiple diffuse scattering caused by many individual scattering centers. While the latter is mostly isotropic ($\langle\cos\vartheta\rangle = 0$), grain boundary scattering can be biased in a certain direction depending on the contrast of the refractive index n between the two grains. If there is no significant contrast, however, scattering will be peaked strongly in the foreward direction ($\langle\cos\vartheta\rangle \geqq 0.95$; $\varepsilon = \alpha$) Pitt and Tozer (1970b) argue that this is the case for upper mantle rocks. This is substantiated by results of Nitsan (1976) who found from measurements on upper mantle rocks the dominant scattering mechanism to be grain boundary scattering, β to be approximately equal to the reciprocal grain size, and $\langle\cos\vartheta\rangle$ to be in the order of 0.98. Therefore, from Equation (4.50), the contribution of β to ε is in the order of 0.2–0.02 cm^{-1} for grain sizes between 1 mm and 1 cm (Shankland *et al.*, 1979) and many authors neglect it (Clark 1957b; Fukao *et al.*, 1968; Fukao, 1969; Pitt and Tozer, 1970b). Nitsan (1976), however, notes that in the lower mantle this may not be the case, and scattering might contribute there as much to opacity as absorption does. Schärmeli (1979) found scattering in sintered olivine to be strongly isotropic ($\langle\cos\vartheta\rangle = 0$; $\varepsilon = \alpha + \beta$). In this case the size of λ_{r2} in Equation (4.48) is strongly dependent on β (Figure 4.24). In rhyolitic obsidian (displaying multiple diffuse scattering) Stein *et al.* (1981) found scattering to be mostly isotropic and to contribute an only weakly temperature and wavelength dependent amount of 6–15 cm^{-1} to the total opacity. They attribute this behaviour to the absence of grain boundaries and to the many small particles in obsidian that cause its dark colour and serve as scattering centers. This interpretation is substantiated by Gable and Shankland (1984) who investigated the optical transmittance of molten, glassy obsidian that did not posses these dark particles. They observed a considerably smaller background scattering of only about 3–5 cm^{-1} in their absorption spectra, and they neglect it. However, in the light of Schärmeli's (1979) results, this small background scatter is still big enough to make λ_{r2} in Equation (4.48) as big as 20 % of λ_{r1} (Figure 4.24); it seems therefore advisable in the isotropic case to account for scattering, no matter how small.

(2) OPACITY ε AND THE ABSORPTION COEFFICIENT α

The Temperature Dependence

The temperature dependence of ε or α was studied very intensively. Since in

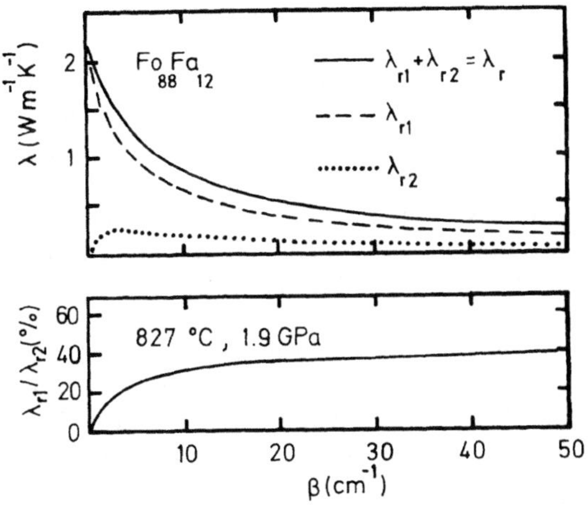

Fig. 4.24. The two components of radiative thermal conductivity as a function of scattering coefficient β as computed from Equations (4.46) and (4.49). The spectrum of the absorption coefficient measured by Fukao *et al.* (1968) was used in combination with Shankland's (1970) results on the pressure dependence (redrawn after Schärmeli, 1977).

absorption measurements on olivines the Fe content will prove a crucial parameter, the relative portions of forsterite $Mg_2[SiO_4]$ and fayalite $Fe_2[SiO_4]$ will always be specified in the following.

Clark (1957a) measured absorption coefficient spectra of gemstone quality minerals at room temperature. Using these results, and assuming ε (or rather α) and n to be independent of temperature he used Equation (4.43) to extrapolate λ_r to temperatures of 2300 °C. However, neglecting an increase in opacity with temperature led him to far too high values for λ_r, due to its T^3 dependence.

Fukao *et al.* (1968), Fukao (1969), Aronson *et al.* (1967, 1970), and Pitt and Tozer (1970b) measured absorption spectra of olivines ($fo_{91}fa_9$ – $fo_{88}fa_{12}$) at elevated temperatures which demonstrated the temperature dependence of ε. They computed ε from Equation (4.43) and found much lower values for λ_r than Clark (1957a). Differences in the results among these authors were later (Shankland *et al.*, 1979) attributed to different oxidation-reduction conditions in the furnace atmospheres: such reactions can drastically alter the absorption spectrum both by changing the defect structure and concentration and by introducing new phases (Nitsan, 1974).

Schatz and Simmons (1972) used a modified Ångström-method to measure under controlled oxigen fugacity simultaneously the radiative and the lattice contributions to the total thermal conductivity of olivines ($fo_{92}fa_8$) and dunite ($fo_{94}fa_6$) from 200–1600 °C: they found a sizable radiative component (about 2 W m^{-1} K^{-1}) for temperatures greater than 500 °C which decreases with increasing fayalite content. However, their data analysis produces a great amount of scatter in their high-temperature radiative conductivities (about ±1 W m^{-1} K^{-1}).

Schärmeli (1977, 1979) made detailed thermal conductivity measurements in a controlled furnace atmosphere on polycrystalline olivine ($fo_{92}fa_8$) from room temperature (λ = 5.2 W m^{-1} K^{-1}) to 1000 °C (λ = 2.5 W m^{-1} K^{-1}) and pressures up to 1.9 GPa. He determined absorption and scattering coefficients from the deviation of measured conductivities from a $1/T$ law using Equations (4.42), (4.48), and (4.49) for a least squares fit of the data. The resulting calculated radiative conductivities are smaller than those determined from absorption measurements on single crystals by a factor of three. However, they are the same size as those that Schatz and Simmons (1972) and Fukao *et al.* (1968) determined for polycrystalline material. A plausible explanation for this systematic difference is grain boundary scattering in polycrystalline samples (or its absence in single crystals) in connection with the different treatments of scattering by various authors.

Shankland *et al.* (1979) measured optical absorption spectra of single and polycrystalline olivine ($fo_{92}fa_8$) up to 1400 °C under oxigen fugacity well within the sample stability field. They used Equation (4.47) to determine λ_r, choosing $\partial n/\partial T$ to $1.8 \cdot 10^{-5}$ K^{-1}, and allowed for the wavelength dependence of n_x by:

$$n_x = n(5.9\ \mu\text{m}) + \partial n/\partial\lambda \cdot (\lambda - 5.9) \quad (4.52a)$$

$$\partial n/\partial\lambda = -0.0205\ \mu\text{m}^{-1} \text{ and } n(5.9\ \mu\text{m}) = 1.66 \text{ at } 22\,°\text{C}. \quad (4.52b)$$

Their results indicate an increase of λ_r from zero at room temperature to about 2.1 W m^{-1} K^{-1} at 1400 °C in good agreement with previous measurements. Figure 4.25 shows the olivine spectra of Shankland *et al.* (1979) as an example for high temperature absorption spectra. Note that there is an increase of λ_r in spite of the rising absorption coefficient. This is due to the shift of the maximum of the spectral radiance (Figure 4.22) into the absorption spectrum's 'transmission window' between about 0.5–6.0 μm. Clark's (1957a) overestimate of λ_r at high temperatures was exactly due to the fact that he did not realize this increase in absorption with temperature.

Obsidian exhibits a different behaviour. Stein *et al.* (1981) measured optical transmittance in rhyolitic obsidian (less than 1.6 % content in iron oxides) from room temperature to 800 °C. Using the same analysis as Shankland *et al.* (1979) they computed λ_r at 800 °C to 0.4 W m^{-1} K^{-1}. Unlike in olivines, there was no noticeable dependence on oxigen fugacity. For the same material Gable and Shankland (1984) extended the temperature range beyond the softening point to investigate absorption in molten obsidian with less than 1% content in iron oxides. Prior to measurements the samples had lost their dark colour due to mechanical and thermal treatment during which they had been molten and crushed several times. With the same analysis as Shankland *et al.* (1979) λ_r was determined from room temperature to 1200 °C. They found that temperature and, most of all, melting considerably increases λ_r, making it comparable in size (1.2 W m^{-1} K^{-1}) with the lattice component λ_p at 700 °C (1.7 W m^{-1} K^{-1}) and dominant at 1145 °C (8.4 W m^{-1} K^{-1}). They therefore assume λ_r to be the dominant component of the total thermal conductivity of silicic, low iron magmas; they point out that this would lead to a faster cooling of silicic magma intrusions than previously believed.

The total amount of iron oxides constitutes an important factor in the determination of the size of λ_r. The Fe^{2+} absorption band at about 1 μm narrows the

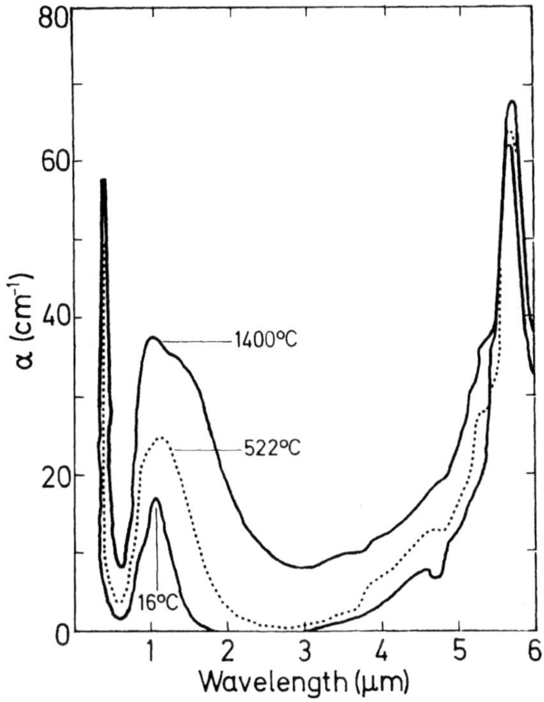

Fig. 4.25. Opacity (absorption) spectrum in single crystal olivine ($fo_{92}fa_8$) measured parallel to the *a*-axis; for these samples the influence of scattering on ε is estimated to be less than 0.5 cm^{-1} (redrawn after Shankland *et al.*, 1979).

infrared pass band for radiation between 0.5 and 2.5 μm depending on the iron content. This pass band is important for heat transfer since the maximum of the blackbody spectral radiance shifts toward smaller wavelengths with increasing temperature. Therefore the findings of Murase and McBirney (1973) substantiate the results of Gable and Shankland (1984): in a controlled furnace atmosphere and for temperatures to 1400 °C they found a strong increase in total thermal conductivity with temperature for rhyolitic obsidian (total content in iron oxides less than 5%) from about 0.8 W m^{-1} K^{-1} at room temperature to 4.0 W m^{-1} K^{-1} at 1400 °C; this increase was not observed, however, for andesitic or basaltic materials with more than 5 % iron oxide content. For these materials there is a continuous decrease in total thermal conductivity with temperature up to about 1200 °C, where λ_r finally becomes dominant and forces the total thermal conductivity to increase again.

Figure 4.26 shows a synopsis of radiative thermal conductivities as a function of temperature for olivines ($fo_{88}fa_{12}$ – $fo_{92}fa_8$) and obsidian, as computed from opacity data by several authors. It is evident that for single crystals λ_r is systematically greater by a factor of about three than for polycrystalline material. Also radiation starts to contribute to the net heat transfer already at temperatures of

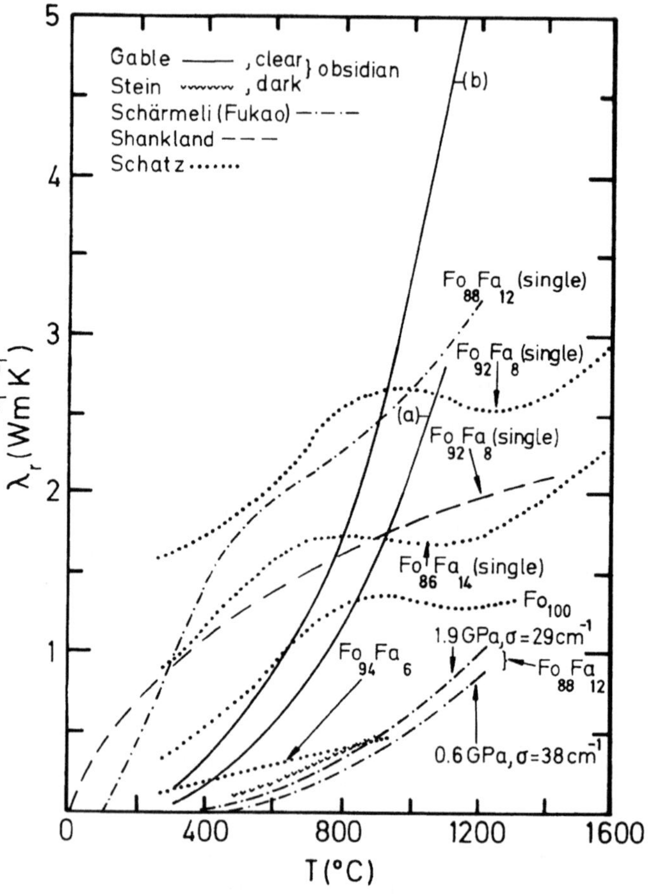

Fig. 4.26. Radiative thermal conductivity as a function of temperature for various single and polycrystalline olivines ($fo_{88}fa_{12}$ – $fo_{92}fa_{8}$), one dunite ($fo_{94}fa_{6}$) and two obsidians. Curves are interpolated from results of : Gable and Shankland (1984) (a: prior to annealing of sample; b: after annealing), Stein *et al.* (1981); Schärmeli (1977) (using the absorption spectra of Fukao *et al.*, 1968), Shankland *et al.* (1979), and Schatz and Simmons (1972).

around 200 °C. Both facts are due to the absence of grain boundary scattering in single crystals. For polycrystals radiation only becomes sizable for temperatures greater than 600 °C. As with single crystals there is an obvious trend of increasing λ_r with decreasing iron (fayalite) content.

The effect of diffuse multiple scattering can be seen on the two obsidian samples: λ_r is relevant only for temperatures greater than 400–600 °C, but becomes quite important for higher temperatures.

Figure 4.27 summarizes the results in terms of the total or effective thermal conductivity $\lambda = \lambda_p + \lambda_r$. The difference between single- and polycrystals is again evident, as well as the increase of the radiative component with decreasing Fe content.

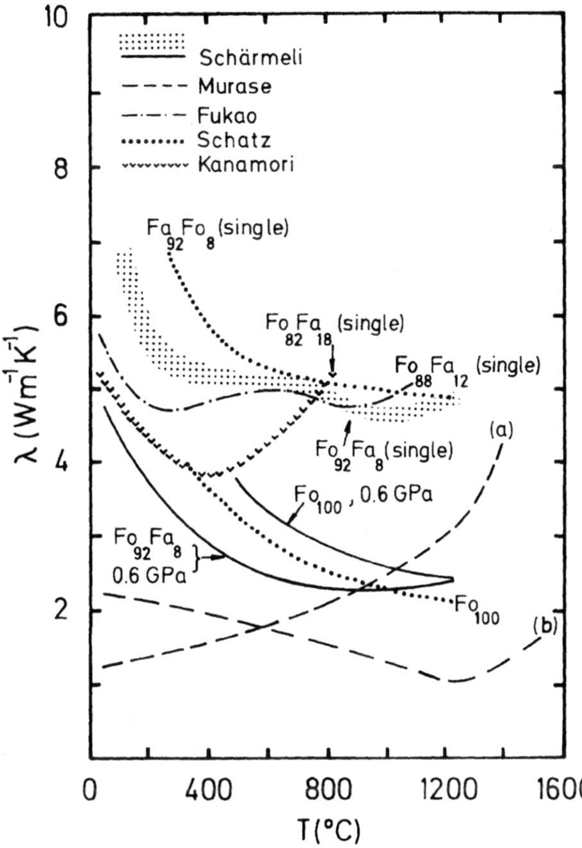

Fig. 4.27. Effective thermal conductivity as a function of temperature for various single and polycrystalline olivines, one Newberry rhyolitic obsidian (a) with 2% iron oxide, and one Galapagos olivine basalt (b) with 12% iron oxides. Curves are interpolated from results of: Kanamori *et al.* (1968); Schatz and Simmons (1972); Fukao *et al.* (1968); Murase and McBirney (1973); and Schärmeli (1977). The dotted area represents conductivities expected by Schärmeli (1977) for the lower crust and upper mantle with no grain boundary scattering.

Figure 4.28 shows two attempts to summarize these results in terms of the thermal conductivity's depth dependence down to the upper mantle: two sets of $\lambda(z)$-curves obtained from oceanic and continental petrophysical models (Schatz and Simmons, 1972) or geotherms (Schärmeli, 1977). While they are similar in trend they are offset by a factor of two. This is due to the different geotherms (or petrophysical models) used and to the fact that Schatz and Simmons (1972) use Equation (4.43) to calculate λ_r while Schärmeli (1977) uses Equations (4.45) and (4.49) and additionally accounts for the pressure dependence from own experimental data. He claims that the curves of Schatz and Simmons (1972) lead to far too high temperatures.

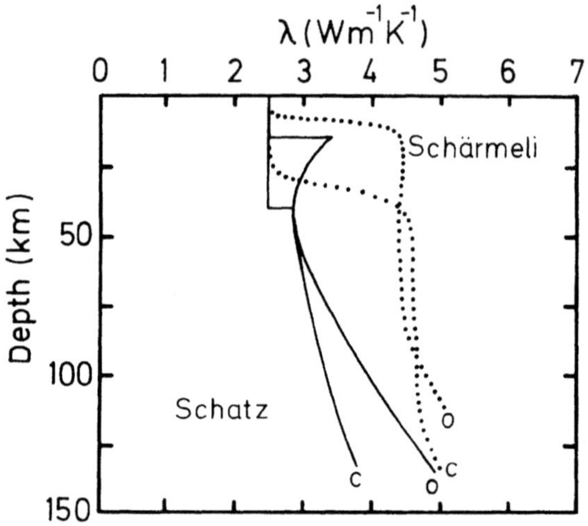

Fig. 4.28. $\lambda(z)$-curves computed for different petrophysical models (Schatz and Simmons, 1972) and geotherms (Schärmeli, 1977); (o) oceanic, (c) continental (redrawn after Schärmeli, 1977).

Finally, in order to show how ε and λ_r can be calculated from the appropriate equations the data set of Fukao *et al.* (1968) on the absorption coefficient of olivine (Table 4.10A) will be used for demonstration purposes. For all equations used this means that a non-trivial integral has to be evaluated, as can be seen from Equations (4.44)–(4.49). Since this would make desk-top calculations very tedious all calculations were done on a computer using a library routine for the numerical integration. The small and (hopefully sufficiently) commented Fortran 77 program is given in the appendix (Chapter 10.5.) so that the results in Table 4.10 can be followed and reproduced by the interested reader. If the numerical library used here should not be available in the specific case any other routine can be used. CPU-time required on a CDC Cyber 825 for the results shown in Table 4.10 is less than two seconds.

First, a blackbody-averaged mean opacity is calculated from Equation (4.45), and results obtained with a wavelength-dependent refractive index are compared with those neglecting this dependence (Table 4.10B); in the latter case Equation (4.45) becomes identical to the formula used by Fukao *et al.* (1968) themselves. As can be seen, the wavelength dependence is not negligible.

Then λ_r is calculated in three ways: Equation (4.43) is used with the two different mean opacities calculated above (Table 4.10B); for the wavelength dependent case this is equivalent to calculating λ_r from Equation (4.46). Moreover, Equation (4.47) is used, which additionally takes into account the temperature dependence of the refractive index. For the wavelength- as well as for the temperature-dependence of the refractive index Equations (4.52a) and (4.52b) are used, as suggested by Shankland *et al.* (1979). All of the results on λ_r are summarized in Table 4.10C. Table 4.10D and Figure 4.29 show the contribution of each wavelength interval to λ_r as calculated from Equation (4.47). It can be

TABLE 4.10

(A) Absorption coefficient of olivine ($fo_{88}fa_{12}$) as a function of wavelength and temperature (data from Fukao *et al.*, 1968). (B) Mean opacity calculated from Equation (4.45) and the data in Table 4.10A; (1) neglecting the wavelength dependence of the refractive index n_x; (2) assuming a wavelength dependence of the n_x as given in Equation (4.52a). (C) Radiative thermal conductivity; (3) and (2) are calcualted from Equation (4.43) using the opacities in Table 4.10B while (3) is calculated from Equation (4.47) assuming a wavelength- and temperature dependence of n_x as given in Equation (4.52a) and (4.52b). (D) Contribution per unit wavelength of each wavelength interval to λ_r calculated from Equation (4.47) (i.e. (3) in Table 4. 10C). The red-shift with increasing temperature is due to Wien's displacement law in the spectral radiance, the weighting function in Equations (4.45), (4.46), and (4.47).

(A) Absorption coefficient in cm^{-1} by Fukao *et al.* (1968)

	Wavelength (10^{-6} m)										
T(K)	0.4	0.5	0.6	1.0	1.4	1.8	2.2	2.5	3.0	4.0	5.0
300	9.80	4.10	2.20	19.40	8.80	0.38	0.10	0.04	0.20	2.50	6.90
400	10.70	4.30	2.30	21.20	12.50	0.75	0.20	0.08	0.25	3.00	7.90
500	11.90	4.80	2.50	17.70	11.10	1.30	0.30	0.17	0.30	3.30	8.30
600	13.80	5.10	2.80	9.10	8.20	2.20	0.50	0.23	0.40	4.00	9.20
700	16.50	5.70	1.90	4.50	4.80	2.20	0.70	0.65	0.70	4.50	9.60
800	21.20	6.40	3.40	3.60	3.40	2.20	1.00	1.00	1.10	4.90	10.00
900	22.80	7.40	3.90	3.90	3.80	3.10	1.70	1.40	1.40	5.40	10.50
1000	30.00	8.20	4.20	5.90	5.70	4.20	2.40	1.90	1.60	5.60	11.50
1100	40.00	9.00	4.50	7.40	6.70	6.20	4.40	2.80	2.30	7.40	13.00
1200	50.00	10.30	5.40	9.00	8.50	7.80	5.60	3.90	3.20	7.60	15.00
1300	60.00	10.30	5.20	10.20	10.10	9.20	6.50	4.10	3.50	7.80	17.00

Table 4.10 (Continued)

(B) Mean opacity

T(K)	Opacity (cm^{-1})	
	(1)	(2)
300	18.66537	17.45526
400	3.51030	3.27042
500	1.71698	1.59776
600	1.29831	1.20612
700	1.78766	1.66002
800	2.17948	2.01974
900	2.63510	2.43924
1000	3.13589	2.90099
1100	4.53128	4.18394
1200	5.99163	5.52068
1300	6.69409	6.15696

(C) Radiative thermal conductivity

T(K)	Radiative thermal conductivity ($W\ m^{-1}\ K^{-1}$)		
	(1)	(2)	(3)
300	0.0122	0.0130	0.0131
400	0.1538	0.1651	0.1652
500	0.6140	0.6598	0.6605
600	1.4032	1.5104	1.5128
700	1.6183	1.7427	1.7464
800	1.9813	2.1380	2.1438
900	2.3333	2.5206	2.5291
1000	2.6895	2.9073	2.9193
1100	2.4774	2.6831	2.6960
1200	2.4324	2.6399	2.6544
1300	2.7681	3.0096	3.0283

Table 4.10 (Continued)

(D) Contribution to (3) in Table 4.10C per 10^{-6} m (10^{-6} W m^{-1} K^{-1})

	Wavelength Intervals (10^{-6} m)									
T(K)	0.4–0.5	0.5–0.6	0.6–1.0	1.0–1.4	1.4–1.8	1.8–2.2	2.2–2.5	2.5–3.0	3.0–4.0	4.0–5.0
300	0.0000000	0.0000000	0.0000000	0.0000000	0.0000000	0.0000058	0.0002213	0.0042467	0.0081830	0.0026752
400	0.0000000	0.0000000	0.0000000	0.0000000	0.0000019	0.0005298	0.0095311	0.0852916	0.1018581	0.0175805
500	0.0000000	0.0000000	0.0000000	0.0000017	0.0000963	0.0064656	0.0845219	0.3400890	0.4118227	0.0506902
600	0.0000000	0.0000000	0.0000011	0.0000866	0.0015995	0.0278690	0.2680763	0.9870211	0.8416800	0.0853209
700	0.0000000	0.0000000	0.0000431	0.0017995	0.0159428	0.1110149	0.6018771	0.8895914	0.9480172	0.1214731
800	0.0000000	0.0000000	0.0002837	0.0127544	0.0825639	0.3039125	0.9619702	1.1294732	0.9759359	0.1546906
900	0.0000000	0.0000006	0.0016728	0.0449067	0.1990631	0.4609747	1.0480876	1.3254463	1.0916158	0.1776807
1000	0.0000000	0.0000067	0.0071329	0.0857755	0.2884398	0.6127961	1.1912257	1.4258152	1.2471431	0.2041756
1100	0.0000003	0.0000496	0.0230907	0.1614741	0.4565908	0.6612875	0.9414374	1.2993478	1.0662352	0.1766469
1200	0.0000019	0.0002462	0.0541042	0.2694149	0.5965144	0.7648122	0.9946238	1.1789916	0.9016321	0.1909456
1300	0.0000095	0.0010623	0.1344669	0.4293559	0.7618327	0.8810320	1.0895121	1.3547011	0.9391364	0.2021796

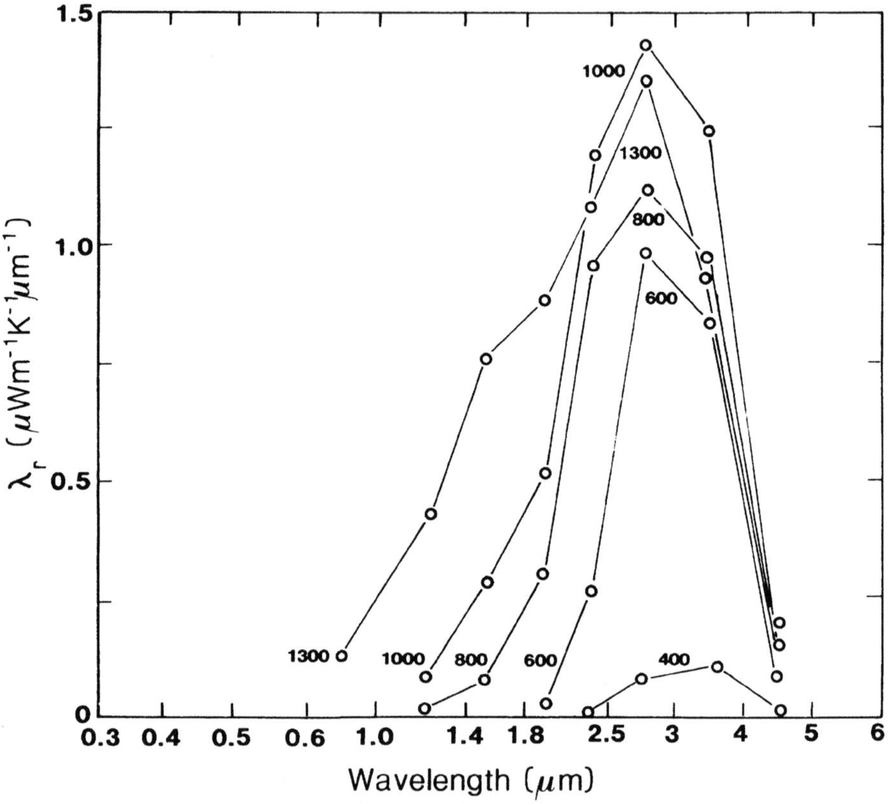

Fig. 4.29. Contribution of each wavelength interval to λ_r, calculated from Equation (4.47) for different absolute temperatures from the olivine absorption coefficients in Table 4.10a (data from Fukao *et al.*, 1968).

seen, that the temperature dependence of the refractive index n is of minor importance; Shankland *et al.* (1979) discuss this in more detail.

If high temperature measurements of thermal conductivity are available, the radiative component and therefore ε can be calculated from Equation (4.43) using residual thermal conductivities, as suggested i.e. by Schärmeli (1977) and MacPherson and Schloessin (1983). Residual conductivities are the differences between measured conductivities and those calculated according to the theory of phonon conduction (Figure 4.30A). This is in most cases a $1/T$ relationship (see MacPherson and Schloessin (1983) for details; also cf. Equation (4.2)). Since ε in Equation (4.43) is a linear function of λ_r and T^3 the mean opacity can be computed from the slope of a plot of these residuals versus T^3 (Figure 4.30B), or from fitting Equation (4.49) to these residuals by choosing an appropriate $\sigma(t)$ (Schärmeli, 1977).

The Pressure Dependence

In the following paragraph pressures will sometimes be converted to depths this

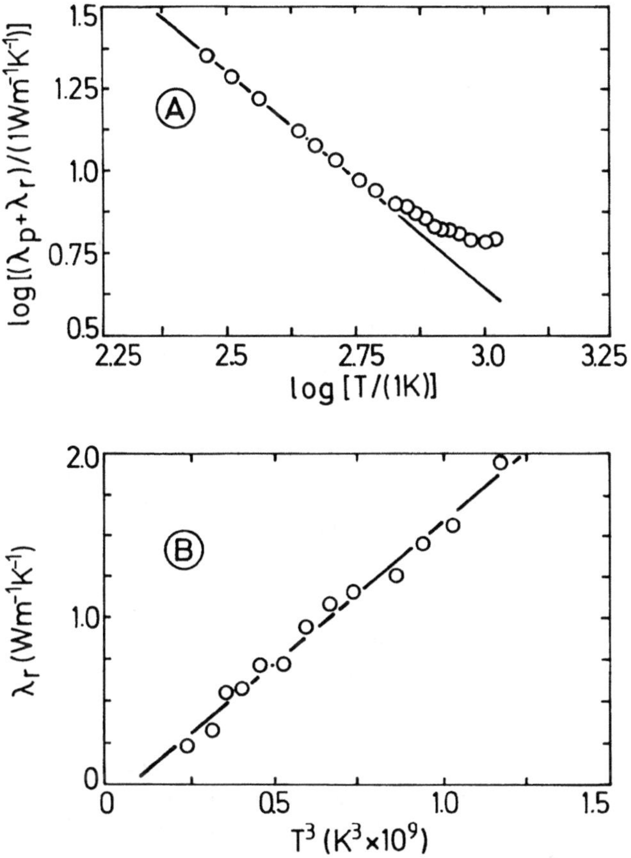

Fig. 4.30. (A) Double logarithmic plot of total thermal conductivity versus absolute temperature for NaCl, showing deviations from the linear extrapolation at high temperatures (redrawn after MacPherson and Schloessin, 1983). (B) λ_r, derived from the residuals determined in Figure 4.30a as a function of T^3 for NaCl at 3 GPa. The straight line is a least-squares fit with a coefficient of determination of 0.985, yielding a mean opacity of 9.29 (redrawn after MacPherson and Schloessin, 1983).

will always be done according to the 'Parametric Earth Model' of Dziewonski *et al.* (1975).

Runcorn (1956) and Balchan and Drickamer (1959) measured the pressure induced red-shift of the ultaviolet (UV) absorption edge in olivines for pressures up to 3.5 and 15 GPa respectively. Pitt and Tozer (1970a, b) investigated the pressure effect in olivine ($fo_{88}fa_{12}$) both on the UV absorption edge and on the infrared (IR) absorption band at 1 μm for pressures to 5 GPa. They combined their results with those of Fukao *et al.* (1968) to demonstrate the combined effect of temperature and pressure on the absorption spectrum (Figure 4.31). Apparently, increasing pressure has two effects: a red-shift of the UV absorption edge, which is moved into the visible part of the spectrum toward greater wavelengths,

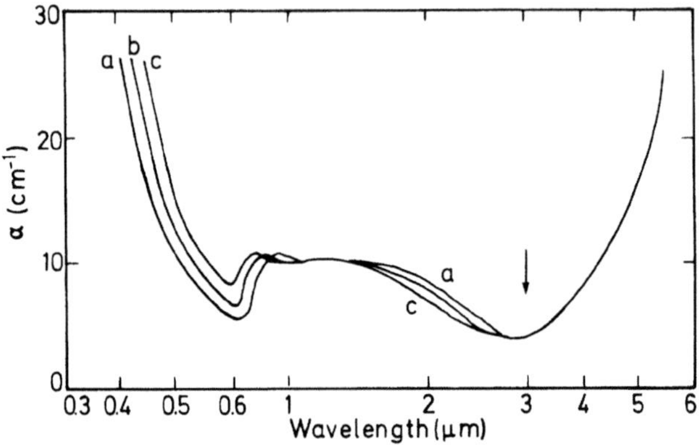

Fig. 4.31. Absorption coefficient calculated from Equation (4.45) for olivine ($fo_{88}fa_{12}$) based on a curve by Fukao *et al.* (1968); for atmospheric pressure (a), for 5 GPa (b), and for 10 GPa (c). The arrow marks the position of the maximum of the weighting function in Equation (4.45), the temperature derivative of the blackbody radiance at 1000 °C; cf. Equation (4.44) (redrawn after Pitt and Tozer (1970b). According to Dziewonski *et al.* (1975) the pressures (b) and (c) correspond to depths in the Earth of 150 km (upper mantle; low velocity zone) and 320 km (mantle below the low velocity zone).

and a blue-shift of the edge of the Fe^{2+} absorption band around 2 μm which is moved toward smaller wavelengths. The latter acts to widen the IR 'transmission window', while increasing temperature tends to close it. Figure 4.31 also demonstrates that the blue-shift is by far the more important pressure effect for radiative heat transfer, at least for pressures below 5 GPa (= 150 km) since very little energy is radiated at wavelengths shorter than 1 μm. For pressures to 12 GPa these results were confirmed by Smith and Langer (1982a, 1983) who made room temperature measurements on olivine ($fo_{90}fa_{10}$), synthetic fayalite, and various garnets (Figure 4.32). Thus it seems that, at least for the lower crust and the upper mantle, increasing pressure enhances radiative heat transfer. For greater depths, such as the lower mantle, this seems no longer to be the case, and there is even some ambiguity in the results among different authors.

Shankland (1970) and Shankland *et al.* (1974) investigated the blue-shift of the IR absorption band for pressures to 5 GPa at ambient conditions in olivines ($fo_{92}fa_8$ and $fo_{90}fa_{10}$), in synthetic spinel, garnets, and in pyroxene. Their results agree well with those mentioned before. However, by extrapolating their results to lower mantle conditions (about 1500 °C and 10 GPa), they calculate that the pressure effect to keep the transmission window open is no longer sufficient to compensate for the contrary temperature effect to broaden the IR absorption band. Hence they infer that radiative transfer is effectively blocked under these pT conditions.

The same conclusion is reached by Mao (1976), but from different evidence: measurements on powdered fayalite (Mao and Bell, 1976) show that for pressures of 20 GPa (= 600 km) or more the red shift of the UV absorption edge sweeps across the visible part of the spectrum into the IR transmission window. Smith and Langer (1982b) found essentially the same results in oriented olivine ($fo_{90}fa_{10}$)

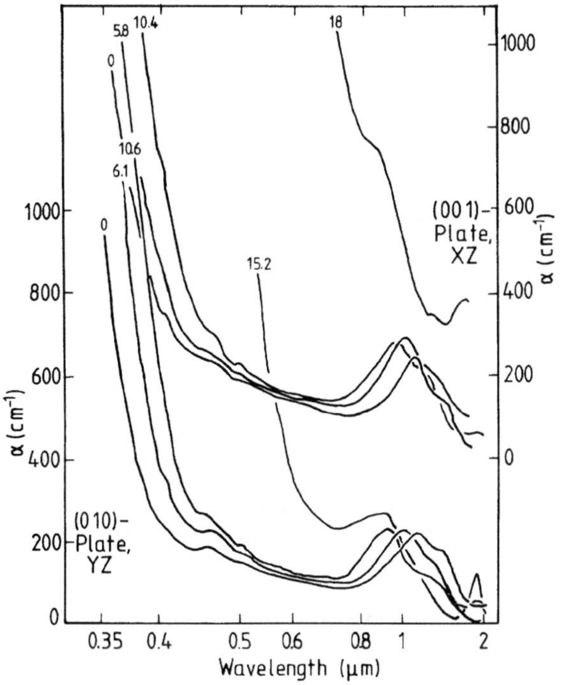

Fig. 4.32. Unpolarized high pressure absorption spectra of two oriented crystal sections of fayalite at room temperature. Curve parameters indicate pressures in GPa (redrawn after Smith and Langer, 1983).

and fayalite single crystals (Figure 4.29; additionally the crystal orientation seems to influence the magnitude of the absorption coefficient). But in agreement with Mao and Bell (1976), their main result, as far as heat radiation is concerned, is that for pressures of 15 GPa (= 500 km) or more the red-shift of the UV absorption edge tends to block radiative transfer.

A totally different interpretation is put forward by Goto *et al.* (1980). They measured opacities in synthetic periclases (MgO with 0.14 and 0.26 mole % Fe^{2+}) for pressures of up to 42 GPa with a shock compression technique, and arrived at very low values of less than 1 cm^{-1}. This is more than two orders of magnitude less than the opacities found by steady state measurements on more Fe^{2+}–rich samples (Mao and Bell, 1976; Smith and Langer, 1982). The fundamental question, whether results of experiments that are transient in the order of microseconds are comparable with those obtained in a steady state, will not be discussed here. It is however conspicuous that the opacities obtained by Goto *et al.* (1980) at zero pressure are smaller by two orders of magnitude than those of, for example, Smith and Langer (1982). This seems to indicate that it might be the difference in the chemical composition of the samples, possibly in Fe^{2+} content, rather than the different experimental techniques used to create high pressures, that is responsible for this divergence in results. Goto *et al.* (1980) argue that all mechanical mechanisms that are different from those under steady state conditions would tend to increase opacity under transient conditions. They too speculate that the Fe^{2+} and Fe^{3+} content may be of importance, especially if the absorption spectra do not scale linearly with the Fe^{2+} concentration.

Goto *et al.* (1980) extrapolate their results for 31 GPa (= 850 km) to a composition corresponding to $(Mg_{0.9}Fe_{0.1})O$, which they feel is typical for the lower mantle, and combine them with those of Shankland *et al.* (1979) to estimate the effect of temperature. They then calculate radiative thermal conductivities for various temperatures using Equation (4.43), and compare these results with those obtained using the pressure dependent opacities of Mao and Bell (1976). Table 4.11 shows this comparison.

TABLE 4.11
Radiative thermal conductivity for $Mg_{0.9}Fe_{0.1}$ versus temperature at 31 GPa calculated from Equation (4.46); (a) opacity scaled from data of Shankland *et al.* (1979) from 5–2.5 μm and $(Mg_{0.78}Fe_{0.22})O$ data of Mao (1976) from 1.75–0.5 μm; (b) opacity scaled from olivine data of Shankland *et al.* (1979) from 5–2.5 μm and 0.26 mole % Fe^{2+} in MgO data of Goto *et al.* (1980) from 1.75–0.5 μm; (after Goto *et al.*, 1980).

T(K)	λ_r(W m^{-1} K^{-1})	
	(a)	(b)
1000	0.13	0.13
2000	0.46	1.30
3000	0.67	3.81

It is clear that small opacities would have a most pronounced effect on the transparency of the lower mantle to heat radiation. It would therefore be important to clarify the high pressure results on opacity in order to provide a consistent theory. Additionally, it should be mentioned at this point that Gaffney (1972), employing crystal field theory, predicts a sizable contribution of radiation to the net heat transfer below 1050 km. This roughly agrees with the depth of 850 km that corresponds to the 31 GPa pressure used by Goto *et al.* (1980) in their calculation, and might be taken as a point in favour of a 'transparent' mantle. On the other hand, since convection is probably the dominant mechanism for heat transport in the lower mantle anyhow, conduction and radiation are most likely to be only second-order effects; the controversy about an opaque or transparent lower mantle may therefore well be of less importance for the net mantle heat transfer.

4.3.4. Summary of Recommendations

In the treatment of radiative heat transfer there are some important points that are suggested independently by the results of severals workers:
1. Scattering should be considered.

Knowledge about the existence and nature of scattering is important when considering the potential contribution of radiation to the overall heat transfer. There is an apparent difference in results between single- and polycrystalline samples which can be attributed to grain boundary scattering. Without it λ_r will start to contribute at temperatures as low as 200–300 °C (as opposed to 600 °C with grain boundary effets) and will be greater by a factor of about three. With multiple diffuse scattering (such as in silicic low-iron magmas) the threshold temperature will be around 400–600 °C, but λ_r will grow very fast with temperature.

2. The Fe^{2+} content should be considered.
   Increasing radiative conductivities can be expected with decreasing Fe-content, since it is mostly the Fe^{2+} content (or the total content in iron oxides respectively) which is responsible for the closing of the IR transmission window with increasing temperature.
3. The pressure effect should be accounted for.
   For pressures of up to 5 GPa, corresponding to a depth of about 150 km, it seems that the effect of pressure is to support radiative transfer. For greater depths (pressures of up to 10 GPa) the contrary and stronger temperature effect to close the IR transmission window, however, seems to make this part of the mantle opaque. For pressures of 15 GPa or more than correspond to depths below the low velocity zone and in the lower mantle, there is an ambiguity: steady state measurements predict total opacity of the material under these pT conditions, while shock compression experiments suggest total transparency. Here more effort would be needed, it seems, either to resolve this ambiguity or to provide a unifying theory. However, in the light of the strong evidence for convection in the mantle (a much more effective heat transport mechanism) this seems not to be a very urgent task.
4. If the results summarized in Figures 4.25 to 4.27 are not sufficient to solve a specific problem, any of the Equations (4.43)–(4.50) can be used to evaluate the radiative thermal conductivity or opacity. The choice will depend on the available data. It is therefore necessary to search in the pertinent literature; if the material in question was already investigated (and this is the case for several potential mantle constituents), numbers for ε (or α and β) can be taken from these papers and inserted in the appropriate equations (as was demonstrated in 4.3.3). Other material parameters related to heat radiation might be found in any of the following data collections: the Consolidated Index (1962), Touloukian and DeWitt (1972), and Raznjevic (1976). Finally, two more rather large review articles should also be mentioned here, the first one especially for the sake of its admirably extensive list of references, a treasure for anyone with special interest in glasses or ceramics: Flynn (1969), and Richmond (1969).

4.3.5 References

Aronson, J. R., Emslie, A. G., McConell, R. K., Eckroad, S. W., and von Thüna, P. C.: 1967, *The Application of High-Temperature Radiative Thermal Conductivity of Minerals and Rocks to a Model of Lunar Volcanism*, A.D. Little Inc., Final Rep., Contract NAS9-5840, NASA CR-916, 67 p.

Aronson, J. R., Bellotti, L. H., Eckroad, S. W., Emslie, A. G., McConnell, R. K., and von Thüna, P. C.: 1970, 'Infrared Spectra and Radiative Thermal Conductivity of Minerals at High Temperatures', *J. Geophys. Res.* **75**(17), 3443–3456.

Balchan, A. S. and Drickamer, H. G.: 1959, 'Effect of Pressure on the Spectra of Olivine and Garnet', *J. Appl. Phys.* **30**, 1446–1447.

Birch, F. and Clark, H.: 1940, 'The Thermal Conductivity of Rocks and its Dependence upon Temperature and Composition, Part 2', *Amer. J. Sci.* **238**(9), 613–635.

Clark, S. P.: 1956, 'Effect of Radiative Transfer on Temperatures in the Earth', *Geol. Soc. Am. Bull.* **67**, 1123–1124.

Clark, S. P.: 1957a, 'Absorption Spectra of Some Silicates in the Visible and Near Infrared', *Am. Mineralogist* **42**, 732–742.

Clark, S. P.: 1957b, 'Radiative Transfer in the Earth's Mantle', *Trans. Am. Geophys. Union* **38**(6), 931–938.

Consolidated Index of Selected Property Values: 1962, *Physical Chemistry and Thermodynamics*, Publication 976, National Academy of Sciences-National Research Council, Washington, D.C. USA, XXIII+274 p.

Dziewonski, A. M., Hales, A. L., and Lapwood, E. R.: 1975, 'Parametrically Simple Earth Models Consistent with Geophysical Data', *Phys. Earth Planet. Interiors* **10**, 12–48.

Flynn, D. R.: 1969, 'Thermal Conductivity of Ceramics', in J. B. Wachtmann (ed.), *Mechanical and Thermal Properties of Ceramics*, Special Publication 303, National Bureau of Standards, Washington, D.C. 20234, USA, 63–123.

Fukao, Y.: 1969, 'On the Radiative Heat Transfer and the Thermal Conductivity in the Upper Mantle', *Bull. Earthquake Res. Inst. Tokyo Univ.* **47**, 549–569.

Fukao, Y., Mizutani, H., and Uyeda, S.: 1968, 'Optical Absorption Spectra at High Temperatures and Radiative Thermal Conductivity of Olivines', *Phys. Earth Planet. Interiors* **1**, 57–62.

Gable, C. W. and Shankland, T. J.: 1984, 'Radiative Heat Transfer in Molten and Glassy Obsidian', *J. Geophys. Res.* **89**(B8) 7107–7110.

Gaffney, E. S.: 1972, 'Crystal Field Effects in Mantle Minerals', *Phys. Earth Planet. Interiors*, **6** 385–390.

Gardon, R.: 1961, 'A Review of Radiant Heat Transfer in Molten Glass', *J. Amer. Ceram. Soc.* **44**, 305–312.

Genzel, L.: 1953, 'Der Anteil der Wärmestrahlung bei Wärmeleitungsvorgängen', *Z. Physik* **135**, 177–195.

Goto, T., Ahrens, T. J., Rossman, G. R., and Syono, Y.: 1980, 'Absorption Spectrum of Shock-Compressed Fe^{2+}-Bearing MgO and the Radiative Conductivity of the Lower Mantle', *Phys. Earth Planet. Interiors* **22**, 277–288.

Kanamori, H., Fujii, N., and Mizutani, H.: 1968, 'Thermal Diffusivity Measurement of Rock-Forming Minerals from 300 K to 1100 K', *J. Geophys. Res.* **73**(2), 595–605.

Kellett, B. S.: 1952, 'Transmission of Radiation through Glass in Tank Furnaces', *J. Soc. Glass Technol.* **36**(169), 115–123.

Lawson, A. W. and Jamieson, J. C.: 1958, 'Energy Transfer in the Earth's Mantle', *J. Geol.* **66**, 540–551.

Lubimova, E. A.: 1967, 'Theory of the Thermal State of the Earth', in T. F. Gaskell (ed.), *The Earth's Mantle*, Academic Press, London, 231–323.

MacPherson, W. R. and Schloessin, H. H.: 1983, 'Apparent, Lattice, and Radiative Thermal Conductivity from 300 to 1500 K and Pressures up to 5.6 GPa: results for MgO and NaCl', *High Temperatures-High Pressures* **15**, 495–509.

Mao, H. K.: 1976, 'Charge Transfer Processes at High Pressure', in R. G. J. Strens (ed.), *The Physics and Chemistry of Minerals and Rocks*, Wiley & Sons, London, 573–581.

Mao, H. K. and Bell, P. M.: 1976, 'Electrical Conductivity and the Red Shift of Absorption in Olivine and Spinel at High Pressure', *Science* **176**, 403–406.

Murase, T. and McBirney, A. R.: 1973, 'Properties of Some Common Igneous Rocks and Their Melts at High Temperatures', *Geol. Soc. Am. Bull.* **84**, 3563–3592.

Nitsan, U.: 1974, 'Stability Field of Olivine with Respect to Oxidation and Reduction', *J. Geophys. Res.* **79**(5), 706–711.

Nitsan, U.: 1976, 'The Effect of Scattering on Radiative Heat Transfer in the Earth's Mantle', *EOS Trans. Am. Geophys. Union* **57**(12), 1005 p.

Nitsan, U. and Shankland, T. J.: 1976, 'Optical Properties and Electronic Structure of Mantle Silicates', *Geophys. J. Royal Astron. Soc.* **45**, 59–87.

Pitt, G. D. and Tozer, D. C.: 1970a, 'Optical Absorption on Natural Ferromagnesian Minerals Subjected to High Pressure', *Phys. Earth Planet. Interiors* **2**, 179–188.

Pitt, G. D. and Tozer, D. C.: 1970b, 'Radiative Heat Transfer in dense Media and its Magnitude in Olivines and some other Ferromagnesian Minerals under Typical Mantle Conditions', *Phys. Earth Planet. Interiors* **2**, 189–199.

Raznjevic, K.: 1976, *Handbook of Thermodynamic Tables and Charts*, McGraw-Hill, New York, VIII+392 p.

Richmond, J. C.: 1969, 'Thermal Radiation Properties of Ceramic Materials', in J. B. Wachtmann (ed.), *Mechanical and Thermal Properties of Ceramics*, Special Publication 303, National Bureau of Standards, Washington, D.C. 20234, USA, 125-137.

Rosseland, S.: 1924, 'Note on the Absorption of Radiation within a Star', *Monthly Notic. Roy. Astron. Soc.* **84**, 525–528.

Rosseland, S.: 1930, 'The Principles of Quantum Theory', in G. Eberhard, A. Kohlschütter, and H. Ludendorff (eds.), Handbuch der Astrophysik, 3(1), Springer, Berlin-New York, 452–474.

Runcorn, S. K.: 1956, 'Experiments on the Displacement of the Ultraviolet Absorption Edge of Olivine at High Pressures', *J. Appl. Phys.* **27**, 598–602.

Schärmeli, G.: 1977, Experimentelle Bestimmung der thermischen Leitfähigkeiten von polykristallinem Olivin FO_{92} bei Drucken bis 25 kb und Temperaturen bis 1500 K und die Identifizierung des Beitrags der Temperatur-Strahlung zum Wärmetransport. Ph.D. Thesis, Ludwig-Maximilians Univ. München, 335 p.

Schärmeli, G.: 1979, 'Identification of Radiative Thermal Conductivity in Olivine up to 25 kbar and 1500 K', in K. D. Timmerhauf and M. S. Barber (eds.), *Proceedings 6th AIRAPT Conf.*, Plenum Press, New York, 60–74.

Schatz, J. F. and Simmons, G.: 1972, 'Thermal Conductivity of Earth Materials', *J. Geophys. Res.* **77**(35), 6966–6983.

Schuster, A.: 1903, 'The Influence of Radiation on the Transmission of Heat', *Phil. Mag., Ser.* **6**, 5, 243–257.

Shankland, T. J.: 1970, 'Pressure Shift of Infrared Absorption Bands in Minerals and the Effect on Radiative Heat Transport', *J. Geophys. Res.* **75**(2), 409–413.

Shankland, T. J., Duba, A. G., and Woronow, A.: 1974, 'Pressure Shifts of Optical Absorption Bands in Iron-Bearing Garnet, Spinel, Olivine, and Periclase', *J. Geophys. Res.* **79**(23), 3273–3282.

Shankland, T. J., Nitsan, U., and Duba, A. G.: 1979, 'Optical Absorption and Radiative Heat Transport in Olivine at High Temperature', *J. Geophys. Res.* **84**(B4), 1603–1610.

Smith, G. and Langer, K.: 1982a, 'High Pressure Spectra of Olivines in the Range 40 000–11 000 cm^{-1}', in W. Schreyer (ed.), High-Pressure Researches in Geoscience, Schweizerbart, Stuttgart, 259–268.

Smith, G. and Langer, K.: 1982b, 'Single Crystal Spectra of Olivines in the Range 40 000–5000 cm^{-1} at Pressures to 200 kbar', *Am. Mineralogist* **67**, 343–348.

Smith, G. and Langer, K.: 1983, 'High Pressure Spectra up to 120 kbars of the Synthetic Garnet End Members Spessartine and Almandine', *N. Jb. Mineralogie Mh.* **12**, 541–555.

Stein, J., Shankland, T. J., and Nitsan, U.: 1981, 'Radiative Thermal Conductivity in Obsidian and Estimates of Heat Transfer in Magma Bodies', *J. Geophys. Res.* **86**(B5), 3684–3688.

Touloukian, Y. S. and DeWitt, D. P.: 1972, 'Thermal Radiative Properties – Nonmetallic Solids', in Y. S. Touloukian and C. Y. Ho (eds.), Thermophysical Properties of Matter, 8, Plenum Press, New York, XXXIV+1856 p.

Van der Held, E. F. M.: 1952, 'The Contribution of Radiation to the Conduction of Heat', *Appl. Sci. Res.* **A3**, 237–247.

5. CONTINENTAL HEAT-FLOW DENSITY

W. G. POWELL and D. S. CHAPMAN
Department of Geology and Geophysics, University of Utah, Salt Lake City, USA

N. BALLING
Laboratoriet for Geofysik, Geologisk Institut, Aarhus Universitet, Denmark

and

A. E. BECK
Department of Geophysics, The University of Western Ontario, London, Canada

5.1. INTRODUCTION

Although on the geological time and space scales the geothermal regime of the Earth is, strictly speaking, both a transient and three dimensional phenomenon, on the global scale by far the most important component of heat transfer is radial. Furthermore, over times of the order of the thermal time constant for the crust, that is a few hundred thousand to a few million years, we consider the thermal regime to be quasi-steady state.

These attitudes are largely a result of the early history of the subject, when it was believed that heat transfer by mass movement and contributions from radiogenic heat sources would be negligible. Therefore, by further arguing that the depth over which we make our thermal measurements, even in deep continental drill holes, was so small compared with the radius of the Earth, and that the area to which those measurements pertained was so small that the curvature of the surface of the Earth was negligible, the parameter of interest to early workers became the amount of heat transferred in the vertical direction across the upper part of the crust by phonon conduction alone; in practice this required corrections for topography to reduce the surface to a plane, and for structure to allow for horizontal inhomogeneity. The problem was therefore simplified to the one-dimensional steady state problem and the parameter to be determined became known as heat-flow density (HFD).

HFD is determined from the product of thermal conductivity λ and temperature gradient ∇T according to Fourier's law, Equation (2.2), $\mathbf{q} = -\lambda \nabla T = -\lambda \mathrm{d}T/\mathrm{d}z$ for the one-dimensional case.

In practice, ∇T is determined by making subsurface temperature measurements in boreholes, mines and tunnels, or in lake sediments (refer to Chapters 3. and 6.); thermal conductivity is most commonly determined on small discrete samples of rock obtained from the drill hole or from nearby localities (refer to Chapter 4.1.). Two methods of combining thermal conductivity and temperature gradient

R. Haenel, L. Rybach, and L. Stegena (eds.)
Handbook of Terrestrial Heat-Flow Density Determination, 167—222.

information are known as the 'interval method' and the 'Bullard method' respectively.

HFD is determined by the simple application of one of these methods to data obtained from measurements in a drill hole a few hundred metres to a few kilometres deep, or within the upper few metres of lake sediments, and is strictly a local value characterizing a region having dimensions of the same order as the depth of the determination. A HFD value determined in this way may contain within it a transient component, sometimes obvious but often so subtle that it goes undetected, due to changing conditions at the Earth's surface ranging from climatic variations to heat transfer by fluid movement. For some purposes this local and sometimes transient HFD may be useful but for many tectonic and geophysical interpretations corrections have to be applied to produce the more useful regional, long-term steady-state (or equilibrium) HFD representative of the longer term thermal state of the crust. Fortunately, many local and transient thermal perturbations can be adequately modelled and hence corrections to the local HFD determination can be applied. In many cases, these corrections have to be applied to the local temperature gradient to produce an equilibrium temperature gradient that is then combined with thermal conductivity information to produce the equilibrium HFD. The bulk of this Chapter is devoted to the discussion of these corrections including corrections for steady-state terrain, climatic history, terrain evolution (erosion and sedimentation), subsurface thermal conductivity structure and groundwater movement.

5.2. DETERMINING CONTINENTAL HEAT-FLOW DENSITY

We assume in this chapter that subsurface temperatures have been measured and corrected for artificial disturbances connected with the drilling of a borehole such as temperature perturbations due to the circulation of drilling mud, free thermal convection within the borehole and induced water flow along a borehole (see Chapters 2.3.3., 3.4. to 3.6.). Thus, the measured but corrected temperatures at this stage represent virgin rock temperatures; however, they may not be used to determine equilibrium temperatures and temperature gradients until the effect, if any, of climatic changes, terrain, regional hydrologic flow and subsurface conductivity variations have been ascertained. Temperatures and temperature gradients at this stage of the analysis are primary input for the calculation of a local HFD value.

5.2.1. Interval Method

In the interval method, a temperature gradient is determined from temperature-depth ($T-z$) data over a specific depth interval and is combined with the representative thermal conductivity for that same interval. Depth intervals are chosen to correspond to major lithologic units intersected by the borehole. If either a single lithologic unit is intersected or if the lithologic units are thin and numerous then the entire borehole can be considered as a single interval. In what follows the term 'specimen' means a single piece of rock prepared for conductivity measurements. The term 'sample' is reserved for its statistical meaning that several specimens make up a sample; the number of specimens making up the sample should be sufficiently large that they are truly representative of the infinite population from which they have been chosen. Unfortunately, as mentioned above, this is not always possible. A depth interval should be considered only if there are both sufficient temperature and conductivity measurements within the interval so that the gradient and average conductivity can be determined within the desired limits of accuracy. The temperature gradient can be determined from a linear least squares regression of T on z (i.e. all error is assumed to be in T) over the selected depth interval; alternatively, it can be determined as the average of gradients determined between consecutive $T-z$ points throughout the interval or from a gradient logging device. A standard error $\alpha = s(N-1)^{-1/2}$ (where s is the standard deviation of the sample of size N) of the temperature gradient should be computed (e.g., Beers, 1962) with the results being reported as the mean gradient $\pm 2\alpha$ which, provided $N > 6$, gives the 95 % confidence limit α_{95}; that is, if we assert a priori that the gradient lies within $\pm \alpha_{95}$ we would be making a correct prediction 95 % of the time.

Thermal conductivity values are generally measured on discrete specimens following procedures described in Chapter 4.1.. For a given depth interval the N discrete values of conductivity should be combined to form the harmonic mean λ_{HM} where;

$$1/\lambda_{HM} = (1/N) \sum_{i=1}^{N} (1/\lambda_i). \qquad (5.1)$$

The harmonic mean is preferred over the arithmetic mean because the temperature change across the interval is directly proportional to thermal resistivity $1/\lambda$ rather than thermal conductivity. Where the entire hole is used as the depth interval because individual lithologic units are too thin the mean conductivity should be weighted by the thickness of lithologies represented by each measured specimen.

If rock were perfectly homogeneous with respect to thermal conductivity, a single specimen would be sufficient to characterize a depth interval. Rocks however exhibit considerable heterogeneity as manifested by thermal conductivity measurements on specimens 1 cm thick and a few centimetres in diameter. It is not uncommon for contiguous specimens taken from drill core to vary by tens of percent in thermal conductivity. A consequence of this heterogeneity for interval HFD determination is the requirement to include a sufficient number of conductivity specimens so that the mean value is determined to the desired limit of accuracy; see also Section 4.1.6.(4).

A guide to the number N of specimens required can be estimated from the following considerations. If the standard error α of the sample mean $\bar{\lambda}$ is to be less than 0.1 $\bar{\lambda}$ then $\alpha = s(N-1)^{-1/2} < 0.1\,\bar{\lambda}$ or, in rearranged form, $N > 1 + (10s/\bar{\lambda})^2$.

A more convenient form of expressing error is the coefficient of variation, V (standard deviation expressed as a percentage of the mean value or $= 100\, s/\bar{\lambda}$ %), of the thermal conductivity; V for most lithologies ranges between 15 % and 30 %, with 20 % a characteristic value. Using $s/\bar{\lambda} = 0.2$, then $N > 4$. For V to be less than 5 % of the mean, the corresponding requirement is $N > 16$. This latter condition is often not satisfied in HFD determinations.

An example of data presentation and heat-flow density determined by the interval method is shown in Table 5.1 and Figure 5.1.

TABLE 5.1
An example of heat-flow density calculated by the interval method in a 570 m deep drill hole, site WSR-1 at Woodside, on the Colorado Plateau, 39°10′43″ N, 110°24′05″ W, elevation 1665 ± 3 m (after Bodell and Chapman, 1982). Depth interval in m, gradient in mK m^{-1}, conductivity in W m^{-1} K^{-1}, heat-flow density in mW m^{-2}.

Depth interval	Thermal gradient	Sample type	Conductivity λ	Interval HFD	Mean value	Corrected value
45–105	15.0 ± 1.0	core	3.96 ± 0.23	60 ± 5		
105–245	18.0 ± 1.0	core	3.43 ± 0.15	62 ± 4		
245–320	24.8 ± 0.7	core	2.75 ± 0.14	68 ± 4	65 ± 4	66
320–455	16.0 ± 0.4	core	4.18 ± 0.15	67 ± 3		
455–515	17.2 ± 0.5	core	4.20 ± 0.14	72 ± 3		
515–575	16.5 ± 0.6	core	3.86 ± 0.09	64 ± 3		

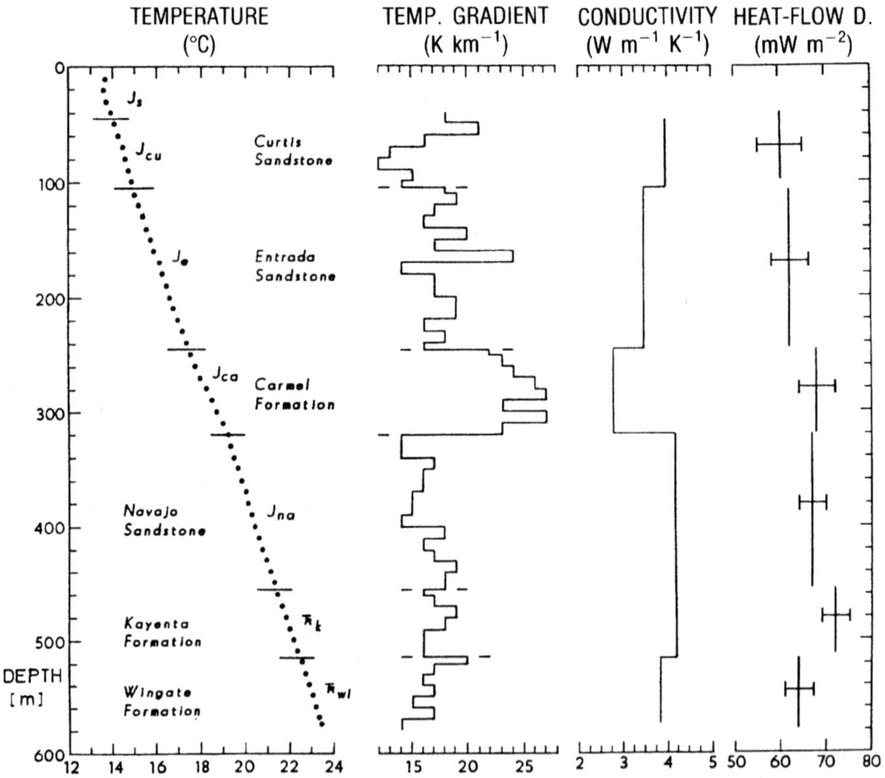

Fig. 5.1. Illustration of heat-flow density determination computed by the interval method. Example shown is site WSR.1 on the Colorado Plateau of the Western USA reported by Bodell and Chapman (1982). Six depth intervals corresponding to lithologic units were chosen. Note the inverse relationship between temperature gradient and thermal conductivity. Details are given in Table 5.1.

5.2.2. Bullard Method

For one dimensional, steady-state, conductive heat flow in a layered medium with negligible heat sources and sinks the subsurface temperature $T(z)$ may be expressed as (see also Equation (7.30a)):

$$T(z) = T_0 + q_0 \sum_{i=1}^{N} (\Delta z_i/\lambda_i), \tag{5.2}$$

where

T_0 = surface temperature
q_0 = constant heat-flow density
λ_i = conductivity over the ith depth interval Δz_i.

The quotient $\Delta z/\lambda$ is formerly called thermal resistance (see Table 2.2) and is summed over the n depth intervals between the surface and the measured temperature; $\Sigma(\Delta z/\lambda)_i$ is often called the thermal depth. A plot of $T(z)$ against $\Delta z/\lambda$ for uniform HFD yields a straight line with a slope of q_0 and intercept T_0.

The use of such a plot by Bullard (1939) to determine heat-flow density in South Africa has given rise to the terminology 'Bullard method'. This procedure is illustrated in Figure 5.2 for a number of sites in the Republic of Zambia reported by Chapman and Pollack (1977). These examples illustrate two cases where use of the Bullard method is appropriate. The first example, is from a flat lying sedimentary sequence of sandstones and mudstones where lithologic units are few, thick, and well characterized by thermal conductivity measurements; the term $\Delta z/\lambda$ in this case is summed over the five formations present. The second example is drawn from broad metamorphic terrain of gneisses and schists in which case it is summed over intervals, each characterized by a single thermal conductivity measurement at the interval midpoint.

5.3. REPORTING UNCERTAINTIES IN HEAT-FLOW DENSITY DETERMINATIONS

Assessing the reliability and quality of the HFD determinations is not as straightforward as it would seem at first sight. Reliability is not synonymous with quality or accuracy; rather, reliability refers to the validity of the method used to calculate the error bounds on the HFD value. If such error bounds have been calculated by a valid statistical technique and are narrow the HFD may be regarded as of both high quality and high reliability; if the error bounds are wide, the value is of low quality but high reliability. If the error bounds are essentially a guess then the HFD is of low reliability and low quality no matter how narrow the error bounds.

In principle assessing uncertainties in local HFD determinations is relatively straightforward, but in practise the situation is quite complicated.

First, although thermal conductivity measurements can readily be made to an imprecision of $V < 2$ % using any one of several standard methods, the small number and small physical size ($\cong$10 mm thickness) of the specimens measured may not sample the rock column adequately, therefore giving a conductivity that

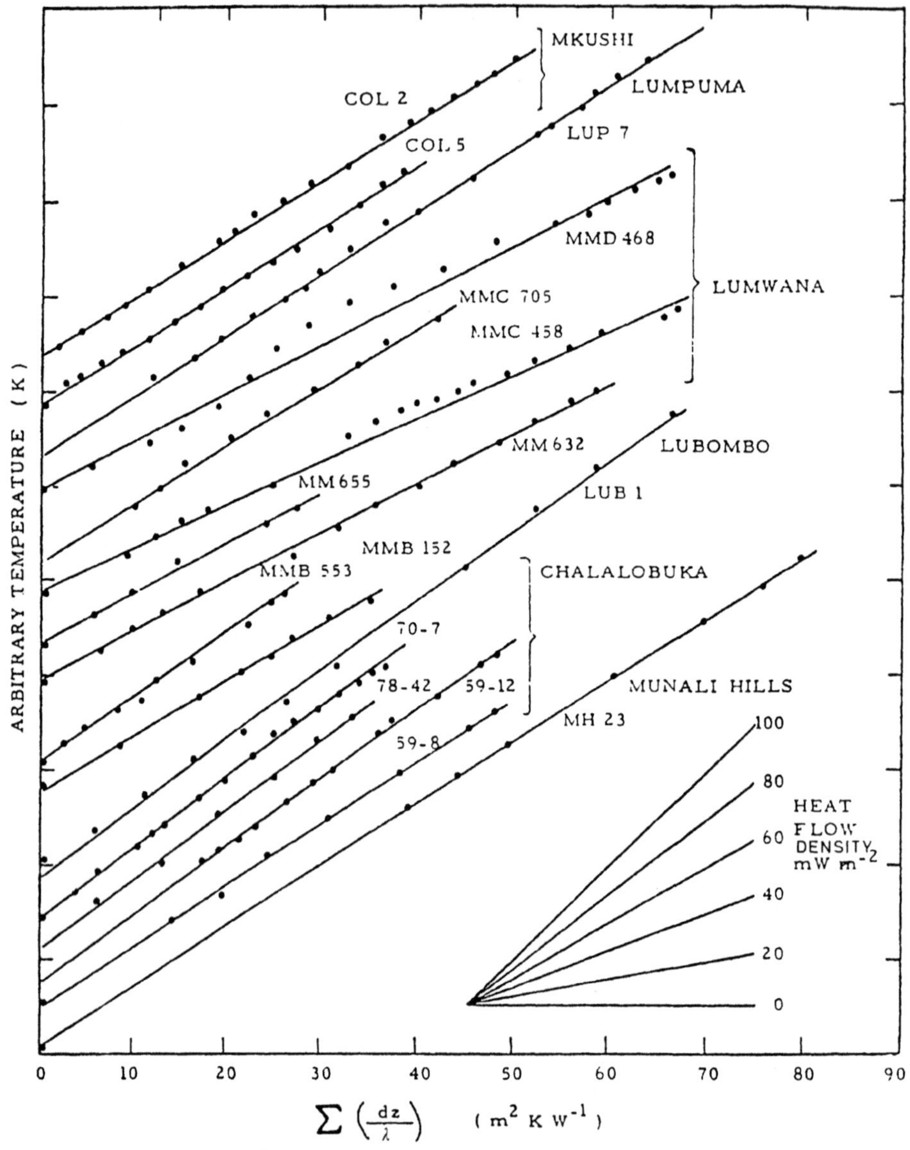

Fig. 5.2. Bullard plot of temperature versus total thermal resistance for a selection of boreholes from the Republic of Zambia. The slope of each curve yields a heat-flow density. After Figure 4 of Chapman (1976).

is inaccurate; the statistical estimate of error of the conductivity measurements is based upon the number of specimens used and, unless complete core is available, there is no satisfactory way of ensuring that the sample mean $\bar{\lambda}$ is the same as the population mean μ or that the standard deviation of the sample s is the same as the standard deviation of the population σ. Furthermore, the problem is complicated by the fact that laboratory measurements may not properly compensate the micro crack and chemical alterations that take place when some of the specimens are removed from the environment; in situ measurements for conductivity are difficult to obtain and in any case may not give conductivity in the direction of heat-flow density that occurs in nature.

Temperature gradients (Γ) can also be determined with low levels of inaccuracy ($< 1\%$) over intervals of tens of metres in an undisturbed borehole but this local gradient may be affected by terrain, transient temperature changes at the Earth's surface, three dimensional subsurface thermal conductivity structure, or slow groundwater percolation. As a result the temperature gradient, while valid locally in space and in time, may not be representative of the long term thermal regime of a region laterally or vertically removed from the measurement site or depth. It is here that assessing the uncertainties of the equilibrium HFD value becomes very difficult because we now have to find a way to combine the uncertainties in conductivity values, local temperature gradient values, and other perturbing effects, in a valid statistical approach. The spatial density of measurements and ancilliary geologic information collected in heat flow studies seldom permits an accurate evaluation of all, or sometimes any, of these effects. As is discussed later, with particular reference to the problem of climatic changes, coupled parameter inversion techniques appear to be the best way of obtaining a maximum likelihood estimate of error, as well as mean value, when there are many interrelated variables with different (large or small) uncertainties.

In spite of these difficulties it is important in heat-flow density studies to report an estimate of the errors involved for each of the principal elements of the HFD calculation. For thermal conductivity measurements at a site, the number of specimens used, the mean and the standard deviation should be reported from which the standard error of the mean may be derived. Either the standard error in conductivity or sometimes twice the standard error (95 % confidence limit) is commonly reported as the conductivity uncertainty (see Chapter 4.1. for more details). Thermal resistivities are also used (Roy *et al.*, 1968); the mean resistivity is seldom very different from the reciprocal of the mean conductivity. For the error in temperature gradient it is common to quote the standard deviation of the slope in the least squares fitting of temperature-depth ($T-z$) data over a depth interval of interest. The *probable* error in HFD is the square root of the summed squares of the fractional errors in conductivity and gradient. If the same confidence limits are used for both λ and Γ, the *maximum* error in HFD to be expected within the same confidence level is the sum of the fractional errors.

These statistical measures of data scatter, however, do not properly address the systematic errors that may arise from terrain, climate, groundwater flow and geological structure effects. An estimate of these possible errors is sometimes provided, by the less formal evaluation of the conductivity and $T-z$ information and the nature of the heat flow site. Sass *et al.* (1971a, b) proposed the following quality classification:

CATEGORY 1 – HIGHEST QUALITY ($V < 10$ %)

$T-z$ profiles are smooth with no sign of hydrologic disturbances below depths of a few tens of metres. Sufficient specimens of rock are available to characterize the effective conductivity of the measured section. No variations in conductivity and gradient that cannot be explained and corrected are apparent.

CATEGORY 2 – MEDIUM QUALITY ($10\ \% < V < 20\ \%$)

$T-z$ profiles have some irregularities. Specimens are too few or conductivity values are too scattered for a good bulk conductivity characterization.

CATEGORY 3 – LOWEST QUALITY ($V > 20\ \%$)

$T-z$ profile highly disturbed. Conductivity poorly determined or estimated. HFD value should be considered only as an estimate on which no definite error bounds can be placed.

A classification scheme based on site characteristics alone such as the Sclater *et al.* (1976) scheme for marine heat-flow density sites has not been proposed for continental sites although consideration of terrain and hydrology could provide a useful reliability filter.

5.4. TERRAIN EFFECTS AND PALAEOCLIMATIC CORRECTIONS

'Terrain effects' are disturbances of the temperature field within the solid Earth caused by the physical processes of heat transfer at the solid Earth's surface. 'Terrain effects' and 'topographic effects' have long been synonymous in the literature of applied geophysics; e.g. Sheriff (1973). However, since it is now recognized that other aspects of the Earth's surface besides topography contribute disturbances to the observations, we see a need to recognize the inherent difference in these terms. Sources of terrain effects include the topography, climatic and microclimatic variation, and variation in thermal and radiative properties of the soil surface. Terrain effects may exhibit time dependence: diurnal and annual variations at shallow depths, slow changes in soil surface temperature due to climatic change, microclimatic changes near the observation site (often caused by human activity) and changes in topography or elevation, including sedimentation and erosion, in tectonically active regions are some temporal phenomena which might disturb temperatures observed in the subsurface. Finally, advective effects of shallow groundwater systems with topography-driven flow fields are also a kind of terrain effect, but because the modelling of hydrologic effects is so involved, these effects are usually considered separately from terrain correction effects. The magnitude of terrain effects seems correlated most strongly with local topography; a site in rugged topography is more likely to require significant correction while the correction is often negligible is subdued topography.

The effect of topography on the near-surface temperature field, illustrated in Figure 5.3A, was recognized late in the nineteenth century; a brief review of the early literature is provided by Birch (1950). Identification of a sources of the more general terrain effects has continued to the present time and concurrent development of numerical techniques for quantitative correction has given the geothermal interpreter a diverse menu of methods applicable in a wide range of circumstances. All terrain correction methods are somewhat tedious to apply and the methods do not have equal value in all situations, so terrain corrections to

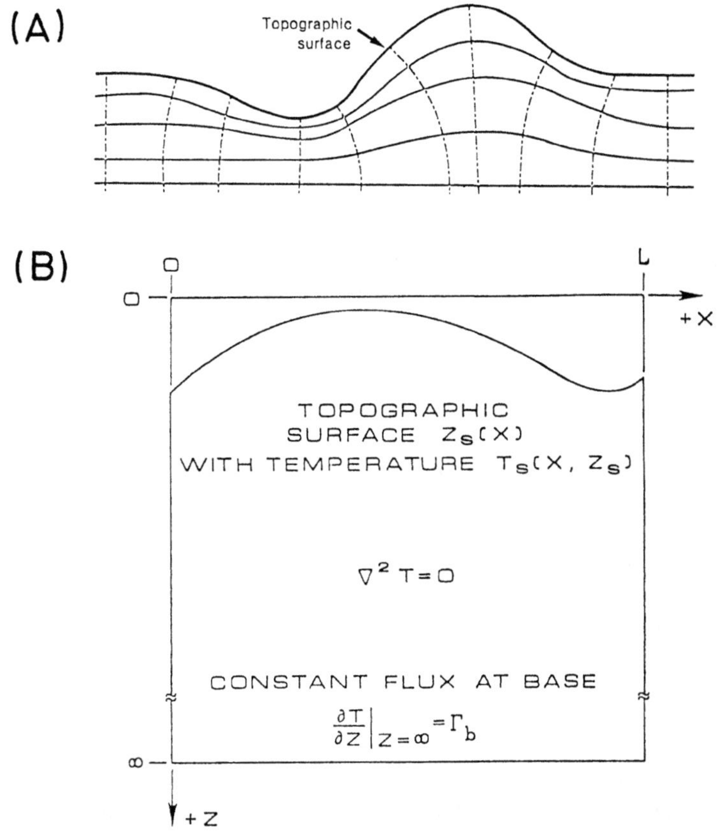

Fig. 5.3. (A) Schematic of perturbation to near surface thermal field due to topography; the solid lines are isogeotherms and the broken lines indicate direction of heat flow. (B) Schematic 2-D steady-state terrain correction problem, illustrating notation used in this chapter. Thermal conductivity is homogeneous, isotropic. No horizontal component of heat flow at lateral boundaries $x = 0$ and $x = L$.

geothermal data are not yet routine. In practice, a semi-quantitative estimate will often suggest that terrain effects are small. When this is the case, corrections are not applied and the effects remain as a component of unknown noise in the observations. The relative importance of each of these effects varies considerably; recognizing the significant sources of disturbance and selecting a technique to correct them requires careful judgement in each case.

The remainder of this section reviews various correction techniques proposed for terrain problems. The discussion concentrates on steady-state problems because they are simpler and more widely applied. Time dependent corrections are briefly covered at the end of the section.

5.4.1. Description of Effects – Steady State

We begin by stating the physical model of terrain disturbances which serves as a starting point for virtually all the correction methods. Although the following description is of a two dimensional model, the extension to three dimensional geometry is simple. Consider an Earth model consisting of a homogeneous isotropic half-space, free of any local heat sources, with a boundary surface modified by some topography (Figure 5.3B). With the additional assumption that heat transport is entirely conductive, the temperature, T, within the Earth must satisfy Laplace's equation, Equation (2.8), $\nabla^2 T = 0$. Corrections are obtained by solving this equation with appropriate boundary conditions. Unfortunately, the boundary conditions of the true Earth are never well known; selection of the appropriate boundary data for topographic correction is an art. Figure 5.3B illustrates the boundaries to be considered and the notation to be followed below.

(1) SUBSURFACE BOUNDARY CONDITIONS

At the base of the space, far enough from the surface to be unaffected by terrain, a constant basal heat-flow density equal to the background heat-flow density in the absence of disturbance is almost always assumed. In a homogeneous space, a constant basal gradient, Γ_b, is an equivalent condition. However, the parameter Γ_b is usually an unknown quantity which is to be determined from the observations. Two practical approaches are available: (1) corrections that do not depend upon Γ_b may be calculated and removed from the observations, with the corrected observations used to determine Γ_b, or (2) a comprehensive thermal model incorporating disturbances and the unknown(s) is constructed and iterative or stepwise adjustments are made in Γ_b until the model fits the subsurface observations in a satisfactory way.

The conditions at the lateral boundaries of the subsurface region under consideration also pose some difficulty; these are rarely physical boundaries but are present because tractable problems are necessarily finite. Temperature or flux conditions on these artificial boundaries are never actually known. Usually, lateral boundaries are assumed to be perfectly insulating (or equivalently, the surface and basal conditions are assumed periodic in x, with even symmetry and period $2L$). This assumption is best when heat flow adjacent to the boundary is most nearly vertical. It is therefore desirable to locate these lateral boundaries where surface relief, surface temperatures and lateral thermal conductivity variations are very small. To minimize the efffect of unavoidable errors at the lateral boundaries on the subsurface region of interest, the distance between the region of interest and these boundaries should be several times the size (the greater of width or depth) of the region of interest.

The effect of these artificial boundaries on the calculated temperature field is easily illustrated. Blackwell *et al.* (1980) used a topographic model after Lees (1910) as a test for topographic correction procedures; we have used the same model as an example here. The parameters of the model are: maximum height of topography, $H = 580$ m, horizontal distance from crest of hill to point at elevation $H/2$, $b = 676$ m, soil surface temperatures follow lapse 4.5 °C km^{-1}; and background gradient $\Gamma_b = 20$ °C km^{-1}. Figure 5.4 compares subsurface temperatures calculated for a simulated borehole at the crest of the hill using finite distances to lateral and basal boundaries with the exact subsurface temperatures determined

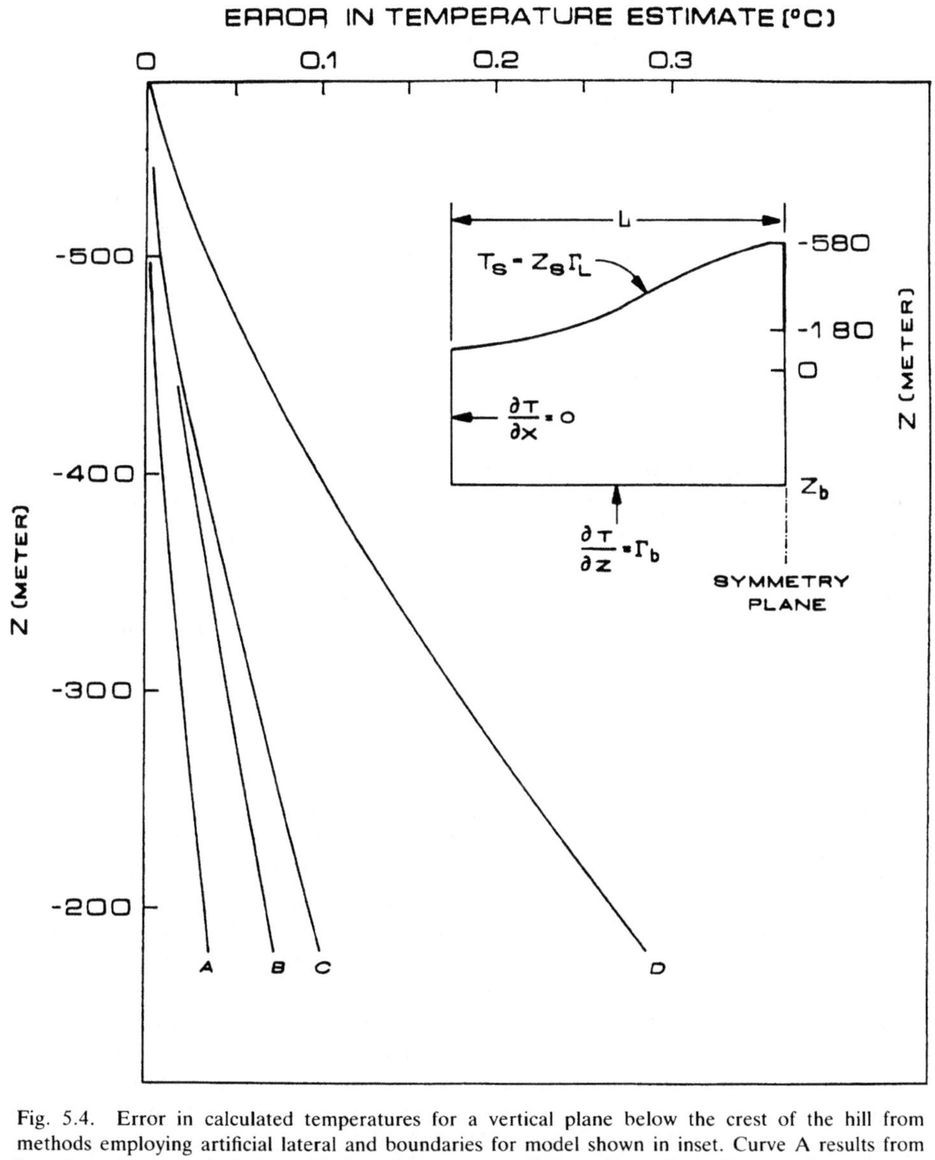

Fig. 5.4. Error in calculated temperatures for a vertical plane below the crest of the hill from methods employing artificial lateral and boundaries for model shown in inset. Curve A results from boundary element calculation with $L = 5000$ m, $Z_b = 6000$ m. Curve B results from linear inverse method with $L = 5000$ m. Curve C results from boundary element method with $L = 5000$ m, $Z_b = 1600$ m. Curve D results from either linear inverse calculation with $L = 1500$ m or boundary element-calculation with $L = 1500$ m and $Z_b = 1600$ m or $Z_b = 6000$ m. See text for details on model and calculation methods.

from the relations given by Lees (1910) for infinitely distant boundaries. Two methods were employed to calculate the effects; the linear inverse or Fourier series method of Blackwell *et al.* (1980), and the boundary element method. Both are described in the following section on correction techniques. The significant difference in the two methods for this example is that the boundary element method assumes a finite depth to the basal boundary for the background heat-flow density boundary condition, while the linear inverse method assumes an effectively infinite depth and therefore incurs no error from this boundary. Although the errors shown in this particular example would be negligible for many practical heat-flow density determinations, the relative importance of the various boundaries is clear from the figure. The result to be emphasized in Figure 5.4 is the importance of locating lateral insulating boundaries sufficiently far from the region of interest.

(2) SURFACE TEMPERATURE BOUNDARY CONDITION

The final boundary represents the source of all terrain effects – the temperature distribution, $T_s(x)$, on the Earth's surface, $z_s(x)$. The surface temperature is controlled by sources outside the solid Earth: soil-atmosphere interactions, solar radiation and the radiative environment of space; the background terrestrial heat-flow density has no significant effect on T_s except in the most extreme cases of anomalous geothermal zones. It should be emphasized that it is $T_s(x)$, the mean annual surface temperature, and not the instantaneous surface temperature, $T_s(x, t)$, which affects geothermal measurements at depths greater than about 20 m. Local mean annual soil temperature can be estimated by extrapolating subsurface temperatures from below the annual wave at the site of a well but an isolated well or even a group of wells usually cannot provide the temperature detail required for topographic correction. There is no alternative but to assume some reasonable surface temperature distribution.

The most popular surface temperature model is a linear decrease of temperature with elevation; usually the rate of decrease is roughly equal to the atmospheric lapse (5 to 10 °C km^{-1}) with the mean annual soil temperature biased 1 to 4 °C higher than the corresponding mean annual air temperature. A significant cause of this bias is persistent snow cover that insulates the soil surface from subfreezing winter air temperatures. Extrapolated surface temperatures from an ensemble of wells in a region tend to cluster about such a relation but deviations at individual sites due to microclimatic effects can be quite large. Figure 5.5 shows the soil temperature lapse observed in a regional study of the Colorado Plateau physiographic province by Bodell and Chapman (1982), and the relationship between mean annual soil temperature and mean annual air temperature. Note the significant deviations, as large as 3 °C, from least squares regression lines, primarily due to microclimatic factors; see also Haenel (1980). The lapse model appears to be effective for corrections to points at a large distance laterally or vertically from the surface area being considered (Birch, 1950; Lachenbruch, 1969); the effects of local variations in microclimate are effectively averaged at large distances.

Another useful surface temperature assumption is made in the insolation model, which relates surface temperature to the solar radiation impinging annually on an element of surface (Blackwell *et al.*, 1980). Other microclimatic factors can result in significant spatial variations in surface temperatures. Differences in vegetation surface properties can lead to sharp differences in surface temperature; soil

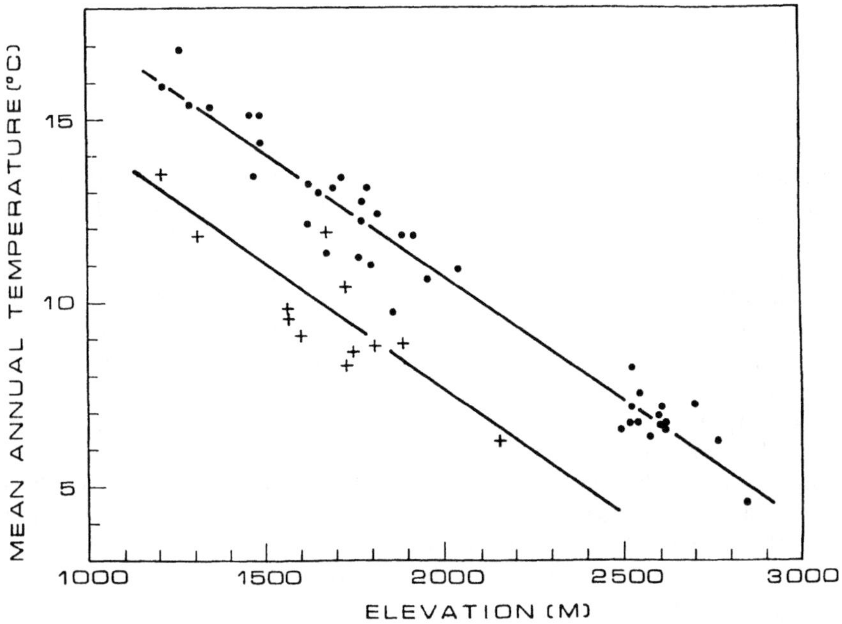

Fig. 5.5. Temperature lapse with elevation in a 10000 km^2 area of Central Utah, USA Soil temperatures (circles) obtained by extrapolation from temperature logs in wells (Bodell and Chapman, 1982). Mean annual air temperatures (crosses) for meteorological stations in the same region compiled from US (Weather Service data by D. Deming. Linear least squares regression for soil and air temperature indicated by lines: lapse is slightly less than 7 mK m^{-1} for both data sets but air temperatures are biased about 3 °C below soil temperatures. Note also the bias of meteorological recording stations towards lower elevations.

temperatures in a forested area are commonly 1 to 2 °C lower than adjacent cleared areas (Geiger, 1965). Local weather patterns, usually topography-related, such as snow cover distribution, wind patterns in mountain valleys, or daily rainfall at a particular time of year and time of day also have pronounced effect on the mean annual soil temperature (Geiger, 1965). The presence of lakes or other shallow bodies of water is another source of surface temperature disturbance, especially well-studied in the case of Alpine lakes in Switzerland; see, for example, Haenel (1970), von Herzen *et al.* (1974) or Finckh (1983). The spatial (and temporal) effects of microclimatic variation on the subsurface thermal regime in permafrost regions has received much recent attention because of the significance for engineering applications; see, for example, Smith (1975) and Lachenbruch (1957a, b). Finally, in accomodating these surface effects, one should remember that factors affecting variation in soil temperature are site-specific and the surface enviroment near the well must be considered when constructing an accurate model of subsurface temperature.

(3) CONCEPTUAL MODEL OF TOPOGRAPHIC DISTURBANCE

In subsequent discussion, the subsurface temperature field is considered as the superposition of three parts:

$$T(x, z) = T_1(x, z, \Gamma_b) + T_2(x, z, z_s, \Gamma_b) + T_3(x, z, z_s, T_s). \qquad (5.3)$$

Note first that T_3 is independent of the unknown background gradient. Since T_s, the surface temperature, is due to sources outside the solid Earth, the effect of these external sources on the subsurface gradient should decrease with depth. Thus T_3 is just the solution to Laplace's equation subject to the boundary conditions $[\partial T_3(x, z)/\partial x]_{z\to\infty} = 0$ and $T_3(x, z_s(x)) = T_s(x)$ which we can readily determine and remove from the data. This is the only part of the temperature field which depends on the surface temperature, so the term 'external' will refer to this component.

Subtracting the external component from the temperature field results in an isothermal topographic surface. This residual temperature field may be further decomposed into a 'background' field, T_1, with uniform gradient (in the absence of unmodelled disturbances), and a 'topographic' field, T_2, representing the well known effect of enhanced gradients beneath valleys and depressed gradients beneath hills. The topographic field is the solution to Laplace's equation with the boundary conditions:

$$[\partial T_2(x, z)/\partial x]_{z\to\infty} = 0 \text{ and } T_2(x, z_s(x)) = -\Gamma_b z_s(x).$$

Any non-zero constant contribution to the subsurface temperatures has been included in the external (i.e. T_3) field. This approach yields $T = 0$ at $z = 0$ in the background (i.e. T_1) field. Actually, the allocation of this constant term among the three components is arbitrary.

The three independent components of a terrain disturbed temperature field are illustrated in Figure 5.6 for the Lees hill example seen previously. It is the background component which we seek through corrections to the observations. The isotherm patterns for the external and topographic fields are similar for this example because the surface temperature in the model follows a lapse rule; more complicated surface temperature effects would change the isotherm pattern for the external component but not the topographic component.

The background and topographic components both depend on the unknown, Γ_b, and as noted previously, the boundary value problem must be solved for several choices of Γ_b either by computing solutions for a range and selecting the choice with the best fit to the observations or by an iterative process which converges on Γ_b. The former approach is discussed by Henry and Pollack (1985) and Bauer and Chapman (1986). Figure 5.7A shows a plot of misfit of model to observations for various choices of Γ_b for the theoretical model of Figure 5.4. The misfit, ε, measured by L_2 or L_1 norms, defined respectively by:

$$\|\varepsilon\|_2 = [\Sigma_{i=1}^{n}(T_i - M_i)^2]^{1/2}$$

and

$$\|\varepsilon\|_1 = \Sigma_{i-1}^{n}|T_i - M_i|,$$

where T_i are temperature observations and M_i are model temperatures calculated

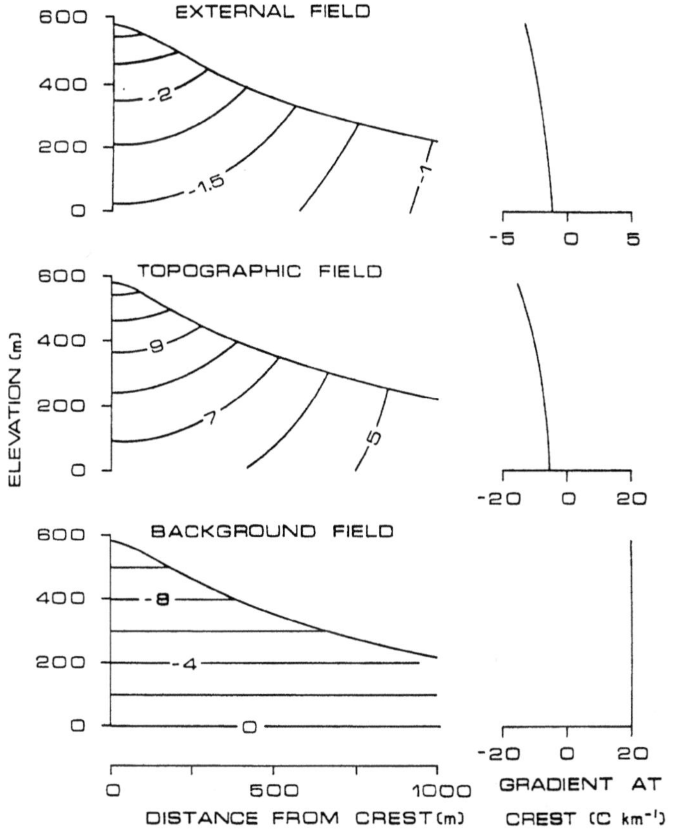

Fig. 5.6. Independent components of terrain-disturbed temperature field for model of Figure 5.4. Surface temperature lapse with z is 4.5 mK m^{-1}, background gradient is 20 mK m^{-1}. Left side of figure shows isotherm pattern in subsurface, right side shows gradient variation for a borehole at the crest of the hill. External field: due to sources outside the solid Earth. This field will vary depending upon the surface temperature distribution. Topographic field: due to interaction of background gradient with the topographic surface. Background field: due to Earth's internal heat sources, without surface disturbances.

at the same locations and L_2 is the common least squares minimization and L_1 minimizes the absolute values of the residuals. In this theoretical case, where only Γ_b is variable and all other model parameters are known exactly, the forms of the misfit functions are simple and a minimum misfit of zero occurs at the true value of Γ_b. In real cases, the minimum becomes broader because of uncertainties in other parameters such as conductivity distribution and surface temperature or because of inadequate data constraints. Figure 5.7B shows an example for a site in the Unita Mountains of Utah (Bauer, 1985; Bauer and Chapman, 1986). Although Henry and Pollack (1985) and Bauer (1985) have used the L_2 norm, the L_1 norm may perform better for some severely noisy cases.

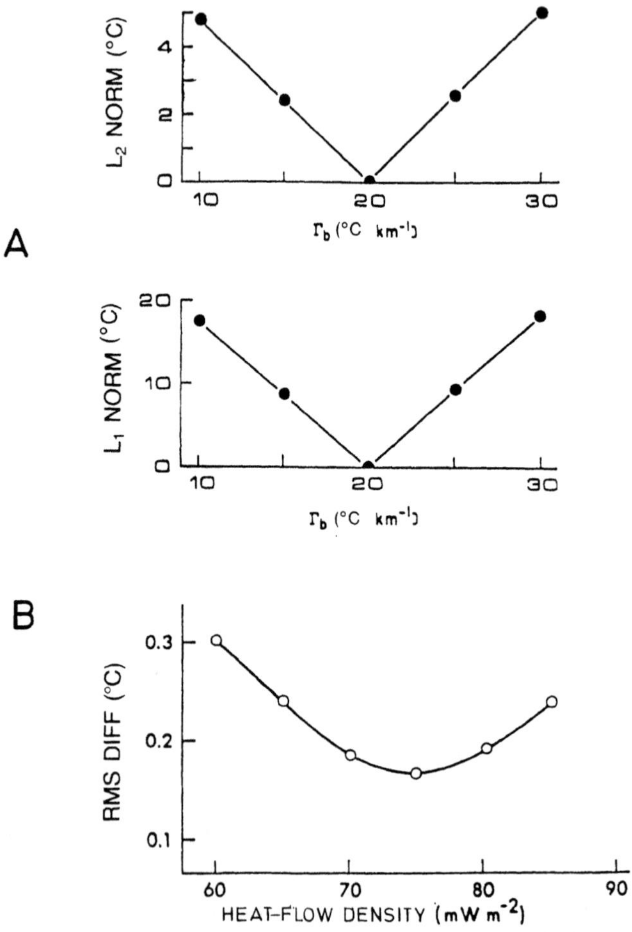

Fig. 5.7. (A) Misfit between observed and calculated subsurface temperatures for noise-free theoretical test case. Minimum misfit is presumed to occur at best choice of background gradient, Γ_b. (B) Misfit between observations and model for data at Upper Stillwater dam site, Uinta Mountains, Utah (from Bauer and Chapman, 1986).

5.4.2. Correction Techniques – Steady State

If we are able to assume that adequate boundary data are available, determining the terrain correction basically requires solving the boundary value problems for disturbance temperature fields T_2 and T_3 shown previously. At least eight numerically distinct methodologies have been applied to the calculation of heat-flow density terrain corrections. Table 5.2 classifies these methods by the mathematical technique employed and illustrates a few of their restrictions. Two broad categories can be identified: the first four methods in Table 5.2 lend themselves to quick but sometimes crude approximation of the expected terrain effects; the last

TABLE 5.2
Steady-state terrain correction methods.

Mathematical technique	Dim.[a]	λ hetero[b]	T_s[c]	Time[d]
Solid angle	2, 3	no	yes	yes
Conformal map	2	no	no	no
Non-linear inverse	2, 3	no	no	no
Linear inverse	2, 3	no	yes	no
Finite difference	2, 3	yes	yes	yes
Monte Carlo	2, 3	yes	yes	yes
Finite element	2, 3	yes	yes	yes
Boundary element	2, 3	yes	yes	yes

[a] Geometric dimensions allowed in model.
[b] Method allows general heterogeneous thermal conductivity.
[c] Method easily handles arbitrary surface temperature.
[d] Method adaptable to time dependent effects.

four are detailed numerical models of the appropriate boundary value problem which allow control of solution accuracy but at considerable cost in computing resources.

Probably the most significant site characteristics to be used in selecting a method are (1) the relationship of magnitude of local relief to depth of measurement being corrected and (2) the number of spatial dimensions required to describe the geometric features of the site. The topographic correction is greatest near sharp breaks in slope and tends to be larger for features with greatest relief. Large variations in surface temperature due to microclimatic effects rather than relief can also require significant correction however. Since 2-D calculations are almost always simpler than 3-D ones, it is important to identify the dimensionality of the problem. Note that 2-D topography might host a more complicated surface temperature or subsurface thermal conductivity requiring full 3-D modelling; these factors must be evaluated before restricting the dimension of the problem. The remainder of this section outlines each of the eight methods for calculating the correction. None of these methods is recommended for all situations. Selection of a method is therefore a new problem for each site; the combination of topography, surface temperature, thermal conductivity distribution, hydrologic characteristics, and accuracy required for the intended application dictates which methods are most suitable.

(1) SOLID ANGLE CORRECTION METHOD

Primarily because of its simplicity and ease of application, this has been the most popular technique. This correction evolved from papers by Jeffreys (1938) and Bullard (1938). A thorough development, including a clear statement of the necessary conditions for its validity, is given by Birch (1950).

In this method, the true topographic surface $z_s(x)$, is replaced by a plane surface, which we identify as $z = 0$, with the z axis defined positive in the downwards direction as in Figure 5.3. The temperature on the topographic surface,

$T_s(x, z_s(x))$, is projected into the reference plane $z = 0$ using the undisturbed background gradient, Γ_b. The temperature on the reference plane $T_0(x, 0)$ is:

$$T_0(x, 0) = T_s(x, z_s(x)) - \Gamma_b z_s(x). \tag{5.4}$$

This projection neglects the divergence or convergence of heat flow beneath topographic features, a significant part of the T_1 and T_2 terms in Equation (5.3). Most published examples have used a lapse model for T_s, although Lachenbruch (1968, 1969) has given examples using an isothermal surface. The failure of projection Equation (5.4) to account for much of the lateral flow of heat near the topographic surface is the most important shortcoming of the method. A useful rule of thumb is if the amplitude of the topographic relief is much less than the depth of the temperature to be corrected, or if the lateral distance to a topographic feature is much more than this depth and more than the amplitude of the relief, the approximation will generally be acceptable (Lachenbruch, 1969).

Now we must determine the subsurface temperature disturbance, $T(x, z)$, in terms of the known boundary data, temperature field $T_0(x, 0)$. Using Green's formula (e.g. Jaswon and Symm, 1977, Chapter 4) it is possible to derive the result:

$$T(x, z) = \frac{1}{\pi}\int_{-\infty}^{\infty} \frac{z}{(x - \xi)^2 + z^2} \cdot T(\xi, 0)\,d\xi. \tag{5.5}$$

The disturbance to the vertical gradient of temperature is easily obtained by formally differentiating this result with respect to coordinate z (Lachenbruch, 1969). The analogous three dimensional results are shown by Birch (1950) and Jaeger (1965). In practice, it is most convenient to rewrite Equation (5.5) in terms of the solid angle, $d\Omega$, occupied by an element $d\xi$ of the surface $z = 0$ as observed from the point being corrected:

$$\begin{aligned} T(x, z) &= \frac{1}{2\pi}\int_0^{2\pi} T_0\,d\Omega \\ &= \frac{1}{\pi}\int_{-\pi/2}^{\pi/2} T_0(x + z\tan\theta, z)\,d\theta, \end{aligned} \tag{5.6}$$

where Ω is solid angle in steradians and θ is plane angle in radians ($\Omega = 2\theta$).

In practical application Equation (5.5) or its three dimensional counterpart is approximated as a finite sum:

$$T(x, z) = (1/\pi)\Sigma_k \bar{T}_k \Delta\theta_k\,, \tag{5.7}$$

where $\Delta\theta_k = (\theta_k - \theta_{k-1})$ and $\bar{T}_k$ is the mean of $T_0(\xi, 0)$ on the interval $x - z\tan\theta_k \leqslant \xi \leqslant x - z\tan\theta_{k-1}$. In the three dimensional case, a cylindrical coordinate system with origin centered at the x and y coordinates of the subsurface measurement is used and the surface is subdivided into annular sectors in concentric rings around the origin. If a sequence of θ_k or annular sectors is established in advance, a template showing the discrete zones can be drawn to the same scale as available topographic maps, allowing the interpreter to estimate rapidly the average elevation for each element of surface. The appropriate solid angle can be easily computed from the surface compartment geometry and the measurement depth.

TABLE 5.3
Topography templates in use for solid angle method. R_0 = outer radius of annulus, n = number of sectors in annulus. Template sources: #1 in use at University of Utah, #2 based on Kappelmeyer and Haenel (1974, Table 3.18), #3 from Birch (1950, Table 8).

Zone	#1 $R_0[m]$	n	#2 $R_0[m]$	n	#3 $R_0[m]$
1	200	4	70	4	200
2	400	6	275	6	400
3	700	8	400	6	600
4	1 000	10	575	8	800
5	1 500	12	1 300	12	1 200
6	2 500	16	1 700	16	1 600
7	4 000	20	2 700	16	2 000
8	6 500	16	3 700	24	2 800
9	10 000	12	5 600	32	4 400
10	15 000	10	6 300	32	8 000
11			11 000	32	16 000
12					35 000
13					65 000

Table 5.3 shows an example of the compartmentalization used for such templates. This scheme is easily adapted to use digital topographic data bases (Bodmer, 1982), although inner zones may still require manual digitization if the digital topography has a coarse sampling interval.

Examples to illustrate the accuracy of this approximate correction technique, using simple models for which an exact expression for the correction can be evaluated, have been given by Birch (1950) and Blackwell *et al.* (1980). Lachenbruch (1968, 1969) has shown the error in the solid angle correction for families of very simple 2-D topography assuming an isothermal surface temperature (Figure 5.8). Where the simple models roughly resemble actual cases, these examples can be used to estimate the size of errors to be expected when using the solid angle correction.

The solid angle method is simple and allows quick estimation of a correction with a minimum of computation. The use of any surface temperature model is possible; both lapse and isothermal models have seen application. Corrections for two or three dimensional topography and T_s are possible and the 3-D case is quite practical. The technique has always been used to model a homogeneous thermal conductivity structure; incorporation of even simple inhomogeneity is likely to complicate the solution enough to make this method unattractive. Time dependent effects (topographic evolution) have been incorporated by Birch (1950) and Jaeger (1965).

(2) CORRECTION BY CONFORMAL TRANSFORMATION

This method of geothermal terrain correction and techniques for applying it to a large class of 2-D models was developed by Lachenbruch (1969), who cites Castoldi (1952) for first applying this method in the special case of a vertical cliff. A conformal map for a different family of transformations, resembling a variety

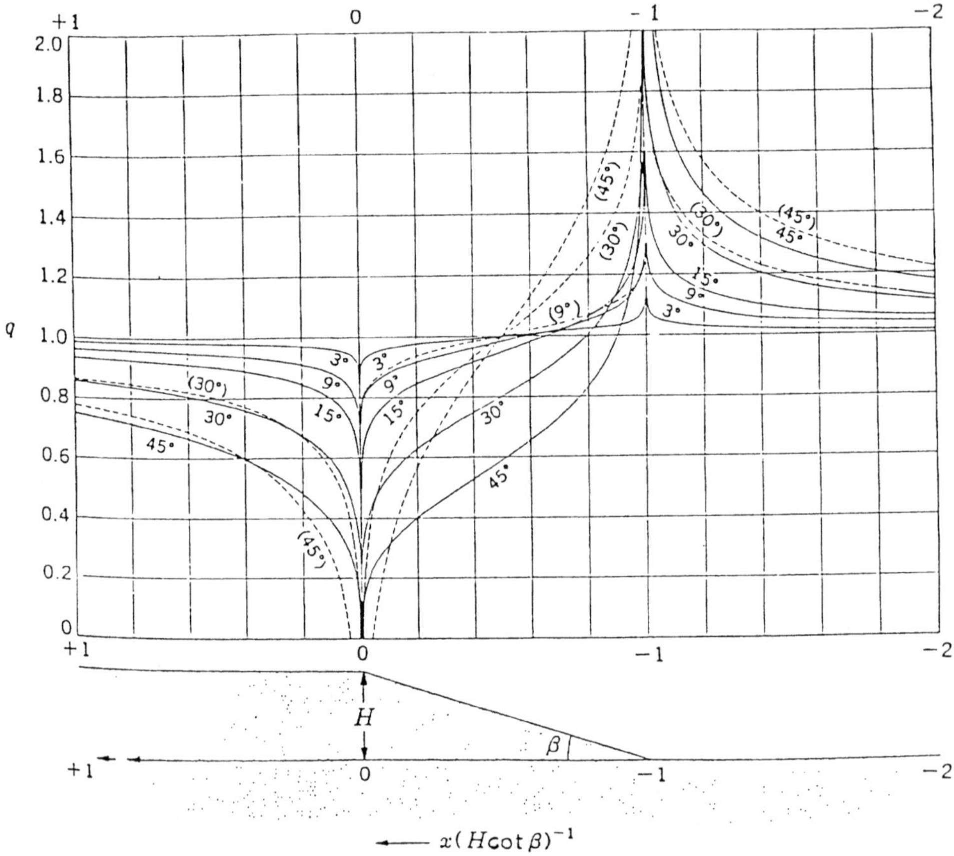

Fig. 5.8. Surface heat-flow density for 2-D plane slope model (solid curves) and surface heat-flow density estimated by solid angle correction method (dashed curves). Geometry of model and meaning of terms toe and brink are shown on Figure 5.9 (from Lachenbruch, 1969, Figure 2).

of symmetric and asymmetric ridges or valleys, was identified and applied to the terrain correction problem by Jobert (1960). Blackwell *et al.* (1980) made the interesting suggestion that corrections for a complicated 2-D topography modelled by any number of linear segments could be calculated using the Schwarz-Cristoffel transformation in the manner of Parker and Klitgord (1972).

Lachenbruch (1968, 1969) noted that the 2-D topographic problem illustrated in Figure 5.9, if the topography is considered a curve in the complex plane, can be transformed by a conformal map to a problem with a horizontal boundary surface for certain special slope angles, $\beta = \pi/n$, where n is an integer.

One of the implicit assumptions in the Lachenbruch application of this method is that the gradient to be corrected has been observed at a sufficiently shallow depth that it is comparable to the gradient at the topographic surface. Lachenbruch's analysis of this assumption can be summarized in some 'rules of thumb':

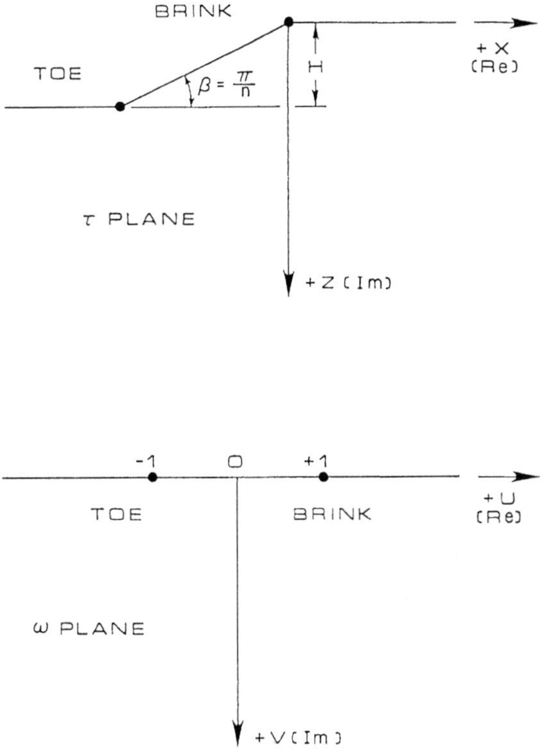

Fig. 5.9. Conformal transformation of simple 2-D topography. Complex plane τ shows the topographic model; angle of slope is π/n where n is restricted to be an integer. Complex plane ω shows topography-free surface when related to τ plane by the transformation $d\tau/d\omega = A[(\omega + 1)/(\omega - 1)]^{1/n}$. Location of points corresponding to breaks in slope are also shown. Figure is based on Figure 1 from Lachenbruch (1969).

(1) the observed gradient should be within roughly 10 % of the gradient calculated at $z = z_s$ when the distance to the relief is more than 5 times the maximum depth of observations; (2) an average gradient from depth to surface should be within 15 % of the calculated surficial gradient when the lateral distance to the relief is more than 2 times the measurement depth. Lachenbruch also notes a fortuitous result that a topographic element at the same lateral distance as measurement depth produces little disturbance. Unfortunately solutions for models of the type shown in Figure 5.9 cannot be superposed to give solutions for more complicated topographies; however, Lachenbruch has shown a method to bound the correction for a flat-bottomed valley, a flat-topped ridge and a horizontal bench interrupting a slope.

Jobert (1960) identified another family of conformal maps relating a topography-free half plane to a half plane distorted by ridge-like or valley-like topography around the origin and tending asymptotically to the topography-free case away from the origin. This method has a major defect: before the transformation can be applied, a shape which provides a reasonable fit to the actual topography has to be selected; this can only be done easily for topography of simple geometric cross-sections.

Conformal transformation thus allows exact determination of the effects of a limited class of 2-D topographic surfaces, often in the form of the disturbance to the vertical gradient of temperature at the topographic surface. The technique is appropriate where the topography is two dimensional and climatic effects are negligible.

(3) NON-LINEAR INVERSE CORRECTION METHOD

This method of correction evolved from the observation of Lees (1910) that certain simple distributions of dipolar heat sources (with moment defined by topographic geometry, background gradient and lapse rate) produce the subsurface temperature field expected beneath a particular ridge-like topography with a surface temperature that obeys a lapse model. Jaeger and Sass (1963) applied this method as a correction for geothermic measurements and derived another family of topographic models from the Lees result. The method is also briefly summarized by Jaeger (1965) and Kappelmeyer and Haenel (1974).

Jaeger and Sass (1963) have remarked that this technique should be used only to estimate the order of magnitude of terrain effects. Real topographic surfaces never exhibit a simple geometric shape as described here. Superposition of many simple models can conceivably match many topographies; the problem is to determine the source locations and moments of all the sources. Determining these source parameters is an infamous ill-posed problem in non-linear inverse theory. Numerical solution of non-linear inverse problems of this type is difficult at best and an automated process to fit a complicated topography will probably never be warranted for the geothermal terrain correction problem. In practice, one or two most significant topographic elements are fitted manually by a trial and error process (J. H. Sass, pers. comm., 1983). Modifying the technique for a surface temperature more complicated than the lapse model is in general impractial. Three dimensional topographic elements are easily modelled by replacing the line of dipoles with a single dipole. The technique is not easily adapted to general inhomogeneity of thermal conductivity nor to time dependent effects. Lees (1910) included the effects of radioactive heat production within the topography; such effects are usually negligible in most practical applications.

(4) LINEAR INVERSE CORRECTION METHOD

A linear inverse formalism was first applied to the geothermic topography correction problem by Ducruix *et al.* (1974b) and another approach is taken by Blackwell *et al.* (1980).

This family of methods addresses the geothermic topographic correction as a problem of continuation of an analytic function. If the surface temperature function is analytic, then its disturbance to subsurface temperatures is also analytic in the source-free region of interest. The common thread in these methods is the construction of a set of orthonormal basis functions describing the temperature field in the subsurface for which the necessary continuation is known explicitly.

The linear combination of basis functions which best approximate the surface temperature $T_s(x)$ on $z_s(x)$ then also gives the best approximation to the sub-surface temperature disturbance; determining this linear combination is a linear inverse problem. Mathematical details for an extension of the method are given in two papers on continuation of potential fields from uneven surfaces in two (Courtillot *et al.*, 1973) and three dimensions (Ducruix *et al.*, 1974a). This method represents the potential field in a basis (ϕ) with elements:

$$\phi_k = \mathrm{Im}\{[\exp(iw) - 1]/w\}, \tag{5.8}$$

where $w = \pi(x - k) + iz$, which reduces to a sync function on the plane $z = 0$. Such a basis can exactly represent a bandlimited harmonic function on $z = 0$. A finite series constructed from this basis is meant to approximate the eigenfunctions for Laplace's equation with Dirichlet conditions on the true boundary z_s. This technique may therefore produce better results than the previous method for some cases, and especially for near-surface correction in rugged topography.

A formal analysis of the problem of continuation of potential fields using finite data has been performed by Huestis and Parker (1979), who have indicated theoretical shortcomings in these methods. Basically, correction techniques in this section construct a possible potential field consistent with the finite data but fail to address the resolution achieved or the nature of non-uniqueness in the solution. Huestis and Parker also give a 2-D away-from-source continuation scheme with explicit consideration of the resolution in the results. The results they derived are not directly applicable to the topographic correction problem but they have indicated the generalization necessary to make them appropriate to our problem.

Determination of topographic corrections via linear inversion can be performed by a variety of techniques which vary widely in complexity and applicability. Once an expansion of the problem in an appropriate basis is accomplished, disturbance temperatures are easily determined anywhere in the subsurface; the difficulty is in determining the basis set and expansion which depend upon the nature of z_s and T_s and on the technique employed. All reasonable surface temperature functions are readily accommodated and three dimensional cases are feasible although they require a considerable increase in computational effort. Since the eigenfunctions for the problem must be known explicitly, accommodation of inhomogeneity in thermal conductivity is impractical, although simple cases such as plane layers may be possible. No attempt has been made to apply these techniques to time dependent problems. Although Huestis and Parker (1979) have found fault with many continuation methods, including those by Henderson and Cordell (1971), Courtillot *et al.* (1973) and Ducruix *et al.* (1974a), these methods have been shown to perform successfully for some cases; a systematic comparison is needed showing where each can be safely applied.

(5) FINITE DIFFERENCE METHOD

The finite difference method (FDM) for approximating and solving partial differential equations was long recognized as potentially applicable to the geothermic terrain correction (see for example Birch, 1950). A brief description of the application to terrain effects is given by Haenel (1970) and Kappelmeyer and Haenel (1974). Sclater *et al.* (1970) and von Herzen *et al.* (1974) show corrections by FDM incorporating thermal conductivity variations. The literature on FDM methods for solving elliptic partial differential equations is substantial; two good basic texts are Ames (1969) and Mitchell and Griffiths (1980). Three recent applications of finite difference modelling to geothermal terrain corrections are by Henry and Pollack (1985) and Bauer and Chapman (1986). The principle is explained in Section 2.4.1.

The FDM allows flexible modelling of subsurface temperatures for many problems with geothermal terrain effects and development of these problems in the calculus of finite differences is conceptually simple. Arbitrary surface geometry and surface temperature functions are easily handled. Quite general thermal conductivity variations can be incorporated with ease. Two or three dimensional models can be constructed; 2-D models are almost always practical but in 3-D cases the number of nodes included is huge and the relative slowness of some algorithms is detrimental. Finite difference approximations are also applied to time dependent problems but restrictions not necessary in the steady state case cause model dependent difficulties. Inability to solve for unknown background gradient is an inconvenience, especially when a large system must be solved several times with trial choices for Γ_b at great expense in computational resources. The ability to model many geological complexities in a simple way is frequently advantageous however, and the method is quite practical for most 2-D cases.

(6) MONTE CARLO CORRECTION METHOD

Monte Carlo modelling techniques were applied to terrain corrections for lunar heat-flow density sites by Langseth *et al.* (1976) and apparently never used again for this purpose. The method for solving thermal conduction problems by Monte Carlo simulation (MCS) is concisely presented by Haji-Sheikh and Sparrow (1967) and basic Monte Carlo procedures are discussed in many texts on simulation techniques or engineering mathematics (for example, Nakamura, 1977).

Because of the efficiency and potential of the method, especially the ability to determine a statistically reliable variance of the temperature, we discuss the method in some detail.

The effect of an element of boundary on a particular subsurface point is related to the probability that a random walk beginning at the subsurface point reaches that element of boundary. In Monte Carlo simulation, this probability is estimated by calculating a large number of random walks starting from the point being considered and counting the proportion reaching each part of the boundary. The random walks may be simulated on a domain represented by a multi-dimensional array of nodes similar to a finite difference grid or in a continuous area or volume. As in the finite difference method, a grid of nodes suffers truncation error so the continuous case is preferred. This section describes the random walk method from the paper by Haji-Sheikh and Sparrow (1967).

Construction of the path of a floating random walk is illustrated by Figure 5.10. The random walker is started from the point where the temperature is to be determined. The initial step from this point and all subsequent steps are performed using the same simple rules until the walk terminates at a boundary. The maximum radius, r, is determined for a circle centered at the walker's present

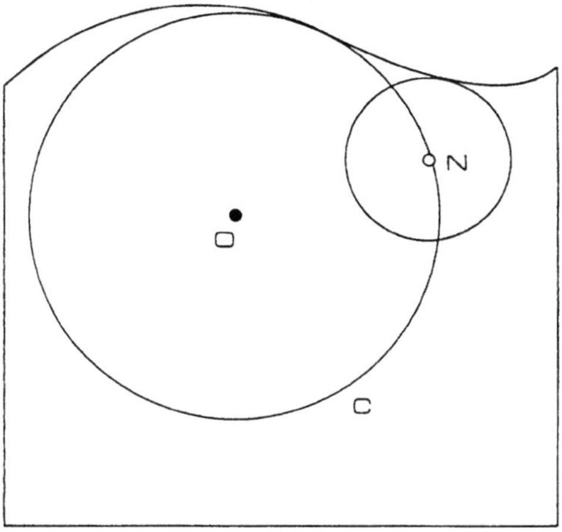

Fig. 5.10. Construction of a step of a floating random walk. Point O (solid dot) is the starting position of the random walker. Circle C, centered on O, is the locus of points which can be reached by this step. Point N (open dot) is position after the step, selected at random from all points of C. Circle centered on N is locus of points for next step.

location and that lies within all the boundaries. This circle is tangent to a boundary at least at one point. The direction of the step, θ, is selected at random from a uniform distribution $0 \leqslant \theta < 2\pi$. A displacement r in the direction θ gives the new location of the walker at the end of the step. A practical difficulty for the method is obvious: the probability of actually reaching the boundary on any step is zero when the circle grazes the boundary at only a finite number of points of tangency. Practical solutions to overcome this are shown later. Another concern, perhaps peculiar to the terrain correction problem, is how to specify the topographic surface $z_s(x)$, in a way that allows the efficient calculation of the maximum radius. Since this calculation must be performed for a new point at every step, it is a major factor determining the total computational time.

When the i'th walker from a point (x, z) reaches a boundary where the temperature is specified, the walk is terminated and the temperature, T_i, encountered at the terminal point is accumulated. If the temperature was specified everywhere on the boundaries of the region, the temperature estimated at (x, z) is:

$$T_e(x, z) = (1/N) \sum_{i=1}^{N} T_i, \tag{5.9}$$

where N is the number of walkers emitted from point (x, z). This temperature estimate converges to the true temperature as $N \rightarrow \infty$. When the i'th walker reaches a boundary where heat flow or gradient normal to the boundary is prescribed, with the normal, n, positive outward, $\partial T/\partial n = \Gamma$, and a slightly more complicated procedure is followed. The temperature at the boundary point is not known explicitly, but a first order Taylor series approximation expanded about an interior point (s_d, z_d), which is a perpendicular distance d from the boundary point being considered, is $T_b = T_d + \Gamma d$. The quantity Γd can be tallied as a part of the contribution of walker i to the desired temperature at (x, z), but since T_d is also unknown, the walk continues from (x_d, z_d). The walker may encounter several prescribed heat flow boundaries before terminating at a prescribed temperature boundary point. In that case, the temperature estimate at (x, z) is:

$$T_e(x, z) = 1/N \sum_{i=1}^{N} \left(T_i + \sum_{j=1}^{m_i} \Gamma_{ij} d \right), \tag{5.10}$$

where m_i is the number of times the i'th walker encounters a prescribed heat flow boundary.

Haji-Sheikh and Sparrow (1967) suggest using one of several methods to allow walkers to actually reach the boundary. Two of these methods are as follows. If the radius for any step is less than some fixed tolerance, r_t, the walker is assumed to have reached the nearest point on the boundary. If the direction chosen is within some fixed tolerance, θ_t, of the point where the circle contacts the boundary, the walker is regarded as arriving at the boundary. When the first method is used, Haji-Sheikh and Sparrow suggest r_t should be less than 1/4 of the node spacing giving an acceptable finite difference or fixed random walk solution. When the second method is used, they have used $\theta_t = 0.10$ radian. Nakamura (1977) suggests performing a fixed random walk on a grid (possibly 1-D) when the walker comes within a specified distance of the boundary. In the fixed walk, a square grid is used and the walker moves from a node to an adjacent node in each step; one of the four nearest neighbour nodes is selected with equal probability for each of the four.

Statistical confidence intervals can be established for the temperature estimates of this method (Haji-Sheikh and Sparrow, 1967, Equation 14). Denoting the estimate from Equations (5.9) or (5.10) by T_e and the actual temperature at the same point by T_a, and assuming each walker makes a contribution with a zero-mean, uncorrelated distribution about the mean, the following interval can be derived (Beck and Arnold, 1977, Equation 6.8.23a). The probability that the estimate is within ε of the actual temperature is:

$$P(|T_e - T_a|) < \varepsilon) = \mathrm{erf}[\varepsilon N/(\sigma\sqrt{2})], \tag{5.11}$$

where σ^2 is the sample variance.

For the examples given by Haji-Sheikh and Sparrow (1967), accuracy of a few percent was obtained using a few thousand random walkers. Despite the large number of walkers used for each point, the Monte Carlo method may be an efficient method for topographic corrections because solutions are only required at a few subsurface points. Since the method is essentially approximating an integral of the temperature field on the boundary, it is important to use a large enough population of walkers to adequately sample all regions of the boundary. Emery, in a comment appended to the paper of Haji-Sheikh and Sparrow (1967), and Langseth *et al.* (1976) have suggested that poor sampling patterns occur in practice; the corrections attempted by Langseth *et al.* appear to be rather poor and they have attributed this to poor sampling of topography at a large distance from the heat flow sites. Haji-Sheikh and Sparrow suggest that biased probabilities could be applied to the walks to improve the sampling of certain important regions. Alternatively, the number density of walkers arriving at each segment of boundary surface could be monitored to assess the adequacy of sampling during the computations.

The MCS method can be adapted to the boundary value problems of the geothermal topographic correction, and it can be applied for any 2D- or 3-D topography and any surface temperature field. Although a large number of calculations must be performed to produce a point estimate of the temperature, the calculations are only needed for the points of interest. Thus this method may be advantageous in terms of computational speed when compared with other numerical modelling techniques, such as finite differences or finite elements, where temperatures must be calculated for points of no interest. A representation of the topographic surface which can be evaluated rapidly is necessary to control processing times for the technique, however, Haji-Sheikh and Sparrow have not given procedures for handling thermal conductivity inhomogeneity, but the extension to this case appears to be simple. The paper by Haji-Sheikh and Sparrow does include an extensive development for transient and moving boundary problems that would also apply to geothermal terrain situations. A comparison of topographic models calculated by this technique and the more widely used finite difference and finite element methods would be of great benefit for evaluating their relative utility for this purpose.

(7) FINITE ELEMENT METHOD

The finite element method (FEM) for approximating solutions to partial differential equations was applied to the 2-D topographic correction problem by Lee and Henyey (1974), primarily because of its facility to accommodate virtually any inhomogeneity in thermal conductivity. An earlier application of FEM to geothermics was an attempt by Geertsma (1971) to model near-surface temperature fields. Van Rooy (1982) presented 3-D modelling of geothermal processes suitable for topographic corrections. The FEM has become one of the most popular techniques of engineering analysis and literature on the method and its applications is voluminous. Among the many useful texts on the technique are those by Livesley (1983), and Zienkiewicz (1977). Again, the principle is explained in Section 2.4.2.

In summary the FEM is a flexible technique adaptable to many situations encountered in geothermal terrain corrections. Quite general surface geometry, surface temperature and subsurface thermal conductivity can be considered, and because of the arbitrary mode distribution these situations are handled with more ease by finite elements than by finite differences. Finite element approximations are also easily adjusted to include heat generation and transient temperature fields (Lee, 1977). Two or three dimensional cases can be computed (a three dimensional formulation using tetrahedral elements is given by Van Rooy, 1982) but as with finite differences, the size of the linear system can become prohibitive in the three dimensional case. The unknown background gradient, Γ_b, can not be found explicitly in this method; the system must be solved several times with trial choices for Γ_b. The FEM has better convergence properties than finite difference solutions for an arbitrary node spacing. A significant disadvantage of FEM models is the large amount of preparation or preprocessing required to construct the approximating linear system. For reasonably simple geometries, however, 2-D models are generally quite practical.

(8) BOUNDARY ELEMENT METHOD

Application of the boundary element method (BEM) for modelling geothermal topographic effects has been proposed by Matsubayashi (1983). The BEM is a recently developed numerical modelling technique with many similarities to the FEM. Relatively few books on BEM have been published but two basic texts are Brebbia (1978) and Brebbia and Walker (1980).

The BEM originates in the weighted residual methods for solving boundary value problems, as does the finite element method. In the FEM, a solution is constructed from a set of basis functions which satisfy the boundary conditions of the problem but satisfy the differential equation only approximately. In the BEM on the other hand, a solution is constructed from a basis set of functions which satisfy the differential equation exactly in the interior of the domain but which only approximate the boundary data. Calculation of a finite element solution reduces to solving a linear system for the value of the approximation at nodal points distributed throughout the domain; calculation of a boundary element solution of the same accuracy reduces to solving a linear system for the approximation values at a much smaller number of nodal points located only on the domain boundaries. Although the linear system of a boundary element problem is much smaller than the corresponding finite element system, there may not be much advantage in terms of computational labor because elements of the boundary element system often must be determined by numerical quadrature and the boundary element system is full, not sparse like the finite element system.

3-D models can be calculated using boundary elements; the linear system in the 3-D case is larger than the 2-D case, but is still probably very much smaller than 3-D finite difference or finite element formulations. The boundary integrals and their numerical evaluation is simple but requires much more computational effort.

Simple inhomogeneities of thermal conductivity can be modelled. The domain is subdivided into several regions of homogeneous conductivity and additional boundary nodes are used to represent the interface between subregions. Each subregion is then a new boundary element problem and continuity of temperature and normal gradient of temperature is imposed between subregions. Intricate thermal conductivity models are probably best approached with finite element techniques however.

The BEM is another quite general numerical modelling technique well-suited to most geothermal terrain correction problems. Although the method requires the computation of many numerical integrals and the solution of a possibly large full linear system of equations, it may be competitive with FDM or FEM because temperatures in the interior of the domain need only be calculated at points of interest. The simplicity of the input parameters is a distinct advantage over finite element and finite difference problems. Two dimensional problems can be solved with the program by Brebbia (1978), 3-D problems are computationally more involved but should still be a practical size. It is possible to generalize the method as outlined here to include heat generation in the subsurface (Brebbia, 1978, pp.

86–89) or transient fields (Brebbia and Walker, 1980, Chapter 6). As with the other numerical modelling techniques, this one cannot directly determine the unknown background gradient, Γ_b; models for the range of Γ_b must be computed and compared to the actual data. Because BEM calculates unknown normal gradients on the boundary, a modification to find Γ_b, at least in some restrictive circumstances, may ultimately be possible. A specific comparison of boundary element terrain calculations with other methods, especially finite difference and finite elements, has not been published but the method appears to offer some promising advantages.

(9) SUMMARY OF STEADY - STATE METHODS

The methods shown have all been applied to geothermal terrain corrections. Some, especially the solid angle correction, are ubiquitous while others such as the Monte Carlo and boundary element techniques have a single reference in the terrain correction literature. Referring again to Table 5.2, one can see at a glance the most significant strengths and weaknesses of the various methods with regard to physical parameters that may be significant in terrain effects. A very brief hierarchy of considerations leading to the choice of method is given here.

Because they are quickly applied without sophisticated computation, the solid angle, conformal map and non-linear inverse techniques are best for initial estimates of topographic effects. A rapid estimate using these techniques would generally use a lapse model for surface temperature and assume a homogeneous Earth. If the topography at the site to be corrected is dominated by a 2-D element, the analysis of Lachenbruch (1968, 1969) or the method of Jaeger and Sass (1963) will almost always be the most convenient. If the topography cannot be considered 2-D, the solid angle correction can be used as long as the scales of topographic amplitude and measurement depth meet the necessary restrictions; otherwise a fit to the topography can be attempted using a small number of dipolar sources in the method of Jaeger and Sass. This rough estimate should suggest at least the order of magnitude of terrain effect expected and serves to determine whether more sophisticated correction attempts are justified for the desired application. Calculations for these initial determinations are simple enough for graphical or hand calculator solution and can therefore be easily performed in the field. These methods are similarly useful in siting studies prior to drilling.

Where additional modelling is required, we can separate two obvious classes of methods. The solid angle (when restrictions are satisfied), linear inverse and boundary element methods are suitable for 2-D or 3-D cases with arbitrary surface temperature but are suitable only if the subsurface thermal conductivity is nearly homogeneous. Calculations of linear inverse and boundary element solutions are intermediate in computational requirements; they are too large for hand computation but can be done with a small portable computer which might even be available in the field. The Monte Carlo, finite difference, finite element and, to some extent, boundary element methods can be applied to 2-D or 3-D cases with arbitrary surface temperature and arbitrary thermal conductivity. These methods require significant computational resources however, especially 3-D cases using finite differences or finite elements. Three dimensional models without interior nodes, such as Monte Carlo, linear inverse, or boundary element methods use, may offer computational advantages.

At the beginning of this section we made the simplifying assumptions of steady-state temperature fields, conductive heat transport, constant background flux and negligible heat generation. There are always actual cases which do not satisfy these assumptions. As noted in the preceding sections, most of the techniques are readily modified to accommodate heat production effects. The more sophisticated modelling techniques (Monte Carlo, finite difference, finite element, boundary element) can also apply lateral variability in the background heat-flow density into the model domain. The problem of advective thermal transport may be significant in many cases and unfortunately the interaction between such effects and the effects described here usually cannot be superposed. If hydrologic effects are significant within the depth range being corrected, they must be included in the correction model. Many methods for modelling these effects have been attempted; Smith and Chapman (1983) have provided an extensive review. Since many of these methods use numerical models, the finite difference, finite element or boundary element techniques described in this chapter can be readily used.

5.4.3. Transient Effects and Corrections

The thermal diffusivities of geological materials are such that many time-varying processes affecting the surface of the Earth can significantly disturb underground temperatures. In the discussion of steady-state effects, the surface temperature and topography appear as the major factors in terrain disturbances. The time dependence of these two factors is similarly the major source contributing to transient thermal effects disturbing the subsurface temperature. The problem of time dependent surface temperature includes the effects of climatic changes, surface disturbances such as construction, changes in vegetation, and changes in the distribution of bodies of water. Time dependent topography involves a variety of tectonic and geomorphic processes which alter the shape and elevation of the Earth's solid surface. These two factors – surface temperature and topography – are described by distinct mathematical models and are discussed separately in the following two sections.

(1) TRANSIENT SURFACE TEMPERATURE EFFECTS AND CORRECTIONS

The diffusion of long period changes in surface temperature conditions into the Earth has a significant effect on the subsurface temperature field. Many attempts have been made to offer quantitative corrections for such effects (for example, Birch, 1948; Beck, 1977; Clauser, 1984) and at a few sites, the inverse problem of recovering palaeoclimatic history from geothermal measurements has been attempted (Hotchkiss and Ingersoll, 1934; Čermák, 1971; Vasseur *et al.*, 1983; Nielsen, 1986). A brief overview of the transient conductive models applied to this correction problem can be found in the standard monographs on geothermics, Jaeger (1965) or Kappelmeyer and Haenel (1974), or in Turcotte and Schubert (1982). As with steady-state corrections, a major difficulty of transient correction theory is finding a suitable model for the (unknown) boundary condition. Beck

(1977) has given a review of this aspect of the problem and identified possible independent sources for information on the history of surface temperature.

The subsurface temperature in a homogeneous, isotropic, source-free Earth with time-varying surface temperature obeys the diffusion Equation (2.6):

$$a\nabla^2 T = \frac{\partial T}{\partial t}, \tag{5.12}$$

where a is the thermal diffusivity of the rock. Because long period changes in surface temperature are expected to be similar throughout a locality, it is conventional to consider only the vertical component of the Laplacian in this equation. The other necessary boundary conditions are as discussed for the steady-state case. If the surface temperature is time-dependent then at any depth z the temperature:

$$T(z) = T_0 + \Gamma z + \Delta T(z, t), \tag{5.13}$$

where $\Delta T(z, t)$ is the temperature perturbation whose form depends on the form of the time dependency of the surface temperature.

Several simple cases are of practical interest. The subsurface temperature and gradient disturbances due to an instantaneous change of surface temperature ΔT_0 at time t before now, Figure 5.11A, are:

$$\Delta T(z, t) = \Delta T_0 \operatorname{erfc}[z/(4at)^{1/2}] \tag{5.14}$$

$$\Delta\Gamma(z, t) = -\Delta T_0(\pi at)^{-1/2} \exp(-z^2/4at). \tag{5.15}$$

For several steps ΔT_{t_n} at times t_n before present, Figure 5.11(B), the corresponding perturbations are:

$$\Delta T(z, t) = \sum_{n=1}^{N} \Delta T_{t_n} \operatorname{erfc}[z/(4at_n)^{1/2}] \tag{5.16}$$

$$\Delta\Gamma(z, t) = \sum_{n=1}^{N} - \Delta T_{t_n}(\pi at_n)^{-1/2} \exp(-z^2/4at_n). \tag{5.17}$$

Table 5.4A gives numerical values for the disturbances to temperature and temperature gradient at selected depths calculated from Equations (5.14) and (5.15) using a single temperature step of 1 K and a thermal diffusivity of 1 mm^2 s^{-1}. Perturbations from single temperature steps of different amplitudes are easily calculated because of the linearity of Equations (5.14) and (5.15). It can be seen from Table 5.4A that for small times the disturbances are restricted to relatively shallow depths; for example, surface temperature changes 10 a ago have no effect at depths of 100 m or below. At greater times the thermal effects penetrate progressively deeper but are damped in amplitude.

What may be more useful is some idea of how deep we have to go to get below a convenient chosen level of inaccuracy. For a single step change, Equation (5.15), with $n = 1$, (or Equation 5.17) with $n = 1$, may be recast to read:

$$z^2 = 4at \ln[\Delta T_0/\{(\pi at)^{1/2}\Delta\Gamma(z, t)\}]. \tag{5.18}$$

For a unit surface change of $\Delta T_0 = 1$ K and with $a = 1$ mm^2 s^{-1} the depths (z_p) below which $\Delta\Gamma(z, t)$ is <1 mK m^{-1} and 0.1 mK m^{-1} are shown in Table 5.4B.

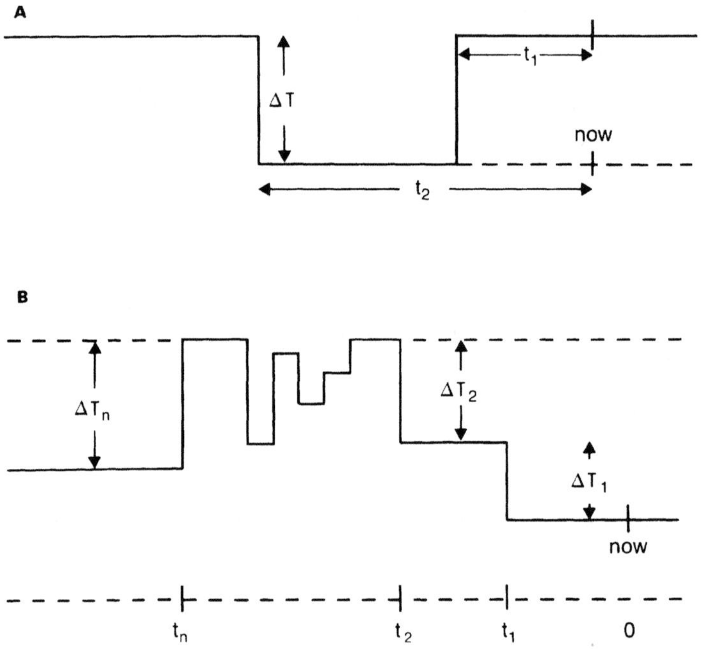

Fig. 5.11. (A) Single step change, Equations (5.14) and (5.15); (B) multiple step change, Equations (5.16) and (5.17); ΔT is temperature change.

TABLE 5.4A

Present perturbations to temperatures ΔT and gradients $\Delta\Gamma$ due to an instantaneous surface temperature change of 1 K at various times prior to the present. Thermal diffusivity = 1 $mm^2\ s^{-1}$, ΔT in mK, $\Delta\Gamma$ in mK m^{-1}.

Depth (m)	1 day		1 month		1 year		1 decade		1 century		1 millenium		10 millenia	
	ΔT	$\Delta\Gamma$	ΔT	$\Delta\Gamma$	ΔT	$\Delta\Gamma$	ΔT	$\Delta\Gamma$	ΔT	$\Delta\Gamma$	ΔT	$\Delta\Gamma$	ΔT	$\Delta\Gamma$
1	16	−110	660	−318	900	−100	970	−32	990	−10	1000	−3	1000	−1
5	0	0	30	−31	530	−82	840	−31	950	−10	980	−3	1000	−1
10	0	0	0	0	210	−45	690	−29	900	−10	970	−3	990	−1
50	0	0	0	0	0	0	50	−4	530	−8	840	−3	950	−1
100	0	0	0	0	0	0	0	0	210	−5	690	−3	900	−1
500	0	0	0	0	0	0	0	0	0	0	50	−1	530	−1
1000	0	0	0	0	0	0	0	0	0	0	0	0	210	−1
5000	0	0	0	0	0	0	0	0	0	0	0	0	0	0

The situation for $n > 1$ is more difficult to generalize, although it is clear that $\Delta\Gamma(z, t)$ must go through zero at some point.

TABLE 5.4B
The depth z_p(m) below which the perturbation to the gradient $\Delta\Gamma$ is less than 1 mK m^{-1} and 0.1 mK m^{-1} after an instantaneous surface temperature change of 1 K, t years ago.

t(a)	for $\Delta\Gamma$ <1 mK m^{-1}	for $\Delta\Gamma$ <0.1 mK m^{-1}
0.5	17.7	21.4
1	24.1	29.5
5	49	62
10	66	85
50	129	177
100	171	241
500	308	490
1000	382	661
5000	471	1294

For the special case of $n = 2$, $\Delta T_{t_1} = -\Delta T_{t_2} = -\Delta T$ it yields:

$$\Delta\Gamma(z, t) = \Delta T(\pi at)^{-1/2} \exp(-x)[\exp(-x) - (1/\sqrt{2})], \tag{5.19}$$

where $x = z^2/8at$. Equation (5.19) is zero at a depth z_m, which is independent of the magnitude of ΔT_0, given by $z_m = \sqrt{4at} \ln 2$.

It can readily be seen that any arbitrary time function of surface temperature change can be modelled by a series of incremental step changes (Beck, 1977). However, instead of using a series of step functions, it may sometimes be convenient to use analytic solutions for special cases.

For a linear change of surface temperature, with T_m constant for time $t < 0$ and changing at a constant rate after $t = 0$ (i.e. $\Delta T_m = \mathrm{bt}$), the half space solutions for temperature (Carslaw and Jaeger, 1959, p. 63) and gradient perturbations are:

$$\Delta T(z, t) = \mathrm{bt}\left[\left(1 + \frac{z^2}{2at}\right)\mathrm{erfc}\left(\frac{z}{\sqrt{4at}}\right) - \frac{z}{\sqrt{\pi at}}\exp\left(\frac{-z^2}{\sqrt{4at}}\right)\right] \tag{5.20}$$

$$\Delta\Gamma(z, t) = \mathrm{bt}\left[\frac{z}{at}\mathrm{erfc}\left(\frac{z}{\sqrt{4at}}\right) - \frac{2}{\sqrt{\pi at}}\exp\left(\frac{-z^2}{\sqrt{4at}}\right)\right]. \tag{5.21}$$

The disturbances to subsurface temperature and gradients due to a sinusoidal surface temperature variation $\Delta T_0 \cos \omega t$ with period $P = 2\pi/\omega$ and setting $A = z(\omega/2a)^{1/2}$, $B = (\omega t - A)$:

$$\Delta T(z, t) = \Delta T_0\, e^{-A} \cos B \tag{5.22}$$

$$\Delta\Gamma(z, t) = -\Delta T_0(A/z)\, e^{-A}[\sin B - \cos B]. \tag{5.23}$$

The wavelength L and propagation velocity S of the perturbations are given by $L^2 = 4\pi aP$ and $S^2 = 4\pi a/P$; when $z = L$, $\Delta T(L, t) = 0.002\Delta T_0 \cos 2\pi[(t/P) - 1]$ which is a maximum of $0.002\Delta T_0$. Table 5.4C shows the perturbations to tem-

TABLE 5.4C
Perturbations to temperature and gradient due to a unit amplitude sinusoidal surface temperature variation; $a = 1\ \mathrm{mm^2\ s^{-1}}$. ΔT in mK and $\Delta\Gamma$ in mK $\mathrm{m^{-1}}$.

		Period											
		1 day		1 month		1 year		1 decade		1 century		1 millenium	
t/P	$z(m)$	ΔT	$\Delta\Gamma$	ΔT	$\Delta\Gamma$	ΔT	$\Delta\Gamma$	ΔT	$\Delta\Gamma$	ΔT	$\Delta\Gamma$	ΔT	$\Delta\Gamma$
0.25	1	2	−10.4	151	−492	693	−290	900	−98.9	968	−31.5	900	−10.0
	10	0	0	0	0.02	−43	13.6	196	−50.8	693	−29.0	901	−9.9
	100	0	0	0	0	0	0	0	0.01	−43	1.4	200	5.1
	1000	0	0	0	0	0	0	0	0	0	0	0	0
0.5	1	0	17.7	297	−161	226	147	90	80.9	31	29.6	10	9.8
	10	0	0	0	−0.02	0	−13.2	310	−11.0	226	14.7	90	8.1
	100	0	0	0	0	0	0	0	0	−1	−1.3	310	−1.1
	1000	0	0	0	0	0	0	0	0	0	0	0	0
0.75	1	−2	10.4	−151	492	−693	290	−900	98.9	−968	31.5	−900	10.0
	10	0	0	0	−0.02	4.3	−13.6	−196	50.8	−693	29.0	−901	9.9
	100	0	0	0	0	0	0	0	−0.01	43	−1.4	−200	5.1
	1000	0	0	0	0	0	0	0	0	0	0	0	0

peratures and gradients at various depths due to sinusoidal surface temperature variations of unit amplitude 1 K for various periods from 1 day to 1000 a, the thermal diffusivity being 1 $\mathrm{mm^2\ s^{-1}}$.

Lachenbruch (1959) and Mundry (1974) give more general solutions for a plane layered Earth. Birch (1948) and many subsequent workers have superposed solutions in the form of Equations (5.14) and (5.15) to match more general climatic histories. Clauser (1984) has given a Laplace transform method for the effect of an arbitrary harmonic surface temperature function on temperature in a plane layered Earth, and Nielsen and Balling (1985) an analytical solution for the effect of an instantaneous surface temperature change also in a plane layered Earth. Wang *et al.* (1986) use an inversion technique in the frequency domain which is valid for layered media in which the thermal conductivity can vary linearly within each layer.

Several other transient surface temperature effects have been modelled, including latent heat effects peculiar to permafrost regions (Lachenbruch, 1982), two dimensional surface temperature functions with the shape of step functions in both space and time (Lachenbruch, 1957a), and the effect of a three dimensional area with a mean temperature which differs from the mean temperature of the surrounding surface, where both vary harmonically in time (Lachenbruch, 1957b). The more flexible numerical schemes for solving partial differential equations introduced previously – finite differences, finite elements and boundary elements – are readily adapted to the diffusion equation and may be useful when rock properties in the subsurface exhibit complicated inhomogeneity or when the surface temperature has complicated spatial variations, requiring two or three dimensional modelling.

The major use of time-varying surface temperature models is in detecting and accounting for palaeoclimatic effects.

The problems and dangers of ignoring what at first sight appear to be insignificant surface temperature variations, and more importantly of using the standard deviation of the local temperature gradient obtained from a linear least squares fit to $T-z$ data as a measure of the reliability and quality of a regional gradient, are illustrated in Figure 5.12. A typical equilibrium temperature gradient of

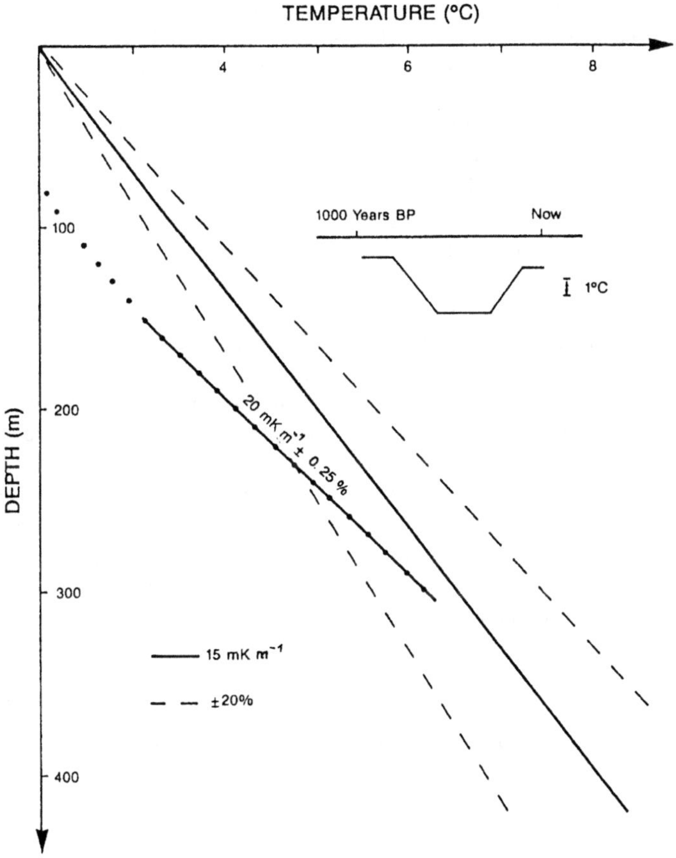

Fig. 5.12. Steady state temperature gradient with 20 % error bounds, together with perturbed temperature depth points for surface disturbance of form shown, and least squares straight line fitted to points from 150–300 m.

15 mK m^{-1} in a PreCambrian shield area is perturbed by the 'Little Ice Age' climatic variation, also shown in Figure 5.12. Temperature measurements made now in a 300 m deep borehole at 10 m intervals would yield $T-z$ points shown; curvature can clearly be seen in the upper part of the hole and, in the absence of any local information on climatic change, careful workers would probably reject data from the upper 150 m and proceed to fit a least squares straight line to the points in the 150–300 m range. This would yield a gradient of 20 mK m^{-1} with a

coefficient of variation of <1 %. Such a result would probably be regarded as a highly accurate and reliable value for temperature gradient, which it is in the local sense; however, it can readily be seen that it departs from the desired equilibrium temperature gradient by more than 25 %. If the hole pentrated another 150 m and only the data from 300–450 m were used, the gradient would be 17 mK m^{-1} + 0.6 %; an improvement but still 12 % different from the true value of 15 mK m^{-1}. Earlier Holocene climatic variations might reduce or increase the error depending upon the particular locality. Clearly, the shallower a borehole that is used for HFD work, the more detailed, and more local, information do we need on past surface temperature variations.

Birch (1948) indicated the importance of palaeoclimatic details over the last 10^6 years for the modern temperature field at depths of a few kilometres, citing examples where step model corrections were 10 to 30 % larger than corrections derived from more detailed Pleistocene climate models. More recently, the importance of Holocene climatic variations on the upper kilometre or two of the subsurface has been considered (Čermák, 1971; Jessop, 1971; Beck, 1977; Balling *et al.* (1981) Vasseur *et al.*, 1983; Clauser, 1984).

Beck (1977) has briefly reviewed the major lines of evidence for inferring global palaeoclimate – ranging from oxygen isotope ratios in continental ice sheets to dendrochronology – and presented a generalized latitude-dependent surface temperature history. His model consists of six step changes in surface temperature and, in the last millenium, two smoother linear changes with time (Table 5.5). Unfortunately, temperatures in a borehole represent a response to temperature variations over a relatively small region of the Earth's surface and models such as this may be quite poor at a particular locality especially if shallow (<300 m) holes are used; Jessop (1971) has indicated some of the variability across Canada. Where possible, local palaeoclimatic evidence must be used along with general global models.

TABLE 5.5
Beck's (1977) synthesis of global climatic changes for heat-flow density correction (see also Figure 10.10).

Latitude:	0–20		20–40		40–60
T[°C]	*t*[YBP]	*T*[°C]	*t*[YBP]	*T*[°C]	*t*[YBP]
- 4.0	97000	- 5.7	81000	−9.0	65000
- 1.3	52000	- 1.9	44000	−3.0	35000
5.3	15000	7.7	12000	12.0	10000
0.7	10000	1.0	8700	1.5	7000
- 0.7	3000	- 1.0	2500	−1.5	2000
0.3	1500	0.5	1200	0.75	1000
- 0.006 °C a^{-1}	1000 to 750	- 0.010 °C a^{-1}	870 to 620	−0.015 °C a^{-1}	700 to 500
0.02 °C a^{-1}	220 to 150	0.032 °C a^{-1}	190 to 125	0.05 °C a^{-1}	150 to 100

Although the uncertainty in time variation of surface temperature at a locality detracts from confidence in a correction and therefore in the HFD value, Beck (1977) offers some procedures to suggest the effectiveness of a correction. The palaeoclimatic effects appear to vary with latitude, thermal diffusivity (i.e. density, specific heat capacity, but principally thermal conductivity) and the depth range of measurements. In a single borehole, HFD may be calculated over several short depth intervals, chosen where the thermal conductivity is well characterized; interval heat-flow density determined in this way will often be different in each depth interval due to various noise sources including conductivity inhomogeneity, refraction of heat, fluid movement, and climatic effects. If a climate correction is carried out, the interval heat-flow densities should tend to converge after correction. Beck noted that correlations of heat-flow density with depth or conductivity, regionally or at a single site, can be due at least in part to climatic effects. Vasseur *et al.* (1983) applied an inverse formalism to interval heat-flow density from a 1 km deep well in a relatively impermeable granitic massif, with continuous core available for sampling thermal conductivity; their results for Holocene surface temperatures agree in gross aspect with the expected climatic changes in global models such as Beck's summary given in this paper. A pragmatic approach to the problem is given by Chapman *et al.* (1984) for the case where several boreholes are available in regions known to have been subjected to similar climatic changes.

In a more recent development Nielsen (1986) used a technique proposed by Tarantola and Vallette (1982a) to jointly invert two or more time separated $T-z$ plots in an attempt to extract not only a HFD value with reliable error bounds but more reliable information on past climatic temperature variations. The method involves a general stochastic approach to weakly non-linear inverse geophysical problems which, for the special case where the errors are Gaussian, simplifies to a conventional linear algebra problem (Tarantola and Vallette, 1982b). The technique incorporates into the inversion process the input of data which may be highly constrained (e.g. if the data values are well known) or virtually unconstrained, the output being a set of desired maximum likelihood parameters and error estimates. Unfortunately, as far as Holocene temperature variations are concerned (i.e. 'periods' of the order of thousands of years) time separation of the $T-z$ profiles should be on the order of hundreds of years; however, the technique has been further developed, as discussed later, for situations where two time separated $T-z$ profiles can be obtained in a reasonable period of time.

The technique of Nielsen (1986) is illustrated in Figure 5.13 by means of a simple model incorporating a homogeneous half-space of thermal conductivity 3 W m^{-1} K^{-1}, thermal diffusivity 12×10^{-7} m^2 s^{-1}, heat generation 3 μW m^{-3}, the HFD = 60 mW m^{-2}. The steady-state subsurface temperature gradient has been perturbed by an instantaneous climate change of 5 K 1000 YBP. Figure 5.13A shows the $T-z$ profile that would be obtained in a 600 m deep borehole if measured now and the difference between that log and one that was obtained 16 a earlier; for convenience, temperatures are shown at approximately 10 m intervals. Figures 5.13B, C, D, and E show the a posteriori results of jointly inverting the profiles using various a priori input data.

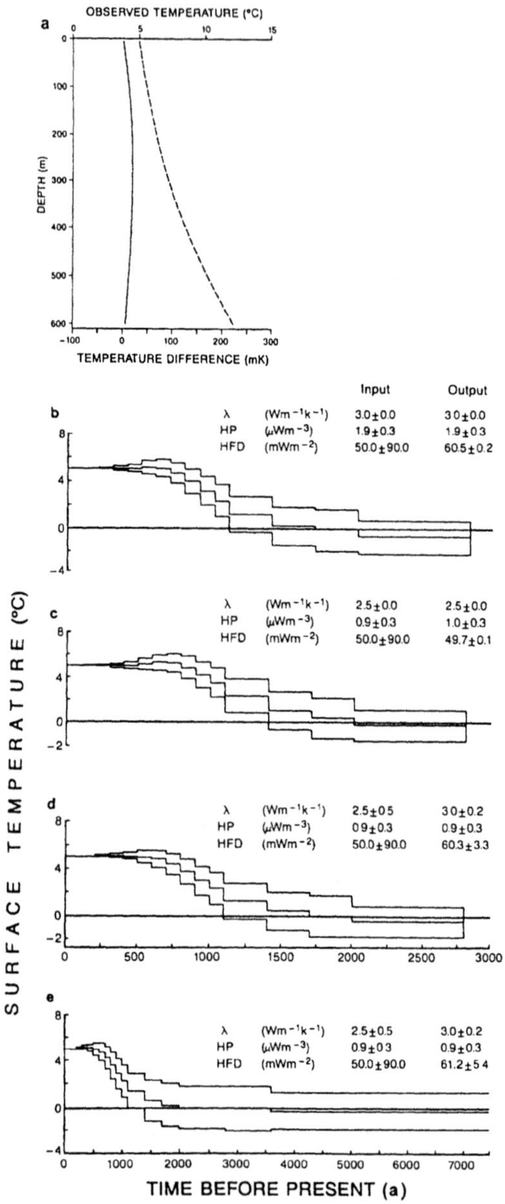

Fig. 5.13. Results of coupled inversion of $T-z$ plots for a homogeneous half space with $\lambda = 3$ Wm^{-1} K^{-1}, $a = 1.2$ mm^2 s^{-1}, HP = 3 μW m^{-3}, HFD = 60 mW m^{-2} the steady state regime being perturbed by a sudden 5 K change of surface temperature 1000 YBP. (A) The $T-z$ profile (---) from temperatures at 10 m intervals measured now in a 600 m deep borehole, together with a profile of the temperature difference (solid line) between the current $T-z$ data and a set gathered 16 years earlier. (B)–(D) Results using input data as shown and constraints on prior climate of 0±2 °C for 2800 YBP to present and 0±0.01 °C prior to this; curves show inferred past climate (central line) with standard deviation (bounding lines). (E) Similar to (D) except that error bounds on climate prior to 2800 YBP are relaxed to 0±2 °C.

In the inverted data a change of 5 K is clearly evident in the past climate but the fact that it is a step change at 1000 YBP is not well resolved; the HFD is well resolved because the thermal conductivity is known and tightly constrained.

If thermal conductivity data are not available a value must be guessed. If an incorrect value is chosen and tightly constrained, the result is an incorrect HFD value as shown in Figure 5.13C; if an incorrect value of thermal conductivity is chosen but loosely constrained, the more reasonable values of thermal conductivity and HFD shown in Figure 5.13D are obtained. The time separation between the two temperature-depth plots is so small that in this case the difference between inverting one and inverting two profiles is not significant.

Although inverse solutions are not unique and are imperfectly resolved, they do have the considerable advantage of giving reliable error bounds. Furthermore, in using the technique to derive past climate from $T-z$ profiles, the correspondence of general features of the solutions with general features of climatic variation known from other evidence, at sites where other noise sources are minimal, strengthens the contention that paleoclimate significantly affects the HFD value, the degree of significance depending on the interval over which HFD has been determined.

A useful application of the techniques of Vasseur *et al.* (1983) and Nielsen (1986) is in the measurement of HFD in lake sediments. Beck *et al.* (1985) and Wang *et al.* (1986) have shown that a one-time probe measurement may need many years of careful and detailed preparatory work, and that one site may not be at all representative of the regional HFD value. Fortunately it may be possible to solve the first problem in a relatively short period of time.

Wang and Beck (1987) have developed the technique of Nielsen (1986) for use in the frequency domain and applied it to the situation of measurements in water covered sediments where the bottom waters have been subjected to significant temperature variations. Basically, the problem is that although due to the heat conduction process the $T-z$ profile must contain information about the past bottom temperature variations as well as the local HFD, with only a single $T-z$ profile having a few data points, the inverse problem of separating bottom temperature variation (BTV) and HFD is so underdetermined that it is impossible to obtain good resolution. However, if there are several $T-z$ profiles separated in time, the problem is much better constrained and it is possible to obtain more useful data for measurements over a relatively short period of time. A frequency inversion is performed on a pair (or more) of $T-z$ profiles to obtain details of the most recent (last one year) BTV and information on earlier longer term trends, together with values of HFD and mean sediment surface temperatures and their maximum likelihood estimates of error. As illustrated in Table 5.6 and Figure 5.14, better *a priori* information leads to more reliable, but not necessarily narrower, a posteriori constraints although the mean value of the HFD may not change significantly.

The model used for Table 5.6 was an isotropic half-space with conductivity $= 0.8\ \mathrm{W\ m^{-1}\ K^{-1}}$, thermal diffusivity $= 2.25 \times 10^{-7}\ \mathrm{m^2\ s^{-1}}$, local HFD $= 60\ \mathrm{mW\ m^{-2}}$, perturbed by six years of BTV as seen, Figure 5.14A, in Lake Greifensee (Finckh, 1981) in which there is a strong annual component, together with some significant higher frequency components; from the record the mean surface tem-

TABLE 5.6

A posteriori HFD and T_0 obtained by joint inversion of two $T-z$ plots separated in time by $(n-1)$ months. The input a priori values of HFD and T_0 are 90 ± 100 mW m^{-2} and 4.5 ± 0.5°C, respectively, in all cases. The *a priori* BTV is set to zero; all the other parameters are accurately known (see text and Figure 5.14 for further details).

n	Profiles used	Separation (days)	*A posteriori* HFD (mW m^{-2})	T_0(°C)
2	1 only	–	24.0 ± 23.2	5.11 ± 0.18
3	1,2	30	44.8 ± 17.9	4.98 ± 0.15
4	1,3	60	55.1 ± 14.7	4.90 ± 0.13
5	1,4	90	61.0 ± 13.4	4.86 ± 0.12
6	1,5	120	61.4 ± 12.8	4.86 ± 0.12
7	1,6	150	59.5 ± 12.7	4.87 ± 0.11
–	1–6		62.4 ± 10.1	4.86 ± 0.10

perature of the sediments is 4.89 °C. Figure 5.14B shows the equilibrium temperature gradient (dashed line) and the $T-z$ profiles that would be measured at time zero (curve 1) and at 30 day intervals prior to that; for curve 1 the complete time series of six years is used whereas for curve 6 the last five months of record are discarded. In the joint inversion process the input *a priori* values are a HFD value of 90 ± 100 mW m^{-2} (i.e. virtually unconstrained). The results of the inversion using several pairs of $T-z$ profiles (and in one case all six $T-z$ profiles) are shown in Table 5.6; corresponding inferred BTV are shown in Figure 5.14C. It can readily be seen that provided the $T-z$ profiles are separated by at least 3 months, the HFD value is recovered albeit with loose constraints, the coefficient of variation being ~20–25 %; increasing the time separation to more than about 3 months (about a quarter period of the dominant variation), or using more than two $T-z$ profiles, does not significantly improve the results. Similarly, the shape of the bottom temperature variations is recovered, in reasonable detail for the previous year but with only an indication of the long term trend for earlier times. Therefore the problem referred to earlier of the considerable time, and detail in the data, required prior to a one-time probe measurement is avoided.

It must be emphasized that this does not eliminate the need to make many such measurements over large areas to determine whether the HFD value recovered from one site is a reliable indicator of the HFD over a region.

The climatic effect has been implicated as a source of disturbance in many HFD determinations. Just as in the steady-state case, difficulties remain in relating most palaeoclimatic data, in the form of air temperature and precipitation estimates, to the mean annual surface temperature. An insidious problem is the possibility of transient groundwater flow patterns of long period. Changes in groundwater circulation accompanying continental glacier retreat might affect greater depths than conductive mechanisms or attenuate surface temperatures more rapidly, depending upon details of flow. The work of Beck (1977) and Vasseur *et al.* (1983) emphasizes the importance of adequately sampling rock properties to allow

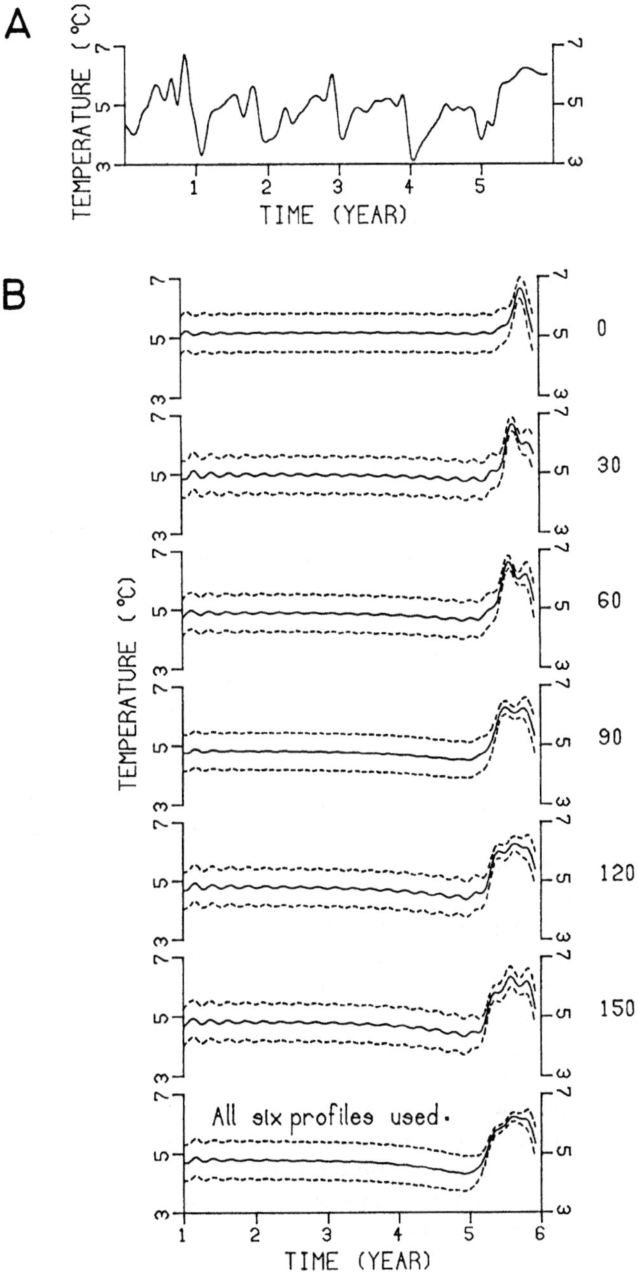

Fig. 5.14. Derived BTV by joint inversion of two $T-z$ profiles separated by times shown in Table 5.6. Model used was a half space of $a = 0.225\ mm^2\ s^{-1}$, mean surface temperature of 4.89 °C, HFD = 60 mW m^{-2}, $\lambda = 0.8\ W\ m^{-1}\ K^{-1}$, perturbed by 6 years of BTV shown in Figure 5.14A.

effective climatic correction. However, if a climatic model and adequate samples are available in a conduction-dominated thermal regime, the corrections are easily calculated and applied. The Earth acts as a filter working to the advantage of the geothermal interpreter in the case of climatic effects: a trade-off exists between measurement depth and uncertainty due to poor knowledge of palaeoclimate. Recent local climatic changes must be known accurately for correction, but they affect only shallow depths and at greater depths, the longer period changes need be known only approximately (Beck, 1977; Balling, 1979; Balling *et al.*, 1981; Haenel, 1983). Palaeoclimatic studies have intensified in recent decades and there is every reason to expect that increasingly sophisticated regional and global models will be available for correcting climatic effects in geothermal data.

(2) TOPOGRAPHIC EVOLUTION

Geologic evidence indicates that changes in the geometry of the Earth's surface and subsurface occur through a variety of mechanisms at a variety of time scales. Three processes are normally considered when accounting for terrain effects: uplift without internal deformation, erosion, which removes surface material and exposes deeper zones to the surface conditions, and sedimentation, which moves the original boundary away from the surface conditions. The general effect of these processes on $T-z$ profiles is illustrated in Figure 5.15.

Other tectonic and geologic processes, such as overthrusting, extension and rifting, intrusion or extrusion of magma etc., can have profound effects on the subsurface temperature regime and have seen detailed attempts at mathematical description in recent years, but these processes are not generally termed terrain effects. The one dimensional moving boundary problem in terrain correction was addressed by Benfield (1949a, b). Carslaw and Jaeger (1959) have given the solution including heat production, and Birch (1950) adapted his steady-state correction to approximate simple temporal evolution with higher geometric dimension. These methods are also outlined briefly in the standard texts: Jaeger (1965), Kappelmeyer and Haenel (1974) and Turcotte and Schubert (1982). England (1978) has explored the interaction of erosion and climatic changes. Various authors have considered the effects of erosion on interpretations of the geological significance of heat-flow density data, such as heat-flow density-age relations for continents (for example, England and Richardson 1980) or heat-flow-density-heat production relations for continental crust (for example, Woodhouse and Birch, 1980).

The case of uplift without erosion or of instantaneous erosion of a finite thickness of material causes a change in surface temperature with time that is just related to the uplift rate and the soil temperature lapse with elevation; this effect is readily estimated with the techniques of the preceding section on climatic change. The thermal effects of erosion or sedimentation are described in general by the advective heat equation. In one space dimension, this equation is (see also Equation (2.15b):

$$a\frac{\partial^2 T}{\partial z^2} = \frac{\partial T}{\partial t} + v\frac{\partial T}{\partial z}, \qquad (5.24)$$

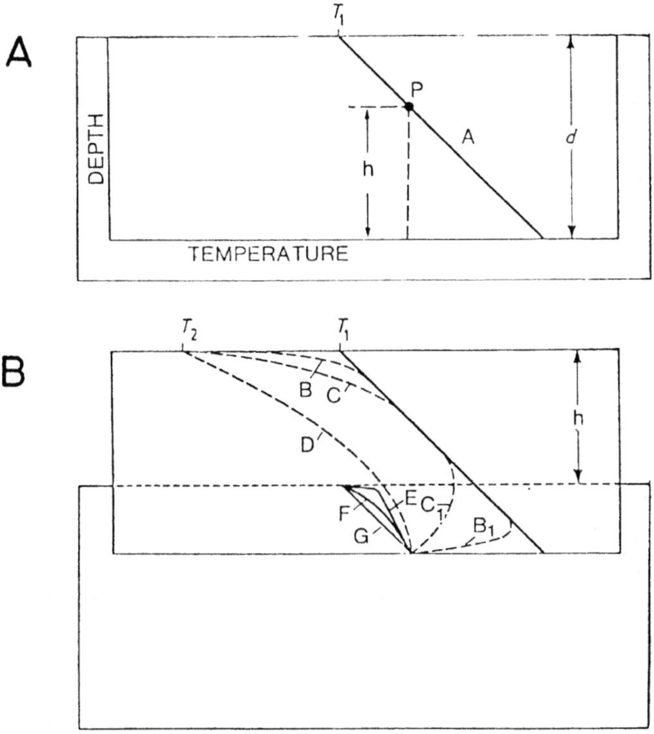

Fig. 5.15. Schematic of perturbation to temperature profiles due to uplift and erosion (after Misener and Beck, 1960). (A) Steady state $T-z$ profile; (B) Rapid uplift has occurred so that bottom of block is at height and temperature of point P in (A); this is equivalent to a sudden decrease in temperature at the block base and at the surface (due to lapse rate). Progression of $T-z$ plots with time will therefore be B, C, D in upper part and B_1, C_1, D in lower part. If there is now sudden erosion to original surface the further progression of $T-z$ profiles will be E, F, G.

where a is thermal diffusivity and v is the velocity of the medium, relative to the boundary surface. Taking z positive down, as in Figure 5.3, positive v corresponds to sedimentation, and negative to erosion. For an equilibrium gradient Γ_b throughout the subsurface prior to initiation of the movement at $t = 0$, and letting Γ_L be the decrease of soil temperature with elevation and T_0 the surface temperature at $t = 0$, $z = 0$, the solution to this problem is:

$$T(z, t) = T_0 + \Gamma_b(z - vt) + \frac{1}{2}\left[\Gamma_b + \frac{\Gamma_b u}{v}\right] \cdot \left[(z + vt) \exp\left(\frac{vz}{a}\right) \operatorname{erfc}\left(\frac{z + vt}{2(at)^{1/2}}\right) - (z - vt) \operatorname{erfc}\left(\frac{z - vt}{2(at)^{1/2}}\right)\right], \quad (5.25)$$

where u is the rate of movement of the boundary with respect to a fixed reference and v is the velocity of the moving medium with respect to a point fixed on the moving boundary. This assumes uplift and boundary changes occur at a constant rate. The corresponding gradient is (Benfield, 1949b; Kappelmeyer and Haenel, 1974):

$$\Gamma(z, t) = \frac{\partial T(z, t)}{\partial z} = \Gamma_b + \frac{1}{2}\left(\Gamma_b + \frac{\Gamma_L u}{v}\right)\left[-\operatorname{erfc}\left(\frac{z - vt}{2(at)^{1/2}}\right) - \frac{z + vt}{(\pi at)^{1/2}} \exp\left(\frac{vz}{a}\right) \exp\left(-\left(\frac{z + vt}{2(at)^{1/2}}\right)^2\right) + \frac{z - vt}{(\pi at)^{1/2}} \exp\left(-\left(\frac{z - vt}{2(at)^{1/2}}\right)^2\right) + \left(1 + \frac{vz}{a} + \frac{v^2 t}{a}\right) \exp\left(\frac{vz}{a}\right) \operatorname{erfc}\left(\frac{z + vt}{2(at)^{1/2}}\right)\right]. \tag{5.26}$$

Figure 5.16 shows graphically the effects of 1-D erosion and sedimentation on the thermal gradient at the Earth's surface for a variety of erosion/sedimentation rates. The thermal gradient disturbances from these processes are all negligible at very short times but grow as the duration of the erosion or sedimentation event increases; the effects are also magnified for high rates. As large scale vertical movements rarely are sustained for longer than a few tens of millions of years, rates of uplift and subsidence less than about 0.1 mm a^{-1} have a negligible effect on the thermal state of the upper crust. At rates greater than 0.1 mm a^{-1}, thermal effects may be significant. For example with an erosion rate of 1 mm a^{-1}, after 10 Ma when 10 km of the crust has been eroded the disturbance to the original gradient is 80 %. Likewise some marine basins have experienced rapid subsidence of several mm a^{-1} for durations of up to a million years; the thermal perturbation in such cases is also severe.

This 1-D analysis is unsuitable for corrections where the topography is significant. Birch (1950) presented a correction allowing for development of the topography from a plane surface over a finite time. Birch's method for topographic evolution is just a generalization of the solid angle method for steady-state topography. As in the steady-state case, the topographic effect is represented approximately by temperature variations projected onto a reference plane, following the suggestion of Jeffreys (1938). Topographic evolution is modelled by allowing the temperature on each element on the reference plane to vary linearly with time. For a 2-D situation, let the temperature on the reference plane $z = 0$ be $tF(x)$ where F is the temperature on the reference plane due to the fully developed topography normalized by the time between initiation of the evolution, $t = 0$, and the present. The temperature disturbance is (Birch, 1950; Clark, 1957):

$$T(x, z) = \frac{t}{\pi}\int_{-\pi/2}^{\pi/2} F(\xi)\left[\frac{R^2}{4at} Ei\left(\frac{-R^2}{4at}\right) + \exp\left(\frac{-R^2}{4at}\right)\right] d\theta, \tag{5.27}$$

where $d\theta$ is the plane angle subtended by reference surface element $d\xi$, $R^2 = (z - \xi)^2 + z^2$ and Ei is the exponential integral $Ei(-s) = -\int_s^\infty \exp(-w)/w \, dw$.

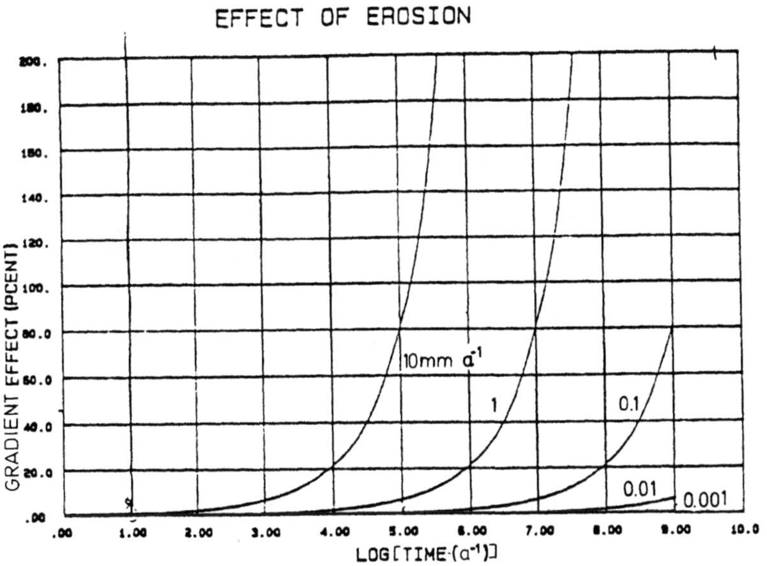

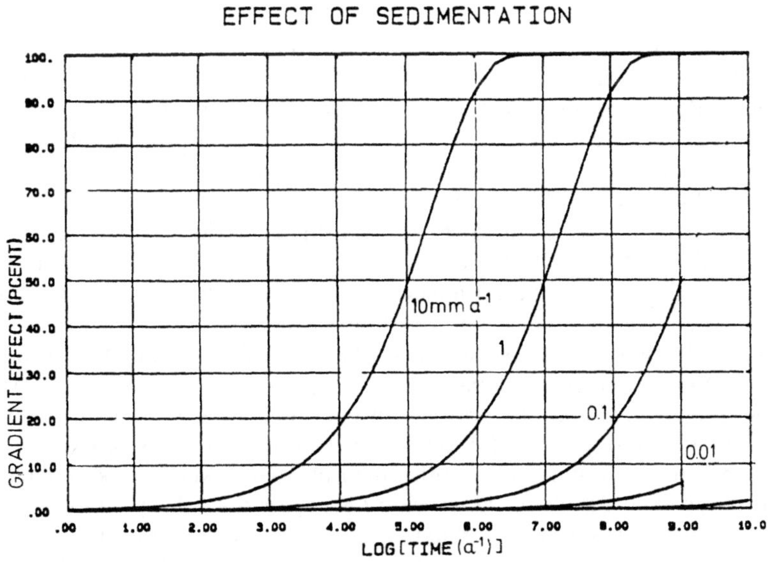

Fig. 5.16. Effects of erosion and sedimentation on the thermal gradient at the Earth's surface. Ordinates in the plots represent the quantity $[(\Gamma(0, t) - \Gamma(0, 0))/\Gamma(0, 0)]$ calculated from Equation 5.26 with $u = 0$ and a thermal diffusivity of 1 $mm^2 s^{-1}$. Curve parameter is erosion and sedimentation rate.

As in the steady-state case, this integral over solid angle θ can be approximated by a finite sum of the integrand averaged on each element $d\theta$. Lachenbruch (1969) has given a rough but easily evaluated estimate of the effect of topographic evolution on the surficial gradient for a 2-D plane slope model as shown in Figure 5.9. The effect of any plane slope at a particular site is the same as the effect due to any of a class of models (called 'equivalent slopes' in Lachenbruch, 1969) with different slopes and locations. Modelling present topography by its equivalent cliff, and assuming the topography formed instantaneously at $t = 0$, the temperature field on the reference plane is just a step change in temperature at the location of the cliff. The effect due to such a surface temperature was given by Lachenbruch (1957a) and the temporal disturbance to the surficial gradient is:

$$\Delta\Gamma(x, t) = \pi^{1/2} \operatorname{ierfc}\frac{x}{2(at)^{1/2}} \cdot \Delta\Gamma(x, t = \infty), \tag{5.28}$$

where

$$\operatorname{ierfc}(s) = \int_s^\infty \operatorname{erfc}(w)\, dw = \pi^{1/2} \exp(-s^2) - s \operatorname{erfc}(s).$$

Here x is the distance from the measurement site to the 'equivalent cliff'. The restrictions on the methods of the steady-state case also apply here. Homogeneous, isotropic rock properties are assumed. Also, since the reference plane projection of Jeffreys (1938) is used, the measurement being corrected must be deep or distant relative to the height of the topography, although the temporal evolution decreases the errors in this approximation.

More complex cases can be handled by the numerical techniques described for the steady-state corrections. Lee (1977) included surface evolution in his finite element model of terrain effects. Haji-Sheikh and Sparrow (1967) illustrate the application of Monte Carlo modelling to moving boundary problems in heat conduction. Boundary element application to moving boundary problems is described by Banerjee and Shaw (1982).

Frequently the history of the local geomorphic evolution is known only poorly, and calculations of this effect probably only suggest its order of magnitude. As with climatic changes, the performance of this correction might be assessed by its tendency to increase the consistency of interval heat-flow density measurements. England (1978) noted that for certain cases, interval heat-flow density might be nearly constant throughout a typical depth range of observation and yet a significant disturbance could still be present. In particular he suggested that erosion of 100–200 metres during the end of the last glaciation could cause a perturbation of about 5 mW m^{-2} throughout a depth range of several kilometres under suitable circumstances. The details of the erosion history of rugged topography also have been shown to be significant (England, 1978); valleys incised during the Holocene can exhibit an augmentation of near-surface heat flow density far beyond that expected due to steady-state topographic effects.

As in the case of climatic change, longer period surface changes affect greater depths, but the surface details need not be closely known, while at shallow depths the details of the geometric changes and their history must be known quite precisely to allow adequate correction. The near surface disturbance is greatest for recently eroded valleys, although significant effects are expected even for planation beneath glaciers (England, 1978). Mathematical tools are available to calculate the effects for most practical situations, but the great sensitivity to the temporal changes at shallow depths means that the required geologic constraints are too incompletely known; once again, it is the limit in our knowledge of the actual boundary conditions of the problem which limit the ability to correct the defect.

5.5. GROUNDWATER DISTURBANCES

The thermal perturbations discussed above and caused by terrain effects, terrain evolution, conductivity heterogeneities, and palaeoclimate variations have all been modelled using the equations of heat conduction. But heat-flow density disturbances can also result from advective heat transfer associated with active groundwater flow systems. A useful summary of papers on the effects of groundwater flow on the thermal regime of the uppermost crust is given in Table 1 of Smith and Chapman (1983).

Groundwater flow that may have a direct effect on heat-flow density determinations can be classified into three systems (Drury, 1984): (a) water movement within the borehole, (b) transient loss or gain of water into or out of fractures during drilling, and (c) water flow through porous or fractured country rock surrounding the borehole. Classes (a) and (b) are discussed in Chapters 2.3, 3.5. and 9.4.3 and (c) in Chapter 2.3, discussion here is limited to thermal perturbations arising from flow in the country rock.

The easiest flow to recognize and correct for is vertical flow of sufficient magnitude to produce a characteristic bowed temperature-depth profile. Bredehoeft and Papadopulos (1965), Section 2.3.2, give an analytic solution of Darcy velocity for 1-D, steady state flow between two points held at constant temperature. Mansure and Reiter (1979) further demonstrated that plots of temperature versus temperature gradient are diagnostic of certain vertical flow regimes and helpful in deducing the flow properties and total heat-flow density.

2-D and 3-D groundwater flow systems are inherently much more complicated than 1-D vertical flows, and their thermal effects are often difficult to recognize, let alone correct for. The problem is acute if heat-flow density determinations are made at isolated sites in relatively impermeable rocks and if there are active flow systems that do not intersect the drillholes. In such cases, the temperature-depth profile and the heat-flow density determination may have all the appearances of a legitimate conductive heat flow site but the local heat-flow density may be significantly above, below, or fortuitiously equal to the regional background heat-flow density. Figure 5.17 shows two model simulations from Smith and Chapman (1983) illustrating various thermal perturbations that could arise from 2-D groundwater flow systems.

In spite of the difficulties in recognizing and correcting for advective disturbances in heat-flow density studies, the following generalizations may be helpful (Domenico and Palciauskas, 1973; Smith and Chapman, 1983):

– For low permeability rocks (intrinsic permeabilities $<\sim 10^{-17}$ m^2 {= 0.01 millidarcy} and gentle topography, groundwater velocities are too small to effect significant redistribution of heat by advection.

– The transition from a conduction to advection dominated thermal regime is

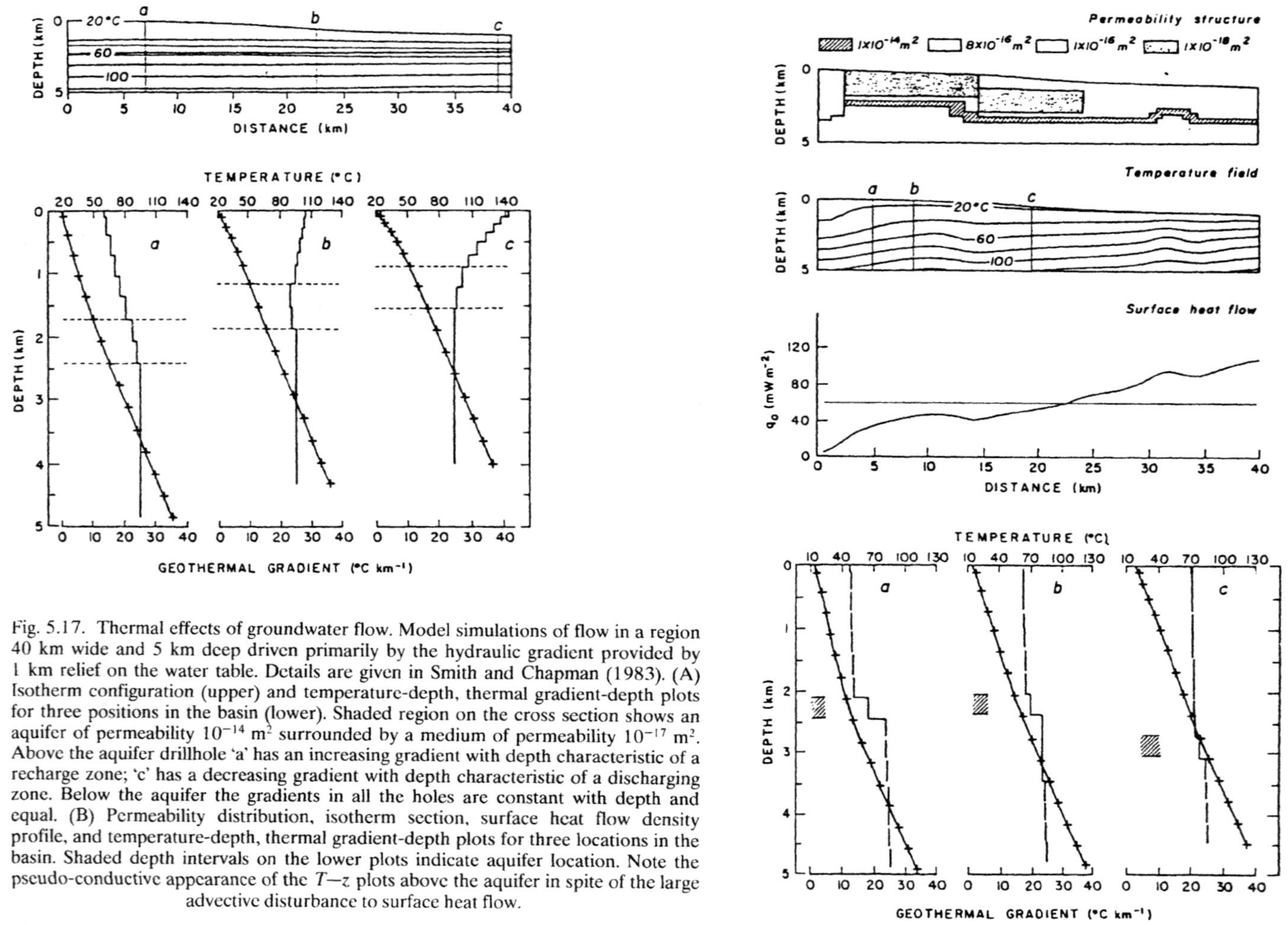

Fig. 5.17. Thermal effects of groundwater flow. Model simulations of flow in a region 40 km wide and 5 km deep driven primarily by the hydraulic gradient provided by 1 km relief on the water table. Details are given in Smith and Chapman (1983). (A) Isotherm configuration (upper) and temperature-depth, thermal gradient-depth plots for three positions in the basin (lower). Shaded region on the cross section shows an aquifer of permeability 10^{-14} m^2 surrounded by a medium of permeability 10^{-17} m^2. Above the aquifer drillhole 'a' has an increasing gradient with depth characteristic of a recharge zone; 'c' has a decreasing gradient with depth characteristic of a discharging zone. Below the aquifer the gradients in all the holes are constant with depth and equal. (B) Permeability distribution, isotherm section, surface heat flow density profile, and temperature-depth, thermal gradient-depth plots for three locations in the basin. Shaded depth intervals on the lower plots indicate aquifer location. Note the pseudo-conductive appearance of the T—z plots above the aquifer in spite of the large advective disturbance to surface heat flow.

sharp. Thus heat-flow density determinations in rocks of permeability $>10^{-15}$ m^2 should be scrutinized for advective effects.

– Thermal regimes, when advective effects are significant, are affected by topographic configuration of the water table, magnitude and spatial distribution of permeability, hydraulic anisotropy and depth of active flow.

– Aquifers produce significant perturbations to the thermal fields of basins, but only if the permeability of the surrounding medium is great enough to permit adequate recharge to and discharge from the aquifer.

– If low permeability rocks overlie part of an aquifer, or a permeable fracture system, the thermal regime within the low permeability rocks will be conduction dominated but the HFD deduced in such a unit may depart significantly from the background flux. Heat-flow density differences in borehole sections above and below a fracture system suspected of having significant flow can be used to compute details of the flow (Drury and Lewis, 1983).

– The thermal regime within a region of moderate to rugged topography and spatially varying permeability will be complex. Caution should be exercised when interpreting heat flow density results in such a region as regional heat flow density values. A series of shallow drill holes along the hydraulic gradient may yield more information about the thermal regime than would one deep drill hole, unless the deep hole penetrated through the deepest aquifer or fracture system.

5.6. PERTURBATIONS DUE TO FREE CONVECTION

There are two types of convection that have to be considered: that which is induced within a borehole by the act of introducing a long cylinder of free standing fluid into the host rock, and that which arises when the regional temperature gradient exceeds the critical temperature gradient in a permeable medium. Induced convection has been discussed in Chapter 3.4. and we deal here only with regional convection.

The critical gradient (i.e. that above which free convection occurs) for a permeable medium depends on two classes of parameters – those of the medium and those of the fluid. These parameters may be usefully combined to give a definitive dimensionless number called the Rayleigh number Ra*. For a given geometry and field conditions there is a first critical Rayleigh number Ra_1^* which defines the onset of convection. In the simplest cases Boussinesq flow is assumed and the physical properties of the formation and convective fluid are also assumed to be constant in space and time. In reality, of course, the permeability of a formation is anisotropic, is a function of pressure and temperature, and may be function of time since chemical reactions frequently take place leading to dissolution and precipitation of formation solids.

For steady state free convection in a permeable layer of infinite horizontal extent and bounded by two perfectly insulating impermeable formations separated by a finite distance $\triangle z$ (i.e. the aspect ratio A^* is large) and with the lower surface maintained at a temperature $\triangle T$ above the upper surface, the first critical Rayleigh number is given by (Lapwood, 1948; Wooding, 1957; Elder, 1965, 1967), see also Table 2.9, Equations (2.25) and (4.25):

$$\mathrm{Ra}_1^* = (k_m \alpha_f g \triangle T \triangle z)/(a_m \nu_f) = 4\pi^2, \tag{5.25}$$

where

g = gravitational acceleration
$a_m = \lambda_m/(\rho c_p)_f$ = effective thermal diffusivity of the medium
λ_m = thermal conductivity of fluid saturated medium
$(\rho c_p)_f$ = specific thermal capacity of the fluid
α_f = volume coefficient of thermal expansion of the fluid
ν = kinematic viscosity of the fluid
k_m = intrinsic permeability of the medium – a property which is independent of the nature of the fluid and of the force field causing fluid movement.

Straus (1974) has defined a second critical Rayleigh number, Ra_2^* above which convection becomes unsteady or turbulent. The temperature gradient does not appear explicitly, partly because it is difficult to think of the gradient when the onset of convection causes it to change with time, and partly because in this case the simple boundary condition requires the bounding surfaces to be maintained at constant temperatures.

Straus and Schubert (1977) have shown that $\mathrm{Ra}_1^* = 4\pi^2$ is a maximum value; in a real case Ra_1^* is considerably influenced by the variable physical properties (such as the temperature dependence of viscosity) and non-Boussinesq behaviour of the fluid; in this case the temperature gradient must be specified. Figure 5.18 shows how Ra_1^* varies with the temperature gradient and thickness of the forma-

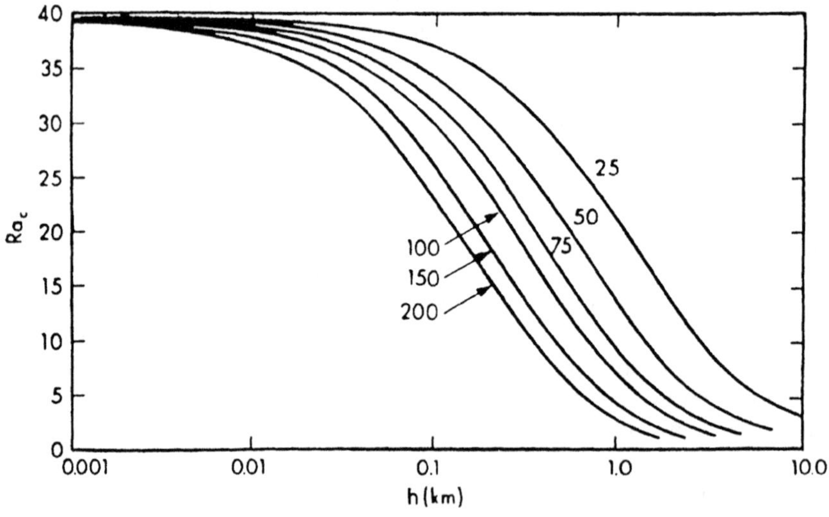

Fig. 5.18. First critical Rayleigh number Ra_c for a geothermal reservoir of thickness h at various temperature gradients in mK m^{-1} and with permeability varying with temperature and pressure (after Straus and Schubert, 1977).

tion in which the thermodynamic and physical properties are temperature and pressure dependent.

Free convection is only likely to be important in geothermal regions. Since in this chapter we are concerned principally with the thermal regime in 'normal' regions the reader is referred to Chapters 2. and 3. and to Cheng (1978) for detailed discussion of the mathematical techniques for analyzing geothermal reservoir data. In the context of equilibrium HFD, once free convection has been recognized there is little one can do with the data except abandon it for HFD work.

5.7. SUMMARY OF RECOMMENDATIONS

1. Continental heat-flow density values are determined by combining temperature gradient and thermal conductivity data directly for specific intervals in a drillhole, or by using a graphical Bullard plot method. All details of the measurement, including estimates of uncertainties in the heat-flow density determination from estimates of errors in conductivity and gradient measurements respectively should be reported.
2. Interval heat-flow density values computed for at least four depth ranges in a borehole serve as a particularly valuable test for the validity of various heat-flow density corrections. Lack of adequate conductivity sampling is often the limiting factor in reducing errors in interval heat-flow density determinations. In this connection, thermal conductivity samples must be obtained and measured in sufficient number so that the uncertainty in interval heat-flow density is much less than the magnitude of the corrections to be tested.
3. Possible perturbations to the regional background heat-flow density arising from uneven terrain surrounding the site, terrain evolution, subsurface conductivity structure, climatic history, and groundwater movement should be assessed, if possible, individually for each heat-flow density site.
4. When well location and drilling program are flexible, (a) select sites to minimize topographic and hydrogeological disturbances, or (b) insure measurement depths much greater than local topographic relief.
5. Approach terrain correction in a hierarchic manner. Use the simplest steady-state methods to estimate the magnitude of effects. Use more sophisticated techniques and transient models when the simpler methods suggest the effects are significant. Dimensionality of the terrain correction model must represent not only topography, but surface temperature and subsurface conductivity variation as well.
6. When interpreting shallow measurements, special attention must be given to local microclimate variations. A sketch map of obvious soil and vegetation variations in the vicinity of a borehole can be invaluable for this purpose.
7. The expected perturbation to heat-flow density arising from past climatic changes should be calculated. If it is not possible to obtain past climate information and HFD's by a maximum likelihood inversion technique, details of local variations in surface temperature where available, should be incor-

porated in the forward method of calculating transient corrections. In the absence of detailed information, the global climatic change model of Beck (1977), repeated as Table 5.4 of this chapter, should be used. A validity check on palaeoclimate models should be made by using the forward method of computing the temperature gradient correction, and hence corrected heat-flow density, for several depth intervals in the drill hole; corrected heat-flow densities should converge towards a single value if the palaeoclimate correction is valid.

8. In regions of moderate to rugged topography and having permeable or fractured country rock, the thermal regime may be perturbed by the advective effects of flowing groundwater. Interval heat-flow density values should be computed above and below any fractures or aquifers suspected of having active flow; differences in the interval heat-flow densities can be used to calculate the flow velocities. If advective effects are suspected for a region, plotting of heat-flow density results in terms of the site location relative to the groundwater flow system would be a useful convention to adopt.
9. In areas, particularly sedimentary basins, where hydraulically driven regional water movement is known to occur, satisfactory elimination of the subtle effects on HFD of the water flow can only be accomplished with data, preferably including permeability values, from a network of several holes distributed across the discharge and recharge zones.

5.8. REFERENCES

Ames, W. F.: 1969, *Numerical Methods for Partial Differential Equations*, Barnes and Noble Inc., New York, 291 p.

Balling, N.: 1979, 'Subsurface Temperatures and Heat Flow Estimates in Denmark', in V. Čermak and L. Rybach (eds.), *Terrestrial Heat Flow in Europe*, Springer-Verlag, Berlin, 161–171.

Balling, N., Kristiansen, J. I., Breiner, N., Poulsen, K. D., Rasmussen, R., and Saxov, S.: 1981, 'Geothermal Measurements and Subsurface Temperature Modelling in Denmark', *Geo rifter* **16**, Department of Geology, Aarhus University, 172 pp.

Banerjee, P. K. and Shaw, R. P.: 1982, 'Boundary Element Formulation for Melting and Solidification Problems', in P. K. Banerjee and R. P. Shaw (eds.), *Developments in Boundary Element Methods – 2*, Applied Science Publishers, London, 1–18.

Bauer, M. S.: 1985, *Heat Flow at the Upper Stillwater Dam Site, Uinta Mountains, Utah*, M.Sc. Thesis, University of Utah, Salt Lake City, 94 p.

Bauer, M. S. and Chapman, D. S.: 1986, 'Thermal Regime at the Upper Stillwater Dam Site, Uinta Mountains, Utah: Implications for Terrain Microclimate and Structural Corrections in Heat Flow Studies', *Tectonophysics* **128**, 1–20.

Beck, A. E.: 1977, 'Climatically Perturbed Temperature Gradients and Their Effect on Regional and Continental Heat Flow Means', *Tectonophysics* **41**, 17–39.

Beck, A. E., Wang, K., and Shen, P. Y.: 1985, 'Sub-Bottom Temperature Perturbations Due to Temperature Variations at the Boundary of Inhomogeneous Lake or Oceanic Sediments', *Tectonophysics* **121**, 11–24.

Beck, J. V. and Arnold, K. J.: 1977, *Parameter Estimation in Engineering and Science*, John Wiley and Sons, New York, 501 p.

Beers, Y.: 1962, *Introduction to the Theory of Error*, Addison-Wesley, Reading, Pa., 66 p.

Benfield, A. E.: 1949a, 'A Problem of the Temperature Distribution in a Moving Medium', *Quart. Appl. Math.* **6**, 439–443.

Benfield, A. E.: 1949b, 'The Effect of Uplift and Denudation on Underground Temperatures', *J. Appl. Phys.* **20**, 66–70.
Birch, F.: 1948, 'The Effects of Pleistocene Climatic Variations upon Geothermal Gradients', *Am. J. Sci.* **246**, 729–760.
Birch, F.: 1950, 'Flow of Heat in the Front Range, Colorado', *Bull. Geol. Soc. Amer.* **61**, 567–620.
Blackwell, D. D., Steele, J. L., and Brott, C. A.: 1980, 'The Terrain Effect on Terrestrial Heat Flow', *J. Geophys. Res.* **85**, 4757–4772.
Bodell, J. M. and Chapman, D. S.: 1982, 'Heat Flow in the North-central Colorado Plateau', *J. Geophys. Res.* **87**, 2869–2884.
Bodmer, Ph.: 1982, 'Beiträge zur Geothermie der Schweiz', Ph. D. Thesis, Eidgenössische Technische Hochschule, Zürich, 201 p.
Brebbia, C. A.: 1978, *The Boundary Element Method for Engineers*, Pentech Press, London, 189 p.
Brebbia, C. A. and Walker, S.: 1980, *Boundary Element Techniques in Engineering*, Butterworth & Co. Ltd., London, 210 p.
Bredehoeft, J. D. and Papadopulos, I. S.: 1965, 'Rates of Vertical Groundwater Movement Estimated from the Earth's Thermal Profile', *Water Resour. Res.* **1(2)**, 325–328.
Bullard, E. C.: 1938, 'The Disturbance of the Temperature Gradient in the Earth's Crust by Inequalities of Height', *Monthly Notices Roy. Astroy. Soc., Geophys. Suppl.* **4**, 360–362.
Bullard, E. C.: 1939, 'Heat Flow in South Africa', *Proc. Roy. Soc. London A*, **173**, 474–502.
Carslaw, H. S. and Jaeger, J. C.: 1959, *Conduction of Heat in Solids*, Oxford Univ. Press, London, 510 p.
Castoldi, L.: 1952, 'Sulla distribuzione della temperatura negli strati superiori della crosta terrestre', *Geofis. Pura et Appl.* **23**, 27–35.
Čermák, V.: 1971, 'Underground Temperature and Inferred Climatic Temperature of the Last Millenium', *Palaeogeogr., Palaeoclimat., Palaeoecol.* **10**, 1–19.
Chapman, D. S.: 1976, 'Heat Flow and Heat Production in Zambia', Ph.D. Thesis, University of Michigan, Ann Arbor, 94 p.
Chapman, D. S. and Pollack, H. N.: 1977, 'Heat Flow and Heat Production in Zambia: Evidence for Lithospheric Thinning in Central Africa', *Tectonophysics* **41**, 79–100.
Chapman, D. S., Howell, J., and Sass, J. A.: 1984, 'A Note on Drillhole Depths Required for Reliable Heat Flow Determinations', *Tectonophysics* **103**, 11–18.
Cheng, P.: 1978, 'Heat transfer in Geothermal Systems', in Irvine, T. F. and Hartnett, J. F. (eds.), *Geothermal Systems*, Advances in Heat Transfer, Vol. 14, Academic Press, New York, 105 p.
Clark, S. P.: 1957, 'Heat Flow at Grass Valley, California', *Trans. Am. Geophys. Union.* **38**, 239–244.
Clauser, C.: 1984, 'A Climatic Correction on Temperature Gradients Using Surface Temperature Series of Various Periods', *Tectonophysics* **103**, 33–46.
Courtillot, V., Ducruix, J., and LeMouel, J. L.: 1973, 'Le prolongement d'un champ du potentiel d'un contour quelconque sur un contour horizontal: une application du la methode de Backus et Gilbert', *Ann. Geophys.* **29**, 361–366.
Domenico, P. A. and Palciauskas, V. V.: 1973, 'Theoretical Analysis of Forced Convective Heat Transfer in Regional Ground-Water Flow', *Geol. Soc. Am. Bull.* **84**, 3803–3814.
Drury, M. J. and Lewis, T. J.: 1983, 'Water Movement within Lac du Bonnet Batholith as Revealed by Detailed Thermal Studies of Three Closely-Spaced Boreholes', *Tectonophysics* **95**, 337–351.
Drury, M. J.: 1984, 'Perturbations to Temperature Gradients by Water Flow in Crystalline Rock Formations', *Tectonophysics* **1984**, 19–32.
Ducruix, J., LeMouel, J. L., and Courtillot, V.: 1974a, 'Continuation of Three Dimensional Potential Fields Measured on an Uneven Surface', *Geophys. Jour. Roy. Astr. Soc.* **38**, 299–314.
Ducruix, J., LeMouel, J. L., and Courtillot, V.: 1974b, 'Une methode simple d'evaluation de la correction topographique dans les problems de flux de chaleur', *Comptes Rendus Acad. Sci. Paris, Series B*, **278**, 841–843.
Elder, J. W.: 1965, 'Physical Processes in Geothermal Areas', in W. H. K., Lee (ed.), *Terrestrial Heat Flow*, Monograph No. 8, A.G.U., Washington, 211–239.
Elder, J. W.: 1967, 'Steady Free Convection in a Porous Medium Heated from Below', *J. Fluid Mech.* **27**, 29–48.

England, P. C.: 1978, 'The Effect of Erosion on Palaeoclimatic and Topographic Corrections to Heat Flow', *Earth Planet. Sci. Lett.* **39**, 427–434.

England, P. C. and Richardson, S. W.: 1980, 'Erosion and the Age Dependence of Continental Heat Flow', *Geophys. J. Roy. Astron. Soc.* **62**, 421–437.

Finckh, P.: 1981, 'Heat Flow Measurements in 17 Perialpine Lakes', *Bull. Geol. Soc. Am.* **92**, 452–514.

Finckh, P.: 1983, 'On the Calibration of Lacustrine Heat-Flow Density Measurements in a Borehole in Lake Zurich, Switzerland', *Zbl. Geol. Palaeont.*, Part 1, 102–117.

Geertsma, J.: 1971, 'Finite-Element Analysis of Shallow Temperature Anomalies', *Geophys. Prospecting* **19**, 662–681.

Geiger, R.: 1965, *The Climate Near the Ground*, Harvard University Press, Cambridge, Mass., 611 p.

Haenel, R.: 1970, 'Eine neue Methode zur Bestimmung der terrestrischen Wärmestromdichte in Binnenseen', *Z. Geophys.* **36**, 725–742.

Haenel, R., (ed.): 1980, *Atlas of Subsurface Temperatures in the European Community*, Hannover, Fed. Repub. Germany (Th. Schäfer), 43 plates.

Haenel, R.: 1983, 'Geothermal Investigations in the Rhenish Massif', in K. Fuchs, K. von Gehlen, H. Mälzer, H. Murawski, and A. Semmel (eds.), *Plateau Uplift*, Springer-Verlag, Berlin, 228–246.

Haji-Sheikh, A. and Sparrow, E. M.: 1967, 'The Solution of Heat Conduction Problems by Probability Methods', *Jour. Heat Transfer, Trans. Am. Soc. Mech. Eng.* **89**, Ser. C., 121–131.

Henderson, R. G. and Cordell, L.: 1971, 'Reduction of Unevenly Spaced Potential Field Data to a Horizontal Plane by Means of Finite Harmonic Series', *Geophys.* **36**, 856–866.

Henry, S. G. and Pollack, H. N.: 1985, 'Heat Flow in the Presence of Topography: Numerical Analysis of Data Ensembles', *Geophysics* **50**, 1335–1341.

Hotchkiss, W. O. and Ingersoll, L. R.: 1934, 'Postglacial Time Calculations from Recent Geothermal Measurements in the Calumet Copper Mines', *J. Geol.* **42**, 113–122.

Huestis, S. P. and Parker, R. L.: 1979, 'Upward and Downward Continuation as Inverse Problems', *Geophys. J. Roy. Astron. Soc.* **57**, 171–188.

Jaeger, J. C.: 1965, 'Application of the Theory of Heat Conduction to Geothermal Measurements', in W. H. K. Lee (ed.), *Terrestrial Heat Flow.*, Geophys. Monographs **8**, Am. Geophys. Union, Washington, D.C., pp. 7–23.

Jaeger, J. C. and Sass, J. H.: 1963, 'Lee's Topographic Correction in Heat Flow and the Geothermal Flux in Tasmania', *Geofis. Pura et Appl.* **54**, 53–63.

Jaswon, M. A. and Symm, G. T.: 1977, *Integral Equation Methods in Potential Theory and Elastostatics*, Academic Press, London, 287 p.

Jeffreys, H.: 1938, 'The Disturbance of the Temperature Gradient in the Earth's Crust by Inequalities of Height', *Monthly Notices Roy. Astron. Soc., Geophys. Suppl.* **4**, 309–312.

Jessop, A. M.: 1971, 'The Distribution of Glacial Perturbation of Heat Flow in Canada', *Can. Journ. Earth Sci.* **8**, 162–166.

Jobert, G.: 1960, 'Perturbation du flux de chaleur dans la croute terrestre due au relief', *Comptes Rendus Acad. Sci., Paris* **250**, 3209–3210.

Kappelmeyer, O. and Haenel, R.: 1974, *Geothermics, with Special Reference to Application*, Gebrüder Borntraeger, Berlin, 238 p.

Lachenbruch, A. H.: 1957a, 'Thermal Effects of the Ocean on Permafrost', *Bull. Geol. Soc. Am.* **68**, 1515–1529.

Lachenbruch, A. H.: 1957b, *Three-Dimensional Heat Conduction in Permafrost Beneath Heated Buildings*, U.S. Geol. Surv. Bull. 1052-B, 51–69.

Lachenbruch, A.H.: 1959, *Periodic Heat Flow in a Stratified Medium with Applications to Permafrost Problems*, U.S. Geol. Surv. Bull. 1083-A, 36 p.

Lachenbruch, A. H.: 1968, 'Rapid Estimation of the Topographic Disturbance to Superficial Thermal Gradients', *Rev. Geophys. and Space Phys.* **6**, 365–400.

Lachenbruch, A. H.: 1969, *The Effect of Two Dimensional Topography on Superficial Thermal Gradients*, U.S. Geol. Surv. Bull. 1203-E, 86 p.

Lachenbruch, A. H.: 1982, 'Permafrost, Heat Flow, and the Geothermal Regime at Prudhoe Bay, Alaska', *J. Geophys. Res.* **87**, B11, 9301–9316.

Langseth, M. G., Keihm, S. J., and Peters, K.: 1976, 'Revised Lunar Heat Flow Values', *Proc. 7th*

Lunar Sci. Conf. 3143–3171.
Lapwood, E. R.: 1948, "Convection of a Fluid in a Porous Medium', *Proc. Camb. Phil. Soc.* **44**, 508–521.
Lee, T. C.: 1977, *Application of Finite Element Analysis to Terrestrial Heat Flow*, Inst. of Geophys. and Planet. Phys., Univ. of California, Riverside, IGPP Rept. No. UCR/GPP-77/15, 64 p.
Lee, T.-C. and Henyey, T. L.: 1974, 'Heat Flow Refraction across Dissimilar Media', *Geophys. J. Roy. Astron. Soc.* **39**, 319–333.
Lees, C. H.: 1910, 'On the Shapes of Isogeotherms under Mountain Ranges in Radioactive Districts', *Proc. Roy. Soc., London, Series A* **83**, 339–346.
Livesley, R. K.: 1983, *Finite Elements: An Introduction for Engineers*, Cambridge University Press, Cambridge, 199 p.
Mansure, A. J. and Reiter, M.: 1979, 'A Vertical Groundwater Movement Correction for Heat Flow', *J. Geophys. Res.* **84**, 3490–3496.
Matsubayashi, O.: 1983, 'Application of Boundary Element Technique to Evaluation of Topographic Effect on Heat Flow Observation', *J. Geothermal Res. Soc., Japan* **5**, 249–257.
Misener, A. D. and Beck, A. E.: 1960, 'The Measurement of Heat Flow over Land', in S. K. Runcorn (ed.), *Methods and Techniques in Geophysics*, Interscience, N.Y., 10–61.
Mitchell, A. R. and Griffiths, D. F.: 1980, *The Finite Difference Method in Partial Differential Equations*, John Wiley and Sons, Chichester, 457 p.
Mundry, E.: 1974, 'Harmonic Temperature Waves in a Horizontally Layered Medium', *J. Geophys.* **40**, 767–773.
Nakamura, S.: 1977, *Computational Methods in Engineering and Science, with Applications to Fluid Dynamics and Nuclear Systems*, John Wiley and Sons, New York, 457 p.
Nielsen, S. B. and Balling, N.: 1985, 'Transient Heat Flow in a Stratified Medium', *Tectonophysics* **121**, 1–10.
Nielsen, S. B.: 1986, 'The Continuous Temperature Log: Principles and Applications', Ph.D. Thesis, University of Western Ontario, London, Canada.
Parker, R. L. and Klitgord, K. D.: 1972, 'Magnetic Upward Continuation from an Uneven Track', *Geophys.* **37**, 662–668.
Roy, R. F., Decker, E. R., Blackwell, D. D., and Birch, F.: 1968, 'Heat Flow in the United States', *J. Geophys. Res.* **72**, 5207–5221.
Sass, J. H., Lachenbruch, A. H. and Jessop, A. M.: 1971a, 'Uniform Heat Flow in a Deep Hole in the Canadian Shield and its Paleoclimatic Implications', *J. Geophys. Res.* **76**, 8586–8596.
Sass, J. H., Lachenbruch, A. H., Munroe, R. J., Green, G. W., and Moses, T. H.: 1971b, 'Heat Flow in the Western United States', *J. Geophys. Res.* **76**, 6376–6413.
Sclater, J. G., Jones, E. J. W., and Miller, S. P.: 1970, 'The Relationship of Heat Flow, Bottom Topography, and Basement Relief in Peake and Freen Deeps, Northeast Atlantic', *Tectonophysics* **10**, 283–300.
Sclater, J. G., Crowe, J., and Anderson, R. N.: 1976, 'On the Reliability of Oceanic Heat Flow Average', *J. Geophys. Res.* **81**, 2997–3006.
Sheriff, R. E.: 1973, *Encyclopedic Dictionary of Exploration Geophysics*, Society of Exploration Geophysicists, Tulsa, 266 p.
Smith, L. and Chapman, D. S.: 1983, 'On the Thermal Effects of Groundwater Flow, 1. Regional Scale Systems', *J. Geophys. Res.* **88**, 593–608.
Smith, M. W.: 1975, 'Microclimatic Influences on Ground Temperatures and Permafrost Distribution, Mackenzie Delta, Northwest Territories', *Can. J. Earth Sci.* **12**, 1421–1438.
Straus, J. M.: 1974, 'Large Amplitude Convection in Porous Media', *J. Fluid Mech.* **64**, 51–63.
Straus, J. M. and Schubert, G.: 1977, 'Thermal Convection in a Porous Medium: Effects of Temperature- and Pressure-Dependent Thermodynamic and Transport Properties', *J. Geophys. Res.* **82**, 325–333.
Tarantola, A. and Valette, B.: 1982a, 'Generalized Non-Linear Inverse Problems Solved Using Least Square Criterion', *Rev. Geophys. Space Phys.* **20**, 191–232.
Tarantola, A. and Valette, B.: 1982b, 'Inverse Problem = Quest for Information', *J. Geophys.* **50**, 159–170.
Turcotte, D. L. and Schubert, G.: 1982, *Geodynamics, Applications of Continuum Physics to Geo-*

logical Problems, John Wiley and Sons, New York, 450 p.

Van Rooy, D.: 1982, *Forward Geothermal Modelling Using the Finite Element Method*, Specialeopgave, Laboratoriet for Geofysik, Aarhus Univ., Aarhus, 88 p.

Vasseur, G., Bernard, Ph., Van de Meulebrouck, J., Kast, Y., and Jolivet, J.: 1983, 'Holocene Paleotemperatures Deduced from Geothermal Measurements', *Palaeogeogr., Palaeoclimat., Palaeoecol.* **43**, 237–259.

Von Herzen, R. P., Finckh, P., and Hsu, K. J.: 1974, 'Heat Flow Measurements in Swiss Lakes', *J. Geophys.* **40**, 141–172.

Wang, K. and Beck, A. E.: 1987, 'Heat Flow Measurement in Lacustrine or Oceanic Sediments Without Recording Bottom Temperature Variations', *J. Geophys. Res.* (In press).

Wang, K., Shen, P. Y., and Beck, A. E.: 1986, 'The Effects of Thermal Properties Structure and Water Bottom Temperature Variation on Temperature Gradients in Lake Sediments', *Can. J. Earth Sci.* **23**, 1257–1264.

Wooding, R. A.: 1957, 'Steady-State Free Thermal Convection of Liquid in a Saturated Permeable Medium', *J. Fluid Mech.* **2**, 273–285.

Woodhouse, J. H. and Birch, F.: 1980, 'Comment on 'Erosion, Uplift, Exponential Heat Source Distribution, and Transient Heat Flux', by T.-C. Lee. *J. Geophys. Res.* **85**, 2692–2693.

Zienkiewicz, O. C.: 1977, *The Finite Element Method*, McGraw-Hill Book Co., New York, 524 pp.

6. OCEANIC HEAT-FLOW DENSITY

EARL E. DAVIS
Pacific Geoscience Centre, Energy, Mines and Resources, Sidney, Canada

6.1. INTRODUCTION

Many marine heat-flow density studies have been completed since the time of pioneering work of Bullard, Revelle, Maxwell, Ewing, and others, and much has been learned about the surficial and deep seated thermal workings of the Earth. Generalized heat-flow density 'mapping' has provided valuable information, but in the future, measurements are likely to be directed more toward solving specific geophysical problems. For these purposes, frequently an instrument must be thoughtfully designed to suit a particular measurement environment or to provide the accuracy necessary to resolve a particular problem. Thus, before any general guidelines for determining marine heat-flow density are suggested or examples of instrumentation given, it is worthwhile to consider briefly a few examples of contemporary problems in marine heat-flow density.

6.1.1. Ridgecrest Problems

Some of the most actively persued problems in marine heat-flow density today concern hydrothermal processes at mid-ocean ridges. Questions which specific observational heat-flow density experiments may yet help to solve concern the depth to which hydrothermal circulation penetrates in the oceanic crust, the nature of 'passive' geothermal gradient-driven circulation, the nature of 'active' circulation which is associated with the cooling of ridge crest magma chambers or hot rock units, the importance of tectonic faulting as an influence on crustal permeability, and the effectiveness of sediment in providing an impermeable barrier to the passage of hydrothermal fluids. The instrumentation requirements for these problems arise primarily from the need for high density station coverage, since major heat-flow density variations occur on a small spatial scale. Also good environmental control is needed at each measurement position, since the relationship between heat-flow density and sea floor topology is important. The measurement accuracy required is generally not high. However, if fluid transport through the sediments occurs, high measurement accuracy and good resolution of the variation of heat-flow density with depth is needed.

6.1.2. Continental Margin Problems

Another group of problems is associated with active and passive continental margins. At active margins there are questions about the way in which water is expelled from rapidly consolidating sediments in accretionary prisms, about migration of pore water up the main subduction thrust boundary, and about the

R. Haenel, L. Rybach, and L. Stegena (eds.)
Handbook of Terrestrial Heat-Flow Density Determination, 223—260.

possibility of hydrothermal circulation in oceanic crust being reinitiated by tectonic faulting on the seaward wall of trenches. The regional variation of heat-flow density from the trench to the arc is also poorly known. This is required to constrain models of the thermal history of sediments in the forearc region. At passive margins, which are generally produced by rifting, heat-flow density determinations are important for improving models of the subsidence and thermal history of sedimentary basins and rifted margins. Knowledge of the heat-flow density at either type of continental margin is of particular importance in constraining models for the maturation of hydrocarbons within the margin basins.

A major requirement for marine heat-flow density equipment employed at continental margins is ruggedness. Surficial sediments are usually coarse grained terrigenous material and thus difficult to penetrate. Where there are no recent deposits, the sediments are commonly indurated. Even fine grained sediments of young age can be semi-consolidated and difficult if not impossible to penetrate with a gravity driven probe. Another requirement for measurements in shallower water depths on continental margins is that the probe penetration be sufficient to avoid thermal disturbances of bottom water temperature variations. The depth of penetration necessary requires careful assessment in any experiment, and this will be treated in a following section. If transients are significant, both deep penetration and good resolution of the variation of the temperature with depth in the sediments are required. The ability to obtain many closely spaced determinations may also be important.

6.1.3. Old Ocean Basin Problems

A third category of geophysical problems about which more marine heat-flow density determinations can provide important information is that concerning the aging of oceanic lithosphere. Existing observations of heat-flow density and topography show that for ages ranging from zero up to 80–100 Ma, oceanic lithosphere cools as a thickening boundary layer; the exact relationships between heat-flow density, topography, and lithospheric age yield parameter kernels which contain the thermal diffusivity, heat capacity, expansion coefficient, and initial temperature. From local variations in the depth-age relationship there are some indications that local differences in the thermal parameters of the lithosphere exist. These variations could be better understood if there were adequate heat-flow density data to show whether similar variations existed in the heat-flow density age relationship. At ages greater than 100 Ma, existing determinations of heat-flow density in old oceanic basins indicate that the heat-flow density through old oceanic lithosphere is too high to be explained by simple boundary layer or plate cooling models. Additional determinations in other old areas should help to reveal the way in which old lithosphere *ceases* to cool, and in some cases is reheated by local hot spots or regional convective heat transfer in the asthenosphere below. For this group of old ocean basin problems, instrumentation must permit the resolution of small magnitude differences in regional heat-flow density, so very accurate (about 5%) determinations are necessary. This accuracy may be achieved in low gradient areas only with instrument design that minimizes physical and thus thermal disturbances to the sediment, adequate depth of penetration, sufficient numbers of determinations, and discriminatory data reduction.

Other problems which may be addressed in future heat-flow density studies include the advection of heat through sediments by pore-water flow driven by supercritical thermal gradients, tectonic dewatering and normal consolidation, chemical and biological heat production in sediments, etc. In each case, close attention must be given to the specific requirements for spacing of measurements, depth of probe penetration, sensor spacing, temperature resolution etc.

6.2. CONTEMPORARY INSTRUMENTATION

Significant improvements have been made in the last few years on the traditional corer-outrigger (e.g. Gerard *et al.*, 1962) and Bullard-probe (e.g. Bullard, 1954) heat flow instruments. In both cases, high resolution digital recording is now commonly used, and real-time data monitoring is often accomplished via digital or analog acoustic telemetry of data to the ship. An improvement on the specific purpose heat-flow density probe was made which allowed multiple penetrations to be made during a single lowering: outrigger thermistors were mounted on a probe which was strong enough to repeatedly resist bending during pullouts (von Herzen and Anderson, 1972). Such instruments only measured the geothermal gradient, however. Further improvements have been incorporated in instruments which follow the design of Lister (described in Hyndman *et al.*, 1978), and they include full resolution digital telemetry, streamlined and rugged violin-bow probe construction, and averaged interval in-situ conductivity using an energy efficient metered or calibrated heat pulse technique (Lister, 1979; Hyndman *et al.*, 1979).

The primary advantage of corer-outrigger instruments lies in their capability to penetrate deeply into sediments. Ten to twelve meters is common in pelagic sediments with corers weighing 800–1000 kg. Unfortunately, the technique has a number of disadvantages: Outriggers and their connecting wires are fragile. The instrument is difficult to set up and deploy and only one penetration can be made on each lowering. Seldom can more than three or four measurements be made in one day. It is also difficult to obtain accurate correspondence between the depth of penetration and the distance along the recovered core used for thermal conductivity; hence the accuracy with which the conductivity can be estimated for a given outrigger gradient interval is severely limited. Significant sediment distortions occur in both the gravity and piston coring process. Gravity cores are shortened relative to the true sediment section penetrated, with maximum recovery limited to about 5–6 m; deeper sediments are usually excluded. Piston corers recover most of the sediment section penetrated, but the core is often severely distorted by unwanted piston motion caused by ship heave and by preferential drawing in of mechanically less competent layers.

Of the various instrument designs, the violin-bow, or Lister-probe can be adapted most easily to the requirements of a wide range of specific heat-flow density problems. It is streamlined and rugged, and it provides a high accuracy complete heat-flow density determination with conductivity measured in situ. Corer-outrigger instruments do allow determination of heat-flow density at locations where bottom water temperature variations prohibit accurate gradient measurements to be made with shorter probes, but they are used best only in a few very specialized situations. Thus, subsequent discussion will be directed primarily toward the measurement of geothermal gradient and thermal conductivity with a Lister-type probe. Application of the discussion to outrigger gradient measurement will be apparent, and some discussion of core sample conductivity measurement will be also included.

Another technique for determining marine heat-flow density involves the use of Deep Sea Drilling Project boreholes. Bottom-hole or insitu sediment temperature measurements are made at intervals during drilling, and thermal conductivity measurements are made on recovered core samples. Instrumentation, techniques used for data reduction, and measurements made to date have been reviewed by Erickson *et al.* (1975), Hyndman *et al.* (1978), Burch and Langseth (1981), Hyndman *et al.* (1983), and others (see references in Hyndman *et al.*, 1983). These measurements have provided unique and valuable information about temperatures deep in marine sediment sections and in the oceanic crust. No discussion of this technique is included in this chapter, however. The reader should refer to the reviews cited above for information about thermal measurements in Deep Sea Drilling Program or other marine boreholes.

Lister Violin-Bow Probes

Probes following the design of Lister have been constructed in a variety of lengths, ranging up to 11 m, and with a variety of mechanical and electronic configurations. Descriptions of several have been given by Hyndman *et al.* (1978), Hyndman *et al.* (1979), and Davis *et al.* (1984); a recent design of Lister (see Davis *et al.*, 1984) will be exemplified here. This instrument, shown in Figure 6.1A, includes a 1 cm diameter, 3.6 m long sensor tube supported nominally 5 cm away from a tapered (5–10 cm) solid strength member. This assembly is gravity-driven into the sea-floor sediments with a set of steel weights totalling 1000 kg. There are seven temperature sensors in the mineral oil-filled probe (Figure 6.1B); four individual thermistors measure temperatures at equally spaced points over an interval totalling 3.42 m, and interspersed groups each comprising sixteen thermistors connected in series-parallel configuration measure the harmonic mean value of temperature between each pair of point sensors. With this sensor string configuration, complete heat-flow density values (gradients with interval mean conductivities) can be determined over three depth intervals. Other instruments are similar. Sensor tubes have ranged from 0.6 to 1.0 cm in diameter, and 2.0 to 11 m in length, and some have used up to 16 individual temperature sensors.

The thermal gradient is measured during the first ten minutes after penetration, with the equilibrium temperatures estimated by extrapolation. The thermal conductivity of the sediment is measured using a heat pulse of about 20 s duration, applied to the full length of the sensor string, and commencing about ten minutes penetration. The technique of using the temperature-time decay of a metered or calibrated heat pulse is described in detail by Lister (1979) and Hyndman *et al.* (1979) and a summary of the reduction scheme is given here. There are several advantages of the pulse-decay technique; the method is more rapid than the traditionally used continuously heated needle-probe technique, it requires less

Fig. 6.1.A Photo of violin-bow multi-penetration instrument described in text. Batteries, electronics, and acoustic transducer are carried in a slot in the weights.

power for the same temperature sensitivity, and it is less sensitive to thermal contact resistance.

In the Lister instrument, temperature data is acoustically transmitted to the ship digitally, with readings taken once each second. The digital telemetry link uses continuous-wave frequency-modulated transmission (11.4–11.6 kHz) that carries the sequential bit stream by means of frequency-shift keying, and the encoding is optimized to allow a simple self-syncronizing receiver to pick up the data rapidly after any signal dropout. Other instruments have used lower accuracy time-interval pingers for real-time telemetry, with accurate digital recording in the instrument. The transmission of accurate digital data is in several ways more convenient than an imprecise transmission coupled with data recording on the sea-floor: the instrument needs to be opened only for repair, and the number of multipenetration heat-flow density measurements is limited only by battery size and by the stamina of shipboard operators. In the newest instruments, this advantage has been gained by the addition of solid state data recording within the instrument that can be interrogated from outside the pressure case. This scheme provides a simple alternate or back-up to high resolution telemetry which is advantageous in the event of poor shipboard acoustics, bad weather etc.

With temperature equivalent readings transmitted once each second in the Lister instrument, a full data cycle (seven sensors and reference resistor) takes

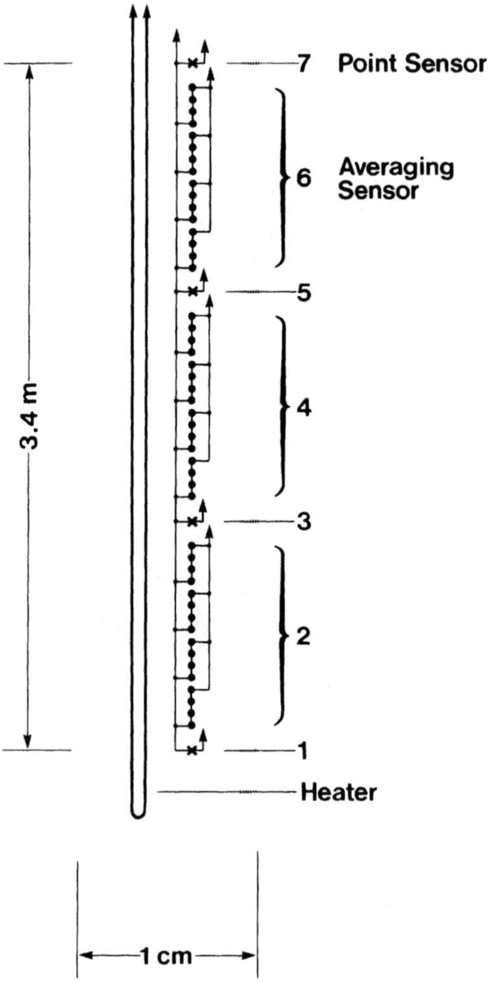

Fig. 6.1B. Schematic diagram of the thermistor distribution in the sensor tube.

eight seconds. Readings are displayed with full 0.001 K resolution graphically and numerically on a 50 cm shipboard graphics recorder. From this display, real-time monitoring of the probe's performance can be accomplished, and data can be selected for heat-flow density reduction.

6.3. SHIPBOARD OPERATION

The normal mode of operating at sea is to make measurements, pogo-probe style, about once an hour while drifting or manouvering at a speed over the bottom of about 0.25 m s^{-1} (0.5 knots). This rate allows time for complete heat-flow density determinations (gradient and conductivity), with time between penetrations for the heavy instrument to swing into an equilibrium position nearly beneath the ship. A pinger mounted some 75 m from the end of the wire allows monitoring of the heat-flow density instrument altitude between penetrations and during lowerings, and maintenance of the optimum amount of slack wire while the instrument is in the bottom. In general, the spacing of penetrations is controlled by the water depth, ship speed, wire diameter, instrument weight, and time required for measurement. A heavy instrument will swing back to an equilibrium towing position after pull-out more quickly than a light one, allowing for more frequent penetrations. Also the equilibrium position will be more directly beneath the ship, ensuring more reliable acoustic signal reception. More frequent penetrations (two to three per hour for the conditions given earlier) can be achieved by reducing the in-bottom time and measuring only gradient at most penetrations. The resulting penetration spacing can thus vary from a few hundred meters to a kilometer or so. Typical lowerings with the Lister instrument shown have included twenty to twenty-five penetrations; during one exceptionally long lowering (2.5 days) eighty-five penetrations were completed.

Navigation is a critical element of multi-penetration heat-flow density surveys. Normal satellite navigation gives excellent accuracy, but fixes are infrequent. Where available, Loran-C corrected by satellite fixes has been employed with most success: the combination provides nearly continuous information with an accuracy of about ±200 m. The heat-flow density instrument can be navigated with respect to the ship to adequate accuracy for most surveys, using the wire direction and by comparing the echo-soundings of the ship with the pinger soundings of the instrument; the accuracy obtainable depends primarily on the nature of the bathymetric relief. Where there is little relief that can be correlated, the instrument position must be estimated using the information given by the wire angle at the surface, the amount of wire out, and the water depth. Where higher accuracy is desired, the instrument can be navigated with the use of sea floor transponders (e.g. Green *et al.*, 1981), or an acoustic system that gives range and bearing from the ship.

6.4. DATA REDUCTION

The various problems associated with reducing marine heat-flow density data have been discussed by a number of authors (e.g. Bullard, 1963; Langseth, 1965; Lister, 1970; Hyndman *et al.*, 1979; Haenel, 1979; Davis *et al.*, 1984). To illustrate the task of marine heat-flow density data reduction, and to point out some of the problems in review, an example from a penetration of the probe described above will be discussed here. A temperature-time record of a typical penetration is shown in Figure 6.2; for simplicity, data from only two of the seven thermistors are shown. Seen clearly on the record are: (1) the time period prior to final descent and penetration, during which 'zero-gradient' reference temperatures are recorded, (2) the thermal decay following the impact of the probe with the sediments, and (3) the thermal decay following the metered heat pulse. A disturbance during the heat pulse thermal decay can be seen late in the record which probably reflects a physical disturbance to the probe caused by a premature pull of the wire. Any physical disturbances such as this, as well as digital transmission

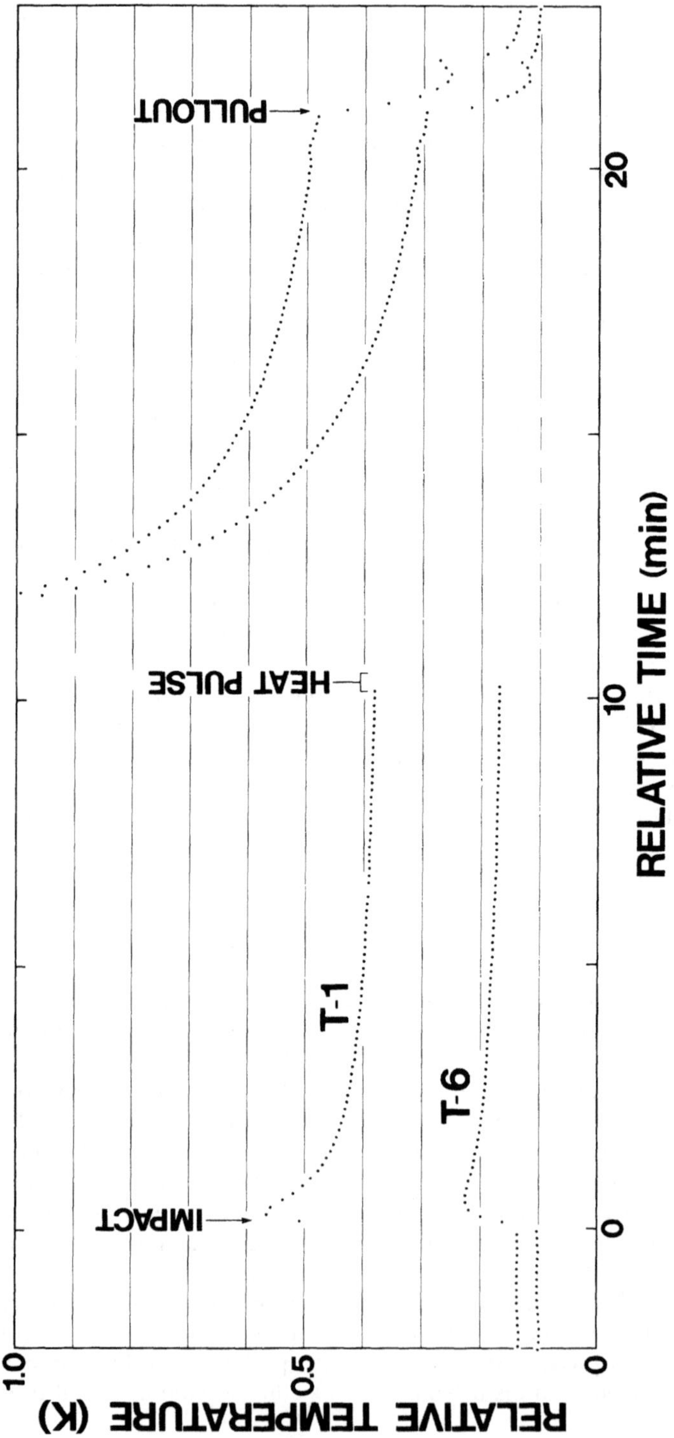

Fig. 6.2. Temperature record from a typical penetration. For simplicity, data from only two thermistors are shown; T-1 is the deepest, a point sensor, and T-6 is a shallow averaging sensor.

or recording disturbances can be seen easily on the high resolution graphical output and excluded by manual selection of data points for later machine reduction. This simple graphical editing step is extremely useful for any recording and reduction scheme since the records often exhibit some disturbance.

6.4.1. Penetration Heating Extrapolation

The gradient is calculated from the recorded data in the following way: the impact decay at each temperature sensor is least squares fitted to the cylindrical decay function $F(\alpha, \tau)$ (see Bullard, 1954; detailed tables in Huppert and Sclater, 1968; Lister, 1979; and Table 6.1) with the origin at the time when impact is first observed. In this way, the data is extrapolated to infinite time in order to establish the temperature rise at a given thermistor above the 'zero gradient' reading (Figure 6.3A) (taken when the probe is hanging stationary in the isothermal near-bottom water prior to penetration), and to the time interval of the heat pulse decay in order to remove the impact frictional heating residual from that part of the temperature record. Initially, an estimate of the thermal time constant $\tau = at/r^2$ (which is the Fourier number Fo, see Table 2.9) is made using the probe radius r, a guess of the conductivity λ, and relationships between λ and diffusivity a and specific heat capacity c_p given by Hyndman *et al.* (1979). A value of 2 is used for α (twice the ratio of the specific heat capacity of the sediment to that of the probe material), estimated from the average probe and sediment properties.

6.4.2. In-Situ Thermal Conductivity

Next, a conductivity is calculated for each of several selected heat pulse decay points (with the impact frictional heating residuals removed) using their temperature rises above the extrapolated infinite time temperature, and a formula derived from the work of Jaeger (1956) (see also Lister, 1979; Hyndman *et al.*, 1979, and following discussion):

$$\lambda = 2\alpha\tau\, F(\alpha, \tau) \cdot \frac{Q}{4\pi\Delta T t}. \tag{6.1}$$

For this probe the effective origin time $t = 0$ has been empirically established as being about 17 s after the center of the heat pulse (see later discussion). The resulting values of λ are then averaged after excluding any points greater than 1.5 times the standard deviation away from the mean. With this new estimate of λ, a new estimate of a is made, and the entire procedure is once repeated. Again, a value of $\alpha = 2$ is used; for typical probe constructions and a reasonably wide range of sediment conductivities (0.75–1.25 W m^{-1} K^{-1}), this assumption introduces less than 1 % error to the conductivity calculated (Hyndman *et al.*, 1979; Lister, 1979). A number of checks have been made with a third iteration, but differences in the results of the second and third calculations were always less than 2 %.

6.4.3. Heat-Flow Density

Figure 6.3 shows the results of a two-iteration calculation. Shown are temperat-

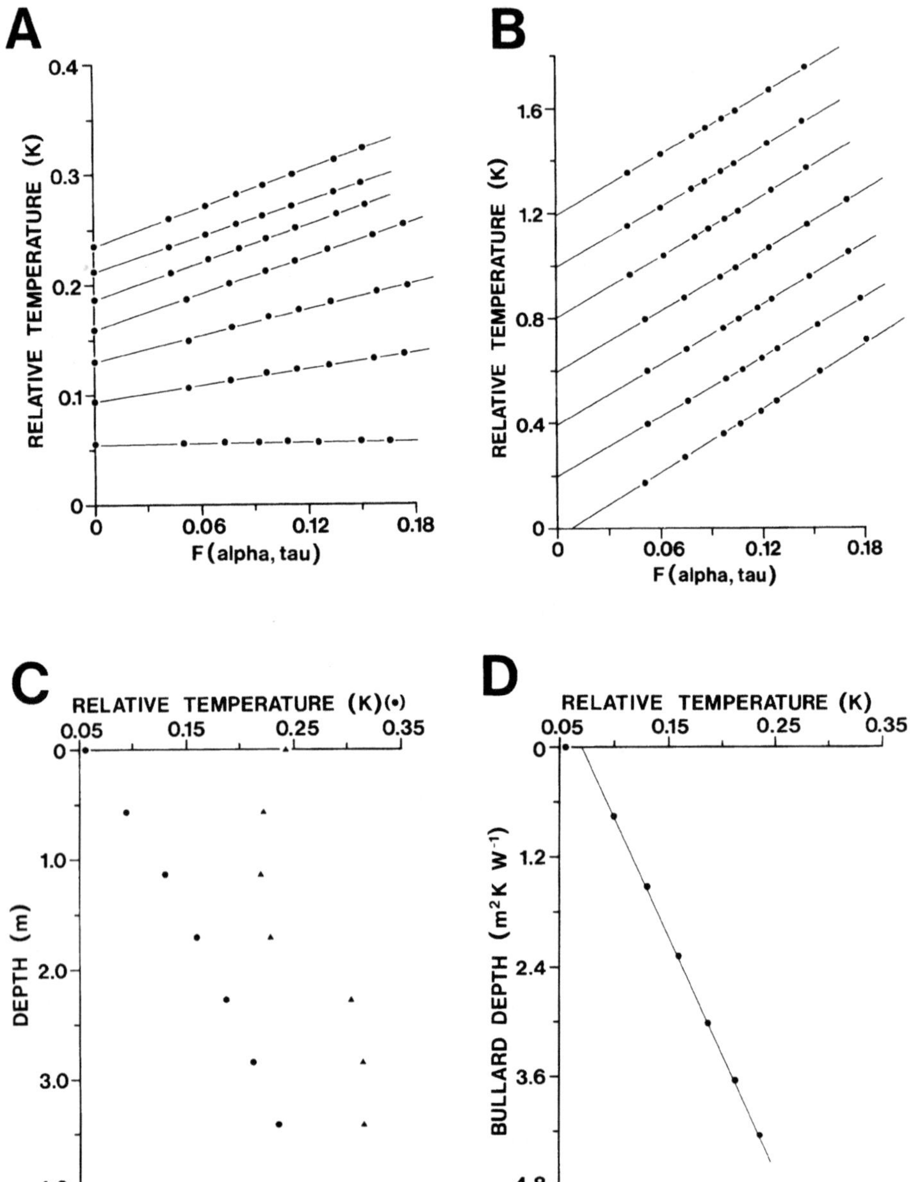

Fig. 6.3. Reduced heat-flow density data: thermal decays after (A) frictional heating and (B) heat pulse, with temperatures plotted against $F(\alpha, \tau)$; (C) temperature rise and conductivity vs depth; (D) temperature rise vs. total vertical thermal resistance (Bullard depth). Temperature of the upper sensor is probably erroneously low as a result of it being surrounded by sediments which have been displaced downward by super-penetration of the large weight stand.

ures during the decay of the frictional heating, relative to the near bottom water temperature, the temperatures during the heat pulse decay relative to the extrapolated equilibrium temperatures; the temperature relative to near bottom water as a function of depth in the sediment, the conductivity as a function of depth in the sediment, and the temperature as a function of the total thermal resistance (Bullard depth). The interval heat-flow density values are simply the product of the interval gradients and the corresponding average conductivity.

6.4.4. Graphical Data Reduction

Data can also be reduced graphically; this approach is sufficiently accurate, and requires no shipboard computer, but is somewhat more tedious. The infinite-time intercept of an $F(\alpha, \tau)$ temperature-time decay is insensitive to errors in α or τ (Lister, 1970), and for extrapolating the thermal decay of impact, single values of α and τ can be assumed with no significant loss in accuracy. A value of $\alpha = 2$ is appropriate for typical probe constructions and most sediment types, and values of τ can be determined assuming a value of $a = 3 \times 10^{-6}$ m^2 s^{-1}.

Graphical reduction of the heat pulse thermal decay can be accomplished in a similar way, although variations in diffusivity cannot be ignored. The relationship between diffusivity and conductivity is accounted for in the iterative way of calculating a summarized above and discussed at length by Lister (1979) and Hyndman *et al.* (1979); graphically it can be accounted for by plotting temperature data on a graph or nomogram with curved time grid lines (see Hyndman *et al.*, 1979). An example of this graphical method is given in Figure 6.4, with the same data as that used in Figures 6.2 and 6.3.

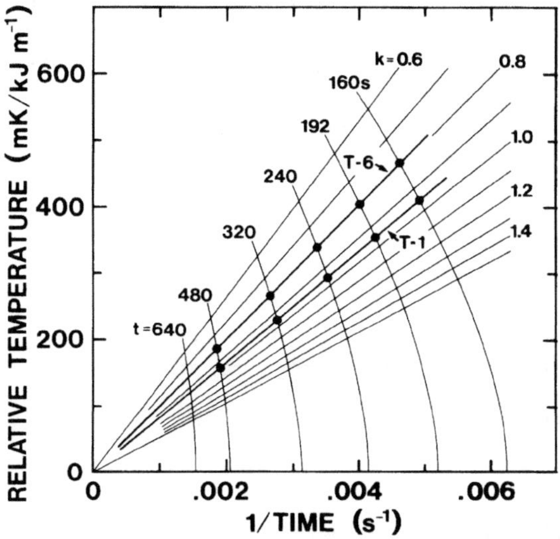

Fig. 6.4. Nomograph for determining thermal conductivity from a heat pulse decay of a 1 cm diameter probe with $\alpha = 2$. Data from Figure 6.2 are plotted as examples. Nomographs for probes of other dimensions can be generated using the mathematical description given in the text and Table 6.1.

6.4.5. Long Time Asymptotic Relations

In the case of the cooling of a smaller diameter probe, such as an outrigger probe measuring sediment temperature or a needle probe measuring sediment thermal conductivity, the thermal decay is often assumed to follow a simple function which is linear with $1/t$, or in the case of continuous heating, with $\ln(t)$. This is appropriate at sufficiently large values of t; in order to evaluate what 'early' and 'late' times are and in order to easily fabricate a graphical or computational reduction scheme, it is worthwhile to consider the nature of the early and late time behavior of the temperature decay following a heat pulse, as well as of the temperature rise during continuous heating.

The observed decay of temperature $T(t)$ in an infinitely conductive probe of initial temperature $T(0)$ is defined by the function cited above:

$$T(t)/T(0) = F(\alpha, \tau). \tag{6.2}$$

For values of α close to 2, the asymptotic behavior of the decay T_a is approached most rapidly, and is described by:

$$T_a/T(0) = \frac{1}{2\alpha\tau} = \frac{Q/T(0)}{4\pi a\tau} \tag{6.3}$$

(see Blackwell, 1954; Hyndman *et al.*, 1979). At a sufficiently long time, the conductivity can be derived from a heat pulse thermal decay using this simple relationship. At earlier times, temperatures can be described by considering the factor $(2\alpha\tau F(\alpha, \tau))^{-1}$ as a 'correction factor', $C(\alpha, \tau)$, so that:

$$T_a = C(\alpha, \tau)\, T(t) \tag{6.4}$$

(see Hyndman *et al.*, 1979; Lister, 1979), and conductivity can be determined using the above asymptotic relationship. The value of $C(\alpha, \tau)$ with $\alpha = 2$, is shown in Figure 6.5 as a function of τ. At a time equivalent to $\tau = 10$, the

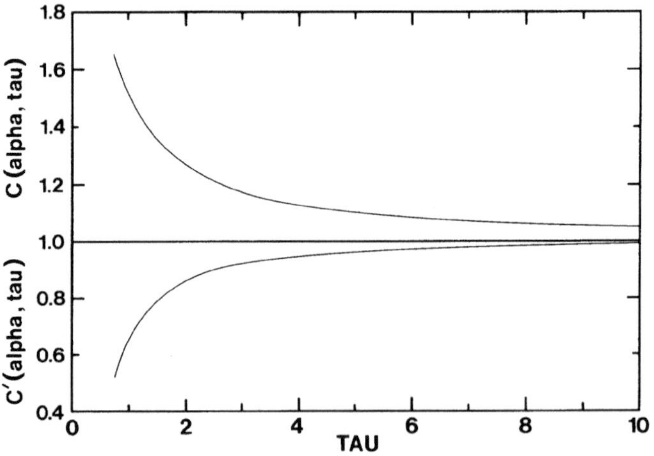

Fig. 6.5. 'Correction factors' $C(\alpha, \tau)$ and $C'(\alpha, \tau)$ vs τ for $\alpha = 2$.

correction becomes small (<5 %), and the asymptotic temperature-time relationship can be used with the observed temperatures to calculate conductivity or to extrapolate frictional decays. At earlier the times, $C(\alpha, \tau)$ correction must be applied iteratively as discussed earlier; at times less than $\tau = 1$ the correction factor exceeds 1.5, and the temperature decay becomes more sensitive to diffusivity, as well as to the probe properties, and reliable values of conductivity cannot be obtained.

A similar 'correction factor' can be used in the description of the temperature rise during the continuous heating of a probe. In that case, the temperature observed at time t is given by:

$$T(t) = (\bar{Q}/\lambda)G(\alpha, \tau), \tag{6.5}$$

where $\bar{Q}$ is heat supplied per unit length per unit time, and $G(\alpha, \tau)$ is a function derived by Jaeger (1956). Tables of values are given by Jaeger (1956) and Sass in Beck (1965) and one is included for convenience here (Table 6.2). As in the case of $F(\alpha, \tau)$, an approximation can be made for $G(\alpha, \tau)$ at long times, and the asymptotic solution for temperature is found to be (see also Equation (4.16)):

$$T_a = \frac{\bar{Q}}{4\pi\lambda}\ln(t) + S. \tag{6.6}$$

Thus at sufficiently long times, the conductivity can be derived from the rate of temperature rise during steady heating using again a simple relationship, and at short times, temperatures can again be described by the asymptotic solution after applying another correction factor:

$$T_a = C'(\alpha, \tau)T(t), \tag{6.7}$$

where

$$C'(\alpha, \tau) = \frac{\ln(2.236\tau)}{4\pi G(\alpha, \tau)}$$

(see Hyndman *et al.*, 1979). Values of $C'(\alpha, \tau)$ are also plotted for various times in Figure 6.5, and it can be seen that the simple asymptotic relationship between temperature and time can be used for calculating conductivity at times greater than about $\tau = 10$. At shorter times, graphical or iterative computational techniques must include corrections for non-asymptotic behavior.

6.5. SOURCES OF ERRORS

Errors in heat-flow density determinations can be considered as falling into two separate categories: there are errors associated with making the measurements of temperature gradient and conductivity, and there are natural errors sometimes inherent in the heat-flow density even when measurements are accurately made. These latter errors are difficult to discuss in any general way; they will not be treated here as 'errors', since in many cases they contain useful information. For example, where sediment permeability is sufficiently high, pore fluids may flow through the sediments and non-uniform heat-flow density depth profiles result. The measurements are unreliable for heat-flow density determinations, although they may provide valuable constraints on the nature and rate of the fluid flow itself (Anderson *et al.*, 1979; Williams *et al.*, 1979, Becker and von Herzen, 1983). Hydrothermal circulation in the crustal rocks beneath the sediments also make individual measurements unreliable for the determination of regional conductive heat-flow density, but they may provide information about the nature of the circulation itself (Lister, 1972; Williams *et al.*, 1974; Davis *et al.*, 1980; Green *et al.*, 1981).

(1) ENVIRONMENTAL SOURCES OF ERROR: SEDIMENTATION

Of a more troublesome nature are those environmental effects whose causes are generally geothermally less interesting; disturbances due to rapid sedimentation, topographic refraction, and bottom water temperature fluctuations are examples. Sedimentation rates in the deep ocean are normally too low to have a significant effect on surface heat-flow density, although this is not always the case. Extreme examples occur in the northeast Pacific where a combination of local tectonic subsidence and high rates of Pleistocene turbidite supply have produced sediment accumulation rates of up to 5 mm a^{-1} and a large reduction in measured heat-flow density (Davis and Lister, 1977; Davis and Riddihough, 1982). The thermal effect of sedimentation has been estimated for the simple single-conductivity conductive situation by von Herzen and Uyeda (1963) and for a convective isothermal bottom boundary case by Davis and Lister (1977). Naturally, the younger the age, the higher the sedimentation rate must be to produce a given thermal effect at the surface. For durations less than the age of most ocean basins, sedimentation rates less than about 0.1 mm a^{-1} have no significant thermal effect.

(2) TOPOGRAPHY AND THERMAL REFRACTION

Topography in most deep ocean environments is relatively subdued and its thermal effect minor. Refraction effects from sediment-basement thermal conductivity contrasts are usually minor also. Where these effects are significant, corrections can be applied using an analytical (e.g. Jeffreys, 1938; Lachenbruch, 1968) or numerical technique (e.g. Sclater and Miller, 1969; Lee and Henyey, 1974). But for few exceptions, however, the variations in oceanic heat-flow density cannot be accounted for by the conductive effects of topography or thermal refraction (e.g. von Herzen and Uyeda, 1963; Sclater *et al.*, 1970; Lister, 1972). Often the relationship between the local variability of heat-flow density and the topography is opposite to that expected from conductive refraction, and this is best explained by the redistribution of heat by hydrothermal circulation (Lister, 1972). Circulation appears to occur in crust of a wide range in age (Sclater *et al.*, 1976; Anderson *et al.*, 1979; Davis *et al.*, 1980) and hence, in most cases it is not appropriate to apply topographic refraction corrections to marine heat-flow density determinations.

(3) BOTTOM WATER TEMPERATURE VARIATIONS

In most deep ocean areas, bottom water temperature variations are small enough to cause no significant disturbance to the geothermal gradient at sub-bottom depths of even less than a meter. Exceptions do occur, however, and it is useful to know roughly what magnitude of variations may occur at a given site in order to plan for adequate penetration. The worst case disturbance can be estimated by considering the maximum effect of a step change in temperature ΔT at a depth z:

$$(\Delta T/z)(2/\pi e)^{1/2} \tag{6.8}$$

(see Carslaw and Jaeger, 1959). In general, however, gains in accuracy should be made more effectively than just linearly with increasing probe length. For example, the disturbance caused by the fluctuation of bottom water temperature will be attenuated exponentially with depth, for a given period.

(4) INSTRUMENTATION ERRORS

The errors from instrumentation should not be significant using an instrument like the one described above, provided the measurement is achieved without disturbance. Relative temperatures can be determined easily to a precision of 1 mK over a wide range of temperatures. Thus gradients as low as 50 mK m^{-1}, can be determined to within 5% over depth intervals in the sediment of only 0.4 m; conductivities can be determined to high accuracy as long as the heat pulse power is sufficient to provide an adequately large temperature change during the measurement interval (i.e. at least 0.1 K). In the example cited in Figure 6.3, a temperature rise of about 0.5 K still remains at the beginning of the useful portion of the decay, at $\tau = 2$. Heater power regulation or monitoring should also be easily accomplished with adequate accuracy, although complete probe calibration in a material of known conductivity is a wise precaution.

(5) PENETRATION HEATING EXTRAPOLATION ERRORS

A significant source of error commonly arises from the inability to extrapolate perfectly the frictional heating decay. Ideally, the decay should follow the $F(\alpha, \tau)$ function described by Bullard (1954), but in practice deviations occur from non-ideal generation of the frictional heat. Because of the limited entry speed of probes simply lowered into the bottom (1–2 m s^{-1}), the length of a given probe (a few meters), and typical data read intervals of a few seconds, the time origin chosen for the frictional heating may be resolved to only several seconds. Furthermore, the nature of penetration is frequently less than ideal. The instrument may settle in soft sediments after initial penetration, and penetration in soft sediments may not be purely axial when the drift rate (0.25 m s^{-1}) is significant in comparison with the penetration velocity (1–2 m s^{-1}). Thus there are differences between the mathematical description (cylindrical thermal decay of a perfectly conductive cylinder of some heat capacity into initially isothermal sediments) and reality (a probe of uniform initial temperature entering warmer sediment, and generating frictional heat at the probe-sediment interface as well as within a probably radially asymmetric region of deformed sediments).

The effects of an origin time error are examined in Figure 6.6 where a particularly long and undisturbed decay is shown with three different origin times. A linear $F(\alpha, \tau)$ decay can be seen over the 10 τ length of record with the origin

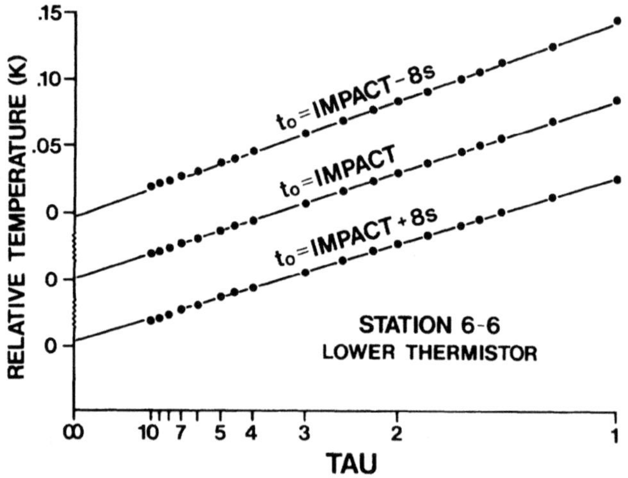

Fig. 6.6. Relative temperature vs $F(\alpha, \tau)$ for an exceptionally long frictional heating decay, showing the effects of a ±8 s timing error of the extrapolated equilibrium value.

chosen at the time of initial disturbance. Plots are also shown for origin times shifted by ±8 s, and although curvature can be seen over the total length of the record, there is negligible curvature in the time interval normally used for data reduction (1 to 5 τ). Thus there is little sensitivity to origin time errors in the form of the temperature decay over the normally used time interval. If origin time errors of this magnitude do exist, Figure 6.6 suggests that the resulting error in extrapolated temperature may be as much as ±5 mK. Clearly, the longer the decay and the smaller the frictional rise, the smaller will be the error in extrapolated temperature. The example has an unusually large frictional temperature rise, and the errors estimated are larger than normal.

Errors in the frictional decay extrapolation also affect the conductivity determination, but only to a small degree. In using as reference level the infinite time extrapolation of the heat pulse delay, only the gradient of the frictional decay is removed as a residual from the heat pulse decay. Using the data in Figure 6.6 as an example, where the extrapolated temperature was found to be in error by ±5 mK, the difference between the temperature drops across the time span equivalent to a heat pulse decay evaluation is only about 1 mK. This error would introduce a negligible error into the conductivity value, since the temperature drop during the portion of heat pulse decay used is about 0.2 K (see Figure 6.3). Referencing the heat pulse decay temperature points to their own extrapolated equilibrium temperature may be of value when obviously disturbed frictional decays are encountered.

(6) HEAT PULSE ORIGIN TIME DECAY

Another potential source of error inherent in the oceanic probe measurement technique arises from the assumptions made when describing the frictional heating and heat pulse thermal decays in a given reduction scheme. A number of these errors has been assessed previously; only a summary will be given here. The temperature decay described by the equations in the previous section is for a perfect cylindrical conductor (thus always isothermal) in perfect contact with an initially isothermal surrounding solid. Heat probes fail to fit this description for a number of reasons: the heat pulse (frictional or electrical) supplied to create the decay is of finite duration, the probe is of finite conductivity, and there may be some contact resistance between the probe and the sediments. The error involved in using the center of a 0.4 τ duration heat pulse as the origin time was shown by Lister (1979) to produce a minor (<1 %) error in the estimate of conductivity at times as early as $\tau = 1$, so this would not appear to be a serious problem. The probe being less than perfectly conducting does appear to be a problem, however; this has been encountered by Lister (1979), Hyndman *et al.* (1979), Davis *et al.* (1984), and Louden (pers. comm. 1983), and dealt with most successfully by allowing for a constant time delay offset in the decay with respect to the idealized origin time. Although this treatment is somewhat ad hoc, it has been justified by the results of a numerical model of the situation and by the fact that only by applying a constant time delay to the decay observations do the data from calibration tests in known conductivity material give the correct value of conductivity (Hyndman *et al.*, 1979; Mojesky, 1981). Values for delays have been established for each thermistor of several probes by finding the minimum standard deviation of conductivities calculated from data at various times along heat pulse decays (Figure 6.7) (Davis *et al.*, 1984), and by determining the best fit to linear $F(\alpha, \tau)$ decays (Hyndman *et al.*, 1979). This is done individually for each thermistor at each penetration for both the penetration and heat pulse decays in a reduction scheme presented by Villinger and Davis (1987a, 1987b).

Resulting values vary from around 16 s for thermistors in 1.0 cm diameter probes designed specifically to provide minimum contact resistance between heater wire and probe to 26 s for thermistors in 0.8 cm diameter probes having a random distribution of thermistors, lead wires, heater wires, and insulation. Significant variations in the optimal time delay occur, however, both from penetration to

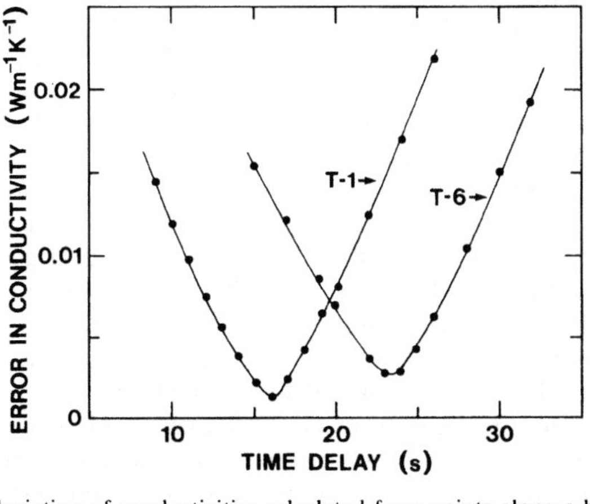

Fig. 6.7. Standard deviation of conductivities calculated from points along a heat-pulse decay, as a function of assumed time delay. Plots for two thermistors are shown.

penetration for a given thermistor and systematically with depth for a given penetration. The source of this variability is not fully understood; it may be in part due to incomplete mineral oil fill within the probe, to variable radial geometry within the probe, and to variable contact resistance between the probe and the surrounding undisturbed sediments. The decay of temperature after the impact frictional heating (concentrated at the sediment-probe interface) follows an $F(\alpha, \tau)$ function surprisingly well with an origin time equal to the time of impact, however, and this would suggest that the time delay is wholly a result of internal contact resistances that are associated primarily with the slow transfer of heat from the interior of the probe to the probe-sediment interface, and that contact resistance between the probe and the sediments is probably unimportant. Since errors in the time delay can introduce significant errors to the conductivity calculation (approximately 1 % s^{-1} of time error for the example discussed here), it is a wise precaution to ensure that the time delay used for each conductivity measurement is correct. This is most easily accomplished by checking that the conductivities calculated during the temperature decay do not vary with time. More thorough investigations of these problems are currently underway (C. Lister, pers. comm.; Villinger and Davis, 1987a).

(7) CORER-OUTRIGGER MEASUREMENT ERRORS

Before this section is concluded, a few sources of errors in making gradient and conductivity measurements with a corer-outrigger instrument should be discussed. In spite of the distortions of sediment caused by both coring techniques, neither appears to alter the thermal properties of the sediments significantly. Comparisons of conductivities measured in situ with those measured in recovered cores have been made in two locations where local variability of conductivity is known to be small (Hyndman *et al.*, 1979; Davis *et al.*, 1984). The correspondence, shown in Figure 6.8, is excellent. Averages of values measured in samples collected with gravity and piston cores using the needle probe technique described by von Herzen and Maxwell (1959) differ from values measured in-situ by less than 10 %. Needle probe measurements used in the comparison have been corrected to account for in situ pressure and temperature conditions following the method outlined by Ratcliffe (1960). Thus it would appear that the primary limitation to the corer-outrigger technique of determining heat-flow density lies in the inadequacy of the sediment sampling, and the highest accuracy should be provided by the piston coring instrument.

The above conclusion cannot be made without noting one qualification regarding the accuracy of the outrigger gradient measurement. Recent measurements were made in the northwestern Atlantic with identical outrigger instrumentation mounted on both piston and gravity (non-piston) corers of identical weight and dimensions (Davis *et al.*, 1984). On the basis of the nature of the frictional heating decays and the temperature-depth profiles, it was concluded that there were significant thermal disturbances caused by the piston coring process which were present at the location of the outriggers, in that case 8 cm from the outside of the 16 cm diameter core barrel. Comparisons of the temperature-time decays and temperature-depth profiles for two of these gravity and piston corer-outrigger

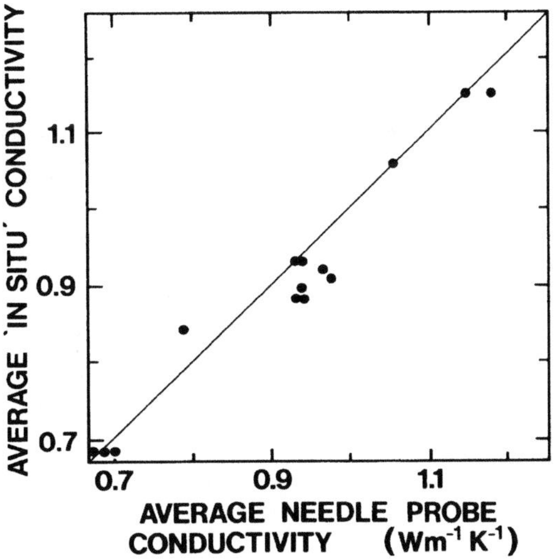

Fig. 6.8. Comparison of thermal conductivity measurements made in-situ and in sediments collected in nearby cores. Data have been selected from areas which have relatively uniform conductivity, from the Blake Bahama Basin, northwestern Atlantic, and from the inlets of southwestern British Columbia.

gradient measurements are shown in Figures 6.9a and 6.9b. It should be noted that the measurements used in this comparison were made in unusually soft sediments, and it may not be appropriate to generalize the conclusion; nevertheless, some caution should be exercised when using the piston corer for making gradient measurements.

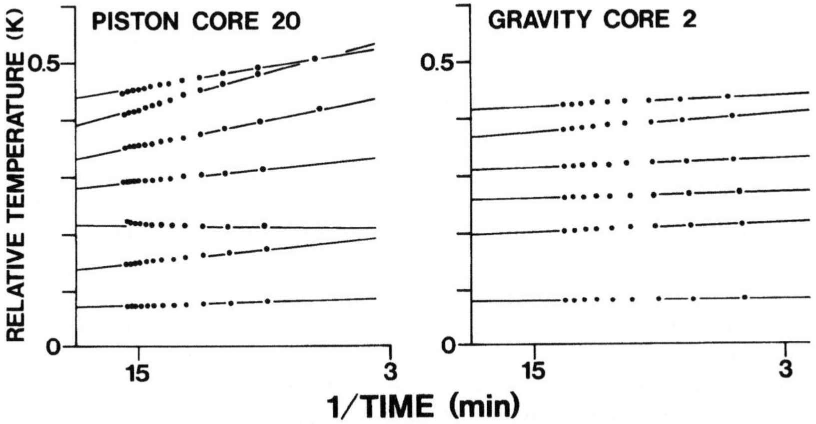

Fig. 6.9A. Thermal decays after frictional heating for outrigger probes mounted on gravity and piston corers, with temperature plotted against inverse time.

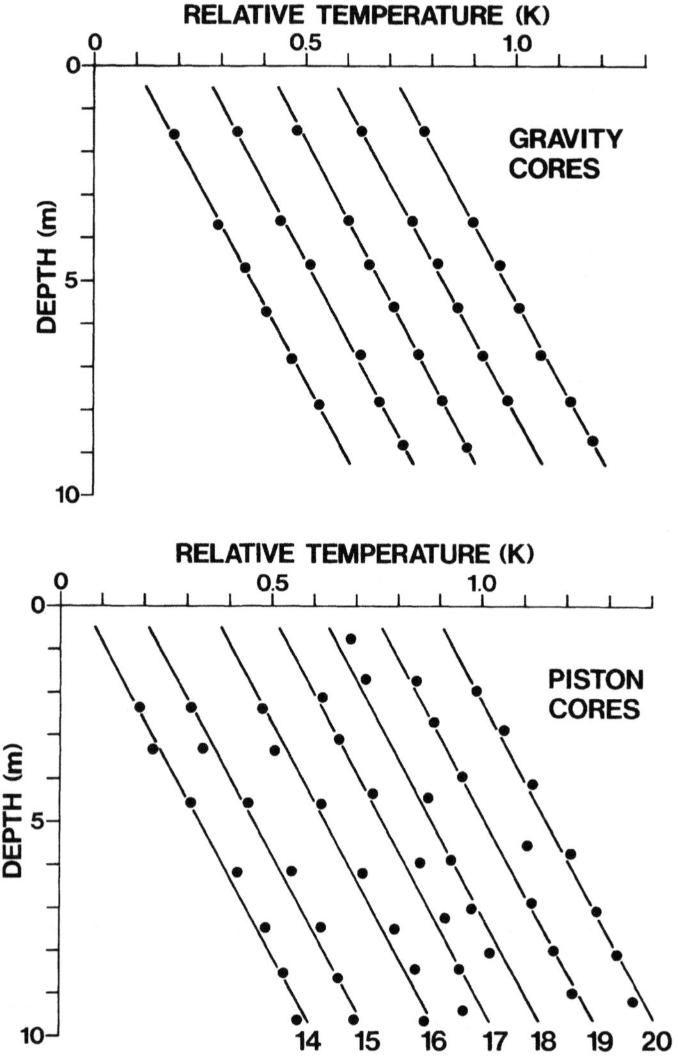

Fig. 6.9B. Temperature vs. depth for gravity- and piston-corer outrigger measurements in the Blake-Bahama Basin, northwestern Atlantic. Sediments at these sites are extremely soft, and the disturbances observed which are presumably due to the piston-coring process are probably much larger than those normally encountered.

Often, conductivity is estimated from water content in a way described by Ratcliffe (1960). The method is successful to a certain degree of accuracy for most sediments, although the second order effects of variable sediment grain conductivities and variable intergranular contact geometry are not accounted for in a single water content-conductivity relationship. These effects have been discussed by Lachenbruch and Marshall (1966), and the potential errors involved have been demonstrated in one area of the northwestern Atlantic by Davis *et al.* (1984). A summary of those results is shown in Figure 6.10, where needle probe conductivity measurements are plotted as a function of corresponding water content values. Previously determined relationships are also shown. Conductivities as estimated from water contents using the Ratcliffe relationship would be correct in some areas (e.g. in this example, for the glacial clays of the northeast Pacific) but in other locations they could be in error by as much as 10–20 % (e.g. north-western Atlantic). There is even some systematic difference between the groups of data at each of four Atlantic sites which are separated from one another by only a few hundred kilometres. Thus very local calibration is obviously necessary before the water content can be used to estimate conductivity with reasonable accuracy.

(8) ERRORS FROM NON-VERTICAL PENETRATION

One additional source of measurement error which can be a problem with any instrument, but particularly with multiple-penetration measurements, is non-vertical penetration. As pointed out previously, drift velocities are commonly significant

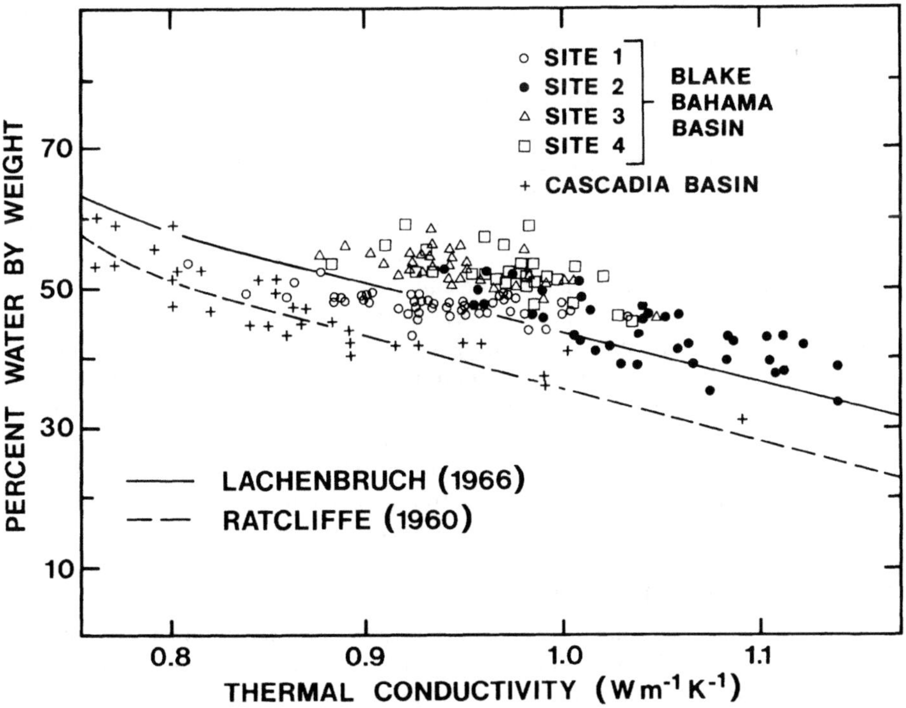

Fig. 6.10. Needle-probe conductivity values vs. water content of cored sediments from four areas in the Blake Bahama Basin, northwestern Atlantic (Goldberg, 1981), and from Cascadia Basin, northeastern Pacific. Shown for comparison are the approximate empirical relationships established by Ratcliffe (1960) (dashed line) and Lachenbruch and Marshall (1966) (solid line).

(25 %) compared to entry velocities, so non-vertical penetrations are bound to result. Tilts of less than 10–15° are of little consequence, so a simple tilt sensor for detecting unacceptable penetration angles is sufficient. In one example, a 15°/30° mercury tilt switch was employed for this purpose.

6.6. SUMMARY OF RECOMMENDATIONS

1. PROBE LENGTH

The most troublesome errors in measuring marine geothermal gradients are caused by short-term bottom water temperature fluxuations, and by the imprecision with which in situ temperatures are measured. Errors from both causes can be reduced by increasing the probe length, although this exercise is only moderately effective. Without knowledge of the nature and timing of bottom water temperature changes in a given situation, the gains in accuracy can only be guaranteed to be made in direct proportion to increasing probe length. Gains in the accuracy of gradient measurement are made even less effectively by trying to reduce the effect of temperature measurement errors with increased probe length. Most of the frictional heat generated during impact is conducted into the probe, so that the longer the probe, the greater will be the frictional heating temperature rise in the lower thermistors. Thus, the gain in accuracy with which the gradient can be measured by increasing the depth interval will be in part offset by a reduction in the accuracy with which the in situ temperature can be measured. Accuracy at the upper sensors may also be reduced because of the greater chance of disturbance of the shallow sediments during penetration. With the benefits of longer probes being thus limited, there is no reason to construct a probe that is too long to be handled conveniently or too long to be robust. Three to five meters seems by experience to be reasonable; greater lengths would not be recommended unless justified by particular circumstances.

This conclusion does not apply to corer-outrigger measurements, however; the rapid response of small diameter outrigger probes prevents frictional heating from causing large errors in temperature-depth measurements, so efficient gains in accuracy of gradient measurements can be made with deep (12 m) penetrations. As discussed previously, the inadequacy of sediment sampling may degrade the quality of the heat-flow density determination, but even as an estimate, the deep heat-flow density value can allow a better assessment of the quality of nearly shallower multiple-penetration determinations.

2. THERMAL CONDUCTIVITY MEASUREMENT

It is rare that marine sediments are sufficiently uniform for conductivity measurements on cores to be relied upon for determining heat-flow density over a given gradient interval. Whether the core and the gradient measurement are taken at the same site, as is the case when a corer-outrigger instrument is used, or at different sites, as is the case when a multi-penetration gradient instrument is used with cores taken nearby, uncertainty is bound to result. Making measurements of conductivity *in-situ* is not a particularly difficult or complex task, and since this is the only reliable way to obtain accurate heat-flow density determinations, it is recommended here. Furthermore, since small scale depth variations in conductivity are often too large to allow a few point measurements to be relied upon, averaging sensors such as those described above should be employed.

3. CALIBRATION

As discussed earlier, some of the errors in measuring conductivity in-situ can be reduced by being certain that the heat pulse power is sufficient to provide a temperature change at a time of about $\tau = 2$ that is large compared to the errors in estimating the ambient temperature conditions. Some guidelines regarding the power can be obtained from Figure 6.4. Errors having their origin in the instrumentation or in a given reduction scheme can be evaluated and hopefully reduced by calibration tests in materials of known conductivity. Some calibration standards are discussed elsewhere in this volume, although none are really suitable for use with a large oceanic heat-flow density probe. Two others are mentioned here. Water, with a small proportion of gelatin added to prevent convection, appears to be an excellent and easily applied primary calibration standard, although its use must be limited to cold-room temperatures to avoid melting. Its conductivity is well known: 0.57 W m^{-1} K^{-1} at 5 °C (Powell, 1958). Glass beads saturated with water provide a reasonable secondary standard; they pack uniformly, but their bulk conductivity must be measured with a needle-probe device. A typical conductivity of this material is 0.8 W m^{-1} K^{-1} (Hyndman *et al.*, 1979). Well sorted sand has also been tried, but uniform packing is difficult to achieve, and 5 % variations in conductivity are unavoidable (Hyndman *et al.*, 1979). Some caution must be used with both water and glass bead calibration materials: For water, α cannot be assumed to be 2, and temperatures during decay require about a −3 % correction; glass beads appear to have a high temperature coefficient of about 1 % per K, so conductivity measurements and probe calibration must be accomplished at the same temperatures.

4. HEAT-FLOW DENSITY SITE PLANNING

Of the factors which contribute to a well designed marine heat-flow density experiment, two of the more important are (1) having an adequate number of individual determinations that are spaced sufficiently closely to reveal the nature of local heat-flow density variability, and (2) having sufficient information about the geological environment at each determination (sediment thickness, sediment distribution, and basement topography) so that the cause of the variability can be understood. Only with this information can the determinations be properly used individually or collectively in solving a local or regional problem. General criteria for site-survey and heat-flow density station planning cannot be given, for every situation and problem to be solved is unique and must be considered individually. Nevertheless, the importance of a thorough site survey (seismic reflection and 3.5 kHz profiling) and careful station siting cannot be emphasized too strongly.

5. DATA PRESENTATION

Suggestions for data presentation are relatively simple for marine heat-flow density. Certainly, a full description of the probe used should be given or referred to, as well as a description of the reduction scheme and the way in which gradient and conductivity errors are estimated. Tabulated information should include penetration location, water depth, penetration depth, gradient intervals, gradients and their error estimates, conductivities and their error estimates, and the resulting heat-flow density values and their error estimates. For convenience, the units presented in Chapter 2. should be used. Finally, any corrections made for topographic effects (rarely important on the sea floor), and sedimentation effects should be noted, with corrected gradients and heat-flow density values listed in addition to the observed values. Clearly, this is a minimum list; in many instances, additional information is desirable or necessary; for more details see Table 2.8. Graphical data display is also valuable to the reader for rapid appraisal of data quality; either summary Bullard plots or temperature- and conductivity-depth plots are suitable for this purpose. Further information are given in Chapter 7.5.

Acknowledgements

The author wishes to thank H. Villinger, R. Haenel, and R. Hyndman for reviewing this chapter, and H. Villinger for performing calculations of $C(\alpha, \tau)$ and $G(\alpha, \tau)$.

6.7. REFERENCES

Anderson, R. N., Hobart, M. A., and Langseth, M. G.: 1979, 'Geothermal Convection Through Oceanic Crust and Sediments in the Indian Ocean', *Science* **204**, 828–832.

Beck, A. E.: 1965, 'Techniques of Measuring Heat Flow on Land', in W. H. K. Lee (ed.), *Terrestrial Heat Flow*, American Geophysical Union, Monograph No. 8, Washington, D.C., 24–67.

Becker, K. and von Herzen, R. P.: 1983, 'Heat Transfer Through the Sediments of the Mounds Hydrothermal Area, Galapagos Spreading Center at 86° W', *J. Geophys. Res.* **88**, 995–1008.

Blackwell, J. H.: 1954, 'A Transient Flow Method for Determination of Thermal Constants of Insulating Materials in Bulk', *J. Applied Physics* **25**, 137–144.

Bullard, E. C.: 1954, 'The Flow of Heat Through the Floor of the Atlantic Ocean', *Proceedings Royal Society London* A, **222**, 408–429.

Bullard, E. C.: 1963, 'The Flow of Heat Through the Floor of the Ocean', in M. N. Hill (ed.), *The Sea*, Vol. III, Interscience, New York, 218–232.

Burch, J. K. and Langseth, M. G.: 1981, 'Heat-Flow Determination in Three DSDP Boreholes near the Japan Trench', *J. Geophys. Res.* **86**, 9411–9419.

Carslaw, H. S. and Jaeger, J. C.: 1959, *Conduction of Heat in Solids*, 2nd edition, Oxford University Press, London, 510 p.

Davis, E. E. and Lister, C. R. B.: 1977, 'Heat Flow Measured over the Juan de Fuca Ridge: Evidence for Widespread Hydrothermal Circulation in a Highly Heat Transportive Crust', *J. Geophys. Res.* **82**, 4945–4860.

Davis, E. E., Lister, C. R. B., Wade, U. S., and Hyndman, R. D.: 1980, 'Detailed Heat Flow Measurements over the Juan de Fuca Ridge System', *J. Geophys. Res.* **85**, 299–310.

Davis E. E. and Riddihough, R. P.: 1982, 'The Winona Basin: Structure and Tectonics', *Canadian Journal of Earth Sciences* **19**, 767–788.

Davis, E. E., Lister, C. R. B., and Sclater, J. G.: 1984, 'Toward Determining the Thermal State of Old Ocean Lithosphere: Heat Flow Measurements from the Blake-Bahama Outer Ridge, Northwestern Atlantic', *Geophys. J. Roy. Astron. Soc.* **78**, **2**, 507–545.

Erickson, A. J., Von Herzen, R. P., Sclater, J. G., Girdler, R. W., Marshall, B. V., and Hyndman, R. D.: 1975, 'Geothermal Measurements in Deep-Sea Drill Holes', *J. Geophys. Res.* **80**, 2505–2528.

Gerard, R., Langseth, M. G. Jr., and Ewing, M.: 1962, 'Thermal Gradient Measurements in the Water and Bottom Sediments of the Western Atlantic', *J. Geophys. Res.* **67**, 785–803.

Goldberg, D. S.: 1981, 'The Physical Properties of Deep Ocean Sediments from the Northern Atlantic: A Comparison of *in situ* and Laboratory Methods', M.S. Thesis, M.I.T., Cambridge, Mass.

Green, K. E., von Herzen, R. P., and Williams, D. L.: 1981, 'The Galapagos Spreading Center at 86° W: A Detailed Geothermal Field Study', *J. Geophys. Res.* **86**, 979–986.

Haenel, R.: 1979, 'A Critical Review of Heat Flow Measurements in Sea and Lake Bottom Sediments', in V. Čermàk and L. Rybach (eds.), *Terrestrial Heat Flow in Europe*, Springer-Verlag, Berlin, 49–73.

Huppert, H. E. and Sclater, J. G.: 1968, *On the In-situ Measurement of the Thermal Conductivity of Deep-Sea Sediments*', Scripps Institution of Oceanography Technical Report, MPL-U-20/68, La Jolla, Calif.

Hyndman, R. D., Rogers, G. C., Bone, M. N., Lister, C. R. B., Wade, U. S., Barrett, D. L., Davis, E. E., Lewis, T. J., Lynch, S., and Seemann, D.: 1978, 'Geophysical Measurements in the Region of the Explorer Ridge off Western Canada', *Canadian J. Earth Sci.* **15**, 1508-1525.

Hyndman, R. D., Davis, E. E., and Wright, J. A.: 1979, 'The Measurement of Marine Geothermal Heat Flow by a Multipenetration Probe with Digital Acoustic Telemetry and *in situ* Conductivity', *Marine Geophys. Res.* **4**, 181–205.

Hyndman, R. D., Langseth, M. G., and Von Herzen, R. P.: 1983, 'A Review of Deep Sea Drilling Project Geothermal Measurements to Leg 70', in R. D. Hyndman, M. H. Salisbury *et al.* (eds.), *Initial Reports of the Deep Sea Drilling Project*, Vol. 78B, U.S. Government printing Office, Washington, D.C., 813–824.

Jaeger, J. C.: 1956, 'Conduction of Heat in an Infinite Region Bounded Internally by a Circular Cylinder of a Perfect Conductor', *Australian J. Phys.* **9**, 167–179.

Jeffreys, H.: 1938, 'The Disturbance of the Temperature Gradient in the Earth's Crust by Inequalities of Height', *Monthly Notices of the Roy. Astron. Soc. Geophys. Suppl.* **4**, 309–312.

Lachenbruch, A. H.: 1968, 'Rapid Estimation of the Topographic Disturbance to Surficial Thermal Gradients', *Rev. Geophys.* **6**, 365–400.

Lachenbruch, A. H. and Marshall, B. V.: 1966, 'Heat Flow Through the Arctic Ocean Floor: The Canada Basin – Alpha Rise Boundary', *J. Geophys. Res.* **71**, 1223–1248.

Langseth, M. G. Jr.: 1965, 'Techniques for Measuring Heat Flow Through the Ocean Floor', in W. H. K. Lee (ed.), *Terrestrial Heat Flow*, Geophysical Monograph Series 8, American Geophysical Union, 58–77.

Lee, T. C. and Henyey, T. L.: 1974, 'Heat Flow Refraction Across Dissimilar Media', *Geophys. J. Roy. Astron. Soc.* **39**, 319-333.

Lister, C. R. B.: 1970, 'Measurement of *in situ* Conductivity by Means of a Bullard-Type Probe', *Geophys. J. Roy. Astron. Soc.* **19**, 521–533.

Lister, C. R. B.: 1972, 'On the Thermal Balance of a Mid-Ocean Ridge', *Geophys. J. Roy. Astron. Soc.* **26**, 515–535.

Lister, C. R. B.: 1979, 'The Pulse-Probe Method of Conductivity Measurement', *Geophys. J. Roy. Astron. Soc.* **57**, 451–461.

Mojesky, T.: 1981, 'Measurement of Sediment Conductivity *in situ* Using a Heat-Pulse Technique', M.Sc. thesis, University of Washington.

Powell, P. W.: 1958, 'Thermal Conductivity and Expansion Coefficients of Water and Ice', *Advances in Physics* **7**, 276–297.

Ratcliffe, E. H.: 1960, 'The Thermal Conductivities of Ocean Sediments', *J. Geophys. Res.* **65**, 1535–1541.

Sclater, J. G. and Miller, S. P.: 1969, *A Machine Method for Computing the Effect of a Two-Layered Conductivity Structure*, Marine physical laboratory of Scripps Institution of Oceanography, Technical Memorandum, **198**, 18 p.

Sclater, J. G., Jones, E. J. W., and Miller, S. P.: 1970, 'The Relationship of Heat Flow, Bottom Topography and Basement Relief in Peake and Freen Deeps, Northeast Atlantic', *Tectonophysics* **10**, 283–300.

Sclater, J. G., Crowe, J., and Anderson, R. N.: 1976, 'On the Reliability of Oceanic Heat Flow Averages', *J. Geophys. Res.* **81**, 2997–3006.

Villinger, H., and Davis, E. E.: 1987a, 'A New Reduction Algorithm for Marine Heat Flow Measurements', *J. Geophys. Res.* **92**, 12846–12856.

Villinger, H., and Davis, E. E.: 1987b, '*HFRED: A Program for the Reduction of Marine Heat Flow Data on a Microcomputer*', Geological Survey of Canada Open File Report 1627, 77 p.

Von Herzen, R. P. and Maxwell, A. E.: 1959, 'The Measurement of Thermal Conductivity of Deep-Sea Sediments by a Needle Probe Method', *J. Geophys. Res.* **64**, 1557–1563.

Von Herzen, R. P. and Uyeda, S.: 1963, 'Heat Flow Through the Pacific Ocean Floor', *J. Geophys. Res.* **68**, 4219–4230.

Von Herzen, R. P. and Anderson, R. N.: 1972, 'Implications of Heat Flow and Bottom Water Temperature in the Eastern Equatorial Pacific', *Geophys. J.* **26**, 427–458.

Williams, D. L., Von Herzen, R. P., Sclater, J. G., and Anderson, R. N.: 1974, 'The Galapagos Spreading Centre: Lithospheric Cooling and Hydrothermal Circulation', *Geophys. J. Roy. Astron. Soc.* **38**, 587–608.

Williams, D. L., Green, K., Van Andel, T. H., Von Herzen, R. P., Dymond, J. R., and Crane, K.: 1979, 'The Hydrothermal Mounds of the Galapagos Rift: Observations with DSRV ALVIN and Detailed Heat Flow Studies', *J. Geophys. Res.* **84**, 7467–7484.

TABLE 6.1
Tabulation of $F(\alpha, \tau)$ for several values of α.

$F(\alpha, \tau)$; $\alpha = 1.5$

	0.00	0.01	0.02	0.03	0.04	0.05	0.06	0.07	0.08	0.09
0.50	0.32793	0.32467	0.32148	0.31836	0.31531	0.31232	0.30939	0.30653	0.30372	0.30096
0.60	0.29826	0.29562	0.29302	0.29047	0.28797	0.28552	0.28311	0.28074	0.27842	0.27614
0.70	0.27390	0.27169	0.26953	0.26740	0.26530	0.26324	0.26122	0.25923	0.25727	0.25534
0.80	0.25344	0.25157	0.24973	0.24792	0.24613	0.24437	0.24264	0.24094	0.23926	0.23760
0.90	0.23597	0.23436	0.23277	0.23121	0.22966	0.22814	0.22664	0.22516	0.22370	0.22226
1.00	0.22084	0.21944	0.21805	0.21669	0.21534	0.21401	0.21269	0.21140	0.21011	0.20885
1.10	0.20760	0.20636	0.20514	0.20394	0.20275	0.20157	0.20041	0.19926	0.19812	0.19700
1.20	0.19589	0.19480	0.19371	0.19264	0.19158	0.19053	0.18950	0.18847	0.18746	0.18646
1.30	0.18547	0.18448	0.18351	0.18255	0.18160	0.18066	0.17973	0.17881	0.17790	0.17700
1.40	0.17611	0.17523	0.17435	0.17349	0.17263	0.17178	0.17094	0.17011	0.16929	0.16847
1.50	0.16766	0.16686	0.16607	0.16529	0.16451	0.16374	0.16298	0.16222	0.16147	0.16073
1.60	0.16000	0.15927	0.15855	0.15783	0.15712	0.15642	0.15573	0.15504	0.15435	0.15368
1.70	0.15301	0.15234	0.15168	0.15103	0.15038	0.14974	0.14910	0.14847	0.14784	0.14722
1.80	0.14660	0.14599	0.14539	0.14479	0.14419	0.14360	0.14301	0.14243	0.14185	0.14128
1.90	0.14072	0.14015	0.13959	0.13904	0.13849	0.13795	0.13740	0.13687	0.13634	0.13581
2.00	0.13528	0.13476	0.13425	0.13373	0.13322	0.13272	0.13222	0.13172	0.13123	0.13074
2.10	0.13025	0.12977	0.12929	0.12882	0.12834	0.12788	0.12741	0.12695	0.12649	0.12603
2.20	0.12558	0.12513	0.12469	0.12425	0.12381	0.12337	0.12294	0.12251	0.12208	0.12166
2.30	0.12123	0.12082	0.12040	0.11999	0.11958	0.11917	0.11877	0.11837	0.11797	0.11757
2.40	0.11718	0.11679	0.11640	0.11601	0.11563	0.11525	0.11487	0.11449	0.11412	0.11375
2.50	0.11338	0.11301	0.11265	0.11229	0.11193	0.11157	0.11122	0.11086	0.11051	0.11016
2.60	0.10982	0.10947	0.10913	0.10879	0.10846	0.10812	0.10779	0.10746	0.10713	0.10680
2.70	0.10647	0.10615	0.10583	0.10551	0.10519	0.10488	0.10456	0.10425	0.10394	0.10363
2.80	0.10332	0.10302	0.10272	0.10242	0.10212	0.10182	0.10152	0.10123	0.10093	0.10064
2.90	0.10035	0.10007	0.09978	0.09950	0.09921	0.09893	0.09865	0.09837	0.09810	0.09782
3.00	0.09755	0.09729	0.09702	0.09675	0.09648	0.09621	0.09595	0.09568	0.09542	0.09516
3.10	0.09490	0.09465	0.09439	0.09413	0.09388	0.09363	0.09338	0.09313	0.09288	0.09263
3.20	0.09239	0.09214	0.09190	0.09166	0.09142	0.09118	0.09094	0.09070	0.09047	0.09023
3.30	0.09000	0.08977	0.08954	0.08931	0.08908	0.08885	0.08863	0.08840	0.08818	0.08796
3.40	0.08773	0.08751	0.08729	0.08708	0.08686	0.08664	0.08643	0.08621	0.08600	0.08579
3.50	0.08558	0.08537	0.08516	0.08495	0.08474	0.08454	0.08433	0.08413	0.08392	0.08372
3.60	0.08352	0.08332	0.08312	0.08292	0.08273	0.08253	0.08233	0.08214	0.08195	0.08175
3.70	0.08156	0.08137	0.08118	0.08099	0.08080	0.08061	0.08043	0.08024	0.08006	0.07987
3.80	0.07969	0.07951	0.07933	0.07914	0.07896	0.07879	0.07861	0.07843	0.07825	0.07808
3.90	0.07790	0.07773	0.07755	0.07738	0.07721	0.07704	0.07687	0.07670	0.07653	0.07636
4.00	0.07619	0.07602	0.07586	0.07569	0.07553	0.07536	0.07520	0.07504	0.07487	0.07471
4.10	0.07455	0.07439	0.07423	0.07408	0.07392	0.07376	0.07360	0.07345	0.07329	0.07314
4.20	0.07298	0.07283	0.07268	0.07252	0.07237	0.07222	0.07207	0.07192	0.07177	0.07162
4.30	0.07148	0.07133	0.07118	0.07104	0.07089	0.07075	0.07060	0.07046	0.07032	0.07017
4.40	0.07003	0.06989	0.06975	0.06961	0.06947	0.06933	0.06919	0.06905	0.06892	0.06878
4.50	0.06864	0.06851	0.06837	0.06824	0.06810	0.06797	0.06783	0.06770	0.06757	0.06744
4.60	0.06731	0.06718	0.06704	0.06692	0.06679	0.06666	0.06653	0.06640	0.06627	0.06615
4.70	0.06602	0.06589	0.06577	0.06564	0.06552	0.06540	0.06527	0.06515	0.06503	0.06490
4.80	0.06478	0.06466	0.06454	0.06442	0.06430	0.06418	0.06406	0.06394	0.06382	0.06371
4.90	0.06359	0.06347	0.06336	0.06324	0.06312	0.06301	0.06289	0.06278	0.06267	0.06255
5.00	0.06244	0.06233	0.06221	0.06210	0.06199	0.06188	0.06177	0.06166	0.06155	0.06144
5.10	0.06133	0.06122	0.06111	0.06100	0.06090	0.06079	0.06068	0.06057	0.06047	0.06036
5.20	0.06026	0.06015	0.06005	0.05994	0.05984	0.05974	0.05963	0.05953	0.05943	0.05932
5.30	0.05922	0.05912	0.05902	0.05892	0.05882	0.05872	0.05862	0.05852	0.05842	0.05832
5.40	0.05822	0.05812	0.05802	0.05793	0.05783	0.05773	0.05764	0.05754	0.05744	0.05735
5.50	0.05725	0.05716	0.05706	0.05697	0.05687	0.05678	0.05669	0.05659	0.05650	0.05641
5.60	0.05632	0.05622	0.05613	0.05604	0.05595	0.05586	0.05577	0.05568	0.05559	0.05550
5.70	0.05541	0.05532	0.05523	0.05514	0.05505	0.05497	0.05488	0.05479	0.05470	0.05462
5.80	0.05453	0.05444	0.05436	0.05427	0.05419	0.05410	0.05402	0.05393	0.05385	0.05376
5.90	0.05368	0.05360	0.05351	0.05343	0.05335	0.05326	0.05318	0.05310	0.05302	0.05294
6.00	0.05285	0.05277	0.05269	0.05261	0.05253	0.05245	0.05237	0.05229	0.05221	0.05213
6.10	0.05205	0.05197	0.05190	0.05182	0.05174	0.05166	0.05158	0.05151	0.05143	0.05135
6.20	0.05128	0.05120	0.05112	0.05105	0.05097	0.05090	0.05082	0.05074	0.05067	0.05060
6.30	0.05052	0.05045	0.05037	0.05030	0.05023	0.05015	0.05008	0.05001	0.04993	0.04986
6.40	0.04979	0.04972	0.04964	0.04957	0.04950	0.04943	0.04936	0.04929	0.04922	0.04915

Table 6.1 (Continued)

	$F(\alpha, \tau)$; $\alpha = 1.5$									
	0.00	0.01	0.02	0.03	0.04	0.05	0.06	0.07	0.08	0.09
6.50	0.04908	0.04901	0.04894	0.04887	0.04880	0.04873	0.04866	0.04859	0.04852	0.04845
6.60	0.04838	0.04831	0.04825	0.04818	0.04811	0.04804	0.04798	0.04791	0.04784	0.04778
6.70	0.04771	0.04746	0.04758	0.04751	0.04745	0.04738	0.04731	0.04725	0.04718	0.04712
6.80	0.04705	0.04699	0.04693	0.04686	0.04680	0.04673	0.04667	0.04661	0.04654	0.04648
6.90	0.04642	0.04635	0.04629	0.04623	0.04617	0.04610	0.04604	0.04598	0.04592	0.04586
7.00	0.04580	0.04573	0.04567	0.04561	0.04555	0.04549	0.04543	0.04537	0.04531	0.04525
7.10	0.04519	0.04513	0.04507	0.04501	0.04495	0.04489	0.04484	0.04478	0.04472	0.04466
7.20	0.04460	0.04454	0.04449	0.04443	0.04437	0.04431	0.04426	0.04420	0.04414	0.04408
7.30	0.04403	0.04397	0.04391	0.04386	0.04380	0.04375	0.04369	0.04363	0.04358	0.04352
7.40	0.04347	0.04341	0.04336	0.04330	0.04325	0.04319	0.04314	0.04308	0.04303	0.04298
7.50	0.04292	0.04287	0.04281	0.04276	0.04271	0.04265	0.04260	0.04255	0.04250	0.04244
7.60	0.04239	0.04234	0.04228	0.04223	0.04218	0.04213	0.04208	0.04202	0.04197	0.04192
7.70	0.04187	0.04182	0.04177	0.04172	0.04167	0.04162	0.04156	0.04151	0.04146	0.04141
7.80	0.04136	0.04131	0.04126	0.04121	0.04116	0.04111	0.04106	0.04102	0.04097	0.04092
7.90	0.04087	0.04082	0.04077	0.04072	0.04067	0.04062	0.04058	0.04053	0.04048	0.04043
8.00	0.04038	0.04034	0.04029	0.04024	0.04019	0.04015	0.04010	0.04005	0.04001	0.03996
8.10	0.03991	0.03987	0.03982	0.03977	0.03973	0.03968	0.03963	0.03959	0.03954	0.03950
8.20	0.03945	0.03941	0.03936	0.03931	0.03927	0.03922	0.03918	0.03913	0.03909	0.03904
8.30	0.03900	0.03896	0.03891	0.03887	0.03882	0.03878	0.03873	0.03869	0.03865	0.03860
8.40	0.03856	0.03852	0.03847	0.03843	0.03839	0.03834	0.03830	0.03826	0.03821	0.03817
8.50	0.03813	0.03808	0.03804	0.03800	0.03796	0.03792	0.03787	0.03783	0.03779	0.03775
8.60	0.03771	0.03766	0.03762	0.03758	0.03754	0.03750	0.03746	0.03742	0.03737	0.03733
8.70	0.03729	0.03725	0.03721	0.03717	0.03713	0.03709	0.03705	0.03701	0.03697	0.03693
8.80	0.03689	0.03685	0.03681	0.03677	0.03673	0.03669	0.03665	0.03661	0.03657	0.03653
8.90	0.03649	0.03645	0.03642	0.03638	0.03634	0.03630	0.03626	0.03622	0.03618	0.03615
9.00	0.03611	0.03607	0.03603	0.03599	0.03595	0.03592	0.03588	0.03584	0.03580	0.03577
9.10	0.03573	0.03569	0.03565	0.03562	0.03558	0.03554	0.03550	0.03547	0.03543	0.03539
9.20	0.03536	0.03532	0.03528	0.03525	0.03521	0.03517	0.03514	0.03510	0.03507	0.03503
9.30	0.03499	0.03496	0.03492	0.03489	0.03485	0.03481	0.03478	0.03474	0.03471	0.03467
9.40	0.03464	0.03460	0.03457	0.03453	0.03450	0.03446	0.03443	0.03439	0.03436	0.03432
9.50	0.03429	0.03425	0.03422	0.03418	0.03415	0.03412	0.03408	0.03405	0.03401	0.03398
9.60	0.03395	0.03391	0.03388	0.03384	0.03381	0.03378	0.03374	0.03371	0.03368	0.03364
9.70	0.03361	0.03358	0.03354	0.03351	0.03348	0.03345	0.03341	0.03338	0.03335	0.03331
9.80	0.03328	0.03325	0.03322	0.03318	0.03315	0.03312	0.03309	0.03305	0.03302	0.03299
9.90	0.03296	0.03293	0.03289	0.03286	0.03283	0.03280	0.03277	0.03274	0.03271	0.03267
10.00	0.03264	0.03281	0.03278	0.03274	0.03271	0.03268	0.03265	0.03262	0.03259	0.03256
10.10	0.03252	0.03249	0.03246	0.03243	0.03240	0.03237	0.03234	0.03231	0.03228	0.03225
10.20	0.03222	0.03219	0.03215	0.03212	0.03209	0.03206	0.03203	0.03200	0.03197	0.03194
10.30	0.03191	0.03188	0.03185	0.03182	0.03179	0.03176	0.03173	0.03170	0.03167	0.03164
10.40	0.03162	0.03159	0.03156	0.03153	0.03150	0.03147	0.03144	0.03141	0.03138	0.03135
10.50	0.03132	0.03129	0.03127	0.03124	0.03121	0.03118	0.03115	0.03112	0.03109	0.03106
10.60	0.03104	0.03101	0.03098	0.03095	0.03092	0.03089	0.03087	0.03084	0.03081	0.03078
10.70	0.03075	0.03073	0.03070	0.03067	0.03064	0.03062	0.03059	0.03056	0.03053	0.03051
10.80	0.03048	0.03045	0.03042	0.03040	0.03037	0.03034	0.03031	0.03029	0.03026	0.03023
10.90	0.03021	0.03018	0.03015	0.03013	0.03010	0.03007	0.03005	0.03002	0.02999	0.02997
11.00	0.02994	0.02991	0.02989	0.02986	0.02983	0.02981	0.02978	0.02975	0.02973	0.02970
11.10	0.02968	0.02965	0.02962	0.02960	0.02957	0.02955	0.02952	0.02950	0.02947	0.02944
11.20	0.02942	0.02939	0.02937	0.02934	0.02932	0.02929	0.02927	0.02924	0.02922	0.02919
11.30	0.02917	0.02914	0.02912	0.02909	0.02907	0.02904	0.02902	0.02899	0.02897	0.02894
11.40	0.02892	0.02889	0.02887	0.02884	0.02882	0.02879	0.02877	0.02874	0.02872	0.02870
11.50	0.02867	0.02865	0.02862	0.02860	0.02857	0.02855	0.02853	0.02850	0.02848	0.02845
11.60	0.02843	0.02841	0.02838	0.02836	0.02833	0.02831	0.02829	0.02826	0.02824	0.02822
11.70	0.02819	0.02817	0.02815	0.02812	0.02810	0.02808	0.02805	0.02803	0.02801	0.02798
11.80	0.02796	0.02794	0.02791	0.02789	0.02787	0.02785	0.02782	0.02780	0.02778	0.02775
11.90	0.02773	0.02771	0.02769	0.02766	0.02764	0.02762	0.02760	0.02757	0.02755	0.02753
12.00	0.02751	0.02748	0.02746	0.02744	0.02742	0.02739	0.02737	0.02735	0.02733	0.02731
12.10	0.02728	0.02726	0.02724	0.02722	0.02720	0.02717	0.02715	0.02713	0.02711	0.02709
12.20	0.02707	0.02704	0.02702	0.02700	0.02698	0.02696	0.02694	0.02691	0.02689	0.02687
12.30	0.02685	0.02683	0.02681	0.02679	0.02677	0.02674	0.02672	0.02670	0.02668	0.02666
12.40	0.02664	0.02662	0.02660	0.02658	0.02655	0.02653	0.02651	0.02649	0.02647	0.02645

Table 6.1 (Continued)

	$F(\alpha, \tau)$; $\alpha = 2$									
	0.00	0.01	0.02	0.03	0.04	0.05	0.06	0.07	0.08	0.09
0.50	0.25969	0.25409	0.25128	0.24855	0.24588	0.24327	0.24072	0.23823	0.23580	0.23342
0.60	0.23108	0.22880	0.22657	0.22439	0.22224	0.22015	0.21809	0.21608	0.21410	0.21217
0.70	0.21027	0.20840	0.20657	0.20478	0.20302	0.20129	0.19959	0.19792	0.19628	0.19467
0.80	0.19309	0.19153	0.19000	0.18850	0.18702	0.18556	0.18413	0.18272	0.18134	0.17997
0.90	0.17863	0.17731	0.17601	0.17473	0.17347	0.17222	0.17100	0.16979	0.16860	0.16743
1.00	0.16628	0.16514	0.16402	0.16291	0.16182	0.16074	0.15968	0.15864	0.15760	0.15659
1.10	0.15558	0.15459	0.15361	0.15265	0.15169	0.15075	0.14982	0.14890	0.14800	0.14710
1.20	0.14622	0.14535	0.14449	0.14364	0.14279	0.14196	0.14114	0.14033	0.13953	0.13874
1.30	0.13795	0.13718	0.13642	0.13566	0.13491	0.13417	0.13344	0.13272	0.13200	0.13130
1.40	0.13060	0.12990	0.12922	0.12854	0.12787	0.12721	0.12656	0.12591	0.12526	0.12463
1.50	0.12400	0.12338	0.12276	0.12215	0.12155	0.12095	0.12036	0.11977	0.11919	0.11862
1.60	0.11805	0.11749	0.11693	0.11638	0.11583	0.11529	0.11476	0.11422	0.11370	0.11318
1.70	0.11266	0.11215	0.11164	0.11114	0.11064	0.11015	0.10966	0.10917	0.10869	0.10822
1.80	0.10775	0.10728	0.10681	0.10635	0.10590	0.10545	0.10500	0.10456	0.10412	0.10368
1.90	0.10325	0.10282	0.10239	0.10197	0.10155	0.10114	0.10073	0.10032	0.09992	0.09952
2.00	0.09912	0.09872	0.09833	0.09794	0.09756	0.09717	0.09680	0.09642	0.09605	0.09568
2.10	0.09531	0.09494	0.09458	0.09422	0.09387	0.09351	0.09316	0.09281	0.09247	0.09212
2.20	0.09178	0.09145	0.09111	0.09078	0.09045	0.09012	0.08979	0.08947	0.08915	0.08883
2.30	0.08851	0.08820	0.08789	0.08758	0.08727	0.08696	0.08666	0.08638	0.08606	0.08577
2.40	0.08547	0.08518	0.08489	0.08460	0.08431	0.08403	0.08374	0.08346	0.08318	0.08291
2.50	0.08263	0.08236	0.08209	0.08182	0.08155	0.08128	0.08102	0.08075	0.08049	0.08023
2.60	0.07998	0.07972	0.07947	0.07921	0.07896	0.07871	0.07846	0.07822	0.07797	0.07773
2.70	0.07749	0.07725	0.07701	0.07677	0.07654	0.07630	0.07607	0.07584	0.07561	0.07538
2.80	0.07515	0.07493	0.07470	0.07448	0.07426	0.06403	0.07382	0.07360	0.07338	0.07317
2.90	0.07295	0.07274	0.07253	0.07232	0.07211	0.07190	0.07169	0.07149	0.07128	0.07108
3.00	0.07088	0.07068	0.07048	0.07028	0.07008	0.06989	0.06969	0.06950	0.06931	0.06911
3.10	0.06892	0.06873	0.06854	0.06836	0.06817	0.06798	0.06780	0.06762	0.06743	0.06725
3.20	0.06707	0.06689	0.06671	0.06653	0.06636	0.06618	0.06601	0.06583	0.06566	0.06549
3.30	0.06532	0.06514	0.06498	0.06481	0.06464	0.06447	0.06431	0.06414	0.06398	0.06381
3.40	0.06365	0.06349	0.06333	0.06317	0.06301	0.06285	0.06269	0.06253	0.06238	0.06222
3.50	0.06207	0.06191	0.06176	0.06161	0.06146	0.06131	0.06116	0.06101	0.06086	0.06071
3.60	0.06056	0.06042	0.06027	0.06013	0.05998	0.05984	0.05970	0.05955	0.05941	0.05927
3.70	0.05913	0.05899	0.05885	0.05871	0.05858	0.05844	0.05830	0.05817	0.05803	0.05790
3.80	0.05776	0.05763	0.05750	0.05737	0.05723	0.05710	0.05697	0.05684	0.05671	0.05659
3.90	0.05646	0.05633	0.05620	0.05608	0.05595	0.05583	0.05570	0.05558	0.05546	0.05533
4.00	0.05521	0.05509	0.05497	0.05485	0.05473	0.05461	0.05449	0.05437	0.05425	0.05414
4.10	0.05402	0.05390	0.05379	0.05367	0.05356	0.05344	0.05333	0.05321	0.05310	0.05299
4.20	0.05288	0.05276	0.05265	0.05254	0.05243	0.05232	0.05221	0.05210	0.05200	0.05189
4.30	0.05178	0.05167	0.05157	0.05146	0.05136	0.05125	0.05115	0.05104	0.05094	0.05083
4.40	0.05073	0.05063	0.05053	0.05042	0.05032	0.05022	0.05012	0.05002	0.04992	0.04982
4.50	0.04972	0.04962	0.04952	0.04943	0.04933	0.04923	0.04914	0.04904	0.04894	0.04885
4.60	0.04875	0.04866	0.04856	0.04847	0.04838	0.04828	0.04819	0.04810	0.04800	0.04791
4.70	0.04782	0.04773	0.04764	0.04755	0.04746	0.04737	0.04728	0.04719	0.04710	0.04701
4.80	0.04692	0.04684	0.04675	0.04666	0.04657	0.04649	0.04640	0.04631	0.04623	0.04614
4.90	0.04606	0.04597	0.04589	0.04581	0.04572	0.04564	0.04556	0.04547	0.04539	0.04531
5.00	0.04523	0.04514	0.04506	0.04498	0.04490	0.04482	0.04474	0.04466	0.04458	0.04450
5.10	0.04442	0.04434	0.04427	0.04419	0.04411	0.04403	0.04395	0.04388	0.04380	0.04372
5.20	0.04365	0.04357	0.04350	0.04342	0.04335	0.04327	0.04320	0.04312	0.04305	0.04297
5.30	0.04290	0.04283	0.04275	0.04268	0.04261	0.04254	0.04246	0.04239	0.04232	0.04225
5.40	0.04218	0.04211	0.04203	0.04196	0.04189	0.04182	0.04175	0.04168	0.04162	0.04155
5.50	0.04148	0.04141	0.04134	0.04127	0.04120	0.04114	0.04107	0.04100	0.04093	0.04087
5.60	0.04080	0.04073	0.04067	0.04060	0.04054	0.04047	0.04041	0.04034	0.04028	0.04021
5.70	0.04015	0.04008	0.04002	0.03995	0.03989	0.03983	0.03976	0.03970	0.03964	0.03957
5.80	0.03951	0.03945	0.03939	0.03933	0.03926	0.03920	0.03914	0.03908	0.03902	0.03896
5.90	0.03890	0.03884	0.03878	0.03872	0.03866	0.03860	0.03854	0.03848	0.03842	0.03836
6.00	0.03830	0.03824	0.03819	0.03813	0.03807	0.03801	0.03795	0.03790	0.03784	0.03778
6.10	0.03773	0.03767	0.03761	0.03756	0.03750	0.03744	0.03739	0.03733	0.03728	0.03722
6.20	0.03717	0.03711	0.03705	0.03700	0.03695	0.03689	0.03684	0.03678	0.03673	0.03667
6.30	0.03662	0.03657	0.03651	0.03646	0.03641	0.03636	0.03630	0.03625	0.03620	0.03615
6.40	0.03609	0.03604	0.03599	0.03594	0.03589	0.03583	0.03578	0.03573	0.03568	0.03563

Table 6.1 (Continued)

	$F(\alpha, \tau)$; $\alpha = 2$									
	0.00	0.01	0.02	0.03	0.04	0.05	0.06	0.07	0.08	0.09
6.50	0.03558	0.03553	0.03548	0.03543	0.03538	0.03533	0.03528	0.03523	0.03518	0.03513
6.60	0.03508	0.03503	0.03498	0.03493	0.03489	0.03484	0.03479	0.03474	0.03469	0.03464
6.70	0.03460	0.03455	0.03450	0.03445	0.03441	0.03436	0.03431	0.04326	0.03422	0.03417
6.80	0.03412	0.03408	0.03403	0.03399	0.03394	0.03389	0.03385	0.03380	0.03376	0.03371
6.90	0.03367	0.03362	0.03357	0.03353	0.03348	0.03344	0.03340	0.03335	0.03331	0.03326
7.00	0.03322	0.03317	0.03313	0.03309	0.03304	0.03300	0.03296	0.03291	0.03287	0.03283
7.10	0.03278	0.03274	0.03270	0.03265	0.03261	0.03257	0.03253	0.03249	0.03244	0.03240
7.20	0.03236	0.03232	0.03228	0.03223	0.03219	0.03215	0.03211	0.03207	0.03203	0.03199
7.30	0.03195	0.03191	0.03186	0.03182	0.03178	0.03174	0.03170	0.03166	0.03162	0.03158
7.40	0.03154	0.03150	0.03146	0.03142	0.03138	0.03135	0.03131	0.03127	0.03123	0.03119
7.50	0.03115	0.03111	0.03107	0.03103	0.03100	0.03096	0.03092	0.03088	0.03084	0.03081
7.60	0.03077	0.03073	0.03069	0.03065	0.03062	0.03058	0.03054	0.03050	0.03047	0.03043
7.70	0.03039	0.03036	0.03032	0.03028	0.03025	0.03021	0.03017	0.03014	0.03010	0.03007
7.80	0.03003	0.02999	0.02996	0.02992	0.02989	0.02985	0.02981	0.02978	0.02974	0.02971
7.90	0.02967	0.02964	0.02960	0.02957	0.02953	0.02950	0.02946	0.02943	0.02939	0.02936
8.00	0.02933	0.02929	0.02926	0.02922	0.02919	0.02915	0.02912	0.02909	0.02905	0.02902
8.10	0.02899	0.02895	0.02892	0.02889	0.02885	0.02882	0.02879	0.02875	0.02872	0.02869
8.20	0.02865	0.02862	0.02859	0.02856	0.02852	0.02849	0.02846	0.02843	0.02839	0.02836
8.30	0.02833	0.02830	0.02827	0.02823	0.02820	0.02817	0.02814	0.02811	0.02807	0.02804
8.40	0.02801	0.02798	0.02795	0.02792	0.02789	0.02786	0.02782	0.02779	0.02776	0.02773
8.50	0.02770	0.02767	0.02764	0.02761	0.02758	0.02755	0.02752	0.02749	0.02746	0.02743
8.60	0.02740	0.02737	0.02734	0.02731	0.02728	0.02725	0.02722	0.02719	0.02716	0.02713
8.70	0.02710	0.02707	0.02704	0.02701	0.02698	0.02696	0.02693	0.02690	0.02687	0.02684
8.80	0.02681	0.02678	0.02675	0.02673	0.02670	0.02667	0.02664	0.02661	0.02658	0.02656
8.90	0.02653	0.02650	0.02647	0.02644	0.02641	0.02639	0.02636	0.02633	0.02630	0.02628
9.00	0.02625	0.02622	0.02619	0.02617	0.02614	0.02611	0.02608	0.02606	0.02603	0.02600
9.10	0.02598	0.02595	0.02592	0.02590	0.02587	0.02584	0.02582	0.02579	0.02576	0.02574
9.20	0.02571	0.02568	0.02566	0.02563	0.02560	0.02558	0.02555	0.02553	0.02550	0.02547
9.30	0.02545	0.02542	0.02540	0.02537	0.02534	0.02532	0.02529	0.02527	0.02524	0.02522
9.40	0.02519	0.02517	0.02514	0.02512	0.02509	0.02507	0.02504	0.02502	0.02499	0.02497
9.50	0.02494	0.02492	0.02489	0.02487	0.02484	0.02482	0.02479	0.02477	0.02474	0.02472
9.60	0.02469	0.02467	0.02465	0.02462	0.02460	0.02457	0.02455	0.02452	0.02450	0.02448
9.70	0.02445	0.02443	0.02441	0.02438	0.02436	0.02433	0.02431	0.02429	0.02426	0.02424
9.80	0.02422	0.02419	0.02417	0.02415	0.02412	0.02410	0.02408	0.02405	0.02403	0.02401
9.90	0.02398	0.02396	0.02394	0.02392	0.02389	0.02387	0.02385	0.02382	0.02380	0.02378
10.00	0.02376	0.02373	0.02371	0.02368	0.02366	0.02364	0.02362	0.02359	0.02357	0.02355
10.10	0.02353	0.02351	0.02348	0.02346	0.02344	0.02342	0.02340	0.02337	0.02335	0.02333
10.20	0.02331	0.02329	0.02327	0.02324	0.02322	0.02320	0.02318	0.02316	0.02314	0.02311
10.30	0.02309	0.02307	0.02305	0.02303	0.02301	0.02299	0.02297	0.02295	0.02292	0.02290
10.40	0.02288	0.02286	0.02284	0.02282	0.02280	0.02278	0.02276	0.02274	0.02272	0.02270
10.50	0.02268	0.02266	0.02263	0.02261	0.02259	0.02257	0.02255	0.02253	0.02251	0.02249
10.60	0.02247	0.02245	0.02243	0.02241	0.02239	0.02237	0.02235	0.02233	0.02231	0.02229
10.70	0.02227	0.02225	0.02223	0.02221	0.02219	0.02217	0.02215	0.02213	0.02212	0.02210
10.80	0.02208	0.02206	0.02204	0.02202	0.02200	0.02198	0.02196	0.02194	0.02192	0.02190
10.90	0.02188	0.02186	0.02185	0.02183	0.02181	0.02179	0.02177	0.02175	0.02173	0.02171
11.00	0.02169	0.02168	0.02166	0.02164	0.02162	0.02160	0.02158	0.02156	0.02154	0.02153
11.10	0.02151	0.02149	0.02147	0.02145	0.02143	0.02142	0.02140	0.02138	0.02136	0.02134
11.20	0.02132	0.02131	0.02129	0.02127	0.02125	0.02123	0.02122	0.02120	0.02118	0.02116
11.30	0.02114	0.02113	0.02111	0.02109	0.02107	0.02106	0.02104	0.02102	0.02100	0.02099
11.40	0.02097	0.02095	0.02093	0.02092	0.02090	0.02088	0.02086	0.02085	0.02083	0.02081
11.50	0.02079	0.02078	0.02076	0.02074	0.02073	0.02071	0.02069	0.02067	0.02066	0.02064
11.60	0.02062	0.02061	0.02059	0.02057	0.02056	0.02054	0.02052	0.02050	0.02049	0.02047
11.70	0.02045	0.02044	0.02042	0.02040	0.02039	0.02037	0.02035	0.02034	0.02032	0.02031
11.80	0.02029	0.02027	0.02026	0.02024	0.02022	0.02021	0.02019	0.02017	0.02016	0.02014
11.90	0.02013	0.02011	0.02009	0.02008	0.02006	0.02005	0.02003	0.02001	0.02000	0.01998
12.00	0.01997	0.01995	0.01993	0.01992	0.01990	0.01989	0.01987	0.01985	0.01984	0.01982
12.10	0.01981	0.01979	0.01978	0.01976	0.01974	0.01973	0.01971	0.01970	0.01968	0.01967
12.20	0.01965	0.01964	0.01962	0.01961	0.01959	0.01958	0.01956	0.01954	0.01953	0.01951
12.30	0.01950	0.01948	0.01947	0.01945	0.01944	0.01942	0.01941	0.01939	0.01938	0.01936
12.40	0.01935	0.01933	0.01932	0.01930	0.01929	0.01927	0.01926	0.01924	0.01923	0.01921

Table 6.1 (Continued)

$F(\alpha, \tau)$; $\alpha = 2.5$

	0.00	0.01	0.02	0.03	0.04	0.05	0.06	0.07	0.08	0.09
0.50	0.20923	0.20672	0.20428	0.20190	0.19959	0.19732	0.19512	0.19296	0.19086	0.18881
0.60	0.18680	0.18484	0.18293	0.18105	0.17922	0.17743	0.17567	0.17395	0.17227	0.17063
0.70	0.16901	0.16743	0.16588	0.16436	0.16287	0.16141	0.15997	0.15857	0.15718	0.15583
0.80	0.15450	0.15319	0.15191	0.15065	0.14941	0.14819	0.14699	0.14582	0.14466	0.14352
0.90	0.14240	0.14130	0.14022	0.13915	0.13811	0.13707	0.13606	0.13506	0.13407	0.13310
1.00	0.13215	0.13121	0.13028	0.12937	0.12847	0.12758	0.12671	0.12585	0.12500	0.12416
1.10	0.12333	0.12252	0.12172	0.12092	0.12014	0.11937	0.11861	0.11786	0.11712	0.11639
1.20	0.11567	0.11495	0.11425	0.11356	0.11287	0.11219	0.11152	0.11086	0.11021	0.10957
1.30	0.10893	0.10830	0.10768	0.10707	0.10646	0.10586	0.10527	0.10468	0.10410	0.10353
1.40	0.10296	0.10240	0.10185	0.10130	0.10076	0.10022	0.09969	0.09917	0.09865	0.09814
1.50	0.09763	0.09713	0.09664	0.09614	0.09566	0.09518	0.09470	0.09423	0.09376	0.09330
1.60	0.09285	0.09239	0.09194	0.09150	0.09106	0.09063	0.09020	0.08977	0.08935	0.08893
1.70	0.08852	0.08811	0.08770	0.08730	0.08690	0.08651	0.08611	0.08573	0.08534	0.08496
1.80	0.08459	0.08421	0.08384	0.08348	0.08311	0.08275	0.08239	0.08204	0.08169	0.08134
1.90	0.08100	0.08066	0.08032	0.07998	0.07965	0.07932	0.07899	0.07867	0.07834	0.07802
2.00	0.07771	0.07739	0.07708	0.07677	0.07647	0.07616	0.07586	0.07556	0.07527	0.07497
2.10	0.07468	0.07439	0.07410	0.07382	0.07354	0.07326	0.07298	0.07270	0.07243	0.07216
2.20	0.07189	0.07162	0.07135	0.07109	0.07083	0.07057	0.07031	0.07005	0.06980	0.06955
2.30	0.06930	0.06905	0.06880	0.06856	0.06831	0.06807	0.06783	0.06759	0.06736	0.06712
2.40	0.06689	0.06666	0.06643	0.06620	0.06598	0.06575	0.06553	0.06531	0.06509	0.06487
2.50	0.06465	0.06443	0.06422	0.06401	0.06380	0.06359	0.06338	0.06317	0.06296	0.06276
2.60	0.06256	0.06236	0.06216	0.06196	0.06176	0.06156	0.06137	0.06117	0.06098	0.06079
2.70	0.06060	0.06041	0.06022	0.06003	0.05985	0.05966	0.05948	0.05930	0.05912	0.05894
2.80	0.05876	0.05858	0.05841	0.05823	0.05806	0.05788	0.05771	0.05754	0.05737	0.05720
2.90	0.05703	0.05686	0.05670	0.05653	0.05637	0.05621	0.05604	0.05588	0.05572	0.05556
3.00	0.05540	0.05525	0.05509	0.05493	0.05478	0.05463	0.05447	0.05432	0.05417	0.05402
3.10	0.05387	0.05372	0.05357	0.05342	0.05328	0.05313	0.05299	0.05284	0.05270	0.05256
3.20	0.05242	0.05228	0.05214	0.05200	0.05186	0.05172	0.05158	0.05145	0.05131	0.05118
3.30	0.05104	0.05091	0.05078	0.05064	0.05051	0.05038	0.05025	0.05012	0.04999	0.04987
3.40	0.04974	0.04961	0.04949	0.04936	0.04924	0.04911	0.04899	0.04887	0.04874	0.04862
3.50	0.04850	0.04838	0.04826	0.04814	0.04802	0.04790	0.04779	0.04767	0.04755	0.04744
3.60	0.04732	0.04721	0.04710	0.04698	0.04687	0.04676	0.04665	0.04653	0.04642	0.04631
3.70	0.04620	0.04609	0.04599	0.04588	0.04577	0.04566	0.04556	0.04545	0.04535	0.04524
3.80	0.04514	0.04503	0.04493	0.04482	0.04472	0.04462	0.04452	0.04442	0.04432	0.04422
3.90	0.04412	0.04402	0.04392	0.04382	0.04372	0.04362	0.04353	0.04343	0.04333	0.04324
4.00	0.04314	0.04305	0.04295	0.04286	0.04277	0.04267	0.04258	0.04249	0.04240	0.04230
4.10	0.04221	0.04212	0.04203	0.04194	0.04185	0.04176	0.04167	0.04158	0.04150	0.04141
4.20	0.04132	0.04123	0.04115	0.04106	0.04098	0.04089	0.04080	0.04072	0.04064	0.04055
4.30	0.04047	0.04038	0.04030	0.04022	0.04014	0.04005	0.03997	0.03989	0.03981	0.03973
4.40	0.03965	0.03957	0.03949	0.03941	0.03933	0.03925	0.03917	0.03909	0.03902	0.03894
4.50	0.03886	0.03879	0.03871	0.03863	0.03856	0.03848	0.03841	0.03833	0.03826	0.03818
4.60	0.03811	0.03803	0.03796	0.03789	0.03781	0.03774	0.03767	0.03760	0.03752	0.03745
4.70	0.03738	0.03731	0.03724	0.03717	0.03710	0.03703	0.03696	0.03689	0.03682	0.03675
4.80	0.03668	0.03661	0.03655	0.03648	0.03641	0.03634	0.03628	0.03621	0.03614	0.03608
4.90	0.03601	0.03594	0.03588	0.03581	0.03575	0.03568	0.03562	0.03555	0.03549	0.03542
5.00	0.03536	0.03530	0.03523	0.03517	0.03511	0.03505	0.03498	0.03492	0.03486	0.03480
5.10	0.03474	0.03467	0.03461	0.03455	0.03449	0.03443	0.03437	0.03431	0.03425	0.03419
5.20	0.03413	0.03407	0.03401	0.03396	0.03390	0.03384	0.03378	0.03372	0.03366	0.03361
5.30	0.03355	0.03349	0.03344	0.03338	0.03332	0.03327	0.03321	0.03315	0.03310	0.03304
5.40	0.03299	0.03293	0.03288	0.03282	0.03277	0.03271	0.03266	0.03260	0.03255	0.03250
5.50	0.03244	0.03239	0.03234	0.03228	0.03223	0.03218	0.03212	0.03207	0.03202	0.03197
5.60	0.03192	0.03186	0.03181	0.03176	0.03171	0.03166	0.03161	0.03156	0.03151	0.03146
5.70	0.03141	0.03136	0.03131	0.03126	0.03121	0.03116	0.03111	0.03106	0.03101	0.03096
5.80	0.03091	0.03087	0.03082	0.03077	0.03072	0.03067	0.03063	0.03058	0.03053	0.03048
5.90	0.03044	0.03039	0.03034	0.03030	0.03025	0.03020	0.03016	0.03011	0.03006	0.03002
6.00	0.02997	0.02993	0.02988	0.02984	0.02979	0.02975	0.02970	0.02966	0.02961	0.02957
6.10	0.02952	0.02948	0.02944	0.02939	0.02935	0.02930	0.02926	0.02922	0.02917	0.02913
6.20	0.02909	0.02905	0.02900	0.02896	0.02892	0.02887	0.02883	0.02879	0.02875	0.02871
6.30	0.02866	0.02862	0.02858	0.02854	0.02850	0.02846	0.02842	0.02838	0.02834	0.02829
6.40	0.02825	0.02821	0.02817	0.02813	0.02809	0.02805	0.02801	0.02797	0.02793	0.02789

Table 6.1 (Continued)

	$F(\alpha, \tau)$; $\alpha = 2.5$									
	0.00	0.01	0.02	0.03	0.04	0.05	0.06	0.07	0.08	0.09
6.50	0.02785	0.02782	0.02778	0.02774	0.02770	0.02766	0.02762	0.02758	0.02754	0.02751
6.60	0.02747	0.02743	0.02739	0.02735	0.02731	0.02728	0.02724	0.02720	0.02716	0.02713
6.70	0.02709	0.02705	0.02702	0.02698	0.02694	0.02691	0.02687	0.02683	0.02680	0.02676
6.80	0.02672	0.02669	0.02665	0.02661	0.02658	0.02654	0.02651	0.02647	0.02644	0.02640
6.90	0.02637	0.02633	0.02630	0.02626	0.02623	0.02619	0.02616	0.02612	0.02609	0.02605
7.00	0.02602	0.02598	0.02595	0.02592	0.02588	0.02585	0.02581	0.02578	0.02575	0.02571
7.10	0.02568	0.02565	0.02561	0.02558	0.02555	0.02551	0.02548	0.02545	0.02542	0.02538
7.20	0.02535	0.02532	0.02529	0.02525	0.02522	0.02519	0.02516	0.02512	0.02509	0.02506
7.30	0.02503	0.02500	0.02497	0.02493	0.02490	0.02487	0.02484	0.02481	0.02478	0.02475
7.40	0.02472	0.02468	0.02465	0.02462	0.02459	0.02456	0.02453	0.02450	0.02447	0.02444
7.50	0.02441	0.02438	0.02435	0.02432	0.02429	0.02426	0.02423	0.02420	0.02417	0.02414
7.60	0.02411	0.02408	0.02405	0.02402	0.02400	0.02397	0.02394	0.02391	0.02388	0.02385
7.70	0.02382	0.02379	0.02376	0.02374	0.02371	0.02368	0.02365	0.02362	0.02359	0.02357
7.80	0.02354	0.02351	0.02348	0.02345	0.02343	0.02340	0.02337	0.02334	0.02332	0.02329
7.90	0.02326	0.02323	0.02321	0.02318	0.02315	0.02312	0.02310	0.02307	0.02304	0.02302
8.00	0.02299	0.02296	0.02294	0.02291	0.02288	0.02286	0.02283	0.02280	0.02278	0.02275
8.10	0.02273	0.02270	0.02267	0.02265	0.02262	0.02260	0.02257	0.02254	0.02252	0.02249
8.20	0.02247	0.02244	0.02242	0.02239	0.02237	0.02234	0.02232	0.02229	0.02227	0.02224
8.30	0.02222	0.02219	0.02217	0.02214	0.02212	0.02209	0.02207	0.02204	0.02202	0.02199
8.40	0.02197	0.02194	0.02192	0.02190	0.02187	0.02185	0.02182	0.02180	0.02177	0.02175
8.50	0.02173	0.02170	0.02168	0.02166	0.02163	0.02161	0.02158	0.02156	0.02154	0.02151
8.60	0.02149	0.02147	0.02144	0.02142	0.02140	0.02138	0.02135	0.02133	0.02131	0.02128
8.70	0.02126	0.02124	0.02121	0.02119	0.02117	0.02115	0.02112	0.02110	0.02108	0.02106
8.80	0.02103	0.02101	0.02099	0.02097	0.02095	0.02092	0.02090	0.02088	0.02086	0.02083
8.90	0.02081	0.02079	0.02077	0.02075	0.02073	0.02070	0.02068	0.02066	0.02064	0.02062
9.00	0.02060	0.02057	0.02055	0.02053	0.02051	0.02049	0.02047	0.02045	0.02043	0.02041
9.10	0.02038	0.02036	0.02034	0.02032	0.02030	0.02028	0.02026	0.02024	0.02022	0.02020
9.20	0.02018	0.02016	0.02014	0.02011	0.02009	0.02007	0.02005	0.02003	0.02001	0.01999
9.30	0.01997	0.01995	0.01993	0.01991	0.01989	0.01987	0.01985	0.01983	0.01981	0.01979
9.40	0.01977	0.01975	0.01973	0.01971	0.01969	0.01968	0.01966	0.01964	0.01962	0.01960
9.50	0.01958	0.01956	0.01954	0.01952	0.01950	0.01948	0.01946	0.01944	0.01942	0.01941
9.60	0.01939	0.01937	0.01935	0.01933	0.01931	0.01929	0.01927	0.01925	0.01924	0.01922
9.70	0.01920	0.01918	0.01916	0.01914	0.01912	0.01911	0.01909	0.01907	0.01095	0.01903
9.80	0.01901	0.01900	0.01898	0.01896	0.01894	0.01892	0.01891	0.01889	0.01887	0.01885
9.90	0.01883	0.01882	0.01880	0.01878	0.01876	0.01874	0.01873	0.01871	0.01869	0.01867
10.00	0.01866	0.01856	0.01854	0.01853	0.01851	0.01849	0.01847	0.01846	0.01844	0.01842
10.10	0.01841	0.01839	0.01837	0.01835	0.01834	0.01832	0.01830	0.01829	0.01827	0.01825
10.20	0.01824	0.01822	0.01820	0.01819	0.01817	0.01815	0.01814	0.01812	0.01810	0.01809
10.30	0.01807	0.01805	0.01804	0.01802	0.01801	0.01799	0.01797	0.01796	0.01794	0.01792
10.40	0.01791	0.01789	0.01788	0.01786	0.01784	0.01783	0.01781	0.01780	0.01778	0.01776
10.50	0.01775	0.01773	0.01772	0.01770	0.01769	0.01767	0.01765	0.01764	0.01762	0.01761
10.60	0.01759	0.01758	0.01756	0.01755	0.01753	0.01751	0.01750	0.01748	0.01747	0.01745
10.70	0.01744	0.01742	0.01741	0.01739	0.01738	0.01736	0.01735	0.01733	0.01732	0.01730
10.80	0.01729	0.01727	0.01726	0.01724	0.01723	0.01721	0.01720	0.01718	0.01717	0.01715
10.90	0.01714	0.01712	0.01711	0.01709	0.01708	0.01706	0.01705	0.01703	0.01702	0.01701
11.00	0.01699	0.01698	0.01696	0.01695	0.01693	0.01692	0.01690	0.01689	0.01688	0.01686
11.10	0.01685	0.01683	0.01682	0.01680	0.01679	0.01678	0.01676	0.01675	0.01673	0.01672
11.20	0.01671	0.01669	0.01668	0.01666	0.01665	0.01664	0.01662	0.01661	0.01659	0.01658
11.30	0.01657	0.01655	0.01654	0.01652	0.01651	0.01650	0.01648	0.01647	0.01646	0.01644
11.40	0.01643	0.01642	0.01640	0.01639	0.01638	0.01636	0.01635	0.01633	0.01632	0.01631
11.50	0.01629	0.01628	0.01627	0.01625	0.01624	0.01623	0.01621	0.01620	0.01619	0.01618
11.60	0.01616	0.01615	0.01614	0.01612	0.01611	0.01610	0.01608	0.01607	0.01606	0.01604
11.70	0.01603	0.01602	0.01601	0.01599	0.01598	0.01597	0.01595	0.01594	0.01593	0.01592
11.80	0.01590	0.01589	0.01588	0.01587	0.01585	0.01584	0.01583	0.01582	0.01580	0.01579
11.90	0.01578	0.01577	0.01575	0.01574	0.01573	0.01572	0.01570	0.01569	0.01568	0.01567
12.00	0.01565	0.01564	0.01563	0.01562	0.01560	0.01559	0.01558	0.01557	0.01556	0.01554
12.10	0.01553	0.01552	0.01551	0.01550	0.01548	0.01547	0.01546	0.01545	0.01544	0.01542
12.20	0.01541	0.01540	0.01539	0.01538	0.01536	0.01535	0.01534	0.01533	0.01532	0.01530
12.30	0.01529	0.01528	0.01527	0.01526	0.01525	0.01523	0.01522	0.01521	0.01520	0.01519
12.40	0.01518	0.01516	0.01515	0.01514	0.01513	0.01512	0.01511	0.01510	0.01508	0.01507

TABLE 6.2
Tabulation of $G(\alpha, \tau)$ for several values of α.

$G(\alpha, \tau)$; $\alpha = 1.5$

	0.00	0.01	0.02	0.03	0.04	0.05	0.06	0.07	0.08	0.09
0.50	0.05781	0.05859	0.05936	0.06012	0.06088	0.06163	0.06237	0.06311	0.06383	0.06456
0.60	0.06527	0.06598	0.06668	0.06738	0.06807	0.06875	0.06943	0.07011	0.07077	0.07143
0.70	0.07209	0.07274	0.07339	0.07403	0.07467	0.07530	0.07592	0.07654	0.07716	0.07777
0.80	0.07838	0.07898	0.07958	0.08017	0.08076	0.08135	0.08193	0.08251	0.08308	0.08365
0.90	0.08422	0.08478	0.08534	0.08589	0.08644	0.08699	0.08753	0.08807	0.08860	0.08914
1.00	0.08966	0.09019	0.09071	0.09123	0.09175	0.09226	0.09277	0.09328	0.09378	0.09428
1.10	0.09478	0.09527	0.09576	0.09625	0.09673	0.09722	0.09770	0.09817	0.09865	0.09912
1.20	0.09959	0.10006	0.10052	0.10098	0.10144	0.10190	0.10235	0.10280	0.10325	0.10369
1.30	0.10414	0.10458	0.10502	0.10546	0.10589	0.10632	0.10675	0.10718	0.10761	0.10803
1.40	0.10845	0.10887	0.10929	0.10970	0.11012	0.11053	0.11094	0.11135	0.11175	0.11215
1.50	0.11255	0.11295	0.11335	0.11375	0.11414	0.11453	0.11492	0.11531	0.11570	0.11608
1.60	0.11646	0.11685	0.11723	0.11760	0.11798	0.11835	0.11873	0.11910	0.11947	0.11983
1.70	0.12020	0.12056	0.12093	0.12129	0.12165	0.12201	0.12236	0.12272	0.12307	0.12342
1.80	0.12377	0.12412	0.12447	0.12482	0.12516	0.12551	0.12585	0.12619	0.12653	0.12687
1.90	0.12720	0.12754	0.12787	0.12821	0.12854	0.12887	0.12920	0.12952	0.12985	0.13017
2.00	0.13050	0.13082	0.13114	0.13146	0.13178	0.13210	0.13241	0.13273	0.13304	0.13335
2.10	0.13367	0.13398	0.13429	0.13459	0.13490	0.13521	0.13551	0.13582	0.13612	0.13642
2.20	0.13672	0.13702	0.13732	0.13761	0.13791	0.13821	0.13850	0.13879	0.13908	0.13938
2.30	0.13967	0.13995	0.14024	0.14053	0.14082	0.14110	0.14138	0.14167	0.14195	0.14223
2.40	0.14251	0.14279	0.14307	0.14335	0.14362	0.14390	0.14417	0.14445	0.14472	0.14499
2.50	0.14526	0.14553	0.14580	0.14607	0.14634	0.14661	0.14687	0.14714	0.14740	0.14766
2.60	0.14793	0.14819	0.14845	0.14871	0.14897	0.14923	0.14949	0.14974	0.15000	0.15025
2.70	0.15051	0.15076	0.15102	0.15127	0.15152	0.15177	0.15202	0.15227	0.15252	0.15277
2.80	0.15301	0.15326	0.15350	0.15375	0.15399	0.15424	0.15448	0.15472	0.15496	0.15520
2.90	0.15544	0.15568	0.15592	0.15616	0.15640	0.15663	0.15687	0.15710	0.15734	0.15757
3.00	0.15781	0.15804	0.15827	0.15850	0.15873	0.15896	0.15919	0.15942	0.15965	0.15988
3.10	0.16010	0.16033	0.16055	0.16078	0.16100	0.16123	0.16145	0.16167	0.16190	0.16212
3.20	0.16234	0.16256	0.16278	0.16300	0.16322	0.16343	0.16365	0.16387	0.16408	0.16430
3.30	0.16451	0.16473	0.16494	0.16516	0.16537	0.16558	0.16579	0.16601	0.16622	0.16643
3.40	0.16664	0.16685	0.16705	0.16726	0.16747	0.16768	0.16788	0.16809	0.16830	0.16850
3.50	0.16870	0.16891	0.16911	0.16932	0.16952	0.16972	0.16992	0.17012	0.17032	0.17052
3.60	0.17072	0.17092	0.17112	0.17132	0.17152	0.17171	0.17191	0.17211	0.17230	0.17250
3.70	0.17269	0.17289	0.17308	0.17328	0.17347	0.17366	0.17385	0.17405	0.17424	0.17443
3.80	0.17462	0.17481	0.17500	0.17519	0.17538	0.17556	0.17575	0.17594	0.17613	0.17631
3.90	0.17650	0.17668	0.17687	0.17705	0.17724	0.17742	0.17761	0.17779	0.17797	0.17816
4.00	0.17834	0.17852	0.17870	0.17888	0.17906	0.17924	0.17942	0.17960	0.17978	0.17996
4.10	0.18014	0.18032	0.18049	0.18067	0.18085	0.18102	0.18120	0.18137	0.18155	0.18172
4.20	0.18190	0.18207	0.18225	0.18242	0.18259	0.18276	0.18294	0.18311	0.18328	0.18345
4.30	0.18362	0.18379	0.18396	0.18413	0.18430	0.18447	0.18464	0.18481	0.18498	0.18514
4.40	0.18531	0.18548	0.18565	0.18581	0.18598	0.18614	0.18631	0.18647	0.18664	0.18680
4.50	0.18697	0.18713	0.18729	0.18746	0.18762	0.18778	0.18794	0.18811	0.18827	0.18843
4.60	0.18859	0.18875	0.18891	0.18907	0.18923	0.18939	0.18955	0.18971	0.18987	0.19002
4.70	0.19018	0.19034	0.19050	0.19065	0.19081	0.19097	0.19112	0.19128	0.19143	0.19159
4.80	0.19174	0.19190	0.19205	0.19220	0.19236	0.19251	0.19267	0.19282	0.19297	0.19312
4.90	0.19327	0.19343	0.19358	0.19373	0.19388	0.19403	0.19418	0.19433	0.19448	0.19463
5.00	0.19478	0.19493	0.19508	0.19522	0.19537	0.19552	0.19567	0.19582	0.19596	0.19611
5.10	0.19626	0.19640	0.19655	0.19669	0.19684	0.19698	0.19713	0.19727	0.19742	0.19756
5.20	0.19771	0.19785	0.19799	0.19814	0.19828	0.19842	0.19857	0.19871	0.19885	0.19899
5.30	0.19913	0.19927	0.19942	0.19956	0.19970	0.19984	0.19998	0.20012	0.20026	0.20040
5.40	0.20054	0.20067	0.20081	0.20095	0.20109	0.20123	0.20137	0.20150	0.20164	0.20178
5.50	0.20191	0.20205	0.20219	0.20232	0.20246	0.20259	0.20273	0.20286	0.20300	0.20313
5.60	0.20327	0.20340	0.20354	0.20367	0.20381	0.20394	0.20407	0.20421	0.20434	0.20447
5.70	0.20460	0.20474	0.20487	0.20500	0.20513	0.20526	0.20539	0.20552	0.20565	0.20578
5.80	0.20592	0.20605	0.20618	0.20630	0.20643	0.20656	0.20669	0.20682	0.20695	0.20708
5.90	0.20721	0.20733	0.20746	0.20759	0.20772	0.20785	0.20797	0.20810	0.20823	0.20835
6.00	0.20848	0.20860	0.20873	0.20886	0.20898	0.20911	0.20923	0.20936	0.20948	0.20961
6.10	0.20973	0.20985	0.20998	0.21010	0.21023	0.21035	0.21047	0.21060	0.21072	0.21084
6.20	0.21096	0.21109	0.21121	0.21133	0.21145	0.21157	0.21170	0.21182	0.21194	0.21206
6.30	0.21218	0.21230	0.21242	0.21254	0.21266	0.21278	0.21290	0.21302	0.21314	0.21326
6.40	0.21338	0.21350	0.21361	0.21373	0.21385	0.21397	0.21409	0.21420	0.21432	0.21444

Table 6.1 (Continued)

	$G(\alpha, \tau)$; $\alpha = 1.5$									
	0.00	0.01	0.02	0.03	0.04	0.05	0.06	0.07	0.08	0.09
6.50	0.21456	0.21467	0.21479	0.21491	0.21502	0.21514	0.21526	0.21537	0.21549	0.21560
6.60	0.21572	0.21584	0.21595	0.21607	0.21618	0.21630	0.21641	0.21652	0.21664	0.21675
6.70	0.21687	0.21698	0.21709	0.21721	0.21732	0.21743	0.21755	0.21766	0.21777	0.21789
6.80	0.21800	0.21811	0.21822	0.21833	0.21845	0.21856	0.21867	0.21878	0.21889	0.21900
6.90	0.21911	0.21922	0.21933	0.21945	0.21956	0.21967	0.21978	0.21989	0.22000	0.22010
7.00	0.22021	0.22032	0.22043	0.22054	0.22065	0.22076	0.22087	0.22098	0.22108	0.22119
7.10	0.22130	0.22141	0.22152	0.22162	0.22173	0.22184	0.22195	0.22205	0.22216	0.22227
7.20	0.22237	0.22248	0.22258	0.22269	0.22280	0.22290	0.22301	0.22311	0.22322	0.22332
7.30	0.22343	0.22354	0.22364	0.22374	0.22385	0.22395	0.22406	0.22416	0.22427	0.22437
7.40	0.22447	0.22458	0.22468	0.22479	0.22489	0.22499	0.22509	0.22520	0.22530	0.22540
7.50	0.22551	0.22561	0.22571	0.22581	0.22591	0.22602	0.22612	0.22622	0.22632	0.22642
7.60	0.22652	0.22663	0.22673	0.22683	0.22693	0.22703	0.22713	0.22723	0.22733	0.22743
7.70	0.22753	0.22763	0.22773	0.22783	0.22793	0.22803	0.22813	0.22823	0.22833	0.22842
7.80	0.22852	0.22862	0.22872	0.22882	0.22892	0.22902	0.22911	0.22921	0.22931	0.22941
7.90	0.22950	0.22960	0.22970	0.22980	0.22989	0.22999	0.23009	0.23019	0.23028	0.23038
8.00	0.23047	0.23057	0.23067	0.23076	0.23086	0.23096	0.23105	0.23115	0.23124	0.23134
8.10	0.23143	0.23153	0.23162	0.23172	0.23181	0.23191	0.23200	0.23210	0.23219	0.23229
8.20	0.23238	0.23247	0.23257	0.23266	0.23276	0.23285	0.23294	0.23304	0.23313	0.23322
8.30	0.23332	0.23341	0.23350	0.23360	0.23369	0.23378	0.23387	0.23397	0.23406	0.23415
8.40	0.23424	0.23433	0.23443	0.23452	0.23461	0.23470	0.23479	0.23488	0.23498	0.23507
8.50	0.23516	0.23525	0.23534	0.23543	0.23552	0.23561	0.23570	0.23579	0.23588	0.23597
8.60	0.23606	0.23615	0.23624	0.23633	0.23642	0.23651	0.23660	0.23669	0.23678	0.23687
8.70	0.23696	0.23705	0.23714	0.23723	0.23731	0.23740	0.23749	0.23758	0.23767	0.23776
8.80	0.23784	0.23793	0.23802	0.23811	0.23820	0.23828	0.23837	0.23846	0.23855	0.23863
8.90	0.23872	0.23881	0.23889	0.23898	0.23907	0.23915	0.23924	0.23933	0.23941	0.23950
9.00	0.23959	0.23967	0.23976	0.23985	0.23993	0.24002	0.24010	0.24019	0.24027	0.24036
9.10	0.24044	0.24053	0.24061	0.24070	0.24078	0.24087	0.24095	0.24104	0.24112	0.24121
9.20	0.24129	0.24138	0.24146	0.24155	0.24163	0.24171	0.24180	0.24188	0.24197	0.24205
9.30	0.24213	0.24222	0.24230	0.24238	0.24247	0.24255	0.24263	0.24272	0.24280	0.24288
9.40	0.24296	0.24305	0.24313	0.24321	0.24329	0.24338	0.24346	0.24354	0.24362	0.24370
9.50	0.24379	0.24387	0.24395	0.24403	0.24411	0.24420	0.24428	0.24436	0.24444	0.24452
9.60	0.24460	0.24468	0.24476	0.24484	0.24492	0.24501	0.24509	0.24517	0.24525	0.24533
9.70	0.24541	0.24549	0.24557	0.24565	0.24573	0.24581	0.24589	0.24597	0.24605	0.24613
9.80	0.24621	0.24629	0.24636	0.24644	0.24652	0.24660	0.24668	0.24676	0.24684	0.24692
9.90	0.24700	0.24708	0.24715	0.24723	0.24731	0.24739	0.24747	0.24755	0.24762	0.24770
10.00	0.24778	0.24786	0.24794	0.24801	0.24809	0.24817	0.24825	0.24832	0.24840	0.24848
10.10	0.24856	0.24863	0.24871	0.24879	0.24886	0.24894	0.24902	0.24909	0.24917	0.24925
10.20	0.24932	0.24940	0.24948	0.24955	0.24963	0.24971	0.24978	0.24986	0.24993	0.25001
10.30	0.25008	0.25016	0.25024	0.25031	0.25039	0.25046	0.25054	0.25061	0.25069	0.25076
10.40	0.25084	0.25091	0.25099	0.25106	0.25114	0.25121	0.25129	0.25136	0.25144	0.25151
10.50	0.25159	0.25166	0.25173	0.25181	0.25188	0.25196	0.25203	0.25210	0.25218	0.25225
10.60	0.25233	0.25240	0.25247	0.25255	0.25262	0.25269	0.25277	0.25284	0.25291	0.25299
10.70	0.25306	0.25313	0.25321	0.25328	0.25335	0.25342	0.25350	0.25357	0.25364	0.25371
10.80	0.25379	0.25386	0.25393	0.25400	0.25408	0.25415	0.25422	0.25429	0.25436	0.25444
10.90	0.25451	0.25458	0.25465	0.25472	0.25479	0.25487	0.25494	0.25501	0.25508	0.25515
11.00	0.25522	0.25529	0.25536	0.25543	0.25551	0.25558	0.25565	0.25572	0.25579	0.25586
11.10	0.25593	0.25600	0.25607	0.25614	0.25621	0.25628	0.25635	0.25642	0.25649	0.25656
11.20	0.25663	0.25670	0.25677	0.25684	0.25691	0.25698	0.25705	0.25712	0.25719	0.25726
11.30	0.25733	0.25740	0.25747	0.25753	0.25760	0.25767	0.25774	0.25781	0.25788	0.25795
11.40	0.25802	0.25809	0.25815	0.25822	0.25829	0.25836	0.25843	0.25850	0.25856	0.25863
11.50	0.25870	0.25877	0.25884	0.25890	0.25897	0.25904	0.25911	0.25918	0.25924	0.25931
11.60	0.25938	0.25945	0.25951	0.25958	0.25965	0.25972	0.25978	0.25985	0.25992	0.25998
11.70	0.26005	0.26012	0.26019	0.26025	0.26032	0.26039	0.26045	0.26052	0.26059	0.26065
11.80	0.26072	0.26078	0.26085	0.26092	0.26098	0.26105	0.26112	0.26118	0.26125	0.26131
11.90	0.26138	0.26145	0.26151	0.26158	0.26164	0.26171	0.26177	0.26184	0.26191	0.26197
12.00	0.26204	0.26210	0.26217	0.26223	0.26230	0.26236	0.26243	0.26249	0.26256	0.26262
12.10	0.26269	0.26275	0.26282	0.26288	0.26295	0.26301	0.26308	0.26314	0.26320	0.26327
12.20	0.26333	0.26340	0.26346	0.26353	0.26359	0.26365	0.26372	0.26378	0.26385	0.26391
12.30	0.26397	0.26404	0.26410	0.26416	0.26423	0.26429	0.26436	0.26442	0.26448	0.26455
12.40	0.26461	0.26467	0.26474	0.26480	0.26486	0.26493	0.26499	0.26505	0.26511	0.26518

Table 6.1 (Continued)

	$G(\alpha, \tau)$; $\alpha = 2.0$									
	0.00	0.01	0.02	0.03	0.04	0.05	0.06	0.07	0.08	0.09
0.50	0.06511	0.06592	0.06673	0.06752	0.06831	0.06909	0.06986	0.07062	0.07137	0.07212
0.60	0.07286	0.07359	0.07432	0.07503	0.07574	0.07645	0.07715	0.07784	0.07852	0.07920
0.70	0.07987	0.08054	0.08120	0.08185	0.08250	0.08315	0.08378	0.08442	0.08504	0.08567
0.80	0.08628	0.08690	0.08750	0.08811	0.08870	0.08930	0.08988	0.09047	0.09105	0.09162
0.90	0.09219	0.09276	0.09332	0.09388	0.09443	0.09498	0.09553	0.09607	0.09661	0.09715
1.00	0.09768	0.09821	0.09873	0.09925	0.09977	0.10028	0.10079	0.10130	0.10180	0.10230
1.10	0.10280	0.10329	0.10378	0.10427	0.10475	0.10523	0.10571	0.10619	0.10666	0.10713
1.20	0.10760	0.10806	0.10852	0.10898	0.10944	0.10989	0.11034	0.11079	0.11123	0.11168
1.30	0.11212	0.11255	0.11299	0.11342	0.11385	0.11428	0.11471	0.11513	0.11555	0.11597
1.40	0.11639	0.11680	0.11722	0.11763	0.11803	0.11844	0.11884	0.11925	0.11965	0.12004
1.50	0.12044	0.12083	0.12122	0.12161	0.12200	0.12239	0.12277	0.12315	0.12353	0.12391
1.60	0.12429	0.12466	0.12504	0.12541	0.12578	0.12615	0.12651	0.12688	0.12724	0.12760
1.70	0.12796	0.12832	0.12867	0.12903	0.12938	0.12973	0.13008	0.13043	0.13078	0.13112
1.80	0.13147	0.13181	0.13215	0.13249	0.13283	0.13316	0.13350	0.13383	0.13416	0.13449
1.90	0.13482	0.13515	0.13548	0.13580	0.13613	0.13645	0.13677	0.13709	0.13741	0.13773
2.00	0.13804	0.13836	0.13867	0.13898	0.13930	0.13961	0.13991	0.14022	0.14053	0.14083
2.10	0.14114	0.14144	0.14174	0.14204	0.14234	0.14264	0.14294	0.14323	0.14353	0.14382
2.20	0.14411	0.14441	0.14470	0.14499	0.14527	0.14556	0.14585	0.14613	0.14642	0.14670
2.30	0.14698	0.14726	0.14754	0.14782	0.14810	0.14838	0.14866	0.14893	0.14921	0.14948
2.40	0.14975	0.15002	0.15029	0.15056	0.15083	0.15110	0.15137	0.15163	0.15190	0.15216
2.50	0.15243	0.15269	0.15295	0.15321	0.15347	0.15373	0.15399	0.15425	0.15450	0.15476
2.60	0.15501	0.15527	0.15552	0.15577	0.15603	0.15628	0.15653	0.15678	0.15702	0.15727
2.70	0.15752	0.15777	0.15801	0.15826	0.15850	0.15874	0.15899	0.15923	0.15947	0.15971
2.80	0.15995	0.16019	0.16043	0.16066	0.16090	0.16114	0.16137	0.16161	0.16184	0.16207
2.90	0.16231	0.16254	0.16277	0.16300	0.16323	0.16346	0.16369	0.16391	0.16414	0.16437
3.00	0.16459	0.16482	0.16504	0.16527	0.16549	0.16571	0.16594	0.16616	0.16638	0.16660
3.10	0.16682	0.16704	0.16726	0.16747	0.16769	0.16791	0.16812	0.16834	0.16856	0.16877
3.20	0.16898	0.16920	0.16941	0.16962	0.16983	0.17004	0.17025	0.17046	0.17067	0.17088
3.30	0.17109	0.17130	0.17150	0.17171	0.17192	0.17212	0.17233	0.17253	0.17274	0.17294
3.40	0.17314	0.17334	0.17355	0.17375	0.17395	0.17415	0.17435	0.17455	0.17475	0.17495
3.50	0.17514	0.17534	0.17554	0.17573	0.17593	0.17612	0.17632	0.17651	0.17671	0.17690
3.60	0.17709	0.17729	0.17748	0.17767	0.17786	0.17805	0.17824	0.17843	0.17862	0.17881
3.70	0.17900	0.17919	0.17938	0.17956	0.17975	0.17994	0.18012	0.18031	0.18049	0.18068
3.80	0.18086	0.18104	0.18123	0.18141	0.18159	0.18177	0.18196	0.18214	0.18232	0.18250
3.90	0.18268	0.18286	0.18304	0.18321	0.18339	0.18357	0.18375	0.18393	0.18410	0.18428
4.00	0.18445	0.18463	0.18481	0.18498	0.18515	0.18533	0.18550	0.18568	0.18585	0.18602
4.10	0.18619	0.18636	0.18654	0.18671	0.18688	0.18705	0.18722	0.18739	0.18756	0.18773
4.20	0.18789	0.18806	0.18823	0.18840	0.18856	0.18873	0.18890	0.18906	0.18923	0.18939
4.30	0.18956	0.18972	0.18989	0.19005	0.19022	0.19038	0.19054	0.19071	0.19087	0.19103
4.40	0.19119	0.19135	0.19151	0.19167	0.19183	0.19199	0.19215	0.19231	0.19247	0.19263
4.50	0.19279	0.19295	0.19311	0.19326	0.19342	0.19358	0.19373	0.19389	0.19405	0.19420
4.60	0.19436	0.19451	0.19467	0.19482	0.19498	0.19513	0.19528	0.19544	0.19559	0.19574
4.70	0.19589	0.19605	0.19620	0.19635	0.19650	0.19665	0.19680	0.19695	0.19710	0.19725
4.80	0.19740	0.19755	0.19770	0.19785	0.19800	0.19815	0.19829	0.19844	0.19859	0.19873
4.90	0.19888	0.19903	0.19917	0.19932	0.19947	0.19961	0.19976	0.19990	0.20005	0.20019
5.00	0.20033	0.20048	0.20062	0.20077	0.20091	0.20105	0.20119	0.20134	0.20148	0.20162
5.10	0.20176	0.20190	0.20204	0.20218	0.20232	0.20246	0.20261	0.20274	0.20288	0.20302
5.20	0.20316	0.20330	0.20344	0.20358	0.20372	0.20385	0.20399	0.20413	0.20427	0.20440
5.30	0.20454	0.20468	0.20481	0.20495	0.20508	0.20522	0.20536	0.20549	0.20563	0.20576
5.40	0.20589	0.20603	0.20616	0.20630	0.20643	0.20656	0.20670	0.20683	0.20696	0.20709
5.50	0.20723	0.20736	0.20749	0.20762	0.20775	0.20788	0.20801	0.20814	0.20827	0.20840
5.60	0.20853	0.20866	0.20879	0.20892	0.20905	0.20918	0.20931	0.20944	0.20957	0.20970
5.70	0.20982	0.20995	0.21008	0.21021	0.21033	0.21046	0.21059	0.21071	0.21084	0.21097
5.80	0.21109	0.21122	0.21134	0.21147	0.21159	0.21172	0.21184	0.21197	0.21209	0.21221
5.90	0.21234	0.21246	0.21259	0.21271	0.21283	0.21296	0.21308	0.21320	0.21332	0.21345
6.00	0.21357	0.21369	0.21381	0.21393	0.21405	0.21417	0.21430	0.21442	0.21454	0.21466
6.10	0.21478	0.21490	0.21502	0.21514	0.21526	0.21538	0.21549	0.21561	0.21573	0.21585
6.20	0.21597	0.21609	0.21621	0.21632	0.21644	0.21656	0.21668	0.21679	0.21691	0.21703
6.30	0.21714	0.21726	0.21738	0.21749	0.21761	0.21772	0.21784	0.21796	0.21807	0.21819
6.40	0.21830	0.21842	0.21853	0.21864	0.21876	0.21887	0.21899	0.21910	0.21921	0.21933

Table 6.2 (Continued)

	$G(\alpha, \tau)$; $\alpha = 2.0$									
	0.00	0.01	0.02	0.03	0.04	0.05	0.06	0.07	0.08	0.09
6.50	0.21944	0.21955	0.21967	0.21978	0.21989	0.22001	0.22012	0.22023	0.22034	0.22045
6.60	0.22057	0.22068	0.22079	0.22090	0.22101	0.22112	0.22123	0.22134	0.22145	0.22157
6.70	0.22168	0.22179	0.22190	0.22200	0.22211	0.22222	0.22233	0.22244	0.22255	0.22266
6.80	0.22277	0.22288	0.22299	0.22309	0.22320	0.22331	0.22342	0.22353	0.22363	0.22374
6.90	0.22385	0.22395	0.22406	0.22417	0.22428	0.22438	0.22449	0.22459	0.22470	0.22481
7.00	0.22491	0.22502	0.22512	0.22523	0.22533	0.22544	0.22554	0.22565	0.22575	0.22586
7.10	0.22596	0.22607	0.22617	0.22628	0.22638	0.22648	0.22659	0.22669	0.22679	0.22690
7.20	0.22700	0.22710	0.22721	0.22731	0.22741	0.22751	0.22762	0.22772	0.22782	0.22792
7.30	0.22802	0.22812	0.22823	0.22833	0.22843	0.22853	0.22863	0.22873	0.22883	0.22893
7.40	0.22903	0.22913	0.22923	0.22933	0.22943	0.22953	0.22963	0.22973	0.22983	0.22993
7.50	0.23003	0.23013	0.23023	0.23033	0.23043	0.23053	0.23062	0.23072	0.23082	0.23092
7.60	0.23102	0.23111	0.23121	0.23131	0.23141	0.23150	0.23160	0.23170	0.23180	0.23189
7.70	0.23199	0.23209	0.23218	0.23228	0.23238	0.23247	0.23257	0.23266	0.23276	0.23286
7.80	0.23295	0.23305	0.23314	0.23324	0.23333	0.23343	0.23352	0.23362	0.23371	0.23381
7.90	0.23390	0.23400	0.23409	0.23418	0.23428	0.23437	0.23447	0.23456	0.23465	0.23475
8.00	0.23484	0.23493	0.23503	0.23512	0.23521	0.23531	0.23540	0.23549	0.23558	0.23568
8.10	0.23577	0.23586	0.23595	0.23605	0.23614	0.23623	0.23632	0.23641	0.23650	0.23660
8.20	0.23669	0.23678	0.23687	0.23696	0.23705	0.23714	0.23723	0.23732	0.23741	0.23750
8.30	0.23759	0.23768	0.23777	0.23786	0.23795	0.23804	0.23813	0.23822	0.23831	0.23840
8.40	0.23849	0.23858	0.23867	0.23876	0.23885	0.23893	0.23902	0.23911	0.23920	0.23929
8.50	0.23938	0.23946	0.23955	0.23964	0.23973	0.23982	0.23990	0.23999	0.24008	0.24017
8.60	0.24025	0.24034	0.24043	0.24051	0.24060	0.24069	0.24078	0.24086	0.24095	0.24103
8.70	0.24112	0.24121	0.24129	0.24138	0.24147	0.24155	0.24164	0.24172	0.24181	0.24189
8.80	0.24198	0.24206	0.24215	0.24223	0.24323	0.24240	0.24249	0.24257	0.24266	0.24274
8.90	0.24283	0.24291	0.24300	0.24308	0.24317	0.24325	0.24333	0.24342	0.24350	0.24358
9.00	0.24367	0.24375	0.24383	0.24392	0.24400	0.24408	0.24417	0.24425	0.24433	0.24442
9.10	0.24450	0.24458	0.24466	0.24475	0.24483	0.24491	0.24499	0.24508	0.24516	0.24524
9.20	0.24532	0.24540	0.24549	0.24557	0.24565	0.24573	0.24581	0.24589	0.24597	0.24606
9.30	0.24614	0.24622	0.24630	0.24638	0.24646	0.24654	0.24662	0.24670	0.24678	0.24686
9.40	0.24694	0.24702	0.24710	0.24718	0.24726	0.24734	0.24742	0.24750	0.24758	0.24766
9.50	0.24774	0.24782	0.24790	0.24798	0.24806	0.24814	0.24821	0.24829	0.24837	0.24845
9.60	0.24853	0.24861	0.24869	0.24877	0.24884	0.24892	0.24900	0.24908	0.24916	0.24923
9.70	0.24931	0.24939	0.24947	0.24955	0.24962	0.24970	0.24978	0.24986	0.24993	0.25001
9.80	0.25009	0.25016	0.25024	0.25032	0.25039	0.25047	0.25055	0.25062	0.25070	0.25078
9.90	0.25085	0.25093	0.25101	0.25108	0.25116	0.25123	0.25131	0.25139	0.25146	0.25154
10.00	0.25161	0.25169	0.25176	0.25184	0.25192	0.25199	0.25207	0.25214	0.25222	0.25229
10.10	0.25237	0.25244	0.25252	0.25259	0.25267	0.25274	0.25281	0.25289	0.25296	0.25304
10.20	0.25311	0.25319	0.25326	0.25333	0.25341	0.25348	0.25356	0.25363	0.25370	0.25378
10.30	0.25385	0.25392	0.25400	0.25407	0.25414	0.25422	0.25429	0.25436	0.25444	0.25451
10.40	0.25458	0.25466	0.25473	0.25480	0.25487	0.25495	0.25502	0.25509	0.25516	0.25524
10.50	0.25531	0.25538	0.25545	0.25552	0.25560	0.25567	0.25574	0.25581	0.25588	0.25596
10.60	0.25603	0.25610	0.25617	0.25624	0.25631	0.25638	0.25645	0.25653	0.25660	0.25667
10.70	0.25674	0.25681	0.25688	0.25695	0.25702	0.25709	0.25716	0.25723	0.25730	0.25737
10.80	0.25744	0.25752	0.25759	0.25766	0.25773	0.25780	0.25787	0.25794	0.25801	0.25807
10.90	0.25814	0.25821	0.25828	0.25835	0.25842	0.25849	0.25856	0.25863	0.25870	0.25877
11.00	0.25884	0.25891	0.25898	0.25905	0.25911	0.25918	0.25925	0.25932	0.25939	0.25946
11.10	0.25953	0.25959	0.25966	0.25973	0.25980	0.25987	0.25994	0.26000	0.26007	0.26014
11.20	0.26021	0.26028	0.26034	0.26041	0.26048	0.26055	0.26061	0.26068	0.26075	0.26082
11.30	0.26088	0.26095	0.26102	0.26109	0.26115	0.26122	0.26129	0.26135	0.26142	0.26149
11.40	0.26155	0.26162	0.26169	0.26175	0.26182	0.26189	0.26195	0.26202	0.26209	0.26215
11.50	0.26222	0.26229	0.26235	0.26242	0.26248	0.26255	0.26262	0.26268	0.26275	0.26281
11.60	0.26288	0.26294	0.26301	0.26307	0.26314	0.26321	0.26327	0.26334	0.26340	0.26347
11.70	0.26353	0.26360	0.26366	0.26373	0.26379	0.26386	0.26392	0.26399	0.26405	0.26412
11.80	0.26418	0.26424	0.26431	0.26437	0.26444	0.26450	0.26457	0.26463	0.26470	0.26476
11.90	0.26482	0.26489	0.26495	0.26502	0.26508	0.26514	0.26521	0.26527	0.26533	0.26540
12.00	0.26546	0.26553	0.26559	0.26565	0.26572	0.26578	0.26584	0.26591	0.26597	0.26603
12.10	0.26609	0.26616	0.26622	0.26628	0.26635	0.26641	0.26647	0.26653	0.26660	0.26666
12.20	0.26672	0.26679	0.26685	0.26691	0.26697	0.26703	0.26710	0.26716	0.26722	0.26728
12.30	0.26735	0.26741	0.26747	0.26753	0.26759	0.26766	0.26772	0.26778	0.26784	0.26790
12.40	0.26796	0.26803	0.26809	0.26815	0.26821	0.26827	0.26833	0.26839	0.26846	0.26852

Table 6.2 (Continued)

$G(\alpha, \tau)$; $\alpha = 2.5$

	0.00	0.01	0.02	0.03	0.04	0.05	0.06	0.07	0.08	0.09
0.50	0.07029	0.07112	0.07194	0.07274	0.07354	0.07433	0.07511	0.07589	0.07665	0.07740
0.60	0.07815	0.07889	0.07962	0.08035	0.08106	0.08177	0.08248	0.08317	0.08386	0.08454
0.70	0.08522	0.08589	0.08655	0.08721	0.08786	0.08850	0.08914	0.08978	0.09040	0.09103
0.80	0.09164	0.09226	0.09286	0.09347	0.09406	0.09465	0.09524	0.09582	0.09640	0.09698
0.90	0.09754	0.09811	0.09867	0.09922	0.09978	0.10032	0.10087	0.10141	0.10194	0.10247
1.00	0.10300	0.10353	0.10405	0.10456	0.10507	0.10558	0.10609	0.10659	0.10709	0.10759
1.10	0.10808	0.10857	0.10905	0.10954	0.11002	0.11049	0.11097	0.11144	0.11190	0.11237
1.20	0.11283	0.11329	0.11375	0.11420	0.11465	0.11510	0.11554	0.11598	0.11642	0.11686
1.30	0.11730	0.11773	0.11816	0.11859	0.11901	0.11943	0.11985	0.12027	0.12069	0.12110
1.40	0.12151	0.12192	0.12232	0.12273	0.12313	0.12353	0.12393	0.12432	0.12472	0.12511
1.50	0.12550	0.12589	0.12627	0.12666	0.12704	0.12742	0.12779	0.12817	0.12854	0.12892
1.60	0.12929	0.12965	0.13002	0.13039	0.13075	0.13111	0.13147	0.13183	0.13219	0.13254
1.70	0.13289	0.13324	0.13359	0.13394	0.13429	0.13463	0.13498	0.13532	0.13566	0.13600
1.80	0.13634	0.13667	0.13701	0.13734	0.13767	0.13800	0.13833	0.13866	0.13898	0.13931
1.90	0.13963	0.13995	0.14027	0.14059	0.14091	0.14122	0.14154	0.14185	0.14216	0.14248
2.00	0.14279	0.14309	0.14340	0.14371	0.14401	0.14432	0.14462	0.14492	0.14522	0.14552
2.10	0.14582	0.14611	0.14641	0.14670	0.14700	0.14729	0.14758	0.14787	0.14816	0.14844
2.20	0.14873	0.14902	0.14930	0.14958	0.14987	0.15015	0.15043	0.15071	0.15099	0.15126
2.30	0.15154	0.15181	0.15209	0.15236	0.15263	0.15291	0.15318	0.15345	0.15371	0.15398
2.40	0.15425	0.15451	0.15478	0.15504	0.15531	0.15557	0.15583	0.15609	0.15635	0.15661
2.50	0.15686	0.15712	0.15738	0.15763	0.15789	0.15814	0.15839	0.15864	0.15890	0.15915
2.60	0.15939	0.15964	0.15989	0.16014	0.16038	0.16063	0.16087	0.16112	0.16136	0.16160
2.70	0.16184	0.16209	0.16233	0.16256	0.16280	0.16304	0.16328	0.16351	0.16375	0.16398
2.80	0.16422	0.16445	0.16468	0.16492	0.16515	0.16538	0.16561	0.16584	0.16607	0.16629
2.90	0.16652	0.16675	0.16697	0.16720	0.16742	0.16765	0.16787	0.16809	0.16832	0.16854
3.00	0.16876	0.16898	0.16920	0.16942	0.16964	0.16985	0.17007	0.17029	0.17050	0.17072
3.10	0.17093	0.17115	0.17136	0.17157	0.17178	0.17200	0.17221	0.17242	0.17263	0.17284
3.20	0.17305	0.17325	0.17346	0.17367	0.17388	0.17408	0.17429	0.17449	0.17470	0.17490
3.30	0.17510	0.17531	0.17551	0.17571	0.17591	0.17611	0.17631	0.17651	0.17671	0.17691
3.40	0.17711	0.17731	0.17750	0.17770	0.17790	0.17809	0.17829	0.17848	0.17868	0.17887
3.50	0.17906	0.17926	0.17945	0.17964	0.17983	0.18002	0.18021	0.18040	0.18059	0.18078
3.60	0.18097	0.18116	0.18134	0.18153	0.18172	0.18191	0.18209	0.18228	0.18246	0.18265
3.70	0.18283	0.18301	0.18320	0.18338	0.18356	0.18374	0.18393	0.18411	0.18429	0.18447
3.80	0.18465	0.18483	0.18501	0.18518	0.18536	0.18554	0.18572	0.18589	0.18607	0.18625
3.90	0.18642	0.18660	0.18677	0.18695	0.18712	0.18729	0.18747	0.18764	0.18781	0.18799
4.00	0.18816	0.18833	0.18850	0.18867	0.18884	0.18901	0.18918	0.18935	0.18952	0.18969
4.10	0.18986	0.19002	0.19019	0.19036	0.19052	0.19069	0.19086	0.19102	0.19119	0.19135
4.20	0.19152	0.19168	0.19185	0.19201	0.19217	0.19234	0.19250	0.19266	0.19282	0.19298
4.30	0.19314	0.19331	0.19347	0.19363	0.19379	0.19395	0.19410	0.19426	0.19442	0.19458
4.40	0.19474	0.19490	0.19505	0.19521	0.19537	0.19552	0.19568	0.19584	0.19599	0.19615
4.50	0.19630	0.19645	0.19661	0.19676	0.19692	0.19707	0.19722	0.19738	0.19753	0.19768
4.60	0.19783	0.19798	0.19813	0.19829	0.19844	0.19859	0.19874	0.19889	0.19904	0.19918
4.70	0.19933	0.19948	0.19963	0.19978	0.19993	0.20007	0.20022	0.20037	0.20051	0.20066
4.80	0.20081	0.20095	0.20110	0.20124	0.20139	0.20153	0.20168	0.20182	0.20197	0.20211
4.90	0.20225	0.20240	0.20254	0.20268	0.20282	0.20297	0.20311	0.20325	0.20339	0.20353
5.00	0.20367	0.20381	0.20395	0.20409	0.20423	0.20437	0.20451	0.20465	0.20479	0.20493
5.10	0.20507	0.20521	0.20534	0.20548	0.20562	0.20576	0.20589	0.20603	0.20616	0.20630
5.20	0.20644	0.20657	0.20671	0.20684	0.20698	0.20711	0.20725	0.20738	0.20752	0.20765
5.30	0.20778	0.20792	0.20805	0.20818	0.20832	0.20845	0.20858	0.20871	0.20884	0.20898
5.40	0.20911	0.20924	0.20937	0.20950	0.20963	0.20976	0.20989	0.21002	0.21015	0.21028
5.50	0.21041	0.21054	0.21067	0.21080	0.21092	0.21105	0.21118	0.21131	0.21143	0.21156
5.60	0.21169	0.21182	0.21194	0.21207	0.21220	0.21232	0.21245	0.21257	0.21270	0.21282
5.70	0.21295	0.21307	0.21320	0.21332	0.21345	0.21357	0.21370	0.21382	0.21394	0.21407
5.80	0.21419	0.21431	0.21443	0.21456	0.21468	0.21480	0.21492	0.21505	0.21517	0.21529
5.90	0.21541	0.21553	0.21565	0.21577	0.21589	0.21601	0.21613	0.21625	0.21637	0.21649
6.00	0.21661	0.21673	0.21685	0.21697	0.21709	0.21721	0.21732	0.21744	0.21756	0.21768
6.10	0.21779	0.21791	0.21803	0.21815	0.21826	0.21838	0.21850	0.21861	0.21873	0.21884
6.20	0.21896	0.21908	0.21919	0.21931	0.21942	0.21954	0.21965	0.21977	0.21988	0.22000
6.30	0.22011	0.22022	0.22034	0.22045	0.22056	0.22068	0.22079	0.22090	0.22102	0.22113
6.40	0.22124	0.22135	0.22147	0.22158	0.22169	0.22180	0.22191	0.22202	0.22214	0.22225

Table 6.2 (Continued)

	$G(\alpha, \tau)$; $\alpha = 2.5$									
	0.00	0.01	0.02	0.03	0.04	0.05	0.06	0.07	0.08	0.09
6.50	0.22236	0.22247	0.22258	0.22269	0.22280	0.22291	0.22302	0.22313	0.22324	0.22335
6.60	0.22346	0.22357	0.22368	0.22379	0.22389	0.22400	0.22411	0.22422	0.22433	0.22444
6.70	0.22454	0.22465	0.22476	0.22487	0.22497	0.22508	0.22519	0.22530	0.22540	0.22551
6.80	0.22561	0.22572	0.22583	0.22593	0.22604	0.22614	0.22625	0.22636	0.22646	0.22657
6.90	0.22667	0.22678	0.22688	0.22698	0.22709	0.22719	0.22730	0.22740	0.22751	0.22761
7.00	0.22771	0.22782	0.22792	0.22802	0.22813	0.22823	0.22833	0.22843	0.22854	0.22864
7.10	0.22874	0.22884	0.22895	0.22905	0.22915	0.22925	0.22935	0.22945	0.22955	0.22966
7.20	0.22976	0.22986	0.22996	0.23006	0.23016	0.23026	0.23036	0.23046	0.23056	0.23066
7.30	0.23076	0.23086	0.23096	0.23106	0.23116	0.23126	0.23135	0.23145	0.23155	0.23165
7.40	0.23175	0.23185	0.23194	0.23204	0.23214	0.23224	0.23234	0.23243	0.23253	0.23263
7.50	0.23273	0.23282	0.23292	0.23302	0.23311	0.23321	0.23331	0.23340	0.23350	0.23360
7.60	0.23369	0.23379	0.23388	0.23398	0.23407	0.23417	0.23426	0.23436	0.23445	0.23455
7.70	0.23464	0.23474	0.23483	0.23493	0.23502	0.23512	0.23521	0.23531	0.23540	0.23549
7.80	0.23559	0.23568	0.23577	0.23587	0.23596	0.23605	0.23615	0.23624	0.23633	0.23643
7.90	0.23652	0.23661	0.23670	0.23680	0.23689	0.23698	0.23707	0.23716	0.23725	0.23735
8.00	0.23744	0.23753	0.23762	0.23771	0.23780	0.23789	0.23799	0.23808	0.23817	0.23826
8.10	0.23835	0.23844	0.23853	0.23862	0.23871	0.23880	0.23889	0.23898	0.23907	0.23916
8.20	0.23925	0.23934	0.23943	0.23951	0.23960	0.23969	0.23978	0.23987	0.23996	0.24005
8.30	0.24014	0.24022	0.24031	0.24040	0.24049	0.24058	0.24066	0.24075	0.24084	0.24093
8.40	0.24101	0.24110	0.24119	0.24128	0.24136	0.24145	0.24154	0.24162	0.24171	0.24180
8.50	0.24188	0.24197	0.24206	0.24214	0.24223	0.24232	0.24240	0.24249	0.24257	0.24266
8.60	0.24274	0.24283	0.24291	0.24300	0.24309	0.24317	0.24326	0.24334	0.24342	0.24351
8.70	0.24359	0.24368	0.24376	0.24385	0.24393	0.24402	0.24410	0.24418	0.24427	0.24435
8.80	0.24444	0.24452	0.24460	0.24469	0.24477	0.24485	0.24494	0.24502	0.24510	0.24519
8.90	0.24527	0.24535	0.24543	0.24552	0.24560	0.24568	0.24576	0.24585	0.24593	0.24601
9.00	0.24609	0.24617	0.24626	0.24634	0.24642	0.24650	0.24658	0.24666	0.24675	0.24683
9.10	0.24691	0.24699	0.24707	0.24715	0.24723	0.24731	0.24739	0.24747	0.24755	0.24763
9.20	0.24771	0.24779	0.24787	0.24795	0.24803	0.24811	0.24819	0.24827	0.24835	0.24843
9.30	0.24851	0.24859	0.24867	0.24875	0.24883	0.24891	0.24899	0.24907	0.24915	0.24923
9.40	0.24930	0.24938	0.24946	0.24954	0.24962	0.24970	0.24977	0.24985	0.24993	0.25001
9.50	0.25009	0.25016	0.25024	0.25032	0.25040	0.25048	0.25055	0.25063	0.25071	0.25078
9.60	0.25086	0.25094	0.25102	0.25109	0.25117	0.25125	0.25132	0.25140	0.25148	0.25155
9.70	0.25163	0.25171	0.25178	0.25186	0.25193	0.25201	0.25209	0.25216	0.25224	0.25231
9.80	0.25239	0.25247	0.25254	0.25262	0.25269	0.25277	0.25284	0.25292	0.25299	0.25307
9.90	0.25314	0.25322	0.25329	0.25337	0.25344	0.25352	0.25359	0.25367	0.25374	0.25381
10.00	0.25389	0.25396	0.25404	0.25411	0.25419	0.25426	0.25433	0.25441	0.25448	0.25455
10.10	0.25463	0.25470	0.25477	0.25485	0.25492	0.25499	0.25507	0.25514	0.25521	0.25529
10.20	0.25536	0.25543	0.25551	0.25558	0.25565	0.25572	0.25580	0.25587	0.25594	0.25601
10.30	0.25608	0.25616	0.25623	0.25630	0.25637	0.25645	0.25652	0.25659	0.25666	0.25673
10.40	0.25680	0.25688	0.25695	0.25702	0.25709	0.25716	0.25723	0.25730	0.25737	0.25744
10.50	0.25752	0.25759	0.25766	0.25773	0.25780	0.25787	0.25794	0.25801	0.25808	0.25815
10.60	0.25822	0.25829	0.25836	0.25843	0.25850	0.25857	0.25864	0.25871	0.25878	0.25885
10.70	0.25892	0.25899	0.25906	0.25913	0.25920	0.25927	0.25934	0.25941	0.25948	0.25955
10.80	0.25961	0.25968	0.25975	0.25982	0.25989	0.25996	0.26003	0.26010	0.26016	0.26023
10.90	0.26030	0.26037	0.26044	0.26051	0.26058	0.26064	0.26071	0.26078	0.26085	0.26092
11.00	0.26098	0.26105	0.26112	0.26119	0.26125	0.26132	0.26139	0.26146	0.26152	0.26159
11.10	0.26166	0.26173	0.26179	0.26186	0.26193	0.26199	0.26206	0.26213	0.26220	0.26226
11.20	0.26233	0.26240	0.26246	0.26253	0.26259	0.26266	0.26273	0.26279	0.26286	0.26293
11.30	0.26299	0.26306	0.26313	0.26319	0.26326	0.26332	0.26339	0.26345	0.26352	0.26359
11.40	0.26365	0.26372	0.26378	0.26385	0.26391	0.26398	0.26404	0.26411	0.26417	0.26424
11.50	0.26430	0.26437	0.26443	0.26450	0.26456	0.26463	0.26469	0.26476	0.26482	0.26489
11.60	0.26495	0.26502	0.26508	0.26515	0.26521	0.26527	0.26534	0.26540	0.26547	0.26553
11.70	0.26560	0.26566	0.26572	0.26579	0.26585	0.26591	0.26598	0.26604	0.26611	0.26617
11.80	0.26623	0.26630	0.26636	0.26642	0.26649	0.26655	0.26661	0.26668	0.26674	0.26680
11.90	0.26686	0.26693	0.26699	0.26705	0.26712	0.26718	0.26724	0.26730	0.26737	0.26743
12.00	0.26749	0.26755	0.26762	0.26768	0.26774	0.26780	0.26787	0.26793	0.26799	0.26805
12.10	0.26811	0.26818	0.26824	0.26830	0.26836	0.26842	0.26849	0.26855	0.26861	0.26867
12.20	0.26873	0.26879	0.26885	0.26892	0.26898	0.26904	0.26910	0.26916	0.26922	0.26928
12.30	0.26934	0.26941	0.26947	0.26953	0.26959	0.26965	0.26971	0.26977	0.26983	0.26989
12.40	0.26995	0.27001	0.27007	0.27013	0.27019	0.27025	0.27031	0.27038	0.27044	0.27050

7. GEOTHERMAL MAPS

V. ČERMÁK
Geofysikalni Ustav, Československa Akademie Ved, Praha, CSSR

and

R. HAENEL
Department of Geophysics, Geological Survey of Lower Saxony, Hannover, FRG

7.1. INTRODUCTION

Geothermal data are commonly shown on maps, usually as the final step of the data acquisition or data compilation. The term geothermal map means the maps displaying the local, regional or global pattern of temperature, temperature gradient, heat-flow density, as well as the maps presenting the distribution of geothermal energy resources and reserves. Together with the geothermal field under consideration, the selected geological, tectonic, hydrogeological and other features are often given in the map as the background (supplementary) information. Geothermal maps may be of different scale, area coverage, isoline interval, projection, size, etc. depending on the purpose for which the particular map will be constructed.

Apart from the use of the geothermal map as a research tool for further general geophysical investigation of the crustal and upper mantle structure, this map may be of great interest in many aspects of applied geophysics, geology and in exploration, such as e.g.:

- Geothermal energy prospecting, since high subsurface temperature is the necessary condition for any economically exploitable geothermal reservoir.
- Oil and gas exploration, because the evolution of the hydrocarbons depends mainly on temperature.
- Mining and engineering geology in order to solve ventilation and other technical problems connected with the construction of deep tunnels and mines.
- Hydrogeology, and thermal water prospection as heat transport is closely related to underground water movements.
- Borehole logging for determination of gas expansion, water inflow, cementation level, etc.
- Mineral exploration, as some economically important deposits may produce an anomalous temperature field.

Maps of temperature gradient (vertical or horizontal) may complement the temperature maps within a certain region to demonstrate better the existing temperature increase or temperature differences for some specific studies. However, the temperature or depth/distance range must be specified.

The term 'geothermal map' should be reserved for isoline patterns with a clear

R. Haenel, L. Rybach, and L. Stegena (eds.)
Handbook of Terrestrial Heat-Flow Density Determination, 261—300.

definition how the contour lines were constructed. Nevertheless, maps, showing only the sites or locations with the values of the obtained data, should accompany the contour maps to give additional information, or substitute it if data density does not enable the construction of contour lines.

To present geothermal data in the form of a contoured map, several principal techniques may be recommended:

(1) Computer interpolation and contouring which include numerous methods that have been employed to interpolate between irregularly spaced data. These methods may be formally divided into two classes (Crain, 1970): mathematical surfaces and numerical surfaces. The former comprises all those that generate surfaces which fit to a closed mathematical expression. The latter are those generated by algorithms whose end product is a numerical array of values. Up to now only the mathematical (functional) representations have been practically used in geothermics, such as the spherical harmonic or polynomial representations. Both of these representations have been applied above all to the heat-flow density data sets to produce global and/or regional heat-flow density patterns. The best fit between the calculated surface and the data set investigated is usually secured by least-squares regression. Different weights can be attributed to the individual data according to their reliability, local importance, density, etc. The resultant contour maps show the calculated analysis of the convenient order.

(2) Construction of the isolines by hand (by 'eye'), i.e. by simple visual interpolation between the individual data points. Generally, such manual interpolation is linear and tends to produce a reasonably smooth and plausible representation. In mapping the heat-flow density field advantage is taken of the experimental relationship between the heat-flow density and the tectonic setting. The isolines of heat-flow density are drawn over a simplified tectonic pattern and the course of the isolines is adapted to the main geological or tectonic boundaries.

7.2. GENERAL STATEMENTS

7.2.1. Computer Contouring

(1) SPHERICAL HARMONIC REPRESENTATION

The Earth's ellipsoid is substituted by a sphere. A set of data (usually covering all the globe or a substantial part of it) $X(\theta_i, \phi_i)$, $i = 1, 2, \ldots, N$, where θ_i is the latitude and ϕ_i is the eastern longitude of the i-th data point (e.g. a heat-flow density station), can be represented by a spherical harmonic expansion of the M-th order (Lee and Uyeda, 1965; Garland, 1979):

$$X(\theta, \phi) = \sum_{n=0}^{M} \sum_{m=0}^{n} [(A_{nm} \cos(m\phi) + B_{nm} \sin(m\phi))] P_{nm} \cos(\theta) \equiv$$

$$\equiv \sum_{j=1}^{k} t_j T_j(\theta, \phi), \tag{7.1}$$

where $k = (M + 1)^2$, and $P_{nm}\cos(\theta)$ are the normalized associated Legendre functions, which satisfy:

$$\int_0^{2\pi}\int_0^{\pi}[P_{nm}\cos(\theta)\,{}^{\sin}_{\cos}(m\phi)]^2 \sin(\theta)\, d\theta\, d\phi = 4\pi \tag{7.2}$$

$T_j(\theta, \phi)$ are the spherical functions, t_j, A_{nm}, B_{nm} are the coefficients of the expansion. To find these coefficients, a system of N equations of k unknown must be solved. Such a system can be solved only if $N \geqslant k$, which is usually satisfied. By the least squares method the system can be reduced to k normal equations and its solution then found by e.g. the matrix inversion:

$$\sum_{j=1}^{k} t_j \sum_{i=1}^{N} T_l(\theta_i, \phi_i) T_j(\theta_i, \phi_i) = \sum_{i=1}^{N} X(\theta_i, \phi_i) T_l(\theta_i, \phi_i)$$

$$l = 1, 2, \ldots, k. \tag{7.3}$$

The individual points are usually irregularly distributed, but if some averaging is done and the characteristic values are calculated over regularly spaced grid elements, the solution is greatly simplified:

$$t_j = \frac{\sum_{i=1}^{N} X(\theta_i, \phi_i) T_j(\theta_i, \phi_i)}{\sum_{i=1}^{N} T_j^2(\theta_i, \phi_i)}, \qquad j = 1, 2, \ldots, k. \tag{7.4}$$

If all the surface of the globe is covered with a regular grid network and the characteristic or mean values of $\bar{X}(\theta, \phi)$ are known in all these grid elements, the numerical integration gives directly (Chapman and Pollack, 1975):

$$\begin{matrix} A_{nm} \\ B_{nm} \end{matrix} = \frac{1}{4\pi}\int_0^{2\pi}\int_0^{\pi} \bar{X}(\theta, \phi) \begin{matrix} \cos(m\phi) \\ \sin(m\phi) \end{matrix} P_{nm}\cos(\theta)\sin(\theta)\, d\theta\, d\phi. \tag{7.5}$$

Summarizing the meaning of the used symbols:

X = the considered value (temperature, heat-flow density etc.)
θ = latitude
ϕ = longitude
A, B, t = coefficients of the spherical harmonic expansion
$P_{nm}\cos(\theta)$ = normalized associated Legendre function
T = spherical function
i = 1, 2, ..., N, number of available values $X(\theta_i, \phi_i)$
j, l = 1, 2, ..., k, number of equations with $N \geqslant k$, $k = (M + 1)^2$
n = 0, 1, ..., M, degree of the expansion
m = 0, 1, ..., n, order of expansion.

The spherical harmonic representation gives a convenient way to illustrate the global heat-flow density distribution. Due to the recent progress in the heat flow studies, several attempts have been made to list all the observations and each such catalogue was supplemented by the derived global heat-flow density pattern. In 1963 Lee and MacDonald reported spherical harmonic coefficients to degree 2 based on 813 data. Subsequently, a revised map appeared based on 987 values which showed the 3rd order expansion (Lee and Uyeda, 1965) and later the 7th order expansion of Horai and Simmons (1969), who used 2812 heat-flow density observations.

As all these spherical harmonic representations were to some extent characterized by unreal distortions caused by the uneven density of data or by lack of observations in several critical areas, Chapman and Pollack (1975) tried to create a synthetic supplement to the existing observations by predicting the heat-flow density in the unsampled regions. Using the empirical relationship of the heat-flow density versus the tectonic age, characteristic heat-flow density values for all 2592 5° × 5° grid elements on the globe were compiled and the 12th degree spherical harmonic coefficients calculated (Chapman and Pollack, 1975) – see Table 7.1. For the time being, their map (Figure 7.1) represents the most complete and updated global heat-flow density representation.

(2) POLYNOMIAL REPRESENTATION

A set of data (usually covering a relatively smaller area of maximum continental size) $X(x_i, y_i)$, $i = 1, 2, \ldots, N$, where x_i, y_i are the coordinates of the i-th data point (not necessarily the geographic coordinates) can be represented by the polynom (Haenel, 1971):

$$X(x, y) = \sum_{m=0}^{M} \sum_{n=0}^{M} C_{mn} x^m y^n, \quad \text{where} \quad C_{mn} = 0 \text{ for all } m + n \leq M. \tag{7.6}$$

M is the order of the expansion, and C_{mn} are its coefficients. To determine these coefficients, we have a system of N equations, which can be again reduced by using the least squares method to a system of k normal equations, $k = \frac{1}{2}(M + 1)(M + 2)$. To apply the least squares method means to find the minimum of the function:

$$F = \sum_{i=1}^{N} [X(x_i, y_i) - \sum_{m=0}^{M} \sum_{n=0}^{M} C_{mn} x_i^m y_i^n]^2, \tag{7.7}$$

which requires to fulfil:

$$\frac{\partial F}{\partial C_{mn}} = 0 \text{ for all } \quad m, n = 0, 1, 2, \ldots\ m + n \leq M. \tag{7.8}$$

The system of k normal equations for k unknown coefficients C_{mn} can then be expressed:

$$\sum_{i=1}^{N} X(x_i, y_i) x_i^m y_i^n = \sum_{p=0}^{M} \sum_{r=0}^{M} C_{pr} \left(\sum_{i=1}^{N} x_i^{m+p} y_i^{n+r} \right), \tag{7.9}$$

TABLE 7.1
The spherical harmonic coefficients of terrestrial heat-flow density up to the order $M = 12$ calculated by Chapman and Pollack (1975). Coefficients are given in mW m^{-2}.

n	m	A_{nm}	B_{nm}	n	m	A_{nm}	B_{nm}
0	0	59.211		9	0	0.568	
1	0	−1.662		9	1	1.651	−0.128
1	1	−2.973	−1.373	9	2	0.387	−0.103
2	0	−0.472		9	3	−1.074	−0.742
2	1	0.997	0.788	9	4	0.101	−0.492
2	2	−3.031	1.396	9	5	0.461	−0.055
3	0	1.415		9	6	1.070	0.926
3	1	−0.479	−0.768	9	7	0.208	−1.247
3	2	0.945	−0.595	9	8	0.342	0.159
3	3	2.004	1.924	9	9	0.769	1.607
4	0	−2.034		10	0	0.468	
4	1	0.328	0.428	10	1	0.577	0.211
4	2	1.204	0.853	10	2	−0.118	0.139
4	3	−0.269	−1.338	10	3	0.835	0.058
4	4	0.351	−2.876	10	4	−0.881	0.224
5	0	1.540		10	5	0.018	0.162
5	1	−0.340	−0.224	10	6	−0.633	0.152
5	2	2.163	0.117	10	7	−0.600	0.470
5	3	1.791	−0.167	10	8	−0.258	0.386
5	4	−2.595	1.363	10	9	1.261	1.036
5	5	−1.473	−1.797	10	10	0.769	−0.832
6	0	−0.891		11	0	−0.463	
6	1	1.213	0.842	11	1	−0.391	0.647
6	2	2.088	−1.637	11	2	0.262	−0.045
6	3	−0.018	−0.118	11	3	0.110	0.115
6	4	−0.850	−1.736	11	4	0.116	−0.505
6	5	−0.354	0.296	11	5	−0.342	1.099
6	6	0.156	1.330	11	6	0.205	−0.235
7	0	0.087		11	7	0.182	−0.126
7	1	−0.419	−0.792	11	8	0.566	0.705
7	2	−0.554	0.576	11	9	−0.139	1.501
7	3	1.001	−0.413	11	10	0.128	−0.029
7	4	0.484	−0.652	11	11	−0.207	−0.605
7	5	−0.025	0.904	12	0	0.310	
7	6	0.269	−0.280	12	1	0.206	−0.039
7	7	0.050	−0.217	12	2	0.199	0.397
8	0	0.062		12	3	0.686	−0.099
8	1	0.140	1.571	12	4	0.441	−0.361
8	2	1.130	1.080	12	5	−0.034	−0.135
8	3	0.092	0.347	12	6	0.245	−0.713
8	4	−0.465	0.664	12	7	−0.421	0.367
8	5	0.865	0.483	12	8	−0.126	1.812
8	6	0.880	2.085	12	9	−0.173	0.388
8	7	0.090	−0.095	12	10	0.715	−0.529
8	8	0.149	−0.289	12	11	−0.741	−0.263
				12	12	−0.640	1.031

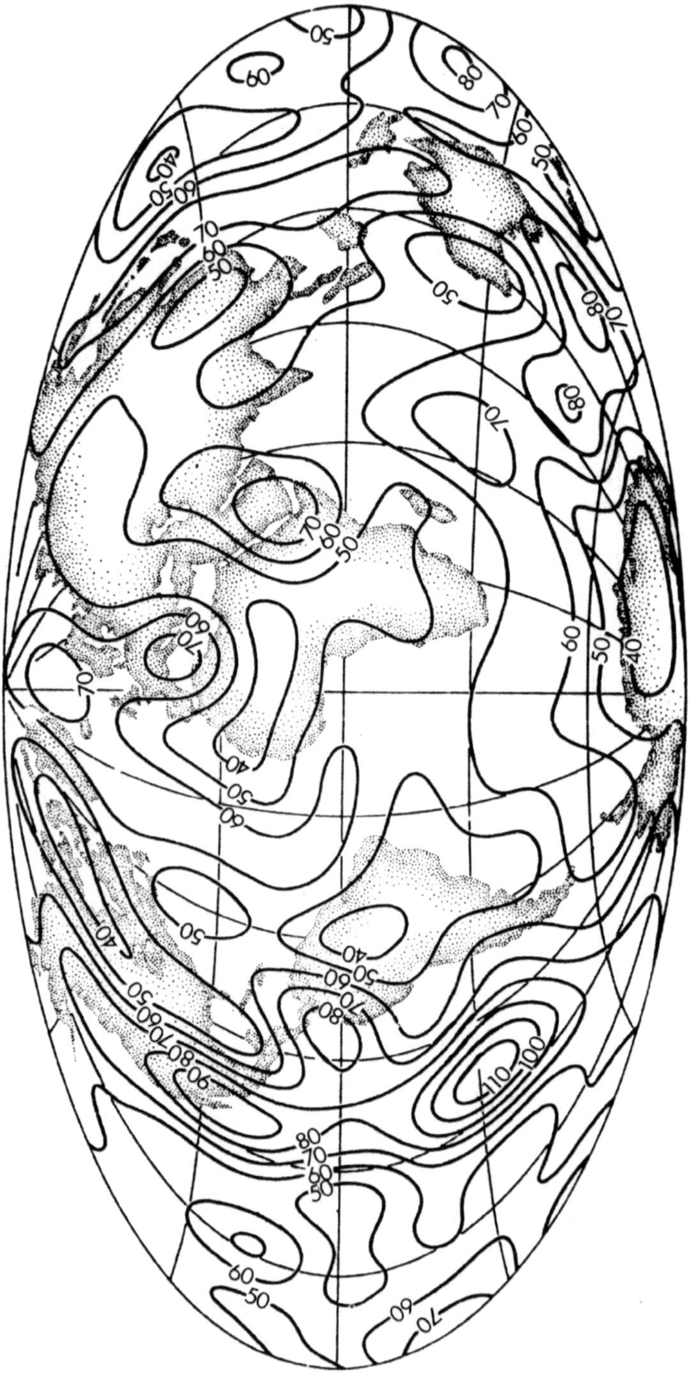

Fig. 7.1. Degree 12 spherical harmonic representation of global heat-flow density (observations supplemented by predictor technique) as proposed by Chapman and Pollack (1975). Isolines are labelled in mW m^{-2}.

where

$$i = 1, 2, \ldots N, \text{ number of measured values } X_i(x_i, y_i)$$
$$m, n = 0, 1, 2, \ldots k, \text{ number of equations with } m + n \leq M$$
$$p, r = 0, 1, 2, \ldots M, \text{ order of expansion with } p + r \leq M.$$

The polynomial representation, the calculated trend of surface data, of the heat-flow density field was proposed for the territory of Germany by Haenel (1971). Haenel and Zoth (1973) published the 5th order representation for Central Europe and later Haenel (1974) used the 7th order representation for all Europe based on 643 data. Similar studies were also performed by Hurtig and Oelsner (1977) and by Čermák and Hurtig (1977), who used the 6th order polynomial representation of the heat-flow density in Europe based in the latter case on about 2400 values.

A good example of the gradual change of the calculated pattern due to the progressive increase of the order of the polynomial trend analysis (from 1 to 8) can be demonstrated on the polynomial heat-flow density maps of France (Vasseur and Nouri, 1980); see Figure 7.2.

(3) OTHER METHODS OF ISOLINE CONTOURING

Modern computer treatment of (mostly non-equispaced) geophysical data often employs some combination of continuous and piece-wise surfaces as part of a purely numerical procedure to obtain their representation. Even though these methods have not yet been routinely used in geothermics, their future use is likely and in some special cases they may be quite promising. An example of numerical heat-flow density contouring was given by Rybach and Bodmer (1983). The critical parameters for such an approach are, besides a sufficient data density, the interpolation radius, the grid spacing, the smoothing factor and the threshold of elimination.

The least squares fitting is generally used as a preliminary step to refine the measured data before the specific method is applied. The numerical methods in the literature may differ considerably, as seen in the review of Crain and Bhattacharyja (1967), Crain (1970). In the following an easy but good approximation is described as presented by Mundry (1970).

To fill the sparely populated area under consideration, additional auxiliary values must be calculated in such a way, that the irregularly measured data set is transformed into a regular grid-point data system.

For that the area under consideration is covered by a fine rectangular grid system. In each grid point a value is calculated using the trend analysis as described below Section (2) but only for the first order (plane of regression). The value is calculated by the measured values in the surroundings, which are provided with weights that are inversely proportional to a certain power of the distance from the considered grid point; here is taken the forth power:

$$w_i = \frac{1}{((x_i - u)^2 + (y_i - v)^2)}, \tag{7.10}$$

where x_i, y_i are the coordinates of the measured value $X_i(x_i, y_i)$, and u, v are the

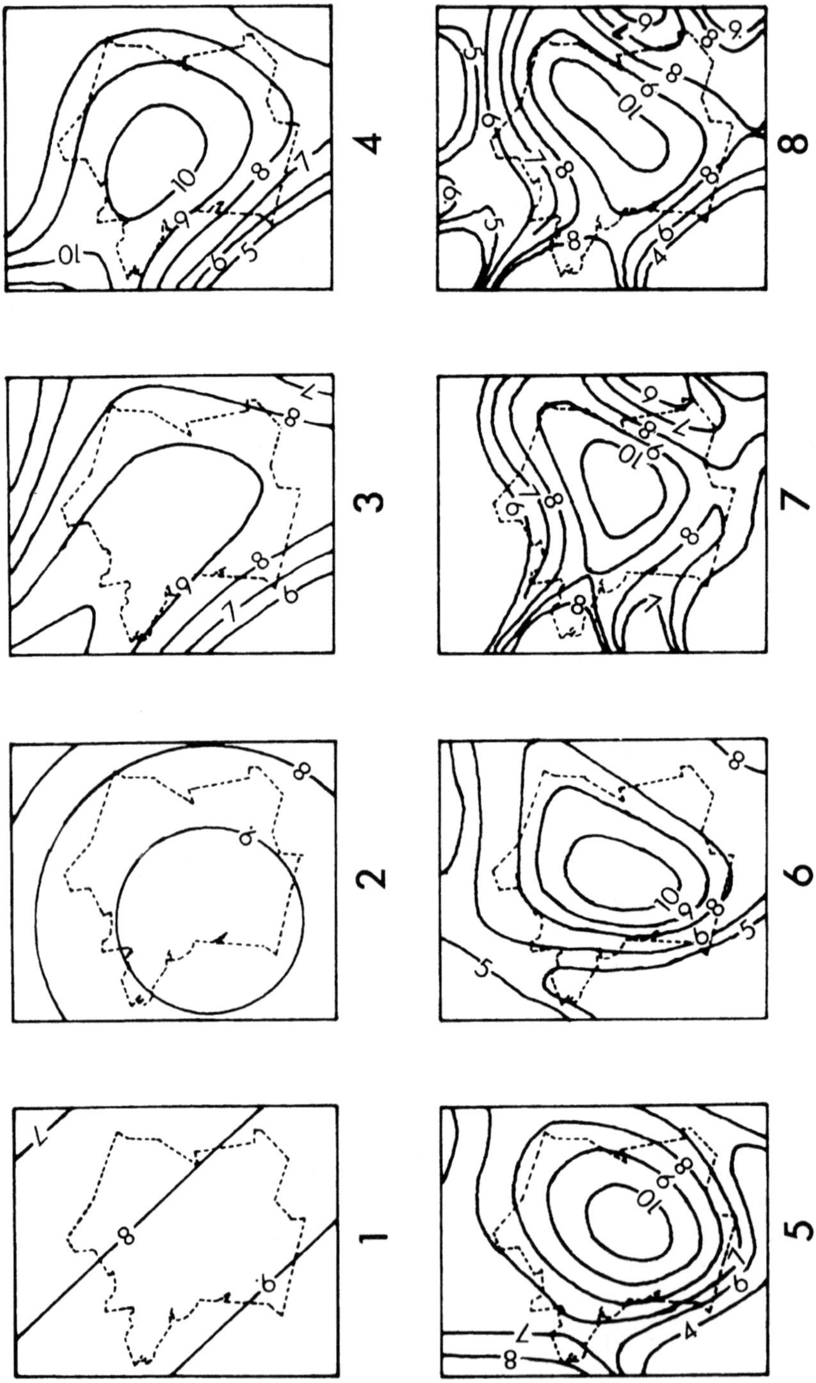

Fig. 7.2. Trends of the heat-flow density field in France calculated by the polynomial analysis (order 1–8), isolines are labelled in 10 mW m^{-2}. Adapted from Vasseur and Nouri (1980).

coordinates of the considered grid point $G(u, v)$.

The weight w must be introduced into Equation (7.9):

$$\sum_{i=1}^{N} w_i X_i x_i^m y_i^n = \sum_{p=0}^{M} \sum_{r=0}^{M} C_{pr} \left(\sum_{i=1}^{N} w_i x_i^{m+p} y_i^{n+r} \right) \tag{7.11}$$

and solved for the order $M = 1$ for the coefficients C_{pr}, keeping in mind that $p, r = 0, 1$ with $p + r \leq M$:

$$\sum_{i=1}^{N} w_i X_i x_i^m y_i^n = C_{00} \sum_{i=1}^{N} w_i x_i^m y_i^n + C_{01} \sum_{i=1}^{N} w_i x_i^m y_i^n + C_{10} \sum_{i=1}^{N} w_i x_i^{m+1} y_i^n,$$

and now solving the equation for $m, n = 0, 1$ with $m + n \leq M$:

$$(m = 0, n = 0)\colon \sum_{i=1}^{N} w_i X_i = C_{00} \sum_{i=1}^{N} w_i + C_{01} \sum_{i=1}^{N} w_i y_i + C_{10} \sum_{i=1}^{N} w_i x_i \tag{7.12}$$

$$(m = 1, n = 0)\colon \sum_{i=1}^{N} w_i X_i x_i = C_{00} \sum_{i=1}^{N} w_i x_i + C_{01} \sum_{i=1}^{N} w_i x_i y_i + C_{10} \sum_{i=1}^{N} w_i x_i^2 \tag{7.13}$$

$$(m = 0, n = 1)\colon \sum_{i=1}^{N} w_i X_i y_i = C_{00} \sum_{i=1}^{N} w_i y_i + C_{01} \sum_{i=1}^{N} w_i y_i^2 + C_{10} \sum_{i=1}^{N} w_i x_i y_i. \tag{7.14}$$

Now, for each grid point a value $G(u, v)$ can be calculated from the above plane of regression, described by the Equations (7.12)–(7.14).

During the next step all squares of the grid system has to be completed by two diagonals (see Figure 7.3) which delivers triangles. In the centre of each square a mean value is calculated from the four neighbouring values. The isolines can be now constructed by means of using a linear interpolation technique between the points of triangles; see also Figure 7.3:

$$\begin{aligned} u_m &= u_1 + \frac{u_2 - u_1}{G_2 - G_1}(\bar{G} - G_1) \\ v_m &= v_1 + \frac{v_2 - v_1}{G_2 - G_1}(\bar{G} - G_1), \end{aligned} \tag{7.15}$$

where

$\bar{G}$ = the looked for isoline value with coordinates u_m, v_m; $\bar{G}$ is given

$G_{1,2}$ = the values of squares and triangles, respectively; the coordinates (u_1, v_1) and (u_2, v_2).

Equation (7.15) can be applied for non-rectangular coordinates too, and of course, also without Equations (7.12) to (7.14).

In cases for which, apart from the measured heat-flow density values, other information exists relating to the course of the isolines (geotherms, tectonic setting), some correction procedure or hand reworking of the computer produced contouring pattern may be necessary.

7.2.2. Visual Contouring

This is the most frequent way of constructing isolines by visual judgment of all

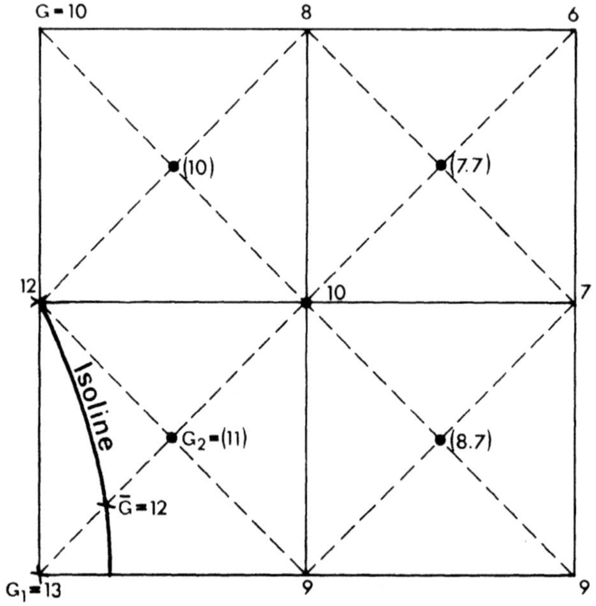

Fig. 7.3. The squares of the grid system are subdivided by diagonals producing triangles. In each grid point the values $G(u, v)$ are given by means of Equation (7.12) to (7.14). The values in brackets are calculated by the four neighbouring values. The isolines are calculated by means of Equation (7.15); here demonstrated for the isoline being 12.

data and taking into account temperature data and/or the main tectonic elements. Different authors used slightly different techniques to construct isolines; in most cases, however, the technique used is not explicitly explained in the corresponding paper. As a matter of fact, there are no general rules for the construction at all. The correctness of the proposed pattern can be checked only by comparing the match of new measurements to the isoline pattern proposed earlier.

Visual contouring requires the knowledge of the nature of the terrestrial heat-flow density, the heat transfer conditions within the crust and the upper mantle, namely the relationship between the specific geothermal parameter and the tectonic structure. The construction of heat-flow density maps usually relates to the dependence of the heat-flow density on the age of the last tectono-thermal event on continents (Polyak and Smirnov, 1968) and/or to the heat-flow density relation to the age of the ocean floor in oceanic regions (Sclater and Francheteau, 1970). More discussion on this topic is given e.g. in Vitorello and Pollack (1980), Pollack (1980) or Rao *et al.* (1982). In areas with a small number of observations the course of the isolines can be fitted to follow the boundaries of the principal tectonic or geological units. Besides of tectonic setting many other geological parameters may be of some help in geothermal mapping, such as geotemperatures, hot spring distribution, fault or fracture systems, major lithological complexes, and crustal thickness.

An original set of data may sometimes be filtered out prior to the contouring procedure in order to exclude anomalous values or data which are likely to be affected by a disturbance not typical of the investigated case. Some authors also claim a minimum number of data (e.g. three points) to close the isoline to show a single anomaly.

Many local heat-flow density patterns and national heat-flow density maps of various countries were published, which were prepared by the visual contouring. So far, however, only a few attempts have been made to construct such a map on a larger or continental scale. Under the International Heat Flow Commission (IHFC) project a preliminary heat-flow density map of Europe to the scale 1:5 000 000 appeared in 1977 (Čermák and Hurtig, 1977) and the final multicolour version was issued in 1979 Čermák and Hurtig, 1979). Another compilation of European heat-flow density data was performed for the area of seven countries of the European Community (Haenel, 1980). The heat-flow density field on the territory of the USSR was shown on the map to the scale of 1:10 000 000 by Smirnov (1980) together with the deep temperature distribution and the heat-flow density on the Mohorovičić discontinuity.

The construction of the heat-flow density map of Europe (Čermák and Hurtig 1979) was based on 3076 existing heat-flow density observations which were supplemented for this purpose by all other suitable information such as deep temperature measurements, heat-flow density estimates and near surface temperature gradients. The interpretation was done on the basis of the regional tectonic pattern. Such a procedure was inevitable at its time in order to produce the most complete representation of the surface geothermal activity. However, to follow the contemporary recommendations of the IHFC not to use estimated heat-flow density data, the future attemps to construct regional or continental heat-flow density patterns should concentrate more on the employment of direct heat-flow density measurements only and all other information should be strictly classified as auxiliary.

7.2.3. Test of Significance

The fit of the calculated mathematical or numerical representation to the original data set can be evaluated by means of the proper statistical criteria such as the significance tests. Such test may be also used to assess the most suitable order of the chosen representation.

The three most common procedures will be briefly described, which are based on the variance:

$$s^2 = \frac{1}{N-1}\sum_{i=1}^{N}(x_i - \bar{x})^2, \tag{7.16}$$

where

x_i = i-th measured value, e.g. the heat-flow density
$\bar{x}$ = mean of all x_i
i = 1, 2, ... N, the number of measured values x_i.

Keeping in mind, the square root of Equation (7.16) is the standard error:

$$s = \left(\frac{\sum_{i=1}^{N} (x_i - \bar{x})^2}{N - 1} \right)^{1/2}. \tag{7.17}$$

(1) THE FIT OF TREND, FIT

A measure of the goodness of fit of the trend can be expressed by the deviation of the data x_i from its mean value $\bar{x}$ as well as by deviation of the data x_i from the calculated trend surface data $x_i^!$; the polynomial representation as shown under 7.2.1 (2). This leads to (Harbaugh and Merriam, 1968):

$$\mathrm{FIT}(\%) = 100 \left(1 - \frac{\sum_{i=1}^{N} (x_i - x_i^!)^2}{\sum_{i=1}^{N} (x_i - \bar{x})^2} \right), \tag{7.18}$$

where

$x_i^!$ = value of trend surface on the place of x_i.

A perfect fit of a trend (a trend surface of high order) to the data points would yield a value of 100 % ($x_i = x_i^!$). This case is uncommon.

If the goodness of FIT is low, e.g. 15–30 %, so the data are not good represented by the choosen order of the trend surface. This is not necssarily bad, but the interpretation of results should be made with this fact clearly in mind. The deviations or residuals, $\varepsilon_i = x_i - x_i^!$, may give geophysical or geothermal importance.

Going with the trend surface from the order M to the next higher order $M + 1$, so a measure of improvement can be represented by (Mundry, 1984):

$$\Delta(\mathrm{FIT}) = \frac{\mathrm{FIT\ of\ }(M + 1)\mathrm{\ order}}{\mathrm{FIT\ of\ }(M)\mathrm{\ order}}. \tag{7.19}$$

(2) THE CORRELATION COEFFICIENT, R

The correlation between x_i and $x_i^!$ is defined as the ratio of the co-variance s to the variances $s_1 = f(x_i, \bar{x})$ and $s_2 = f(x_i^!, \bar{x})$ as shown by Kreyszig (1967):

$$s = \sum_{i=1}^{N} (x_i - \bar{x})(x_i^! - \bar{x})/(N - 1) \tag{7.20}$$

$$s_1 = \left(\sum_{i=1}^{N} (x_i - \bar{x})^2/(N - 1) \right)^{1/2} \tag{7.21}$$

$$s_2 = \left(\sum_{i=1}^{N} (x_i^! - \bar{x})^2/(N - 1) \right)^{1/2} \tag{7.22}$$

and:

$$r = \frac{s}{s_1 s_2} = \frac{\sum_{i=1}^{N} (x_i - \bar{x})(x_i^t - \bar{x})}{\left(\sum_{i=1}^{N} (x_i - \bar{x})^2 \cdot \sum_{i=1}^{N} (x_i^t - \bar{x})^2\right)^{1/2}}, \tag{7.23}$$

r ranges between -1 and 1. In the case $x_i^t \equiv x_i$, the correlation coefficient equals to 1. So that means, the higher r the better the approximation by the trend surface is.

The improvement from trend surface of order M to order $M + 1$ can be expressed again by:

$$\Delta(r) = \frac{r \text{ of } (M + 1) \text{ order}}{r \text{ of } (M) \text{ order}}. \tag{7.24}$$

(3) THE STATISTICAL SIGNIFICANCE OF A TREND SURFACE

The objective is to determine if components of a trend function are statistically significant or not. For that the mean sum of squares due to trend function is considered (Harbaugh and Merriam, 1968):

$$V_1 = \frac{\sum_{i=1}^{N} (x_i^t - \bar{x})^2}{M}, \tag{7.25}$$

and the mean sum of squares due to deviations:

$$V_2 = \frac{\sum_{i=1}^{N} (x_i - x_i^t)^2}{N - M - 1}, \tag{7.26}$$

where the quantities $M = f_1$ and $N - M - 1 = f_2$ are the degress of freedom. The ratio V_1/V_2 is called the Fisher function or the F-value:

$$F = F(f_1, f_2) = \frac{V_1}{V_2}. \tag{7.27}$$

F is a known probability distribution, which is associated with the two values f_1 and f_2 for degrees of freedom. In the case the components of trend function have a normal distribution with a statistical reliability of $1 - \alpha$ (e.g. 95 %), the F-value must be higher (better) than a threshold-value $F_s(f_1, f_2, 1 - \alpha)$:

$$F(f_1, f_2) > F_s(f_1, f_2, 1 - \alpha). \tag{7.28}$$

The F_s-values are tabulated and given in Table 7.2.

To have a comparable procedure of improvement from one trend surface to the higher one as given above, the ratio has to be calculated:

$$\Delta(F) = \frac{F \text{ of } (M + 1) \text{ order}}{F \text{ of } (M) \text{ order}}. \tag{7.29}$$

TABLE 7.2
The threshold values $F_s(f_1, f_2, 1 - \alpha)$ for $1 - \alpha = 95$ % of F-distribution

f_2 \ f_1	1	2	3	4	5	6	7	8	9	10	11	12	13	14	15
1	161	200	216	225	230	234	237	239	241	242	243	244	245	245	246
2	18.5	19.0	19.2	19.2	19.3	19.3	19.4	19.4	19.4	19.4	19.4	19.4	19.4	19.4	19.4
3	10.1	9.55	9.28	9.12	9.01	8.94	8.89	8.85	8.81	8.79	8.76	8.74	8.73	8.71	8.70
4	7.71	6.94	6.59	6.39	6.26	6.16	6.69	6.04	6.00	5.96	5.94	5.91	5.89	8.87	5.86
5	6.61	5.79	5.41	5.19	5.05	4.95	4.88	4.82	4.77	4.74	4.70	4.68	4.66	4.64	4.62
6	5.99	5.14	4.76	4.53	4.39	4.28	4.21	4.15	4.10	4.06	4.03	4.00	3.98	3.96	3.94
7	5.59	4.74	4.35	4.12	3.97	3.87	3.79	3.73	3.68	3.64	3.60	3.57	3.55	3.53	3.51
8	5.32	4.46	4.67	3.84	3.69	3.58	3.50	3.44	3.39	3.35	3.31	3.28	3.26	3.24	3.22
9	5.42	4.26	3.86	3.63	3.48	3.37	3.29	3.23	3.18	3.14	3.10	3.07	3.05	3.03	3.01
10	4.96	4.10	3.71	3.48	3.33	3.22	3.14	3.07	3.02	2.98	2.94	2.91	2.89	2.86	2.85
11	4.84	3.98	3.59	3.36	3.20	3.09	3.01	2.95	2.90	2.85	2.82	2.79	2.76	2.74	2.72
12	4.75	3.89	3.49	3.26	3.11	3.00	2.91	2.85	2.80	2.75	2.72	2.69	2.66	2.64	2.62
13	4.67	3.81	3.41	3.18	3.03	2.92	2.83	2.77	2.71	2.67	2.63	2.60	2.58	2.55	2.53
14	4.60	4.74	3.34	3.11	2.96	2.85	2.76	2.70	2.65	2.60	2.57	2.53	2.51	2.48	2.46
15	4.54	3.68	3.29	3.06	2.90	2.79	2.71	2.64	2.59	2.54	2.51	2.48	2.45	2.42	2.40
16	4.49	3.63	3.24	3.01	2.85	2.74	2.66	2.59	2.54	2.49	2.46	2.42	2.50	2.37	2.35
17	4.45	3.59	3.20	2.96	2.81	2.70	2.61	2.55	2.49	2.45	2.41	2.38	2.35	2.33	2.31
18	4.41	3.55	3.16	2.93	2.77	2.66	2.58	2.51	2.46	2.41	2.37	2.34	2.31	2.29	2.27
19	4.38	3.52	3.13	2.90	2.74	2.63	2.54	2.48	2.42	2.38	2.34	2.31	2.28	2.26	2.23
20	4.35	3.49	3.10	2.87	2.71	2.60	2.51	2.45	2.39	2.35	2.31	2.28	2.25	2.22	2.20
22	4.30	3.44	3.05	2.82	2.66	2.55	2.46	2.40	2.34	2.30	2.26	2.23	2.20	2.17	2.15
24	4.26	3.40	3.01	2.78	2.62	2.51	2.42	2.36	2.30	2.25	2.21	2.18	2.15	2.13	2.11
26	4.23	3.37	2.98	2.74	2.59	2.47	2.39	2.32	2.27	2.22	2.18	2.15	2.12	2.09	2.07
28	4.20	3.34	2.95	2.71	2.56	2.45	2.36	2.29	2.24	2.19	2.15	2.12	2.09	2.06	2.04
30	4.17	3.32	2.92	2.69	2.53	2.42	2.33	2.27	2.21	2.16	2.13	2.09	2.06	2.04	2.01
32	4.15	3.29	2.90	2.67	2.51	2.40	2.31	2.24	2.19	2.14	2.10	2.07	2.04	2.01	1.99
34	4.13	3.28	2.88	2.65	2.49	2.38	2.29	2.23	2.17	2.12	2.08	2.05	2.02	1.99	1.97
36	4.11	3.26	2.87	2.63	2.48	2.36	2.28	2.21	2.15	2.11	2.07	2.03	2.00	1.98	1.95
38	4.10	3.24	2.85	2.62	2.46	2.35	2.26	2.19	2.14	2.09	2.05	2.02	1.99	1.96	1.94
40	4.08	3.23	2.84	2.61	2.45	2.34	2.25	2.18	2.12	2.08	2.04	2.00	1.97	1.95	1.92
50	4.03	3.18	2.79	2.56	2.40	2.29	2.20	2.13	2.07	2.03	1.99	1.95	1.92	1.89	1.87
60	4.00	3.15	2.76	2.53	2.37	2.25	2.17	2.10	2.04	1.99	1.95	1.92	1.89	1.86	1.84
70	3.98	3.13	2.74	2.50	2.35	2.23	2.14	2.07	2.02	1.97	1.93	1.89	1.86	1.84	1.81
80	3.96	3.11	2.72	2.49	2.33	2.21	2.13	2.06	2.00	1.95	1.91	1.88	1.84	1.82	1.79
100	3.94	3.09	2.70	2.46	2.31	2.19	2.10	2.03	1.97	1.93	1.89	1.85	1.82	1.79	1.77
200	3.89	3.04	2.65	2.42	2.26	2.14	2.06	1.98	1.93	1.88	1.84	1.80	1.77	1.74	1.72
300	3.87	3.03	2.63	2.40	2.24	2.13	2.04	1.97	1.91	1.86	1.82	1.78	1.75	1.72	1.70
500	3.86	3.01	2.62	2.39	2.23	2.12	2.03	1.96	1.90	1.85	1.81	1.77	1.74	1.71	1.69
1000	3.85	3.00	2.61	2.38	2.22	2.11	2.02	1.95	1.89	1.84	1.80	1.76	1.73	1.70	1.68
∞	3.84	3.00	2.60	2.37	2.21	2.10	2.01	1.94	1.88	1.83	1.79	1.75	1.72	1.69	1.67

Table 7.2 (Continued)

16	17	18	19	20	24	30	40	50	60	80	100	200	500	∞	f_1 / f_2
246	247	247	248	248	249	250	251	252	252	252	253	254	254	254	1
19.4	19.4	19.4	19.4	19.4	19.5	19.5	19.5	19.5	19.5	19.5	19.5	19.5	19.5	19.5	2
8.69	8.68	8.67	8.67	8.66	8.64	8.62	8.59	8.58	8.57	8.56	8.55	8.54	8.53	8.53	3
5.84	5.83	5.82	5.81	5.80	5.77	5.75	5.72	5.70	5.69	5.67	5.66	5.65	5.64	5.63	4
4.60	4.59	4.58	4.57	4.56	4.53	4.50	4.46	4.44	4.43	4.41	4.41	4.39	4.37	4.37	5
3.92	3.91	3.90	3.88	3.87	3.84	3.81	3.77	3.75	3.74	3.72	3.71	3.69	3.68	3.67	6
3.49	3.48	3.47	3.46	3.44	3.41	3.38	3.34	3.32	3.30	3.29	3.27	3.25	3.24	3.23	7
3.20	3.19	3.17	3.16	3.15	3.12	3.08	3.04	3.02	3.01	2.99	2.97	2.95	2.94	2.93	8
2.99	2.97	2.96	2.95	2.94	2.90	2.86	2.83	2.80	2.79	2.77	2.76	2.73	2.72	2.71	9
2.83	2.81	2.80	2.78	2.77	2.74	2.70	2.66	2.64	2.62	2.60	2.59	2.56	2.55	2.54	10
2.70	2.69	2.67	2.66	2.65	2.61	2.57	2.53	2.51	2.49	2.47	2.46	2.43	2.42	2.40	11
2.60	2.58	2.57	2.56	2.54	2.51	2.47	2.43	2.40	2.38	2.36	2.35	2.32	2.31	2.30	12
2.51	2.50	2.48	2.47	2.46	2.42	2.38	2.34	2.31	2.30	2.27	2.26	2.23	2.22	2.21	13
2.44	2.43	2.41	2.40	2.39	2.35	2.31	2.27	2.24	2.22	2.20	2.19	2.16	2.14	2.13	14
2.38	2.37	2.35	2.34	2.33	2.29	2.25	2.20	2.18	2.16	2.14	2.12	2.10	2.08	2.07	15
2.33	2.32	2.30	2.29	2.28	2.24	2.19	2.15	2.12	2.11	2.08	2.07	2.04	2.02	2.01	16
2.29	2.27	2.26	2.24	2.23	2.19	2.15	2.10	2.08	2.06	2.03	2.02	1.99	1.97	1.96	17
2.25	2.23	2.22	2.20	2.19	2.15	2.11	2.06	2.04	2.02	1.99	1.98	1.95	1.93	1.92	18
2.21	2.20	2.18	2.17	2.16	2.11	2.07	2.03	2.00	1.98	1.96	1.94	1.91	1.89	1.88	19
2.18	2.17	2.15	2.14	2.12	2.08	2.04	1.99	1.97	1.95	1.92	1.91	1.88	1.86	1.84	20
2.13	2.11	2.10	2.08	2.07	2.03	1.98	1.94	1.91	1.89	1.86	1.85	1.82	1.80	1.78	22
2.09	2.07	2.05	2.04	2.03	1.98	1.94	1.89	1.86	1.84	1.82	1.80	1.77	1.75	1.73	24
2.05	2.03	2.02	2.00	1.99	1.95	1.90	1.85	1.82	1.80	1.78	1.76	1.73	1.71	1.69	26
2.02	2.00	1.99	1.97	1.96	1.91	1.87	1.82	1.79	1.77	1.74	1.73	1.69	1.67	1.65	28
1.99	1.98	1.96	1.95	1.93	1.89	1.84	1.79	1.76	1.74	1.71	1.70	1.66	1.64	1.62	30
1.97	1.95	1.94	1.92	1.91	1.86	1.82	1.77	1.74	1.71	1.69	1.67	1.63	1.61	1.59	32
1.95	1.93	1.92	1.90	1.89	1.84	1.80	1.75	1.71	1.69	1.66	1.65	1.61	1.59	1.57	34
1.93	1.92	1.90	1.88	1.87	1.82	1.78	1.73	1.69	1.67	1.64	1.62	1.59	1.56	1.55	36
1.92	1.90	1.88	1.87	1.85	1.81	1.76	1.71	1.68	1.65	1.62	1.61	1.57	1.54	1.53	38
1.90	1.89	1.87	1.85	1.84	1.79	1.74	1.69	1.66	1.64	1.61	1.59	1.55	1.53	1.51	40
1.85	1.83	1.81	1.80	1.78	1.74	1.69	1.63	1.60	1.58	1.54	1.52	1.48	1.46	1.44	50
1.82	1.80	1.78	1.76	1.75	1.70	1.65	1.59	1.56	1.53	1.50	1.48	1.44	1.41	1.39	60
1.79	1.77	1.75	1.74	1.72	1.67	1.62	1.57	1.53	1.50	1.47	1.45	1.40	1.37	1.35	70
1.77	1.75	1.73	1.72	1.70	1.65	1.60	1.54	1.51	1.48	1.45	1.43	1.38	1.35	1.32	80
1.75	1.73	1.71	1.69	1.68	1.63	1.57	1.52	1.48	1.45	1.41	1.39	1.34	1.31	1.28	100
1.69	1.67	1.66	1.64	1.62	1.57	1.52	1.46	1.41	1.39	1.35	1.32	1.26	1.22	1.19	200
1.68	1.66	1.64	1.62	1.61	1.55	1.50	1.43	1.39	1.36	1.32	1.30	1.23	1.19	1.15	300
1.66	1.64	1.62	1.61	1.59	1.54	1.48	1.42	1.38	1.34	1.30	1.28	1.21	1.16	1.11	500
1.65	1.63	1.61	1.60	1.58	1.53	1.47	1.41	1.36	1.33	1.29	1.26	1.19	1.13	1.08	1000
1.64	1.62	1.60	1.59	1.57	1.52	1.46	1.39	1.35	1.32	1.27	1.24	1.17	1.11	1.00	∞

An example of trend analysis including a test of significance will be studied. For the trend analysis the Equation (7.9) is applied and the results are shown in Figure 7.4. The calculations were kindly supported by a computer programme of Mundry (1984). The results of the significance test are compiled in Table 7.3,

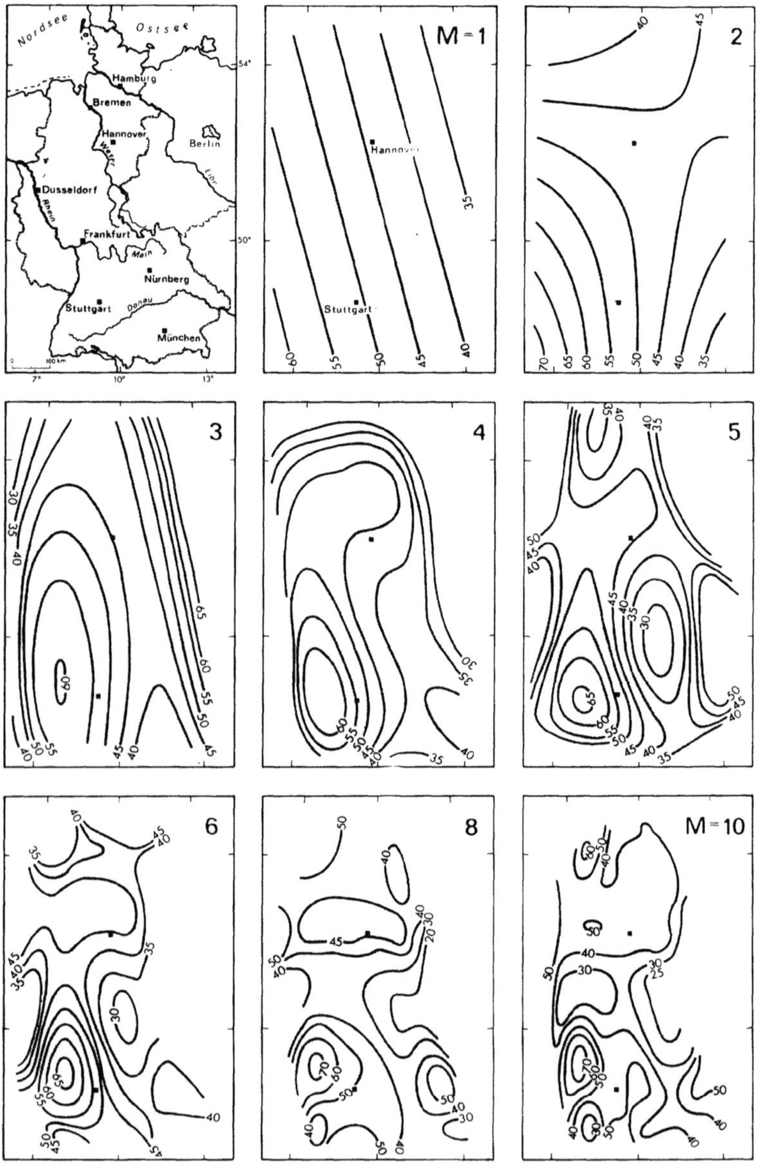

Fig. 7.4. Trends of the temperature field at 1000 m depth below ground level in the German Federal Republic calculated by the polynomial analysis; Order $M = 1$ to $M = 10$.

TABLE 7.3
Calculation for the best fit of trend based on 1005 temperature values from the area of Federal Republic of Germany at 1000 m depth below ground level. Formulas used for calculation: FIT (7.18), r (7.23), F (7.27), Δ (FIT) (7.19), $\Delta(r)$ (7.24), $\Delta(F)$ (7.29).

Order M	FIT(%)	r	F-value	F_s	Δ(FIT)	$\Delta(r)$	$\Delta(F)$
1	14.1	0.376	14.60	3.85			
2	20.7	0.455	27.55	3.00	1.47	1.20	1.89
3	24.0	0.495	14.12	2.61	1.16	1.09	0.51
4	28.8	0.537	10.84	2.38	1.20	1.11	1.81
5	31.3	0.559	5.81	2.22	1.09	1.04	1.54
6	36.0	0.600	10.24	2.11	1.15	1.07	1.76
7	39.2	0.626	6.41	2.02	1.08	1.04	0.63
8	42.2	0.650	5.55	1.95	1.08	1.04	0.87
9	44.1	0.644	3.22	1.89	1.04	1.02	0.58
10	45.8	0.677	2.77	1.84	1.04	1.02	0.87

using the Equations (7.18), (7.23), (7.27) and Table 7.3 for the F_s-values; further the Equations (7.19), (7.24) and (7.29) for the measure of improvement from order M to order $M + 1$.

From the FIT, r, F, and F_s value (Table 7.3) it can be concluded that all values deliver a good trend surface. Because the FIT and r should be as high as possible, the trend surface of order 10 would be the best. Following the ratio-equations, expressed by Δ, the best trend surface is given if all three Δ-values have a maximum. This is the case for the trend surface of order $M = 2$; and the next best trend surface is of order $M = 4$. Keeping in mind that the FIT and the correlation coefficient r should be as high as possible, the trend surface of order $M = 4$ is to prefer.

7.2.4. Comparison of Individual Techniques

Both the spherical and the polynomial analyses reveal purely mathematical representations, thus giving pictures not influenced by any other geophysical or geological information or by a personal opinion of the interpreter. However, the number and the density of the data and their quality can greatly limit the reliability of such a picture. A certain smoothing effect may tend to reduce the extreme values in the centres of the anomalies, on the other hand, some diffusion of the local anomalies may be caused. Moreover, the polynomial representation may quite often yield strong gradients near the margins, which may not be representative at all, unless the boundary conditions are prescribed (e.g. by assumed, estimated or predicted data), which, on the other hand, may influence the objectivness of the data distribution as well.

Specific problems may arise also from the absence of data in some areas, where unreal distortions of the computed field may appear. This difficulty was partly overcome by Chapman and Pollack's (1975) attempt to complement the original set of data by a set of predicted (estimated) data from the unsampled regions, i.e. to ensure a regular data coverage. This method was successfully used to propose a global heat flow density pattern (Figure 7.1), but may pose problems if used in some other cases of constructing geothermal maps. Neither the use of the empirical predictor needs to reflect all the fineness of the studied relationship. However, the future advance in the prediction technique and in the establishment of similar experimental relationships may help considerably in both, large scale and/or local scale geothermal mapping.

Any harmonic or polynomial representation filters out the short wavelength features with dimensions smaller than a certain characteristic value, which depends on the order of the representation used. Some local, but structurally important anomalies may not appear and to increase the order of the expansion for their recognition has little sense unless a greater volume of observational material is ensured.

Contouring by the visual method is not so vulnerable to the density of data nor to the size of individual anomalies. Anyhow, the main objection to this technique is its strong dependence on the used auxiliary data or model. Once there are no strictly mathematically defined rules of the interpolation between the individual data, there may be an unlimited number of resulting patterns corresponding to the same original set of observations. When preparing the heat-flow density map of Europe, Čermák and Hurtig (1979) were aware of this problem and they tried to minimize this effect by involving a number of specialists into the team of co-authors. The individual subjective view was thus replaced by a collective view. It is believed that such approach is the only one possible if it concerns a map of a large territory, the interpretation of which can be hardly governed by a single person.

Three maps of the heat-flow density in Europe are shown, which were produced by different methods (Figures 7.5, 7.6 and 7.7). As all of them cover the same territory and were based on an approximately identical material, they may serve well for the mutual comparison of the different mapping techniques. Figure 7.5 shows the 6th order polynomial representation of the heat-flow density in Europe prepared by Hurtig and based on 2605 data points (Čermák and Hurtig 1977). The similar heat-flow density pattern over the European continent was derived from the global analysis of Chapman and Pollack (1975). For this purpose the original data base was complemented by additional information from Europe and the 18th order spherical harmonic coefficients were calculated (Figure 7.6) (Chapman *et al.*, 1979). The map area is roughly a square 4500 km on a side, for which the characteristic dimension represented by the 6th order polynomial analysis is about 750 km. This is not greatly different from the 1100 km dimension represented by the spherical harmonics of order 18.

Figure 7.7 shows a simplified version of the heat-flow density map of Europe prepared by the visual contouring (Čermák and Hurtig 1979). Even though the latter map is much more complicated and 'broken' into a series of smaller anomalies, the principal regional features of all three heat-flow density patterns are generally in good agreement.

The short wavelength features such as the anomalies of the Pannonian Basin, the Upper Rhine Graben, the Tyrrhenian Sea, etc. do not appear in the polynomial or the spherical harmonic map, being too small compared to the characteristic dimensions of 750 and/or 1100 km, respectively. The vulnerability of the

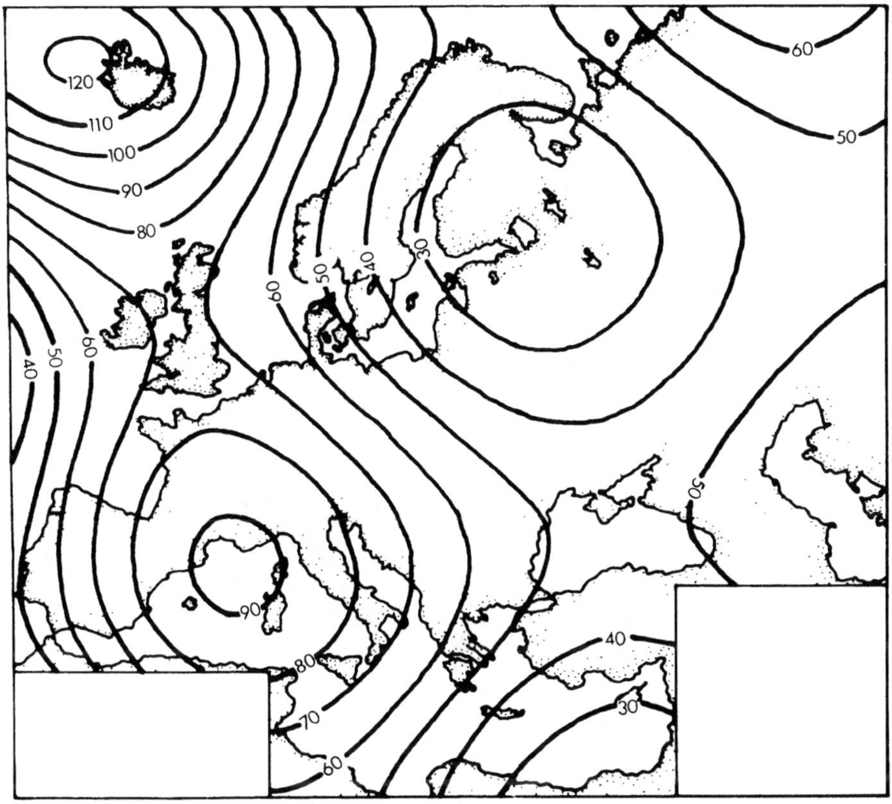

Fig. 7.5. Degree 6 polynomial representation of the heat-flow density field in Europe as prepared by E. Hurtig, isolines are labelled in mW m $^{-2}$. Adopted from Čermák and Hurtig (1977).

computer produced map due to the paucity of data in some critical areas, resulting then in unreal distortions of the field, may be well demonstrated in the case of Turkey. In both computer produced maps (Figures 7.5 and 7.6) this area is of low heat-flow density, contrary to the matter of fact that this region has a young geological history and undoubtedly high heat flow. No direct heat-flow density measurements from Turkey have been reported yet, but the preliminary isoline pattern based on hydrogeological survey and applied geothermal prospecting (Tezcan, 1979) clearly confirms a pronounced geothermal high in this region. Moreover several high heat-flow density values were found in the Aegean Sea (Jongsma, 1974) and in Greece (Fytikas and Kolios 1979).

7.2.5. Corrections in Geothermal Mapping

Geothermal maps that show the subsurface temperature or the temperature gradient distribution are usually based on the directly measured (observed) values, which apart from technical and steady-state corrections need not be corrected prior to their use. Their main purpose is to present the really existing conditions

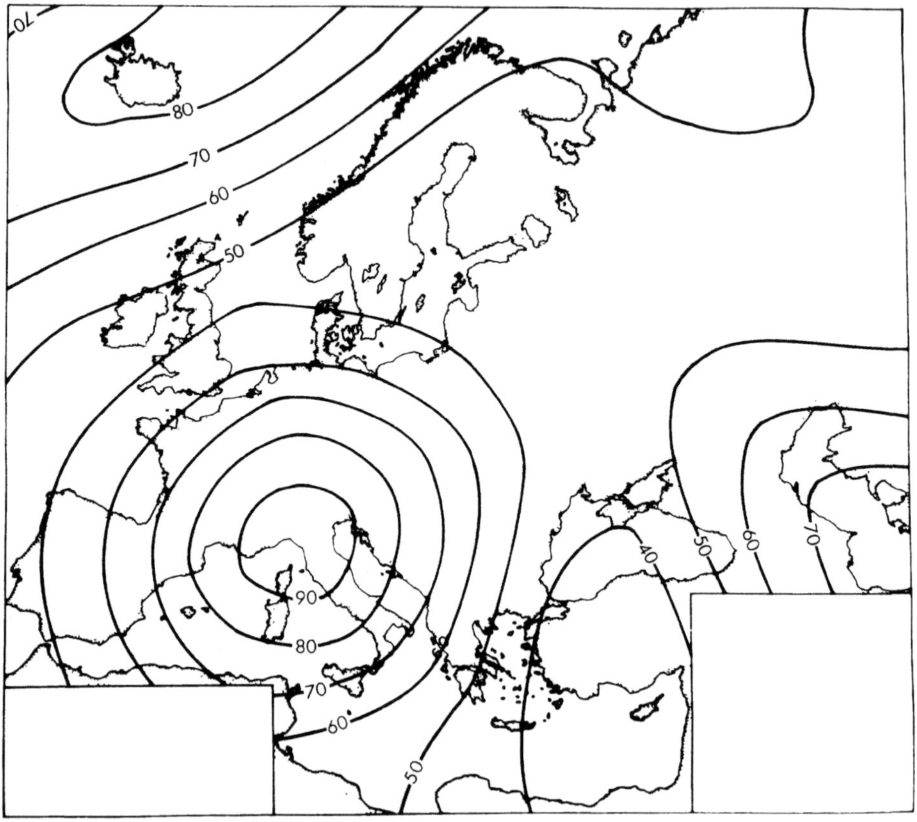

Fig. 7.6. Degree 18 spherical harmonic representation of the global heat-flow density field shown for the European region, isolines are labelled in mW m^{-2}. Adopted from Chapman *et al.* (1979).

of the temperature field at depth. On the other hand, heat-flow density maps or maps showing deep crustal temperatures should be based on corrected data. The local or regional environmental factors (topography, hydrogeology, past climate, etc.) as well as the geological history (sedimentation, uplift, erosion) may, to some extent, have distorted the temperature field in the uppermost layers, in which the temperature loggings take place. To get the genuine value of the heat-flow density from the depth, all the above effects must be evaluated and the measurements corrected accordingly (see Chapters 3 to 6).

Topographic corrections and the effect of the local disturbances are nowadays adopted by most authors and thus pose no serious problems, but a systematic application of corrections for the long-term climatic variations and for the geological history is still rather uncommon. For any compilation of data to be mapped, the authors must be sure not to mix together data to which different types of corrections were applied. Usually the problem of the use of corrected and/or uncorrected data is not so critical in smaller territories and if all data were obtained by a single interpreter or if there was a uniform interpretation tech-

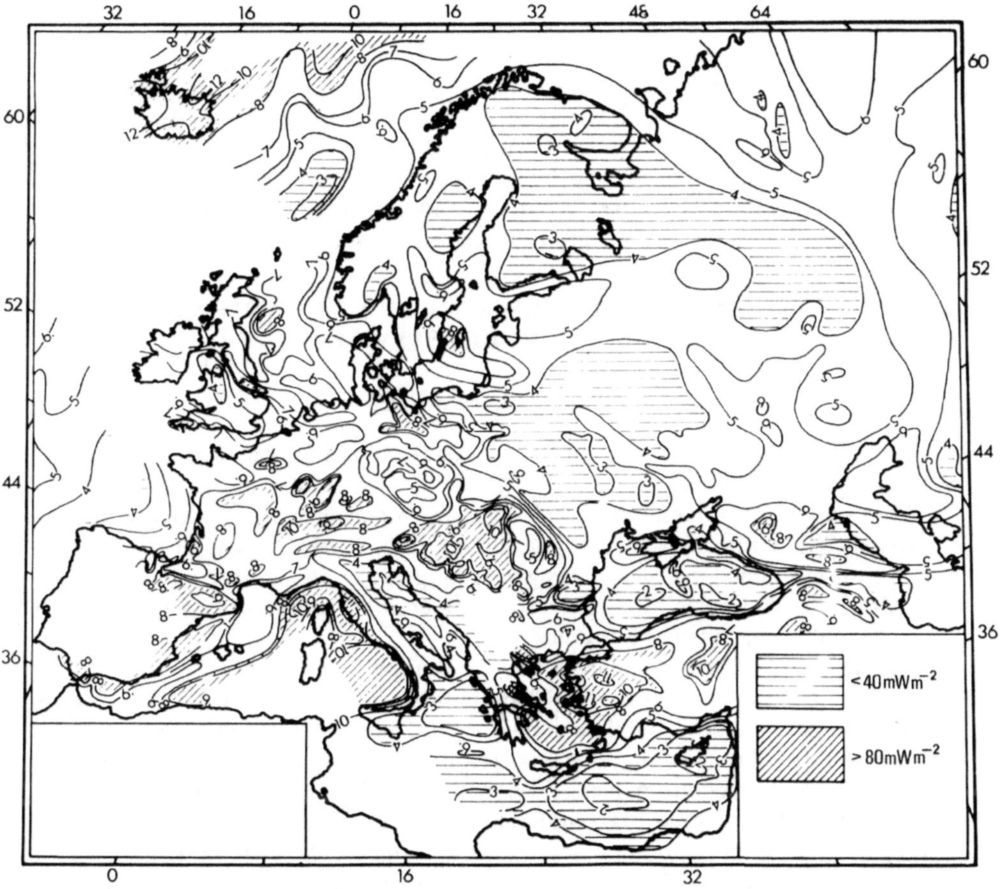

Fig. 7.7. Simplified heat-flow density map of Europe, isolines are labelled in 10 mW m^{-2}. Adopted from Čermák and Hurtig (1979).

nique, which is the case of most national heat-flow density maps. However, in producing heat-flow density maps of continental size or in preparing derived maps, such as the heat-flow density or the temperature distribution at the Mohorovičić discontinuity, etc., only uniform data sets should be used and the nature and quality of data must be mentioned. Several categories of data may be proposed: (a) all non-corrected data, (b) all data corrected for the topographic effect only, (c) all data corrected for the topographic and palaeoclimatic effects, and (d) all effect evaluated. At present, probably, item (c) is the optimum which can be fulfilled on the regional scale.

In mapping the geothermal parameters to assess the geothermal energy resources, one may quite often encounter anomalous values which are connected to local structures such as hydrothermal manifestations (Figure 7.8), water movements, etc., which are certainly not characteristic of the deep crustal structure. It depends very much on the purpose for which the map is constructed. Some anomalous values which are to be neglected in the global or regional representation, on the contrary, must be specially shown in other specific cases. Sometimes, the local pattern may be so intricate or disturbed that the isolines cannot be given at all, an special symbols are to be used. Having the tectonic/geological background and contouring the map by hand, one can easily recognize such cases. Numerical methods are more vulnerable and before the data set is fed to the computer, a careful selection of all the extreme values is necessary.

7.3. TEMPERATURE, TEMPERATURE GRADIENT AND HEAT-FLOW DENSITY MAPS

7.3.1. Temperature Maps

Temperature and pressure control the physical and chemical processes and the behavior of rocks.

The knowledge of temperature is thus of importance for geology and geophysics. While the maps of the subsurface temperature distribution are used above all in applied geophysics, engineering geology and in mining and drilling activities, derived maps of deep crustal temperatures are mostly used in general geophysical and geodynamic studies.

Subsurface temperature maps show the temperature field up to the depths reached by the deepest holes and measurements in mines and tunnels. Sometimes, these data are complemented by continuous temperature logging, the temperature deduced from drill stem tests and pumping tests in holes and by temperatures derived from the heat-flow density data from shallow holes. Preference should be given to the temperature data obtained in an equilibrium state, while all other information may serve only to complete the general pattern and its correctness should be checked. The most common way is to used the Earth's surface as a reference level and to give subsurface temperatures at certain depths below the surface, such as 100, 500, 1000, 2000, 3000 m etc. (see Chapter 3.9.). A map of mean annual ground temperature may complete the above set of subsurface temperature maps. In special cases, the sea level may also be used as a reference level and the subsurface temperature are then related to the specific depths below or above sea level. This method needs the additional knowledge of the topography, which renders it generally impractical. However, in terrains of rough topography the use of a deep reference surface related to the sea level and believed to have temperature independent on the surface topography, may facilitate the prediction of the underground temperature field (Bodmer, 1982). Still another variant presents the regional patterns of depths (map of isobaths) at which the subsurface temperature attains a selected value, such as 50, 100, 200 °C (Čermák and Safanda 1982). These maps can be used in planning drilling activity or in geothermal energy exploration. For similar studies and/or to evaluate the existing geothermal conditions at a certain geologic boundary (such as on the surface of a crystalline basement, i.e. at the base of the sedimentary cover) the maps showing the corresponding temperature distribution may be also constructed (e.g. Motyl-Rakowska *et al.*, 1977).

Numerous subsurface temperature maps have been produced for various purposes. As a few examples one can mention a set of subsurface temperature maps on the territory of the Federal Republic of Germany (Wohlenberg, 1979) using almost 5000 data points and applying the sliding average technique over a grid network, temperature maps of Hungary (Stegena, 1976a; Horvath *et al.*, 1981); Poland (Wesierska, 1970; Majorowicz and Plewa, 1979), and many others.

Though of historical value nowadays, the first attempt to summarize data from several countries was undertaken by Stegena (1972, 1976b), who published the 1:10000000 map showing the temperature distribution at 1 km depth in Central and Eastern Europe. A much more elaborate compilation of such data presents the Atlas of Subsurface Temperatures in the European Community (Haenel, 1980). The Atlas covers the territory of seven EEC countries and includes a series of 1:5000000 temperature maps down to a depth of 5 km (with the increment of 500 m). These general maps are complemented by 1:1500000 to 1:2000000 maps of all the member states and by local and regional maps of several selected areas of special interest and sufficient data coverage at the scales varying from 1:50000 to 1:800000 and within the depth range of 500 to 3000 m.

The solution of the equation of heat conduction enables, in the absence of convective heat transfer e.g. by deep groundwater movements, the surface heat-flow density to be converted into deep temperatures. For this purpose the knowledge of the surface heat-flow density (q_0) must be complemented by a model of the distribution of the thermal conductivity, $\lambda = \lambda(z)$, and the distribution of the heat sources (heat generation), $H = H(z)$.

It is customary to represent the temperature field in one dimension by 'geotherms': they describe, for a given location, the variation of temperature (usually increase) with depth, $T(z)$.

For a purely conductive, steady-state regime the heat Equation (2.7) to consider temperature-dependent thermal conductivity $\lambda(T)$ and depth-dependent heat production $H(z)$ can be written as:

$$H(z) + \frac{\mathrm{d}}{\mathrm{d}z}\left[\lambda(T)\frac{\mathrm{d}T}{\mathrm{d}z}\right] = 0.$$

Analytical $T(z)$ solutions with the boundary conditions $T_0 = T_{\text{surface}}$ and $q_0 = \lambda(\mathrm{d}T/\mathrm{d}z)_{z=0}$ are e.g.:

– Several (N) layers with negligible heat production ($H(z) = 0$), thermal conductivity layerwise constant (thickness of the i-th layer Δz_i, λ_i = const):

$$T(z) = T_0 + q_0 \sum_{i=1}^{N} \Delta z_i/\lambda_i. \qquad (7.30a)$$

– One single layer with Constant H and λ ($H(z) = H_0$, $\lambda(T) = \text{const} = \lambda_0$):

$$T(z) = T_0 + (q_0/\lambda_0)z - (H_0/2\lambda_0)z^2. \qquad (7.30b)$$

– One layer with H = constant = H_0, λ depends on temperature as $\lambda(T) = \lambda_0/(1 + CT)$ (C is in the order of 10^{-3} K^{-1}):

$$T(z) = (1/C)\{(1 + CT_0)\exp[(C/\lambda_0)(q_0 z - H_0 z^2/2)] - 1\}. \qquad (7.30c)$$

– One layer with exponential depth dependence of H as $H(z) = H_0\exp(-z/D)$

(D: logarithmic decrement of the exponential function, in the order of 10 km), λ = const. = λ_0:

$$T(z) = T_0 + (q_0 - H_0 D)z/\lambda_0 + H_0 D^2[1 - \exp(-z/D)]/\lambda_0. \qquad (7.30d)$$

– One layer with depth dependence of $H(z) = H_0 \exp(-z/D)$, temperature dependence of $\lambda(T) = \lambda_0/(1 + CT)$:

$$T(z) = (1/C)[(1 + CT_0)\exp\{(C/\lambda_0)[H_0 D^2(1 - \exp(-z/D)) - H_0 Dz + q_0 z]\} - 1]. \qquad (7.30e)$$

– Several (N) layers, heat production H_i and thermal conductivity λ_1 layerwise constant, layer thickness: Δz_i:

$$T(z) = T_0 + \left(q_0 - \sum_{i=1}^{N-1} H_i \Delta z_i\right)\left(z - \sum_{i=1}^{N-1} \Delta z_i\right)/\lambda_1 - H_n\left(z - \sum_{i=1}^{N-1} \Delta z_i\right)^2/2\lambda_1. \qquad (7.30f)$$

The regional heat-flow density patterns may thus be used to project the maps of deep (crustal) temperature distribution including e.g. a map of the temperature at the Mohorovičić depth (see also Chapter 8.1). Several attempts of this kind were made in the studies of the heat-flow density field in Europe, see e.g. the temperature distribution in Central Europe (Oelsner, 1978) and in Eastern Europe (Lubimova, 1979) or detailed three-dimensional studies of the geothermal anomaly below the Pannonian Basin, for which Bodri and Bodri (1982) calculated the depths to the surfaces of equal temperatures of 500 and 800°C. Similar works showing the calculated temperature distribution at the Mohorovičić discontinuity were reported by Sokolova (1978) for Siberia and by Smirnov (1980) for the whole territory of the USSR.

The global distribution of the temperatures at depths of 25 and 50 km based on the order 10 spherical harmonic representation of heat-flow density data and the conversion of heat-flow density into deep temperatures was reported by Lubimova and Suetnova (1982). Unfortunately, the latter attempt presents the typical example, when the final isoline pattern partly suffers from both, the uneven data coverage and the resolution factor given by the order of the representation (2000 km in this case), as well as from a certain schematicity of the model used, which evidently cannot reflect the specific features of all the tectonic units.

For the subsurface temperature maps the isoline interval depends on the accuracy of the available data, the scale of the map, and the depth of the represented data. For shallow depths (several hundred meters) where sufficient data control exists, an isoline spacing of 10 °C can be used. For maps showing the temperature distribution at relative great depths (greater than 3000 m) which are often based on values extrapolated from shallower holes, the isoline interval of 20 °C seems to be appropriate. For the patterns of even deeper temperatures (crustal temperatures, Mohorovičić temperatures, etc.) the interval of 200 °C may be approved if the maximum uncertainty of the calculated temperature is less than 200 °C. The isoline interval of 100 °C which is sometimes also given for deep temperature patterns seems to be rather problematic especially if applied to a map of the regional scale and with diverse tectonic settings.

It may be of great help for the users of temperature maps if the characteristic temperature-versus-depth profiles for selected regions and/or the families of the calculated geotherms for the specific tectonic units are shown. A brief, but complete description of the model used for the calculation in the latter case would be essential.

7.3.2. Maps of Vertical and Horizontal Temperature Gradient

The knowledge of not only the temperature field but also of the rate of its increase with depth or its lateral changes (gradient) may be of interest. Actually under the term 'gradient map' we understand rather a map showing the quotient of temperature difference over certain depth/distance interval. Vertical gradient is thus usually replaced by temperature increase from the surface to either certain depth or from surface to certain geological boundary. To exclude near surface distortion of the temperature field the reference level may be chosen e.g. at 100 m. Since the temperature gradient depends on the thermal conductivity of rocks, which varies from point to point, the construction of a gradient map is justified only within a certain depth range and within a limited area.

Quite a lot of gradient maps were prepared and published in the early days of the initial interest in regional geothermal research some 15–20 years ago. If the thermal conductivity is relatively constant, the gradient map may to a certain extent substitute the information given by the heat-flow density distribution. At that time when routine measurements of the thermal conductivity of rocks were still missing in some countries, to produce the gradient map was the only way to furnish the material needed for further geophysical studies. Among these maps one can rate e.g. Boldizsar's (1968, 1973) patterns of the temperature gradient distribution in the Vienna and Little Hungarian Basins and/or the gradient map of Hungary, Boyadgieva's *et al.* (1973) 1 : 1 000 000 gradient map of Bulgaria, which showed the gradient values related to the depth of 500 m, Makarenko's *et al.* (1972) geothermal map of the USSR, Majorowicz's (1973) gradient map of Poland, etc. In some regions studied in great detail, such as e.g. the Upper Rhine Graben, a series of subsurface temperature and gradient patterns were prepared, see Delattre *et al.* (1970), Werner and Doebl (1974). Several gradient maps were prepared recently for the area of conterminous USA (Guffanti and Nathenson, 1980; Kron and Heiken 1980).

At the present time, when the conductivity determinations have become generally available, the above gradient maps are usually less important for basic geophysical studies. Once the conductivity of rocks is known and the heat-flow density pattern can be constructed, such a pattern will certainly surpass any gradient map.

Nevertheless, some use of the gradient maps can be justified in large sedimentary basins, where the conductivity does not change much on a relatively large scale. Most of these maps are widely used in the oil or mineral exploration, when within the framework of a regional survey numerous temperature loggings or bottom-hole-temperature data are obtained. These data are usually referred to as the 'file data', their accuracy may be poorer, for not all holes were in equilibrium prior to the logging and only very rarely are temperatures accompanied by relevant information on the conductivity. However, as the oil companies are more openminded at present to disclose their measurements, formerly often classified as confidential, to the scientific community, enormous volume of information contained in their files may present an extremely valuable data base for many 'purely' scientific studies.

Regional examples of the temperature gradient maps based on the data obtained in the oil fields include e.g. Harper's (1971) approximate gradient map of the North Sea or the gradient map of Southeast Asia (SEAPEX, 1977). On a local scale a similar gradient map was prepared e.g. by Thamrin *et al.* (1980) for a basin in Central Sumatra. In this case, the temperature data were complemented by the pattern showing the thermal conductivity distribution as well as the local heat-flow density map (see also Carvalho *et al.*, 1980).

Subsurface temperature and temperature gradient maps may be very important in the hydrogeological assessments of the hydraulic conditions in the basinal areas, see e.g. the gradient pattern of the Western Canada Prairies Basin (Majorowicz and Jessop, 1981). Jones and Lamb's (1983) geothermal gradient contour map of one locality in Northwestern Alberta, Canada, may serve as a good example of the computer produced local pattern of the gradient data based on numerous bottom-hole-temperature data in an area of intensive drilling.

The maps showing the horizontal temperature gradients may be constructed in order to assess the horizontal (lateral) component of the temperature field for special studies, such as the tension or stress distribution in the crust e.g. for the appraisal of the release mechanism of seismic energy (Stegena, 1983).

The magnitude and the direction of the maximum temperature increase can be found by the numerical derivation of the isoline pattern in the corresponding temperature map. Alternatively, to evaluate the existing lateral temperature differences, the area under investigation can be covered by a regular grid network and the difference between the maximum and minimum value in each such grid element characterizes to some extent also the horizontal variability of the temperature field. Here again the depth range of validity must be specified.

7.3.3. Maps of Heat-Flow Density

The knowledge of the heat-flow density is of fundamental importance in geophysics and the maps showing its distribution present the final output of the field or compilation work for further interpretation. As mentioned above, the heat-flow density maps show either the mathematical or visual representation of the individual data and their preparation usually requires some insight into the relation of the pattern to be mapped to the basic tectonic features of the area in question. In constructing the heat-flow density isolines, great attention should be paid to the preparatory selection of data and to correcting them properly.

The isoline interval for contouring should equal or preferably be greater than the accuracy of the heat-flow density determination. Nowadays the use of the isoline interval of 10 mW m^{-2} has been generally adopted, in some older maps also the interval of 0.5 μcal cm^{-2} s^{-1}(= 20 mW m^{-2}) was commonly applied in addition to the interval of 0.2 μcal cm^{-2} s^{-1} (= 8 mW m^{-2}).

As the accuracy of the present routine heat-flow density determinations amounts to 10–15 %, the use of the isoline interval of 10 mW m^{-2} may be generally recommended for further heat-flow density mapping, except in the areas of very high heat flow or in the areas of highly variable heat-flow density, where the construction of detailed isolines may be dubious.

The labeling of isolines should be carried out as for example for topographic hights. The top of the lettering points into the direction of higher values and the bottom to lower values. This would avoid misunderstanding whether an extreme is a minimum or a maximum, additional signs as plus or minus are not any more necessary.

The major global and/or continental heat-flow density maps were mentioned earlier. Besides them a number of local and regional heat-flow density patterns and maps were constructed and presented in literature. These maps usually cover a territory of a single country and characterize the progress in the heat-flow density studies obtained there.

For studies of crustal dynamics and the geothermal conditions of the upper mantle, the knowledge of the outflow of heat from below a certain depth may be important. Using the linear experimental relationship between the surface heat-flow density and the radiogenic heat production together with the empirical relation between the reduced heat-flow density and the mean surface heat-flow density for continents and the estimate of heat liberated by cooling of the oceanic crust, being the function of the age, Pollack and Chapman (1977) proposed the global order 12 spherical harmonic representation of mantle heat-flow density derived from their 1975 representation of the surface heat-flow density. Alternatively, using the surface heat-flow density and the crustal thickness map of Europe together with deep seismic sounding data (seismic velocity-depth distribution) converted into heat production, Čermák (1982) proposed the continental map of the heat-flow density on the Mohorovičic discontinuity below Europe. To avoid the dependence on the resolution factor of the mathematical representation, preference was given to the visual contouring of averaged 4° × 4° grid element data of the mantle heat-flow density. The calculation have been carried out by means of:

$$q_M = q - \int_0^{z_M} H(z)\,\mathrm{d}z \tag{7.31}$$

where

q_M = heat-flow density at Mohorovičić discontinuity
q = surface heat-flow density
z_M = crustal thickness
$H(z)$ = heat production.

q and z_M were taken from corresponding maps, and $H(z)$ was assessed for a layered crustal model according to the conversion relation between seismic velocity and heat generation reported by Rybach (1973).

Other heat-flow density patterns on the Mohorovičić discontinuity in Europe were constructed by Oelsner (1978) for Central Europe, by Lubimova (1979) for Eastern Europe, and by Bodri and Bodri (1982) for the Pannonian Basin. For the USSR territory the similar pattern was proposed by Smirnov (1980).

7.4. GEOTHERMAL RESOURCE AND RESERVE REPRESENTATION

7.4.1. Objectives

The objective of a geothermal resource/reserve assessment is to estimate the available heat in place in the subsurface and to quantify that part of this heat in place which can be extracted economically. This would provide user-oriented groups (such as scientists working in this field as well as public and private decision makers) with a data base for long-term planning.

For this the given natural hydrogeothermal subsurface conditions must be known and they have to be transformed into the energy available at well head. The delimination and representation of that part which can be economically used also depends on the present legal conditions and the technology of surface processes, such as the exploitation technique, the heat distribution system, and the (possible) consumer. Furthermore, resource/reserve assessment not only includes the quantities that could be produced under economic conditions, but also comprises the quantities which have not been discovered yet or which might be produced with an improved technology or under different economic conditions. The assessment of geothermal energy also includes estimates of the amounts of byproducts, such as salt, metals, gases and the water itself, which might be produced and used economically along with the thermal energy. Finally, the extracted fluid can be used successively by several consumers at different temperature levels; this is called energy cascading; see Haenel (1985).

As the assessment is a statement made at a given time using a given set of data and assumptions which can change rapidly with exploration activities and technology developments, the assessment is only of transitory value and must be brought up-to-date periodically.

The most important terms are 'resource' and 'reserve', where the

RESOURCE is that part of geothermal energy which *could be* extracted economically and legally at a specified time in the near future; and the

RESERVE is the resource known and characterized by drilling or by geochemical, geophysical and geological evidence and which *can be* extracted economically at present time.

The reserves are further subdivided with regard to their certainty into proven, probable, and possible reserves; see Figure 2.1 and Table 2.7 in Chapter 2.

Fundamental papers for resource/reserve assessment are presented by Muffler and Cataldi (1978), Muffler and Guffanti (1979), Cataldi and Celati (1983), White and Williams (1975), and Haenel and Staroste (1988).

7.4.2. Assessment and Mapping in Low Enthalpy Areas

Data on the subsurface are required for an evaluation of geothermal resources and reserves as given in Table 7.4. These data should be presented in such a way (see Chapters 7.1 and 7.2) that the reader can also carry out the transformation of these hydrogeothermal data into geothermal resources and reserves as demonstrated in the following.

The heat in place (Q_0 in J) contained within a given aquifer can be determined using a volume model of heat extraction (Muffler and Cataldi, 1978):

$$Q_0 = ((1 - P)\rho_m c_m + P\rho_w c_w)(T_t - T_0)A\Delta z, \tag{7.32}$$

where

P = effective porosity, unitless
ρ = mean density of the rock column, kg m^{-3}
c = mean specific heat capacity, J kg^{-1} K^{-1}
T_t = temperature at the top of the aquifer, °C
T_0 = temperature at the Earth's surface, °C
A = surface area under consideration, m^2
Δz = net thickness of the aquifer, m
m = rock matrix, and
w = water.

It is obvious that only part of Q_0 can be recovered. This is expressed by a term called the recovery factor R. The product of R and Q_0 is the identified resources Q_1, commonly abbreviated to resources. Because the water in most cases is highly saline, it has to be reinjected. The identified resources at wellhead for a doublet (injection well and extraction well) is given empirically as follows (Lavigne 1978):

$$Q_1 = R_0 Q_0 \tag{7.33}$$

$$R_0 = 0.33\frac{T_t - T_r}{T_t - T_0}, \tag{7.34}$$

where

T_r = temperature of the reinjected water (set by the EC expert group at 25 °C in order to be able to compare calculations).

TABLE 7.4
Required hydrogeothermal aquifer data.

- Extension:
  area and thickness (i.e. net pay), depth of the aquifer top

- Geology:
  stratigraphy, lithology, geochemistry (e.g. salinity)

- Physical parameters:
  effective porosity, temperature at the top of the aquifer, permeability (hydraulic conductivity), transmissibility (transmissivity), potentiometric level, yield

- isoline maps, cross sections and tables.

If only a production well without reinjection is considered (a singlet), the recovery factor is (Gringarten, 1979):

$$R_0 \approx 0.1. \tag{7.35}$$

Only a part of the geothermal energy resources can be exploited economically at present. This can also be expressed by a recovery factor R. The proven reserves at wellhead, commonly abbreviated to reserves, can then be calculated as follows:

$$Q_2 = R_1 Q_1. \tag{7.36}$$

The recovery factor R_1 is difficult to determine. It depends on the site-specific geological conditions, as well as on the cost of installation of a singlet or doublet. Therefore, determination is left to each working group. A general rule is given by Koppe *et al.* (1983). Following their procedure, the recovery factor R_1 is defined as the ratio of the extractable energy E minus the installation costs I to the available identified resources Q_1:

$$R_1 = \frac{E - I}{Q_1} \tag{7.37}$$

$$E = Q_v(T_t - T_r) \cdot \rho_w c_w \Delta t \tag{7.38}$$

$$Q_v = \frac{2\pi T_v s}{\ln(R/r)} \quad \text{or} \quad Q_v = 2\pi r z \frac{k_f}{15} \qquad (7.39a), (7.39b)$$

$$I = \frac{I_\$ C}{P}, \tag{7.40}$$

where

E = extractable geothermal energy, J
I = installation cost of singlet or doublet expressed in heat, J
$I_\$$ = installation cost expressed in \$ (contains the cost of drilling, well completion, extraction and injection pumps, and heat exchanger; it includes no pipeline costs),
Q_v = groundwater yield from confined aquifer, $m^3\ s^{-1}$; (Equation 7.39a) is after Thiem-Dupuit, Equation (7.39b) after Sichardt,
Δt = life-time of the plant, usually taken to be 30 yr,
$T_v = k_f z$ = transmissivity, $m^2\ s^{-1}$
$k_f = k\rho_w g/\mu$ = hydraulic conductivity, $m\ s^{-1}$
k = rock permeability, m^2
g = gravitation constant, $m\ s^{-2}$
μ = dynamic viscosity of water, $kg\ m^{-1}\ s^{-1}$
R = radius of the drawdown cone, m
r = radius of the borehole, m
s = drawdown of the water level in the well, m
C = specific heat content of oil, $J\ t^{-1}$
P = oil price at a specific time, $\$\ t^{-1}$
z = thickness of aquifer, m.

The yields calculated with Equations (7.39a) and (7.39b) are different, but the lower value is that which should be used for Equation (7.38). Equation (7.40) can be used to express the installation cost in joules. Substituting Equation (7.37) into Equation (7.36) leads to:

$$Q_2 = E - I \quad \text{for } Q_1 > E \tag{7.41}$$

or

$$Q_2 = Q_1 - I \quad \text{for } E > Q_1. \tag{7.42}$$

Thus the expression for proven reserves at wellhead is simplified to:

reserves = extractable energy − installation cost.

Neglecting natural water flow and recharge area of an aquifer, the available water supply Y within an area A is given by the expression:

$$Y = PzA \tag{7.43}$$

and the number of singlets within an area is:

$$a = \frac{Y}{Q_v \Delta t} = \frac{P \Delta z A}{Q_v \Delta t}, \tag{7.44}$$

where

Δt = life-time of the system, normally assumed to be 30 a.

Equations (7.38) and (7.40) now have to be multiplied by the factor a of Equation (7.44).

Equation (7.44) is not necessary for a doublet, because the extracted water is always reinjected. The water volume necessary for a doublet is governed mainly by the amount of heat to be extracted per unit time. Model calculations are necessary to determine this volume.

The spacing D of the two wells of a doublet has a critical effect on the determination of reservoir life-time. A simplified formula is given by Landel and Sauty (1978):

$$D^2 = \frac{3}{\pi} \frac{Q_v t_b}{z} \frac{\rho_w c_w}{(1 - P)\rho_m c_m + P \rho_w c_w}, \tag{7.45}$$

where t_b = breakthrough time, i.e. the time it takes the cold, reinjected water to reach the extraction borehole; z = net thickness of the aquifer.

But it should be kept in mind that when a system of doublets is used in which the doublets are not isolated from the influence of each other, the recovery factor can be increased and the life-time of the reservoir (Gringarten, 1979), too.

If the natural flow of water in the aquifer is also considered, a model calculation must be made. Numerical and analytical models, using the standard equations of geothermics and hydraulics, are presented by Gringarten and Sauty (1975), Landel and Sauty (1978), and Schulz (1985) (see Figure 7.8).

The heat supply from the terrestrial heat flow, in percentage of terrestrial heat-

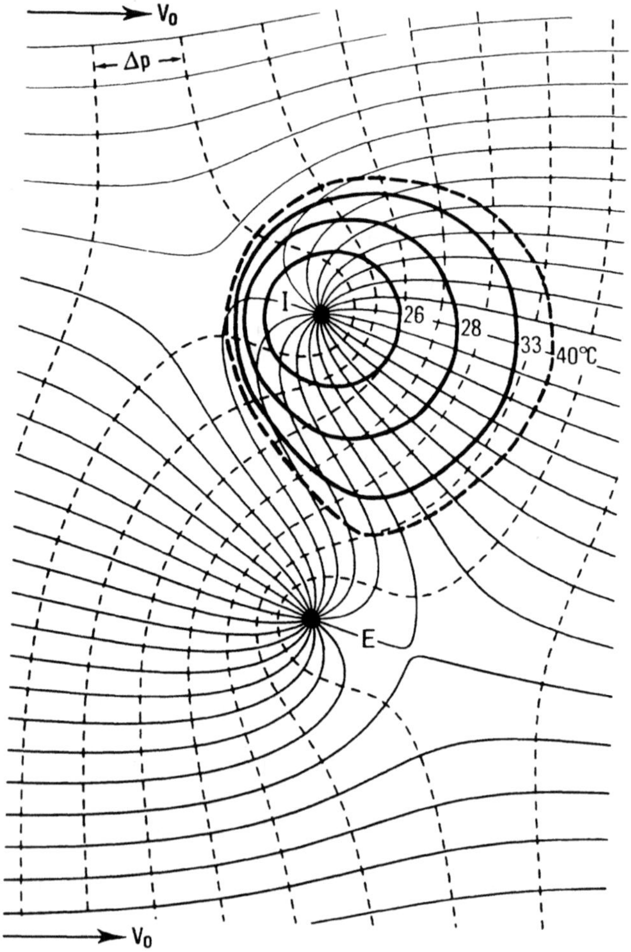

Fig. 7.8. Pattern of isolines for water flow, pressure and temperature after three years of operation of a doublet (Schulz, 1985). The flow lines of the undisturbed field were straight and parallel, as indicated by the arrows at left ($v_0 = 10^{-6}$ m s^{-1}; distance between extraction well (E) and injection well (I) = 400 m; extraction and injection rate = 30 l s^{-1}; undisturbed aquifer temperature = 40°C; temperature of reinjected water = 25°C; transmissivity = 3×10^{-3} m^2 s^{-1}); Δp = pressure difference, flow lines ——, temperature isolines ——, pressure isolines - - - -.

flow density, depends on the aquifer temperature. The lower the aquifer temperature, the greater the heat supply.

Minimum values for economic utilization of low enthalpy geothermal energy have not yet been clearly defined. In any case, a temperature anomaly is not necessarily a favorable site. A more important parameter is the expected production rate (l s^{-1}), which is a function of transmissivity, temperature, and proximity to the consumer. Based on known projects, an attempt of definition is made in Figure 7.9.

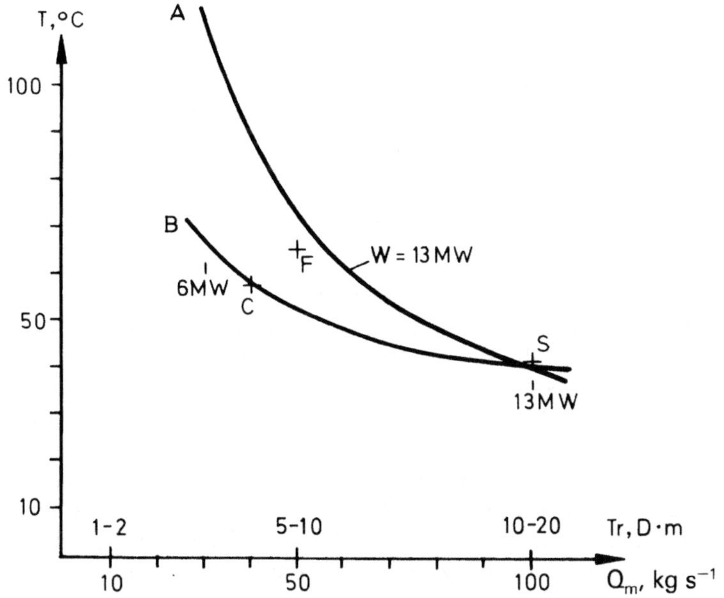

Fig. 7.9. Possible minimum values for economic utilization of geothermal water (Haenel, 1982). Point S is the geothermal heating power W that can be economically extracted in the Saulgau project in Germany; $W = Q_m c(T - T_0)$, where Q_m is in kg s^{-1}, c is in J kg^{-1} K^{-1}, and T_0 = 12°C. Point C is the corresponding value for the Crail plant in France and F is the mean of 5 geothermal plants in France. Curve A represents W = 13 MW and curve B is the interpolation of points S and C.

A constant production rate for a 30 a period and a maximum drawdown of 150–200 m are generally assumed for the evaluation (Sorey *et al.*, 1983; Gringarten, 1979).

A reasonable scheme for the assessment of geothermal energy potential at well head is:

$$\text{Name of area} \quad \underset{(\text{km}^2)}{A} \quad \underset{(\text{km})}{z} \quad \underset{(\text{km}^3)}{V_m} \quad \underset{(\%)}{P} \quad \underset{(\text{km}^3)}{V_w} \quad \underset{(°\text{C})}{T_t} \quad \underset{(°\text{C})}{T_0} \quad \underset{(\text{J})}{Q_0} \quad \underset{(\text{J})}{Q_1} \quad \underset{(\text{J})}{Q_2} \tag{7.46}$$

where

A = area under consideration
z = aquifer thickness
V = volume of aquifer; m = matrix, w = water
P = effective porosity
T_t = temperature at the top aquifer
T_0 = mean annual temperature at the Earth's surface
Q_0 = heat in place
Q_1 = geothermal resources at wellhead
Q_2 = geothermal reserves at wellhead.

7.4.3. Assessment and Mapping in High Enthalpy Areas

The mapping of the hydrogeothermal data should be carried out as explained in

Chapter 7.3., and the transformation of these data into geothermal resources and reserves as presented below.

(1) Following the procedure given by Cataldi *et al.* (1978) for a volume model of heat extraction, the identified resources at wellhead area:

$$Q_3 = R_2 Q_0 \tag{7.47}$$

and the proven reserves at wellhead are:

$$Q_4 = R_3 Q_3. \tag{7.48}$$

The heat in place Q_0 is given by Equation (7.32), in which T_0 is now 60°C.

The recovery factor R_2 correlates directly with the effective porosity. It has been found that an ideally permeable reservoir of hot fluid (porosity $P = 20$ %) would have a recovery factor of 0.5 (Cataldi *et al.*, 1978); see Figure 7.10.

The determination of the recovery factor R_3 for calculating the proven reserves requires a very detailed investigation of the reservoir. Therefore, only estimates can be made. A primary factor for the economic feasibility of extracting geothermal energy is the drilling cost. Thus, the recovery factor has been expressed in terms of drilling depth, as shown in Figure 7.11. The maximum drilling depth is 3 km (see also Figure 2.1) below which no economic utilization can be expected at present.

(2) Another procedure is presented by Brook *et al.* (1979). It is based on experience gained with the vapour-dominated system of the Geysers. The identified resources at wellhead are given by:

$$Q = m(h - h_r) \tag{7.49}$$

$$m = 0.5 x P \rho_w V, \tag{7.50}$$

where

m = mass of steam produced at wellhead, kg
x = fraction of pores filled with liquid water, unitless
P = effective porosity, unitless

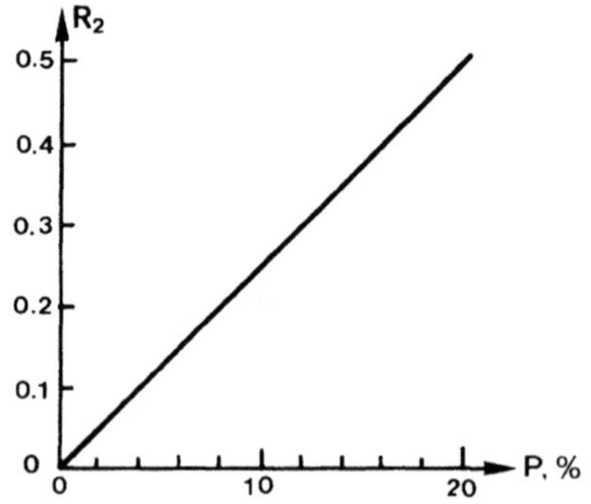

Fig. 7.10. Graph showing variation of geothermal resource recovery factor R_2 as a function of effective porosity P (in %) for a reservoir producing by an intergranular flow mechanism (Cataldi *et al.*, 1978).

ρ_w = density of water, kg m^{-3}
V = reservoir volume, m^3
h = specific enthalpy at wellhead, J kg^{-1}
h_r = specific enthalpy of saturated water at the reference temperature of 15 °C, J kg^{-1}.

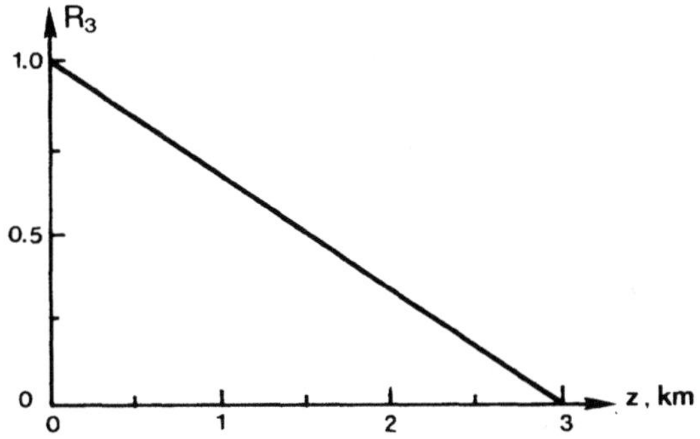

Fig. 7.11. Graph showing variation in the reserve recovery factor R_3 as a function of depth (Cataldi *et al.*, 1978).

The conversion of part of the resources calculated with Equation (7.49) into electricity equals the proven reserves at well head. The work W available for this conversion can be written as follows:

$$W = \Delta H - T\Delta S = m(h - h_0 - T_0(s - s_0)) \qquad (7.51)$$
$$h = h_R - gz,$$

where
H = enthalpy, J
S = entropy, J K^{-1}
T = temperature of the rock into which the water is reinjected, K
h_0 = specific enthalpy of the fluid after heat extraction, J kg^{-1}
T_0 = temperature of the reinjected water, K
s = specific entropy at wellhead, J K^{-1} kg^{-1}
s_0 = specific entropy of the fluid after heat extraction, J K^{-1} kg^{-1}
h_R = specific enthalpy of the fluid in the reservoir, J kg^{-1}
z = depth to the middle of the reservoir, m
g = gravity constant, m s^{-2}.

The electrical energy E can be obtained by:

$$E = \eta W, \qquad (7.52)$$

where

η = utilization factor (Nathenson, 1975); see Figure 7.12.

And again, a similar scheme for the representation of the geothermal potential in terms of resources and reserves is recommended as in Chapter 7.4.2. (see also Muffler and Cataldi, 1978):

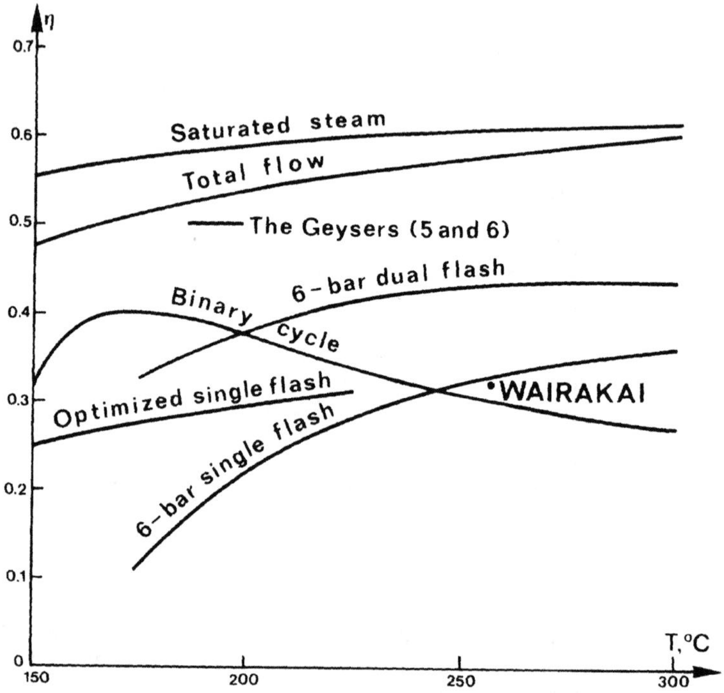

Fig. 7.12. Utilization factors for electric power generation after Brook *et al.* (1979) calculated for several conversion technologies as a function of temperature (based on Nathenson, 1975). For saturated steam, the temperature is wellhead temperature. For hot-water systems, the temperature corresponds to wellhead enthalpy, treating the fluid as liquid water. All the cycles assume 40°C condensing temperature. Single- and dual-flash cycles assume 6 bars for the first separation and are thus not appropriate at temperatures lower than about 200°C. Optimized single flash is calculated by iterating wellhead separation pressure to obtain maximum work. Many cycles involving various optimization schemes are available in the literature; the various cycles shown here illustrate the general behavior.

$$\begin{array}{lcccccccccc} \text{Name of area} & A & z & V_m & P & V_w & T_t & T_0 & Q_0 & Q_3 & Q_4 \\ & (\text{km}^2) & (\text{km}) & (\text{km}^3) & (\%) & (\text{km}^3) & (^\circ\text{C}) & (^\circ\text{C}) & (\text{J}) & (\text{J}) & (\text{J}) \end{array} \quad (7.53)$$

or

$Q_3 = Q$ (Equation (7.49)), $Q_4 = W$ and E, respectively (Equation (7.59), (7.52)).

7.5. SUMMARY OF RECOMMENDATIONS

The following rules were proposed by the International Heat Flow Commission, discussed and accepted by many investigators during the workshops as mentioned in Chapter 1:

1. The use of SI–units only as already stressed in Chapter 2.
2. Equal areas or conformal projections, and the use of the Greenwich system (latitude – longitude marks) are proposed. A scale is necessary. Further, a simplified tectonic pattern will certainly increase the informative value of any map of this kind.

3. Compilations showing the regional or continental heat-flow density patterns should use the internationally accepted projection, usually that used for the tectonic map of the respective territory. The choise of the background tectonic map, its scale, projection, etc. may be consulted with the Commission for the Geological Map of the World, Paris.
4. It is generally recommended not to use any estimated values when constructing geothermal maps.
5. Maps showing the underground temperature distribution should be refered to the Earth's surface as a reference level. To facilitate the future use and comparison of these maps, the following sequence of depths is recommended: 50, 100, 500, 1000, 1500, 2000, 3000, 4000, 5000 m and for greater depths 10, 20, 30, 40 km etc. plus selected depths such as the Mohorovičić discontinuity.
6. Complete list of all the references (primary sources of information) is indispensable with any geothermal map, as well as brief description of the mapping technique used.
7. Temperature gradient maps can be accepted only if the depth range is specified.
8. To produce heat-flow density maps and patterns only uniform data sets should be used. e.g.:
   - all non-corrected data only,
   - data corrected for the topographic effect only,
   - data corrected for the topographic and palaeoclimatic effect,
   - data corrected by all necessary and feasible corrections, respectively.
9. To close an anomaly a minimum number of three values should be taken.
10. The top of lettering on isolines should point to the direction of higher values and the bottom to the lower values.
11. The best trend surface should always be selected by means of the fit of trend.
12. The terms geothermal resources and reserves should be used as defined in Table 2.7.
13. For comparison the assessment of resources and reserves should be carried out as presented in the text until new and better procedures will be given and accepted.

7.6. REFERENCES

Bodmer, Ph.: 1982, 'Beiträge zur Geothermie der Schweiz', Ph.D Thesis, ETH Zürich Nr. 7034.

Bodri, L. and Bodri, B.: 1982, 'Geothermal Model of the Heat Anomaly of the Pannonian Basin', in V. Čermák and R. Haenel (eds.), *Geothermics and Geothermal Energy*, E. Schweizerbart'sche Verlagsbuchhandlung, Nägele u. Obermiller, Stuttgart, 37–43.

Boldizsár, T.: 1968, 'Geothermal Data from the Vienna Basin', *J. Geophys. Res.* **73**, 613–618.

Boldizsár, T.: 1973, 'Positive Heat Flow Anomaly in the Carpathian Basin', *Geothermics* **2**, 61–67.

Boyadgieva, K. T., Velinov, T., and Georgiev, D.: 1973, *Skhematicheskaya karta N. R. Bolgarii geotermicheskogo gradienta na glubine 500 m ot zemnoy proverkhnosti, 1:1 000 000*, IGP i GK, Sofia.

Brook, C. A., Mariner, R. H., Mabey, D. R., Swanson, J. R., Guffanti, Marianne, and Muffler, L. J. P.: 1979, 'Hydrothermal convection systems with Reservoir Temperatures 90 °C', *U.S. Geological Survey Circular* **790**, 18–85.

Carvalho, H. S., Purwoko, Siswoyo, Thamrin, M., and Vacquier, V.: 1980, 'Terrestrial Heat Flow in the Tertiary Basin of Central Sumatra, *Tectonophysics* **69**, 163–188.

Cataldi, R., Lazzarotto, A., Muffler, P., Squarci, P., and Stefani, G.: 1978, 'Assessment of Geothermal Potential of Central and Southern Tuscany', *Geothermics* **7**, 91–131.

Cataldi, R. and Celati, R.: 1983, 'Review of Italian Experience in Geothermal Resource Assessment', in R. Haenel and M. Gupta (eds.), *Results of the First Workshop on Standards in Geothermics*, Zbl. Geol. Paläont., I, 1/2, 2–4, Stuttgart, 168–184.

Čermák, V. and Hurtig, E.: 1977, *Preliminary Heat Flow Map of Europe (1:5 000 000). Map and Explanatory Text*, Geofyz. Úst. Czechosl. Acad. Sci., Praha and Zentralinst. f. Physik d. Erde, Potsdam, 58 p.

Čermák, V. and Hurtig, E.: 1979, 'Heat Flow Map of Europe', in V. Čermák and L. Rybach (eds.), *Terrestrial Heat Flow in Europe, Colour Enclosure*, Springer-Verlag, Berlin, Heidelberg, New York.

Ćermák, V.: 1982, 'Crustal Temperature and Mantle Heat Flow in Europe', *Tectonophysics* **83**, 123–142.

Čermák, V. and Šafanda, J.: 1982, 'Subsurface Temperature Distribution in Western Czechoslovakia and Its Mapping for Appraising the Exploitable Sources of Geothermal Energy, in V. Čermák and R. Haenel (eds.), *Geothermics and Geothermal Energy*, E. Schweizerbart'sche Verlagsbuchhandlung, Nägele u. Obermiller, Stuttgart, 265–270.

Chapman, D. S. and Pollack, H. N.: 1975, 'Global Heat Flow: A New Look', *Earth. Planet. Sci. Lett.* **28**, 23–32.

Chapman, D. S., Pollack, H. N., and Čermák, V.: 1979, 'Global Heat Flow with Special Reference to the Region of Europe', in V. Čermák and L. Rybach (eds.), *Terrestrial Heat Flow in Europe*, Springer-Verlag, Berlin, Heidelberg, New York, 41–48.

Crain, I. K. and Bhattacharyja P. K.: 1967, 'Treatment of Non-Equispaced Two-Dimensional Data with a Digital Computer', *Geoexploration* **5**, 173–194.

Crain, I. K.: 1970, 'Computer Interpolation and Contouring of Two-Dimensional Data: A Review', *Geoexploration* **8**, 71–86.

Delattre, J. N., Hentinger, R., and Lauer, J. P.: 1970, 'A Provisional Geothermal Map of the Rhinegraben (Alsacian part)', in J. H. Illies and St. Mueller (eds.), *Graben Problems. Intern. Upper Mantle Proj., Scient. Rep. 27*, E. Schweizerbart'sche Verlagsbuchhandlung, Nägele u. Obermiller, Stuttgart, 107–110.

Fytikas, M. D. and Kolios, N. P.: 1979, 'Preliminary Heat Flow Map of Greece', in V. Čermák and L. Rybach (eds.), *Terrestrial Heat Flow in Europe*, Springer-Verlag, Berlin, Heidelberg, New York, 197–205.

Garland, G. D.: 1979, *Introduction to Geophysics (Mantle, Core and Crust)*, W. B., Saunders Comp., Toronto, 494 p.

Gringarten, A. C. and Sauty, J. P.: 1975, 'A Theoretical Study of Heat Extraction from Aquifer with Uniform Regional Flow', *J. Geophys. Res.* **80**, 35, 49–56.

Gringarten, A. C. and Sauty, J. P.: 1975, 'Influence de la réinjection sur la tempétature d'un réservoir géothermique utilisé pour le chauffage urbain', 2. UN Symp. on the Development and Use of Geothermal Energy, San Francisco, USA, 20–29 May 1975, U.S. Government Printing Office, Washington, D.C. 20402, 1365–1374.

Gringarten, A. C.: 1979, *Reservoir Lifetime and Heat Recovery Factor in Geothermal Aquifers Used for Urban Heating*, Pageoph., **117**, 297–308.

Guffanti, M. and Nathenson, M.: 1980, 'Preliminary Map of Temperature Gradients in the Conterminouns United States', *Trans. US Geotherm.* Res. Counc. **4**, 53–56.

Haenel, R.: 1971, 'Heat Flow Measurements and a First Heat Flow Map of Germany', *J. Geophys.* **37**, 119–134.

Haenel, R.: 1974, 'Heat Flow Measurements in Northern Italy and Heat Flow Maps of Europe', *J. Geophys.* 367–380.

Haenel, R. and Zoth, G.: 1973, 'Heat Flow Measurements in Austria and Heat Flow Maps of Central Europe', *J. Geophys.* **39**, 425–439.

Haenel, R. (ed.): 1980, 'Atlas of Subsurface Temperatures in the European Community', Th. Schäfer Druckerei GmbH, Hannover, 36 p. + 43 maps.

Haenel, R.: 1982, *Die Möglichkeiten der Nutzung von geothermischer Energie in der Bundesrepublik Deutschland*, Statusreport 1982, Geotechnik und Lagerstätten, KFA (PLE) Jülich, 19–38.

Haenel, R.: 1985, 'EC project on the Evaluation of the Community Potential of Geothermal Energy', in A. S. Strub and P. Ungemach (eds.), *European Geothermal Update*, D. Reidel Publ. Co., Dordrecht, Holland, 21–38.

Haenel, R. and Staroste, E. (eds.): 1988, Atlas of Geothermal Resources in the European Community, Austria and Switzerland. Th. Schäfer Druckerei GmbH, Hannover, no plates.

Harbaugh, J. W. and Merriam, D. F.: 1968, *Computer applications in Stratigraphic Analysis*, J. Wiley & Sons, Inc., New York, London, Sydney, 282 p.
Harper, M. L.: 1971, 'Approximate Geothermal Gradients in the North Sea Basin', *Nature* **230**, 235–236.
Horai, K. and Simmons, G.: 1969, 'Spherical Harmonic Analysis of Terrestrial Heat Flow', *Earth Planet. Sci. Lett.* **6**, 386–394.
Horváth, F., Dövényi, P., and Liebe, P.: 1981, 'Geothermics of the Pannonian Basin', *Earth Evolution Sci.* **3–4**, 285–291.
Hurtig, E. and Oelsner, Ch.: 1977, 'Heat Flow, Temperature Distribution and Geothermal Models in Europe: Some Tectonic Implications', *Tectonophysics* **41**, 147–156.
Jones, F. W. and Lamb, H. L.: 1983, 'Geothermal Gradients in the Steen River Area of Northwestern Alberta', *Tectonophysics*, **103**, 1–4, 263–272.
Jongsma, D.: 1974, 'Heat Flow in the Aegean Sea', *Geophys. J. Roy. Astron. Soc.* **37**, 337–346.
Koppe, I., Kleefeldt, M. and Haenel, R.: 1983, *Evaluation of Geothermal Energy Resources and Reserves in Selected Areas of the Federal Republic of Germany*, Report, NLfB, Hannover (FRG), Archive No. 95, 314, 10 p.
Kreyszig, E.: 1967, *Statistische Methoden und ihre Anwendungen*, Vandenhoeck & Ruprecht, Göttingen, 421 p.
Kron, A. and Heiken, G.: 1980, *Geothermal Gradient Map of the Conterminous United States*, Los Alamos National Laboratory Mem LA-8476-MAP.
Landel, P. A. and Sauty, J. P.: 1978, *Etude de l'influence des caracteristiques physiques de l'aquifere et roches encaissantes sur la temperature de l'eau an puits de production d'un doublet hydrothermique*, Report BRGM (Orleans, France), No. 78 SGN 405 GTH.
Lavigne, J.: 1978, *Les resources géothermiques francaises possibilities de mise en valeur*, Ann. des Minees, Avril, 16 p.
Lee, W. H. K. and MacDonald, G. J. F.: 1963, 'The Global Variations of the Terrestrial Heat Flow', *J. Geophys. Res.* **68**, 6481–6492.
Lee, W. H. K. and Uyeda, S.: 1965, 'Review of Heat Flow Data', in W. H. K. Lee (ed.), *Terrestrial Heat Flow*, AGU Geophys. Monograph. Washington, D.C., 8, 87–190.
Lubimova, E. A.: 1979, 'Kompleksnoye izucheniye teplovogo potoka, epeyrogenezisa i seismicheskikh proyavlenii na territorii Vostochno-Evropeyskoy platformi i ee obramleniya', in M. P. Volarovich (ed.), *Eksperimentalnoy i teoreticheskoye izucheniye teplovikh potokov*, Izd. Nauka, Moskva (in Russian), 5–33.
Lubimova, E. A. and Suetnova, E. I.: 1982, 'Global Thermal Regime of the Lithosphere with Its Different Permeability', in V. Čermák and R. Haenel (eds.), *Geothermics and Geothermal Energy*, E. Schweizerbart'sche Verlagsbuchhandlung, Nägele u. Obermiller, Stuttgart, 11–16.
Majorowicz, J.: 1973, 'Heat Flow in Poland and Its Relation to the Geological Structure', *Geothermics* **2**, 24–28.
Majorowicz, J. and Plewa, S.: 1979, 'Study of Heat Flow in Poland with Special Regard to Tectonophysical Problem', in V. Čermák and L. Rybach (eds.), *Terrestrial Heat Flow in Europe*, Springer-Verlag, Berlin, Heidelberg, New York, 240–252.
Majorowicz, J. and Jessop, A. M.: 1981, 'Regional Heat Flow Patterns in the Western Canadian Sedimentary Basin', *Tectonophysics* **74**, 209–238.
Makarenko, F. A., Velyugo, A. Ya., Gavlina, G. B., Mavritsky, B. F., Pokrovsky, V. A., Polyak, B. G., and Smirnov, Ya. B.: 1972, *Geotermicheskaya karta SSSR*, Nauka, Moskva (in Russian).
Motyl-Rakowska, J., Plewa, S., and Schoeneich, K.: 1977, 'Relationships Between Terrestrial Heat Flow and Structure of the Earth's Crust in Poland', *Geothermics* **6**, 99–106.
Muffler, L. J. P. and Cataldi, R.: 1978, 'Methods for Regional Assessment of Geothermal Resources', *Geothermics* **7**, 53–89.
Muffler, L. J. P. and Guffanti, M.: 1979, *Assessment of Geothermal Resources of the United States – 1978*, U.S. Geological Survey Circular 790, 1–7.
Mundry, E.: 1970, 'Zur automatischen Herstellung von Isolinienplänen', *Beih. Geol. Jb.* **98**, Hannover, 77–93.
Mundry, E.: 1984, 'Personal Communication', Address: NLfB, Stilleweg 2, Hannover, FRG.
Nathenson, M.: 1975, *Physical Factors Determining the Fraction of Stored Energy Recoverable from Hydrothermal Convection Systems and Conduction-Dominated Areas*, US Geological Survey Open-File Report 75–525, 35 p.

Oelsner, Ch.: 1978, 'Wärmestrom und Temperatur an der Moho in Mitteleuropa', *Rev. Roum. Géol. Géophys. Géogr.-Géophys.* **22**, 49–58.
Pollack, H. N. and Chapman, D. S.: 1977, 'Mantle Heat Flow', *Earth Planet. Sci. Lett.* **34**, 174–184.
Pollack, H. N.: 1980, 'The Heat Flow from the Earth', in P. A. Davies and S. K. Runcorn (eds.), *Mechanism of Continental Drift and Plate Tectonics*, Academic Press, New York, 183–192.
Polyak, B. G. and Smirnov, Ya. B.: 1968, 'Svyaz glubinnogo teplovogo potoka s tektonicheskim stroyeniem kontinentov', *Geotektonika* **4**, 3–19 (in Russian).
Rao, R. U. M., Rao, G. V., and Reddy, G. K.: 1982, 'Age Dependence of Continental Heat Flow – Fantasy and Facts', *Earth Planet. Sci. Lett.* **59**, 288–302.
Rybach, L.: 1973, *Wärmeproduktionsbestimmungen an Gesteinen der Schweizer Alpen*, Beitr. Geol. Schweiz. Geotechn. Ser., 51, 43 p.
Rybach, L. and Bodmer, Ph.: (1983), 'Processing and Representation of Heat Flow Density Maps. Part II. Construction and Contouring of Heat Flow Density Maps', in R. Haenel and M. Gupta (eds.), *Results of the First Workshop on Standards in Geothermics*, Zbl. Geol. Paläontol., Teil I, Stuttgart, 87–92.
Schulz, R.: 1985, *Analytische Modellberechnungen für die Hydrogeothermik in einem gespannten Aquifer*, Report, NLfB, Hannover (FRG), Archive No. 98 556, 40 p.
Sclater, J. G. and Francheteau, J.: 1970, 'The Implication of Terrestrial Heat Flow Observations on Current Tectonic and Geochemical Models of the Crust and Upper Mantle of the Earth', *Geophys. J. R. Astr. Soc.* **20**, 509–542.
SEAPEX: 1977, *Geothermal Gradient Map of Southeast Asia*, Southeast Asia Petroleum Exploration Society, Cities Service Oil Co., Singapore.
Smirnov, Ya. B. (ed.): 1980, *Karta teplovogo potoka i glubinnikh temperatur territorii SSSR i sopredelnikh rayonov (mashtab 1:10 000 000)*, Glavnoye upravleniye geodezii i kartografii, Moskva, 150 pp. and 2 enclosures (in Russian with English Abstract).
Sokolova, L. S.: 1978, 'Raschet temperaturi zemnoy kori na primere oblastey yuga Sibiri', *Rev. Roum. Géol., Géophys. Géogr.-Géophys.* **22**, 99–105 (in Russian).
Sorey, M. L., Nathenson, M., and Smith, Chr., 1983: *Methods for Assessing Low-Temperature Geothermal Resources*, US Geological Survey, Circular 892, 17–30.
Stegena, L.: 1972, 'Geothermal Map of Eastern Europe', *Geothermics* **1**, 140.
Stegena, L.: 1976a, 'The Variation of Temperature with Depth in the Pannonian Basin', in A. Ádám (ed.), *Geoelectric and Geothermal Studies (East-Central Europe and Soviet Asia)*, Adadémiai Kiadó, Budapest, 425–438.
Stegena, L.: 1976b, 'Geothermal Temperature Map of Central and Eastern Europe', in A. Ádám (ed.), *Geoelectric and Geothermal Studies (East-Central Europe and Soviet Asia)*, and enclosure, Akadémiai Kiadó, Budapest, 381–383.
Stegena, L.: 1983, 'Geothermics and seismicity in Central Europe', *Earthquake Prediction Research*, 17–23.
Tezcan, A. K.: 1979, 'Geothermal Studies, Their Present Status and Contribution to the Heat Flow Contouring in Turkey', in V. Čermák and L. Rybach (eds.), *Terrestrial Heat Flow in Europe*, Springer-Verlag, Berlin, Heidelberg, New York, 283–292.
Thamrin, M. and Siswoyo, Prayito: 1980, *Heat Flow Measurement in the Tertiary Basin of North Sumatra, Indonesia*, Report submitted to the 17th CCOP Meeting, Bangkok.
Vasseur, G. and Nouri, Y.: 1980, 'Trend of Heat Flow in France: Relation with Deep Structure', *Tectonophysics* **65**, 209–223.
Vitorello, I. and Pollack, H. N.: 1980, 'On the Variation of Continental Heat Flow with Age and the Thermal Evolution of Continents', *J. Geophys. Res.* **85(B2)**, 983–995.
Werner, D. and Doebl, F.: 1974, 'Eine geothermische Karte des Rheingraben-Untergrundes', in J. H. Illies and K. Fuchs (eds.), *Approaches to Taphrogenesis*, IGC Scient. Rep. No. 8, Stuttgart, 182–191.
Wesierska, M.: 1970, 'Results of Geothermal Temperature Measurements in Poland', *Mat. i Prace, Warszawa* **34**, 25–28.
White, D. E. and Williams, D. L. (eds.): 1975, *Assessment of Geothermal Resources of the United States – 1975*, US Geological Survey, Circular 726, 155 p.
Wohlenberg, J.: 1979, 'The Subsurface Temperature Field of the Federal Republik of Germany', *Geol. Jb.* **E15**, 3–29.

8. CRUST AND MANTLE GEOTHERMICS

8.1. THERMAL REGIME OF THE LITHOSPHERE

H. N. POLLACK
Department of Geological Sciences, The University of Michigan, Ann Arbor, USA

and

J. H. SASS
United States Department of the Interior, Geological Survey, Flagstaff, USA

The Earth's lithosphere may be specified in a number of ways depending on the physical or chemical characteristics being used to define it. Thermally the lithosphere is probably best defined as that portion of the Earth above a zone separating predominantly conductive heat transfer from predominantly non-conductive (convective and radiative) heat transfer (see e.g. Morgan 1984). This definition is somewhat oversimplified in that non-conductive heat transfer may dominate within the lithosphere in tectonically active regions for significant time intervals; however, the definition suitably characterizes most of the Earth's lithosphere, both continental and oceanic.

For most of the oceans (and to a lesser extent, marginal basins), the lithosphere may be defined explicitly in terms of a thermal boundary layer resulting from the conductive cooling of hot crustal and upper mantle material introduced at the axis of mid-ocean spreading (McKenzie, 1967; Parsons and Sclater, 1977). In areas of very young crust and non-existent or discontinuous sedimentary cover, the surface of this boundary layer is profoundly affected by hydrothermal convection (e.g. Davis and Lister, 1977); however, away from these areas, the thermal regime of oceanic lithosphere may be characterized adequately by the time elapsed since its creation at a spreading center. Heat-flow density is inversely proportional to the square root of age, and subsidence (thermal contraction) is directly proportional to the same parameter. The thermal relaxation of the oceanic lithosphere from its initial hot condition occurs in a characteristic time on the order of 10^8 a.

Two analytical models have commonly been used to describe the thermal regime of the oceanic lithosphere. The first is simply a half-space initially at a high temperature which loses heat by conduction across a cold upper surface. The governing equation is:

$$T(z, t) = T_0 \operatorname{erf}[z/(4at)^{1/2}], \qquad (8.1)$$

where

T_0 = initial high temperature
a = thermal diffusivity
t = time after cooling began.

R. Haenel, L. Rybach, and L. Stegena (eds.)
Handbook of Terrestrial Heat-Flow Density Determination, 301—308.

The second model is a slab of thickness h initially at a high temperature, which loses heat by conduction across a cold upper surface, while maintaining the base of the slab at the initial temperature T_0. The governing equation is:

$$T(z, t) = T_0 z/h + (4T_0/\pi) \sum_{n=0}^{\infty} \frac{1}{2n+1} \sin \frac{(2n+1)\pi z}{h} \exp \frac{-(2n+1)^2\pi^2 at}{h^2} - (2T_0/\pi) \sum_{n=1}^{\infty} \frac{(-1)^{n+1}}{n} \sin \frac{n\pi z}{h} \exp \frac{-n^2\pi^2 at}{h^2}. \quad (8.2)$$

Figure 8.1 shows geotherms calculated from Equations (8.1) and (8.2). As is apparent, they are identical for times up to 50 Ma, at which time the basal boundary condition of Equation (8.2) begins to take effect, maintaining slightly higher temperatures in the lower part of the slab and slightly greater heat flow at the surface.

For continents, the situation is not nearly so simple. The creation and reworking of continental lithosphere involves both heating and cooling processes of varying intensity and duration. The tectono-thermal age of a given area may vary considerably depending on the criteria (and the geochronometers) used to define it. When viewed on a global basis, the continental heat-flow density data reveal a suggestion of a decrease of heat-flow density with age, albeit with considerable scatter. The average of an extremely heterogeneous Cenozoic and Mesozoic population of heat-flow density age pairs is higher than the global heat-flow density average for continents, and heat-flow density from Archean rocks is significantly lower than the average (cf. Polyak and Smirnov, 1968; Chapman and Furlong, 1977; Vitorello and Pollack, 1980, Sclater *et al.*, 1980). There apparently is not, however, a robust statistical correlation between heat-flow density and age for continents (Rao *et al.*, 1982).

From the above discussion, we may conclude that while it is relatively easy to construct a geotherm for an area of ocean whose age is well known, for an analogous continental region tectonic age is only one factor to be evaluated for the specific area under consideration. An analytical representation of a continental geotherm requires a solution to the heat-conduction equation that accounts for the surface heat-flow density, subsurface distribution of radioactive heat-producing elements, vertical and lateral variation of thermal conductivity, magmatic and metamorphic history, uplift and erosion, surface temperature history, and a host of local phenomena such as microclimates and the shallow hydrologic regime (see Chapters 2. to 5. for details). In practice, however, an exact characterization of many if not most of these variables is simply not feasible, and, in some instances, may result in a mathematically intractable formulation of the problem. Simplifying assumptions must therefore be made both for analytical convenience and because of lack of data on the variation of some parameters. An example of simple, steady state, conductive geotherms is shown in Figure 8.2 (after Lachenbruch and Sass, 1977). The governing equation is (see also Section 7.3.1.):

$$T(z) = [(q_0 - DH_0)z + D^2H_0\,(1 - \exp(-z/D))]/\lambda, \quad (8.3)$$

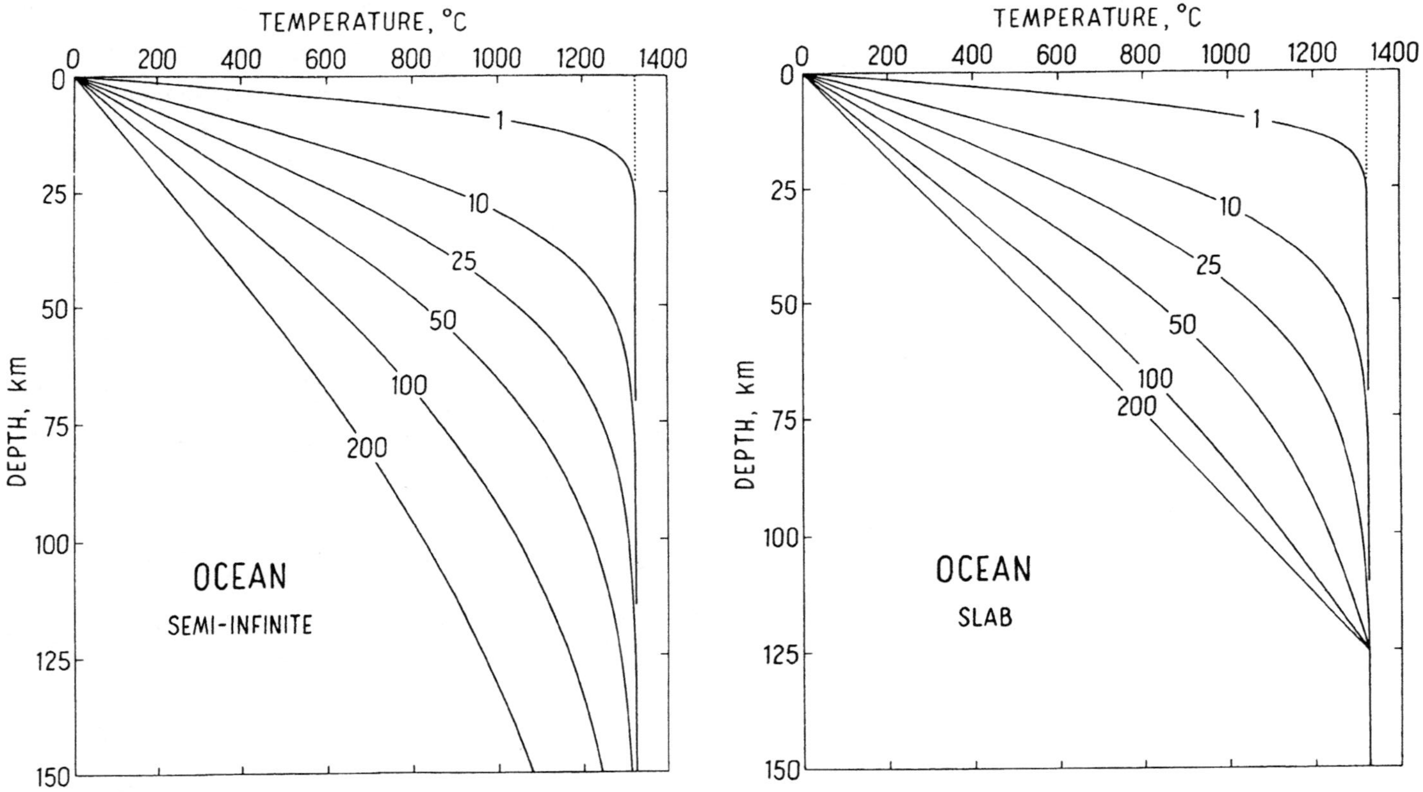

Fig. 8.1. Geotherms for oceanic lithosphere cooling from initial high temperature, half-space model (left) and constant temperature lower boundary model (right). Lithospheric thickness is 125 km, initial temperature 1325 °C and thermal diffusivity 32 km^2 Ma^{-1}. Numbers on curves indicate time in millions of years after cooling began.

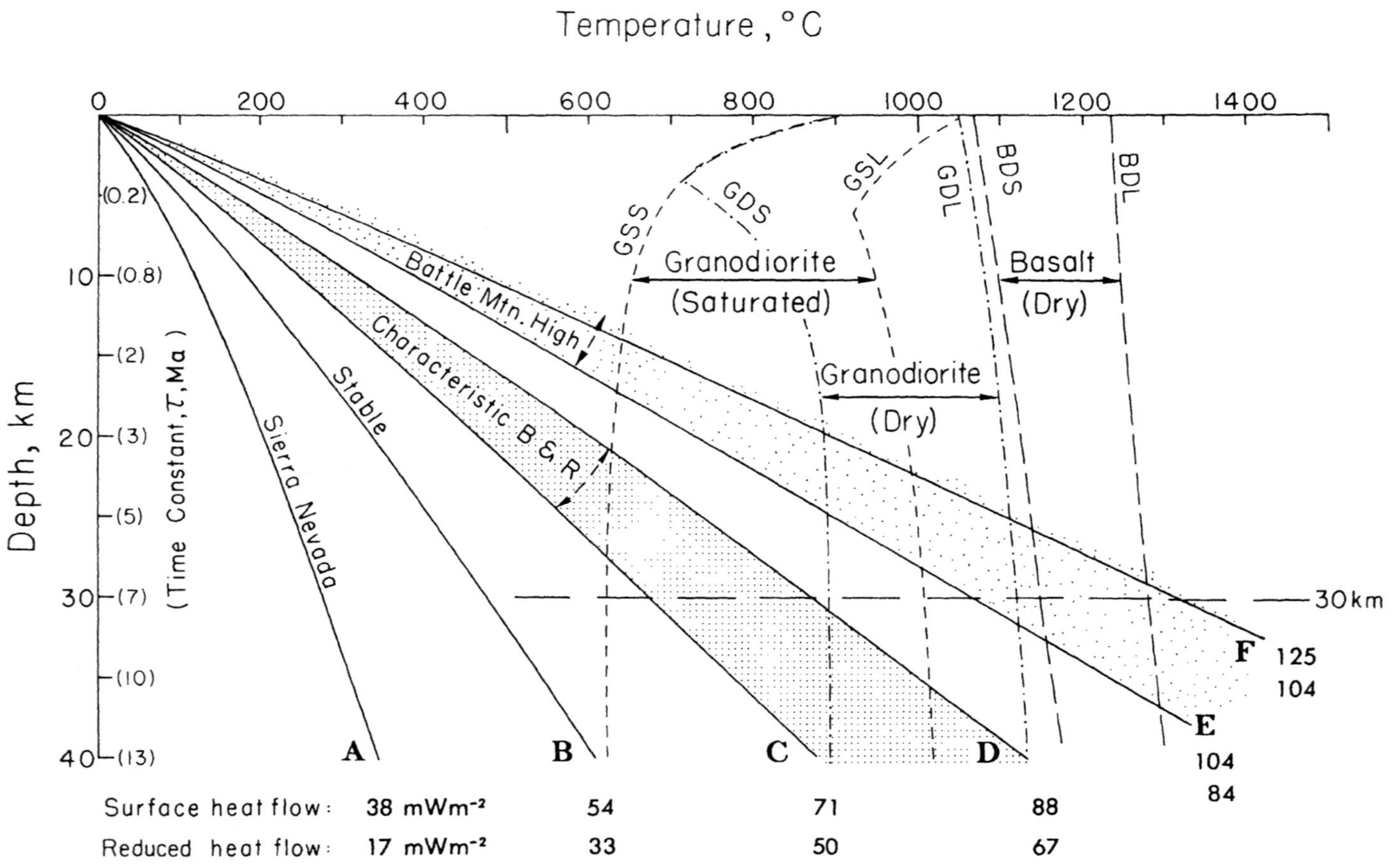

Fig. 8.2. Generalized steady-state conductive temperature profiles for various heat-flow density provinces in the United States (Lachenbruch and Sass, 1977). Surface heat-flow density and reduced heat-flow density are shown at the bottom of each curve. Solidus and liquidus curves for dry and saturated granodiorite (GDS, GDL, GSS, GSL) and basalt (BDS, BDL) are also shown (after Wyllie, 1971).

where

T = temperature
q_0 = surface heat-flow density
H_0 = volumetric radiogenic heat generation at the surface
D = parameter characterizing the depth distribution of heat sources
λ = thermal conductivity.

The thermal time constant $\tau = z^2/4a$ is a measure of the relaxation time for a thermal perturbation extending to depth z. The geotherms in Figure 8.2 are constrained by surface heat-flow density, zero surface temperature and a surface radiogenic heat generation of 2 $\mu W\ m^{-3}$. For analytical convenience, the mean λ of the crust was assumed constant at 2.5 $W\ m^{-1}\ K^{-1}$ and the heat source distribution was assumed to decrease exponentially with depth with a logarithmic decrement D of 10 km. The reduced heat-flow densities (q_r) shown in Figure 8.2 are calculated by simply stripping off an approximation (DH_0) of the upper crustal radiogenic component according to:

$$q_r = q_0 - DH_0. \tag{8.4}$$

The crustal portions of these geotherms illustrate the problems of attempting to characterize with steady-state conductive thermal models areas subject to contemporary or Neogene tectonism. Low heat-flow density in the Sierra Nevada of the western United States is the result of thermal transients associated with Miocene subduction of the terrane (Roy *et al.*, 1972; Henyey and Lee 1976). Thus, at some depth within the lithosphere, near the base of the subducted block, there will occur a rapid increase in temperature and geotherm 'A' will become invalid. Geotherm 'B' is probably an adequate representation of conditions in the stable eastern United States. Steady-state conductive geotherms for the Basin and Range province (geotherms C–F, Figure 8.2), most particularly for the large subprovince of heat-flow density >100 $mW\ m^{-2}$ (the 'Battle Mountain high'), have an implicit 'melting paradox', as pointed out by Lachenbruch and Sass (1977). These geotherms predict temperatures high enough to cause widespread melting of intermediate and even basaltic rocks in the lower crust – a prediction considerably at odds with geological and other geophysical observations.

Lachenbruch and Sass (1978, see also Lachenbruch, 1978) demonstrated that the melting paradox for the Basin and Range province could be resolved by constructing geotherms based on simple, one dimensional, convective models involving stretching of the lithosphere (by normal faulting and/or ductile flow) accompanied by some combination of basaltic underplating or pervasive intrusion of basalt along fractures induced by tectonic extension. These geotherms are cooler by several hundred degrees in the lower crust and uppermost mantle than the purely conductive ones owing to the more efficient heat transfer resulting from the combination of solid and magmatic convection accompanying distributed extension. For extension rates consistent with geological observations (e.g. displacements on normal faults) these models predict a thermal regime compatible with seismically derived estimates of lithosphere thickness and with pT conditions inferred from petrologic data.

The continuation of continental geotherms into the upper mantle in principle requires no new concepts; specification of the mantle heat source distribution and thermal conductivity enables the continuation of the temperature profile downward from the base of the crust. However, the petrology of the mantle contrasts markedly with that of the crust, and in particular the upper mantle is strongly depleted of radioisotopes, and therefore heat generation is virtually absent. In the absence of heat generation, the extended geotherms in a medium of uniform conductivity would be linear with depth, in a cartesian coordinate system. However, two factors, one physical and one geometric, lead to a departure from linearity. First is the temperature dependence of the thermal conductivity. This effect, described more fully in Chapters 4.1 and 10.1, generally enhances conductivity in the temperature range of the upper mantle, leading to lower temperature gradients than would occur with a uniform conductivity. The second factor is geometric, arising from the Earth's sphericity. Conduction of heat through a uniform conductivity spherical shell with no heat sources follows a geotherm that is inversely proportional to radius, thus yielding temperature gradients higher than those across an equivalent planer layer. These two factors have opposing effects but the conductivity effect dominates. From the base of the continental crust to 200 km depth the net difference in temperature (as compared to a planar layer of uniform conductivity characteristic of the uppermost mantle) is a decrease of approximately 250 and 500 °C, respectively, for reduced heat-flow density (surface heat-flow density minus crustal radiogenic component) of 20 and 40 mW m^{-2}.

The ultimate depth to which conductive geotherms can be extended is entwined with the thermal definition of the lithosphere as the zone of predominantly conductive heat transfer. Thus the depth at which conditions promote significant solid-state creep also is the probable lower boundary of the thermal lithosphere. Just as in the crustal environment, the presence (or absence) of volatiles, principally H_2O and CO_2, plays a very important role in establishing the onset of creep. The significant parameter is the ratio of the ambient temperature T to the solidus temperature T_m, with the latter strongly affected by even minor amounts of volatile components. When the ratio T/T_m exceeds about 0.85, solid-state creep with its accompanying enhanced heat transfer increasingly overshadows conduction in significance. Figure 8.3 (after Chapman and Pollack, 1977) shows a family of continental geotherms, parametric in surface heat-flow density, superimposed on some generalized mantle solidi representative of a range of volatile environments. It is clear that conductive geotherms suitably characterize only the outer few tens of kilometers in regions of high heat-flow density. In contrast, conduction may extend to several hundred kilometers beneath low heat-flow density provinces such as the Precambrian shields, particularly if the upper mantle beneath shields is depleted in volatiles.

Tectonism and magmatism in various settings can lead to transient perturbations of steady-state geotherms. Indeed, Vitorello and Pollack (1980) considered a 90 mW m^{-2} geotherm as a 'perturbed' 45 mW m^{-2} geotherm with the temperature difference between the two an initial condition superimposed on the 'steady state' lower geotherm. The decay of this perturbation they interpreted as the decaying

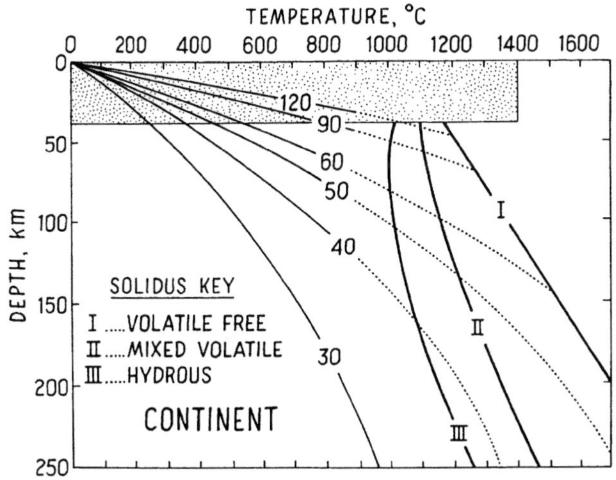

Fig. 8.3. Conductive geotherm family for continental terranes (after Pollack and Chapman, 1977; and Chapman and Pollack, 1977). Numerical parameter on each geotherm is the corresponding surface heat-flow density in mW m^{-2}. Generalized solidus curves for periodotite in different volatile environments are also shown. Stippled area above 40 km corresponds to depth and temperature range of Figure 8.2. Geotherms are shown with solid lines where conduction is the principal mode of heat transfer, and with dotted lines where other modes of heat transfer may have increasing significance.

thermal signature of a tectonic event. Similar elevated geotherms result from simple extension models (e.g. McKenzie, 1978) who interpreted the subsequent decay and accompanying thermal contraction in terms of sedimentary basin formation. Crustal thickening under compression by contrast might actually lead to an initially cooler geotherm which recovers to a warmer steady state. Compressional events involving thrusting have been modeled as initial value problems characterized by thermal repetition accompanying the stratigraphic repetition (Oxburgh and Turcotte, 1974). Such initial value models have been investigated principally in the context of temperature histories of crustal terranes, with implications for metamorphism, changing rheology, radiometric dates, acquisition of palaeomagnetism, and other systems involving thermal activation and closure. Crustal thermal histories are, of course, significantly influenced by the erosional/depositional history following a tectono-thermal event (England and Richardson, 1977).

It is impossible in a volume of this scope to provide a general solution of the heat-flow equation that will describe adequately the Earth's thermal regime in all tectonic situations, particularly for continents. The construction of continental geotherms requires a knowledge of many factors, both regional and local. In general, the uncertainty of the geotherm increases with depth because of corresponding uncertainties in the mineralogic composition, thermal conductivity, and abundances of heat-producing elements. We have attempted by means of a few simple examples to illustrate the general process of geotherm calculation, and to emphasize some of the pitfalls and difficulties in what amounts to a downward continuation of near-surface heat-flow data. As more data are collected in regions and tectonic settings now sparsely represented in the world data base, it should be possible to refine regional continental geotherms and to place needed constraints on tectonic and geochemical models for the Earth's deep interior.

8.2. REFERENCES

Chapman, D. S. and Furlong, K.: 1977, *Continental Heat Flow – Age Relationships*, EOS Trans. Am. Geophys. Union 58, 1240 p.

Chapman, D. S. and Pollack, H. N.: 1977, 'Regional Geotherms and Lithospheric Thickness', *Geology* **5**, 265–268.

Davis, E. E. and Lister, C. R. B.: 1977, 'Heat Flow Measured over the Juan de Fuca Ridge: Evidence for Widespread Hydrothermal Circulation in a Highly Heat Transportive Crust', *J. Geophys. Res.* **82**, 4845–4860.

England, P. C. and Richardson, S. W.: 1977, 'The Influence of Erosion Upon the Mineral Facies of Rocks from Different Metamorphic Environments', *J. Geol. Soc. Lond.* **134**, 201–213.

Henyey, T. L. and Lee, T. C.: 1976, 'Heat Flow in Lake Tahoe, California-Nevada, and the Sierra Nevada – Basin and Range Transition', *Geol. Soc. America Bull.* **87**, 1179–1187.

Lachenbruch, A. H.: 1978; 'Heat Flow in the Basin and Range Province and the Thermal Effects of Tectonic Extension', *Pageoph* **117**, 34–50.

Lachenbruch, A. H. and Sass, J. H.: 1977, 'Heat Flow and the Thermal Regime of the Crust', in: J. G. Heacock (ed)., *The Earth's Crust, Its Nature and Physical Properties*, American Geophysical Union, Washington, D. C., 625–675.

Lachenbruch, A. H. and Sass, J. H.: 1978, 'Models of an Extending Lithosphere and Heat Flow in the Basin and Range Province', in R. B. Smith and G. P. Eaton (eds.), *Cenozoic Tectonics and Regional Geophysics of the Western Cordillera*, Geological Society of America, Memoir 152, Boulder, Colorado, 209–250.

McKenzie, D. P., 1967: 'Some Remarks on Heat Flow and Gravity Anomalies', *J. Geophys. Res.* **72**, 6261–6273.

McKenzie, D. P.: 1978, 'Some Remarks on the Development of Sedimentary Basins', *Earth Planet. Sci. Lett.* **40**, 25–32.

Morgan, P.: 1984, 'The Thermal Structure and Thermal Evolution of the Continental Lithosphere', *Physics and Chemistry of the Earth* **15**, 107–193.

Oxburgh, E. R. and Turcotte, D. L.: 1974, 'Thermal Gradients and Regional Metamorphism in Overthrust Terrains with Special Reference to the Eastern Alps', *Schweiz. Miner. Petrogr. Mitt.* **54**, 641–662.

Parsons, B. and Sclater, J. G.: 1977, 'An Analysis of the Variation of Ocean Floor Bathymetry and Heat Flow with Age', *J. Geophys. Res.* **82**, 803–827.

Pollack, H. N. and Chapman, D. S.: 1977, 'On the Regional Variation of Heat Flow, Geotherms, and the Thickness of the Lithosphere', *Tectonophysics* **38**, 279–296.

Polyak, B. G. and Smirnov, Ya.B.: 1968, 'Relationship Between Terrestrial Heat Flow and the Tectonics of Continents', *Geotectonics, Engl. Trans.* **4**, 205–213.

Rao, R. U. M., Rao, G. V., and Reddy, G. K.: 1982: 'Age Dependence of Continental Heat Flow–Fantasy and Fact', *Earth Planet. Sci. Lett.* **59**, 288–302.

Roy, R. F., Blackwell, D. D. and Decker, E. R.: 1972, 'Continental Heat Flow, in E. C. Robertson (ed.) *The Nature of the Solid Earth*, New York, McGraw-Hill Book Co., 506–543.

Sclater, J. G., Jaupart, C., and Galson, D.: 1980, 'The Heat Flow Through Oceanic and Continental Crust and the Heat Loss from the Earth', *Rev. Geophys. Space Phys.* **18**, 269–311.

Vitorello. I. and Pollack, H. N.: 1980; 'On the Variation of Continental Heat Flow with Age and the Thermal Evolution of Continents', *J. Geophys. Res.* **85**, 983–995.

Wyllie, P. J.: 1971, 'Experimental Limits for Melting in the Earth's Crust and Upper Mantle', in J. G. Heacock *The Structure and Physical Properties of the Earth's Crust*, American Geophysical Union, Washington, D. C., 279–301.

8.3. DOWNWARD CONTINUATION

D. STROMEYER
Central Institute for Physics of the Earth, Telegrafenberg, Potsdam, GDR

8.3.1. Introduction

This paper reviews the most important mathematical methods dealing with inverse problems in connection with downward continuation in geothermics. Only linear inverse problems are taken into consideration because their theory is well developed and powerful numerical techniques are known.

8.3.2. Inverse Problems for Steady-State Heat Conduction Models

A simple but often reasonable approximation of the heat transport mechanism of the Earth's crust and upper mantle is the steady-state heat conduction model described by the following linear differential equation:

$$-\nabla(\lambda \nabla T) = H, \quad \nabla = \left(\frac{\partial}{\partial x}, \frac{\partial}{\partial y}, \frac{\partial}{\partial z}\right), \tag{8.5}$$

where

T = temperature
λ = thermal conductivity
H = heat production rate.

Equation (8.5) can be solved if parameters λ, H and appropriate boundary conditions are known. From Equation (8.5) follows:

$$-\frac{\mathrm{d}}{\mathrm{d}z}\left(\lambda \frac{\mathrm{d}}{\mathrm{d}z} T\right) = H, \tag{8.6}$$

if the density of heat flow is considered only in the vertical direction z. From Equation (8.6) it is easy to derive the well-known linear relation:

$$q = q_M + \int_0^{z_M} H(z)\, \mathrm{d}z \tag{8.7}$$

between measured terrestrial heat-flow density $q = -\lambda\ \mathrm{d}T/\mathrm{d}z$ at the Earth's surface, the heat production rate $H(z)$ down to the depth z and the heat-flow density q_M at the depth z_M. The quantities H and q_M are unknown. Therefore, the estimation of H and q_M from near-surface q measurements represents the inverse problem of the steady-state heat conduction equation. Obviously, the solution of this problem is an ambiguous one. The number of heat production rate distributions $H(z)$ which produce the same q is infinitly large even if q_M is given.

R. Haenel, L. Rybach, and L. Stegena (eds.)
Handbook of Terrestrial Heat-Flow Density Determination, 309—315.

In true cases the distribution of H with depth can be estimated by some more or less speculative standard models (see also Chapter 4.2.). Then the determination of q_M can be carried out by means of Equation (8.7) in a straightforward manner. In connection with downward continuation of surface geothermal data not only q_M and H are to be estimated but also the distribution of thermal conductivity λ with depth. Similar to the determination of H, this problem has no unique solution. However, in most cases the relationship between q and λ is nonlinear (e.g. if $\lambda = \lambda(T)$ is a function of the temperature; see Chapter 10.1.) and therefore this problem will not be treated.

The simple Equation (8.7) for measured data and unknown model parameters gets a more complicated character if the density of heat flow is not restricted to the vertical direction. Moreover, a new difficulty arises while treating two- and three-dimensional models. The determination of q_M from q becomes an ill-posed problem even if a distribution of the heat production rate H is given because the solution q_M does not depend continuously on data q (Stromeyer and Hurtig, 1983). In the following this will be shortly demonstrated for steady-state two-dimensional temperature calculations.

The solution of Equation (8.5) for a two-dimensional temperature section depends on the parameters λ and H and appropriate boundary conditions respectively. At the Earth's surface, these conditions are real measurements. At all other boundaries of the section reasonable assumtions of the thermal state are to be introduced. Commonly the following conditions are used (Figure 8.4):

- temperature values $T_0\,(x)$ at the surface;
- vanishing horizontal heat-flow density components at the ends of the section;
- estimated temperature and/or heat-flow density distribution along the lower boundary. For crustal models these values represent the temperature and/or heat-flow density at the Mohorovičić level (T_M, q_M), respectively.

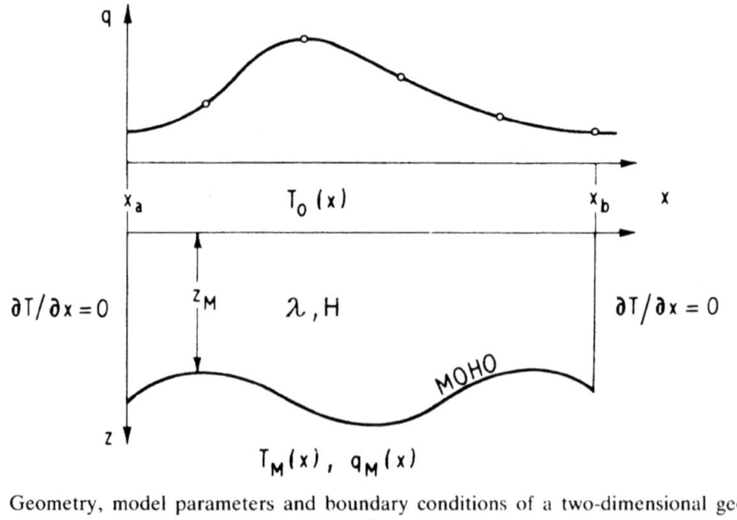

Fig. 8.4. Geometry, model parameters and boundary conditions of a two-dimensional geothermal profile.

The main imperfection of this approach is that the calculated temperature field $T(x, z)$ does not depend on the measured terrestrial heat-flow density q at the surface. Therefore, q can be used to improve models of λ and H or to estimate q_M or T_M. Only the heat source intensity H and the boundary condition q_M (or T_M) are linearly related with q.

These connections can be expressed in general form by means of an integral equation as follows:

$$q(x') = \int_0^{z_M} \int_{x_a}^{x_b} F(x'; x,z)H(x,z)\, dxdz + \int_{x_a}^{x_b} Q(x'; x)q_M(x)\, dx. \qquad (8.8)$$

Here the integration is carried out over the whole investigated profile section and its horizontal extension respectively.

For non-uniform heat conductivity λ however, it is impossible to construct the integral kernels F and Q explicitly and discrete approximations are necessary. This results in linear equations of the form:

$$\mathbf{q} = \mathbf{FH} + \mathbf{Qq}_M. \qquad (8.9)$$

Here $\mathbf{q}$, $\mathbf{A}$ and $\mathbf{q}_M$ are vectors and $\mathbf{F}$ and $\mathbf{Q}$ matrices. There is a simple way to calculate $\mathbf{F}$ and $\mathbf{Q}$ if effective procedures are available to solve Equation (8.5) numerically by replacing the differential operator ∇ by finite difference (or finite element) approximations. Then the columns of $\mathbf{F}$ and $\mathbf{Q}$ are formed by the heat-flow density values at the surface produced by a sequence of models where one component of the vectors $\mathbf{H}$ or $\mathbf{q}_M$ is equal to 1 while others are equal to 0.

Equation (8.9) is also obtained if special heat source models with free parameters are investigated, e.g. point sources distributed in a half space (Brott *et al.*, 1981) or a block model of a layer with uniform thermal conductivity (Huestis, 1979).

8.3.3. Linear Inverse Problems

Now, algorithms will be considered to solve the linear equation (8.9) from a general mathematical point of view.

The standard linear inverse problem may be expressed in the parametric form:

$$\mathbf{y} = \mathbf{Bx}. \qquad (8.10)$$

Here $\mathbf{y}$ is a N vector of observations, $\mathbf{x}$ is a M vector of unknown parameters and $\mathbf{B}$ is a known $N \times M$ matrix. Difficulties connected with Equation (8.10) are the non-uniqueness ($N < M$) and the instability of its solution for many real cases. For overcoming this situation a priori information for $\mathbf{x}$ must be used. Then the type of this information controls the different approaches to the solution of Equation (8.10). The following methods are restricted to least squares solutions of (8.10) and solutions with general lateral constraints respectively because for these techniques well proven numerical procedures and computer routines are available.

In recent years many papers have been published concerning linear inverse problems. For geophysical problems, the contributions by Sabatier (1977), Jackson (1978), and Parker (1979) must be strongly recommended.

(1) UNCONSTRAINED LEAST SQUARES PROBLEMS

The standard methods to solve Equation (8.10) are procedures for the least-norm solution and the least squares solution of (8.10). They are available as routines at each well established computer center. The first is defined by:

$$\min_{\mathbf{x} \in K} \|\mathbf{x}\| \qquad K = \{\mathbf{x}: \mathbf{B}\,\mathbf{x} = \mathbf{y}\} \tag{8.11}$$

and the second for an inconsistent system by:

$$\min_{\mathbf{x}} \|\mathbf{B}\,\mathbf{x} - \mathbf{y}\|.$$

These solutions are strongly related with the Moore-Penrose-pseudoinverse of a rectangular matrix and with a special approach of the Backus-Gilbert formalism (Sabatier, 1977). Both minimization problems are ill-posed. Therefore they must be realized numerically by a regularization algorithm (Tikhonov, 1963; Marquardt, 1963) of the form:

$$\min_{\mathbf{x}} (\|\mathbf{B}\,\mathbf{x} - \mathbf{y}\|^2 + \alpha\,\|\mathbf{x}\|^2). \tag{8.12}$$

α is a small positive value, the so called regularization parameter. Equation (8.11) can be generalized to:

$$\min_{\mathbf{x} \in K} \|\mathbf{D}\,(\mathbf{x} - \mathbf{x}_0)\|.$$

This is the problem of finding $\mathbf{x}$ close to an initial solution $\mathbf{x}_0$ where the difference $\mathbf{x} - \mathbf{x}_0$ is determined by a $L \times M$ matrix $\mathbf{D}$ (theory of weighted pseudoinverses, Eldén, 1975). Introducing this into (8.12) it follows:

$$\min_{\mathbf{x}} (\|\mathbf{B}\,\mathbf{x} - \mathbf{y}\|^2 + \alpha\,\|\mathbf{D}\,(\mathbf{x} - \mathbf{x}_0)\|^2). \tag{8.13}$$

(2) CONSTRAINED LEAST SQUARES PROBLEMS

To make the problem related to the solution of (8.10) wellposed one can introduce some a priori information in form of a bound ω of the norm of $\mathbf{D}(\mathbf{x} - \mathbf{x}_0)$. This results in a constrained least squares problem:

$$\min_{\mathbf{x} \in K} \|\mathbf{B}\,\mathbf{x} - \mathbf{y}\|, \quad K = \{\mathbf{x}: \|\mathbf{D}(\mathbf{x} - \mathbf{x}_0)\| \leq \omega\}. \tag{8.14}$$

In practice $\mathbf{y}$ often contains measurement errors and it is not meaningful trying to satisfy Equation (8.11) exactly. The assumption that the solution $\mathbf{x}$ is smooth in some sense leads to the following minimization problem:

$$\min_{\mathbf{x} \in K} \|\mathbf{D}(\mathbf{x} - \mathbf{x}_0)\|, \quad K = \{\mathbf{x}: \|\mathbf{B}\,\mathbf{x} - \mathbf{y}\| \leq \varepsilon\} \tag{8.15}$$

for some value of ε related to the statistical distribution of the errors in $\mathbf{y}$. Eldén (1975) has proved that in all interesting cases the minima are attained on the boundaries of K. Using the method of Lagrange multipliers (8.14) and (8.15) are equivalent to solving the nonlinear equations:

$$\|\mathbf{D}(\mathbf{x}_\alpha - \mathbf{x}_0)\|^2 - \omega^2 = 0$$

and

$$\|\mathbf{B}\ \mathbf{x}_\alpha - \mathbf{y}\|^2 - \varepsilon^2 = 0,$$

respectively, where $\mathbf{x}_\alpha$ is the solution of (8.13). All numerical algorithms discussed above were implemented in a computer program and published by Eldén (1979).

It should be mentioned that the use of a priori information in least squares problems is restricted to average properties of $\mathbf{x}$ and $\mathbf{y}$. Additional constraints in form of upper and lower bounds for the components of $\mathbf{x}$ and $\mathbf{y}$ can only be taken into account by removing the semi-linear character of the corresponding minimizing problems.

8.3.4. Lateral Constraints in Linear Inverse Problems

In 1977 Sabatier introduced a method which completely solves linear inverse problems:

$$y_i = \sum_j B_{ij}\, x_j \qquad i = 1,\ldots,N;\ j = 1,\ldots,M \tag{8.16}$$

with positive constraints:

$$x_i \geq 0. \tag{8.17}$$

The conditions (8.16) and (8.17) are very general because all lateral constraints, e.g. $\underline{x}_i < x_i < \bar{x}_i$ or $|x_i^\circ - x_i| < \delta_i$ can be reduced to this form.

The analysis of Sabatier is based on the fact that all $\mathbf{x} = (x_1,\ldots,x_M)$ satisfying (8.16) and (8.17) form a convex set which can be described by its extremals. A 'best solution' of this set is characterized by the minimum of a linear functional of $\mathbf{x}$. It can be found in a straightforward manner by means of linear programming. In the following some problems will be reviewed which can be solved in terms of this method.

– Minimize the norm:

$$\sum_i w_i\, x_i, \tag{8.18}$$

where
w_i is a (non-negative) weight vector.

– Determine a solution which is the closest to an initial vector $\mathbf{x}^\circ$ by minimizing:

$$\max_i |w_i(x_i - x_i^\circ)|. \tag{8.19}$$

– Solve (8.16) and (8.17) when (8.18) is replaced by:

$$\sum_i |w_i\,(x_i - x_i^\circ)|. \tag{8.20}$$

– Parker's ideal body theory (Parker, 1975) is equivalent to the minimization of the functional:

$$\max_i |x_i|. \tag{8.21}$$

In all cases the effect of experimental errors can by studied simply if (8.16) is replaced by:

$$|\sum_j B_{ij} x_j - y_i| < \varepsilon_i, \tag{8.22}$$

which can be transformed in the standard form using $2N$ non-negative slack variables. An effective package of computer routines for linear inverse problems with lateral constraints based on linear convex analysis was published by Cuer and Bayer in 1980. The first geothermal application of this method was given by Huestis (1979) in the case of uniform λ to calculate extremal temperature bounds in the presents of upper and lower constraints for the heat production rate H.

8.3.5. Determination of Mantle Heat-Flow Density

In comparison with other geophysical branches there are only few publications interpreting geothermal measurements by means of inversion techniques. Therefore, there are only few experiences in connection with the discussed linear methods in geothermics and our recommendation must be preliminary.

To simplify matters for a moment the relation between q and q_M in Equation (8.9) will be investigated in absence of heat sources. Then (8.9) reduces to $\mathbf{q} = \mathbf{Q}\mathbf{q}_M$. What inversion method may be appropriate to give a good estimation of q_M? Two conditions seem to be resonable. Upper and lower bounds may be introduced or the mean lateral variation of q_M must be bounded. The first condition leads to the linear convex methods of Sabatier, the second to a constrained least squares problem defined Equation (8.14). Choosing the discrete differential operator $\mathrm{d}/\mathrm{d}x$ for the matrix $\mathbf{D}$ in Equation (8.14) the mean value of the first derivative of q_M is bounded and the solution has a good smoothness. This method was used by Stromeyer (1984) to calculate a reasonable mantle heat-flow density q_M for a two-dimensional temperature profile crossing Central Europe.

8.3.6. Optimal Heat Source

Similar to the determination of q_M from measured data q, optimal heat source models can be calculated by means of constrained least squares methods. On this occasion attention must be paid to the fact that this problem is not only ill-posed but also an ambiguously one. If lateral variations of heat production rate $H(x,z) = H(x)$ are of interest it can be shown that H is determined uniquely by surface heat-flow density data.

For a two-dimensional geothermal profile some optimal heat production rate models $H(x)$ were calculated by Stromeyer (1984) for the granitic layer by means of Equation (8.14). The same techniques was used by Hurtig and Stromeyer (1985) to estimate the heat production rate $H(x,y)$ within the granitic upper crust for a three-dimensional crustal model of Central Europe and adjacent areas.

It should be mentioned that the method of Sabatier for linear inverse problems with lateral constraints is also a powerful tool to calculate optimal heat source models because a priori information in form of an initial source model H_0 with strong upper and lower bounds is often available. But if the interest is concentrated on details the number of unknown parameters becomes large. Today there are no constrained reasons in geothermics to prefer one inversion method and reject another one. Because of the general properties of the methods it can be expected that least squares procedures give a good information about average features of the investigated parameters and the method based on linear convex analysis about extremal solutions.

Acknowledgment

I am thankful to Prof. L. Stegena for his suggestions to improve the final version of this paper.

8.4. REFERENCES

Brott, C. A., Blackwell, D. D., and Morgan, P.: 1981, 'Continuation of Heat Flow Data: A Method to Construct Isotherms in Geothermal Areas', *Geophysics* **46**, 1732–1744.

Cuer, M. and Bayer, R.: 1980, *A Package of Routines for Linear Inverse Problems*, Cahiers mathematiques 18, Montpellier

Eldén, L.: 1975, *A Note of Weighted Pseudoinverse With Application to the regularization of Fredholm Integral Equations of First Kind*, Report of Linköping University, LTH–MAT–R–75–11.

Eldén, L.: 1979, *A Program for Interactive Regularization*, Report of Linköping University, Lith–MAT–R–79–25.

Huestis, S. P.: 1979, 'Extremal Temperature Bounds from Surface Gradient Measurements', *Geophys. J. Roy. Astron. Soc.* **58**, 249–260.

Hurtig, E. and Stromeyer, D.: 1985, 'Three-Dimensional Modelling of Crustal Temperature and MOHO Heat Flow Density in Central Europe and Adjacent Areas', *J. of Geodynamics* **4**, 63–73.

Jackson, D. D.: 1979, 'The Use of *a priori* Data to Resolve Non-uniqueness in Linear Inversion', *Geophys. J. Roy. Astron Soc.* **57**, 137–157.

Marquardt, D. W.: 1963, 'An Algoritm for Least-Squares Estimation of Nonlinear Parameters', *J. Soc. Ind. Appl. Math.* **11**, 431–441.

Parker, R. L.: 1975, 'The Theory of Ideal Bodies for Gravity Interpretation', *Geophys. J. Roy. Astron. Soc.* **42**, 315–334.

Parker, R. L.: 1979, 'Understanding Inverse Theory', *Ann. Rev. Earth Planet. Sci.* **5**, 35–64.

Sabatier, P. C.: 1977, 'On Geophysical Inverse Problems and Constraints', *J. Geophys.* **43**, 115–137.

Stromeyer, D. and Hurtig, E.: 1983, 'Parameter Estimation for Steady-State Heat Conduction Models', *Zbl. Geol. Paläont. Teil* I, 1–2, 93–101.

Stromeyer, D.: 1984; 'Downward Continuation of Heat Flow Data by Means of the Least Squares Method', *Tectonophysics* **103**, 55–66.

Tikhonov, A. N.: 1963, 'Solution of Incorrectly Formulated Problems and the Regularization Method', *Soviet Math. Dokl.* **4**, 1035–1038.

9. IMPLICATIONS

9.1. GEODYNAMICS

S. UYEDA

Earthquake Research Institute, University of Tokyo, Japan
and
Department of Geophysics, Texas A&M University, College Station, USA

9.1.1. Introduction

The Earth is often regarded as a heat engine because many of the important phenomena of the Earth are caused by its internal thermal energy. The phenomena include such a large scale one as global plate tectonics and small scale ones as local seismicity, crustal movements, volcanism and hydrocarbon maturation. To elucidate how this giant heat engine operates, the nature and distribution of its heat sources, internal temperature and mode of heat transfer must be clarified. Through such studies, the mechanisms of Earth's processes are inferred and the thermal history and tectonic evolution of the Earth can be reconstructed. However, the path of such research is not an easy one, because the problem is essentially an inverse one both in time and space with non-unique solutions.

Modern measurement of terrestrial heat-flow density was initiated during the 1930's in land areas and 1950's in oceanic areas. When Birch (1954) reviewed the available data, the total number of measurements was only about sixty, but now the number exceeds 10000 (D. Chapman and H. Pollack, personal communication). Table 9.1 shows the time change of the number of measurements. Figure 9.1 shows the global distribution of heat-flow density data in 5° × 5° grid as of 1979. North America and Australia are examples of well studied continents. Recently monumental compilations of heat-flow density data have been published for Europe (Čermák and Rybach, 1979) and the USSR (Smirnov, 1980). However, as can be observed in Figure 9.1, there are still many areas where only small number of measurements have been reported, e.g. South America (Vitorello *et*

TABLE 9.1
Numbers of published heat-flow density measurements (after Pollack, 1982).

Year of compilation	Continental	Oceanic	Reference
1954	43	20	Birch (1954a)
1965	131	913	Lee and Uyeda (1965)
1970	597	2530	Lee (1970)
1976	1699	3718	Jessop *et al.* (1976)
1979	2808	4409	Chapman and Pollack (1980)

R. Haenel, L. Rybach, and L. Stegena (eds.)
Handbook of Terrestrial Heat-Flow Density Determination, 317—351.

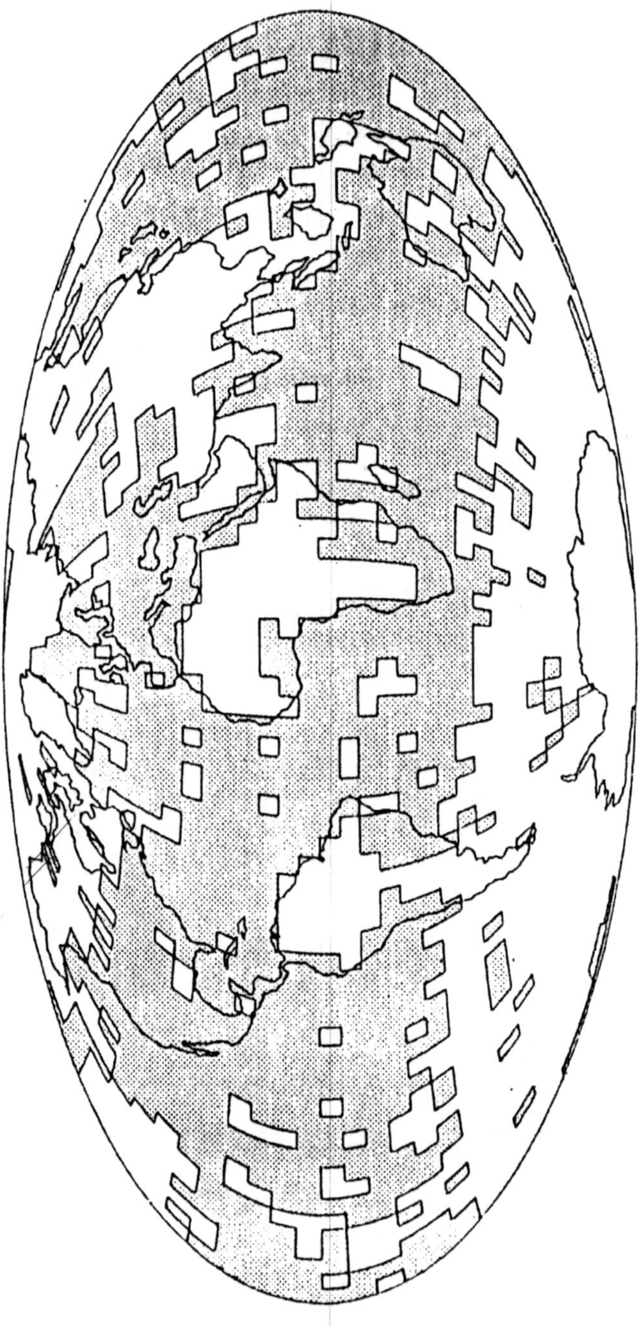

Fig. 9.1. Global heat-flow density data distribution in 5° × 5° geographic grid. Shaded areas contain at least one observation (after Pollack, 1982).

al., 1980; Uyeda and Watanabe, 1982), Africa (Degens *et al.*, 1973, Chapman and Pollack, 1977), Central and Southeast Asia (Verma, 1969; Uyeda, 1980) and the polar regions (Lachenbruch and Marshall, 1966: Decker and Burcher, 1977).

Since the early stage of heat-flow density studies, two important issues have been discussed from time to time. One was the apparent equality of heat-flow density on land and in oceanic areas. Because oceanic crust is believed to contain much less radioactive heat sources than continental crust, equality of heat flow led to speculations on the differences in the thermal state of the upper mantle under oceans and continents (e.g. Bullard *et al.*, 1956). The other issue was the apparent equality of present heat output from the Earth with the rate of heat production from chondritic meterorite with the size of the Earth ('chondrite coincidence'). This again induced speculations regarding the meteoritic origin of the Earth (e.g. Birch, 1958). However, both of these issues appear to have lost their importance in the present thinking of Earth scientists, as will be explained later.

Review of heat-flow density data from a global viewpoint first started in the 1960's (Lee and MacDonald 1963; Lee and Uyeda, 1965, Simmons and Horai, 1968). It was found that heat-flow density generally decreases from young orogenic zones to Pre-Cambrian shields on land and from mid-oceanic ridges to older basins in the sea. From about that time, a revolutionary progress in solid Earth sciences came about as represented by the theories of sea-floor spreading and plate tectonics. Interpretation of heat-flow density data naturally was deeply influenced by these new thoughts (e.g. Sclater and Francheteau, 1970). In this article, the geodynamic implications of heat-flow density will be reviewed in terms of these new thoughts.

9.1.2. Global Heat Output

Early estimates of global heat output were made by simple averaging of measured values (Lee and Uyeda, 1965, Simmons and Horai, 1968; Lee, 1970; and a more advanced one by Kappelmeyer and Haenel, 1974). The results was the equality of average heat-flow density over continents and oceans, the average being about 60 mW m^{-2}, giving a total of global heat output of ca. 3×10^{13} W. Now, the situation has changed. In the framework of plate tectonics, more definite physical models may be devised and more meaningful average values may be determined. Some representative estimates based on geodynamics models are given in Table 9.2.

As can be observed from this table, estimates of oceanic heat output drastically increased after the later 1970's. This was due to the recognition of powerful hydrothermal circulation in the oceanic crust. Taking the non-conductive component into account, the average oceanic heat-flow density is now estimated to be almost 1.8 times that of the continental average, making the ocean-continent equality totally obsolete. Considering the areas covered by oceans and continents, the ratio of heat output from the oceans and continents is about 3 : 1. In plate tectonic models, oceanic plates are the surface boundary layer of the convecting mantle and are playing the role of cooling fins of the global heat engine. The latest estimates of total global heat output are over 4×10^{13} W (Table 9.2).

TABLE 9.2
Estimates of average heat-flow density for continents and oceans, and total heat output from the Earth.

Average heat-flow density[a]			Total heat output	
Continental	Oceanic	Global	(10^{13} W)	Reference
61 (1.46)	61 (1.46)	61 (1.46)	3.11	Lee (1970)
53 (1.27)	62 (1.48)	59 (1.41)	3.01	Chapman and Pollack (1975)
61 (1.46)	87 (2.08)	77 (1.84)	3.92	Langseth and Anderson (1979)
53~61 (1.27~1.46)	100 (2.39)	81~84 (1.93~2.01)	4.15~4.30	Davies (1980a)
57 (1.37)	99 (2.36)	82 (1.95)	4.2	Sclater *et al.* (1980)

[a] In mW m^{-2}; values in brackets: 1×10^{-6} cal cm^{-2} s^{-1}.

9.1.3. Oceanic Heat-Flow Density

In plate tectonics models, oceanic plates are generated at mid-oceanic ridges and spread laterally. As they spread, cooling from above penetrates deeper and the plates increase their thickness. In these models, oceanic lithosphere is the thermal boundary layer of the convective mantle. During spreading, the thickness of oceanic plate and the depth of sea floor are both predicted theoretically to increase as $t^{1/2}$, where t is the age of plate (Yoshii, 1973; Oldenburg, 1975). This model, successful in many ways was, however, not able to explain at all the observed distribution of heat-flow density in the oceanic areas. For example, Figure 9.2 shows the observed heat-flow density and water depth across the Pacific Ocean from the East Pacific Rise to the Japan Sea. Heat-flow density is certainly high at the East Pacific Rise, but has a large scatter and decreases very rapidly on the flanks of the rise. It rises again at a distance from the rise crest and then decreases more slowly. This behavior is at odds with the behavior predicted from heat conduction theory based on cooling plate models (e.g. Parsons and Sclater, 1977), namely:

$$q(t) = 473\ t^{-1/2}, \tag{9.1}$$

where q is heat-flow density in mW m^{-2} and t age of the lithosphere in millions of years. Water depth, on the other hand, was found to follow the theoretical prediction quite well.

This phenomenon was also observed in the Atlantic and Indian Oceans. Figure 9.3 is a compilation for various oceanic areas, in which average heat-flow density values are plotted against age of the lithosphere. In all cases, heat-flow density values are much lower than the expected theoretical curve at younger ages and approach the theoretical curve only after age becomes greater than certain values. This disagreement between theory and observation was a puzzle for sometime in

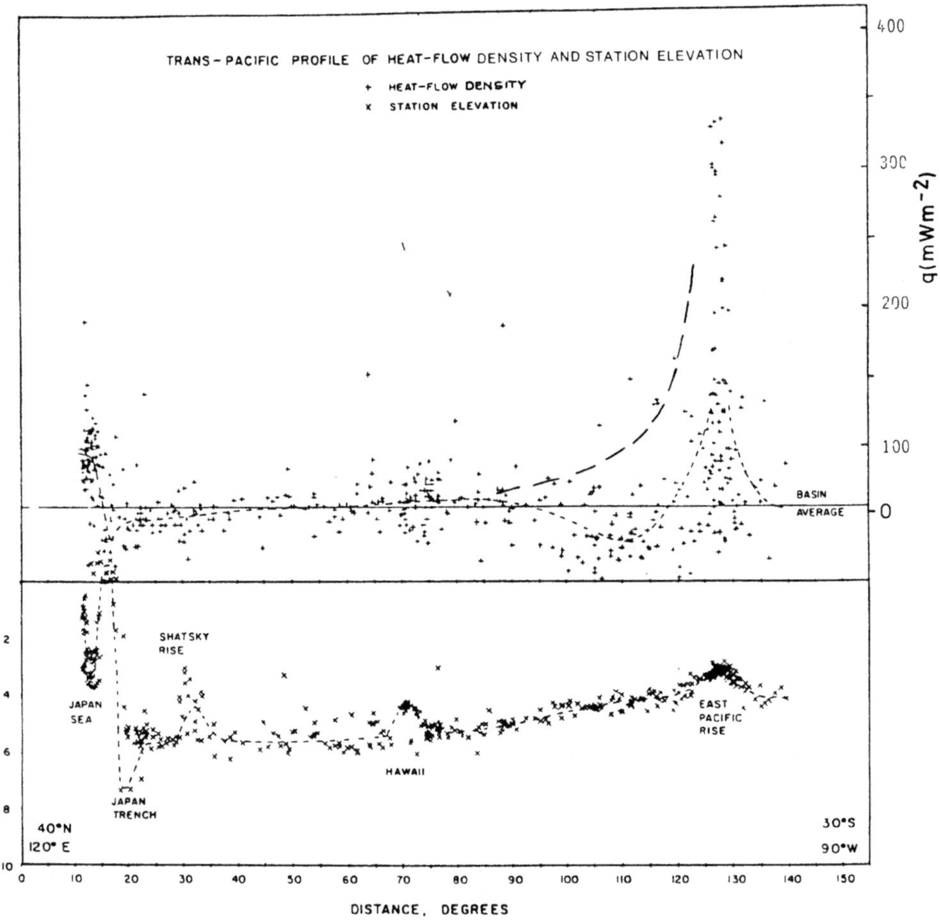

Fig. 9.2. Profile of heat-flow density (top) and station elevation (bottom) from Japan Sea across the Pacific to the East Pacific Rise (after Langseth and Von Herzen, 1970). Broken lines are averages of observed values. Thick broken line shows approximate heat-flow density/age relation predicted by the plate model.

the heat flow community. At present, however, this disagreement is attributed to the existence of non-conductive heat transfer by hydrothermal circulation in the oceanic curst; the ordinary probe method (see Chapter 6.) can measure only a fraction of heat output (see e.g. Anderson and Skilbeck, 1980). This means that the igneous oceanic crust is so highly permeable that seawater can penetrate it and circulate actively. As age increases an impermeable sedimentary cover thickens to some 200–300 m and sea water circulation is shut off. After this, heat flow follows the theoretical curve.

This important idea was first introduced explicitly by Lister (1972, 1980) and was unequivocally verified through detailed heat-flow density measurements in

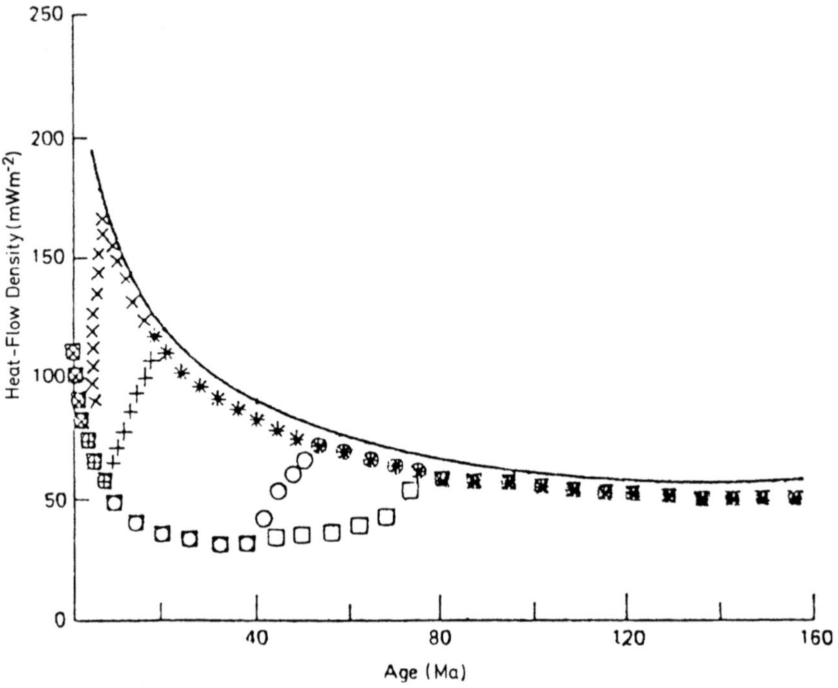

Fig. 9.3. Variation in mean heat-flow density vs. age in each of the major mid-ocean ridge segments. (−) theoretical; (∗) Galapagos; (+) East Pacific Rise; (○) Indian Ocean; (□) Atlantic Ocean (after Anderson and Skilbeck, 1980).

the Galapagos area with the multiple penetration type of heat flow probe (Williams *et al.*, 1974). With this new device, oceanic heat-flow density measurements can be made with extremely close station spacing as shown in Figure 9.4. Heat-flow density was shown to change more or less periodically with a wave length of about 10 km near the Galapagos spreading center. Such a local change can not be explained without hydrothermal circulation through the several km thick oceanic crust. Similarly dramatic local heat-flow density variations have been found in the area around the alleged back-arc spreading center of the Mariana Trough (Figure 9.5, Hobart *et al.*, 1979).

From these observations, it can be inferred that the ocean floor is indeed releasing heat as the theoretical curve in Figures 9.2–9.4 predicts. The estimates of oceanic heat output cited in the previous section were based on this inference.

Spectacular manifestations of hydrothermal activity have been discovered by deep diving manned submersibles in a series of dives since the late 1970's at the crests of mid-oceanic spreading centers where such activity is believed to be most vigorous (e.g. Corliss *et al.*, 1979; Spiess *et al.*, 1980). Along the crest of the East Pacific Rise, many vents have been found with hot water jetting out at over 350 °C, together with eco-systems having exotic live creatures and hydrothermal

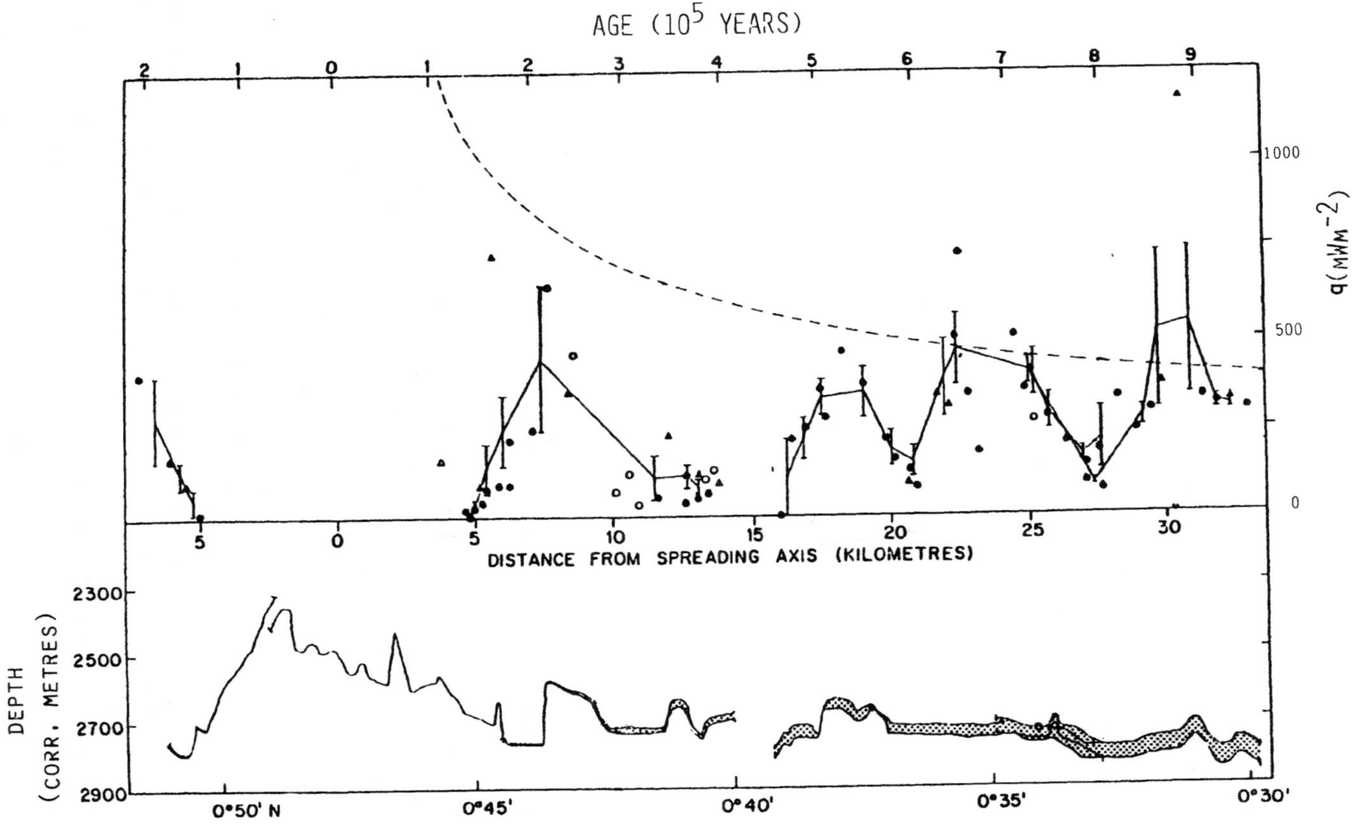

Fig. 9.4. Heat-flow density, topography and sediment thickness (stippled pattern) over Galapagos spreading center. Closed circles: measurements within 2 km of the profile, open circles: measurements more than 2 km away from the profile, triangles: minimum heat-flow density values, dashed line: theoretical heat-flow density: solid lines connect average values over 1 km intervals (after Williams *et al.*, 1974).

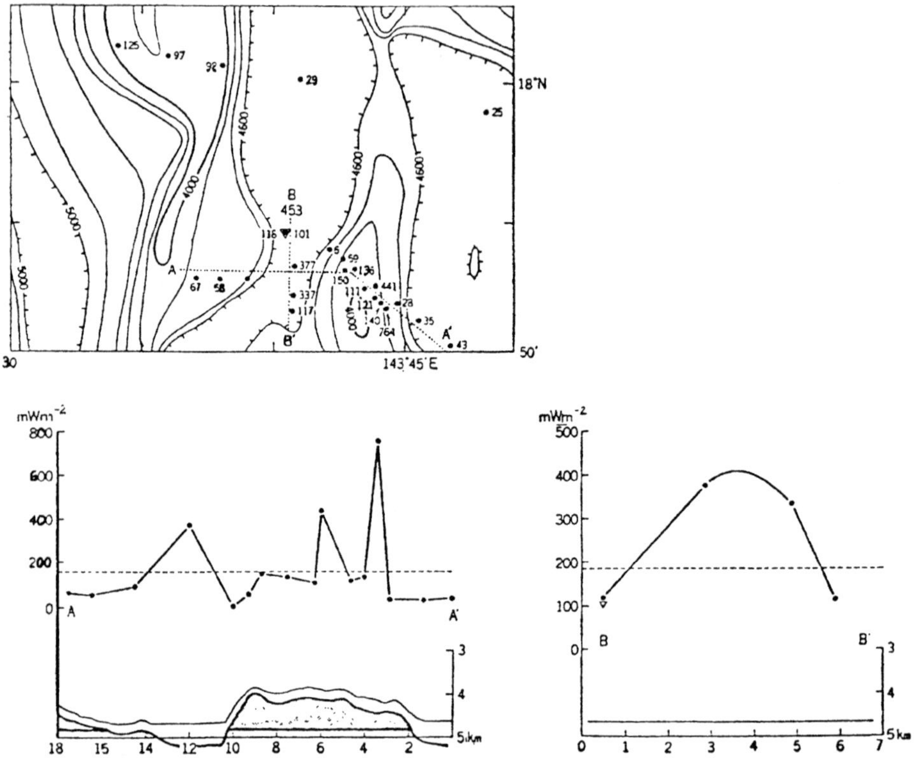

Fig. 9.5. Heat-flow density in the Mariana Trough near DSDP site 453. A–A′ and B–B′ in the lower figures correspond to tracks A–A′ and B–B′ of the chart (after Hobart *et al.*, 1979). Broken lines are theoretically expected values of heat-flow density for 5–6 Ma old lithosphere.

ore bodies. A single 'black smoker' is estimated to be releasing a few hundred watts of thermal energy (MacDonald *et al.*, 1980). Submersible studies in other types of spreading centers, such as the Mariana and Okinawa Troughs, Lau Basin, Andaman Sea, and Bonin back-arc depressions are likely to be very important both from scientific and resources points of view.

9.1.4. Continental Heat-Flow Density

Existence of age dependence of heat-flow density in continental areas was noticed by Lee and Uyeda (1965), but it was more explicitly demonstrated by Polyak and Smirnov (1968) and Hamza and Verma (1969). This relationship has since been repeatedly and variously postulated (e.g. Verma *et al.*, 1970; Sclater and Francheteau, 1970; Vitorello and Pollack, 1980; Sclater *et al.*, 1980a). The representations tend to take different forms, because 'age' is not unequivocally definable in the case of continents. Figure 9.6 illustrates some of these, as compiled by Rao *et al.* (1982). It appears that the time constant involved in continental heat-flow

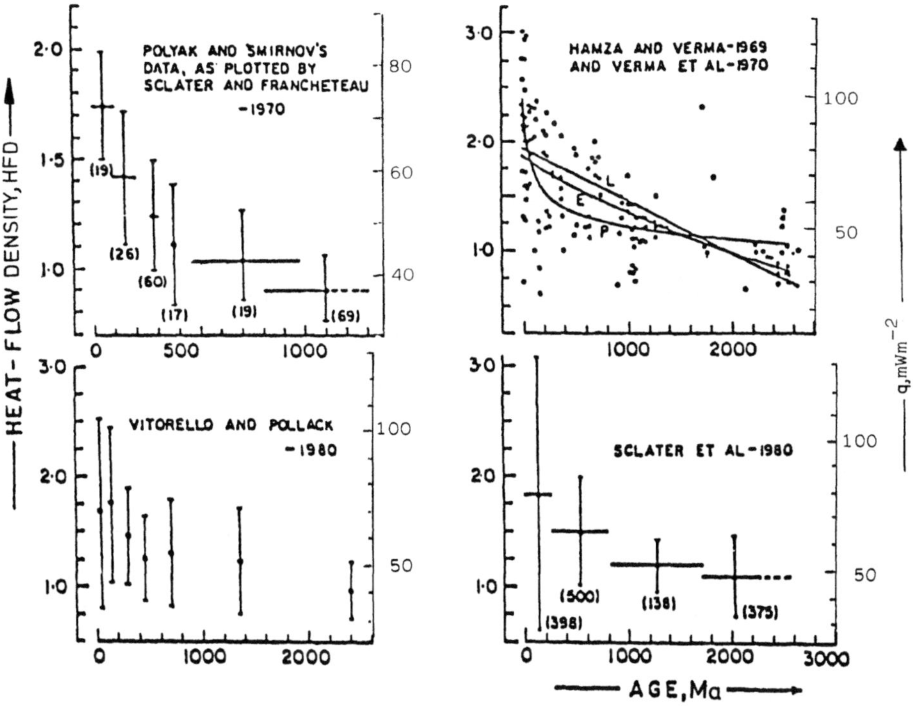

Fig. 9.6. Heat-flow density vs age relations for continental areas. Vertical bars indicate one standard deviation. Horizontal bars represent age range. Number in parentheses indicates the number of heat-flow density values for the data set. L, E, and P (for the Hamza and Verma, 1969 and Verma *et al.*, 1970 data) are linear, exponential and power-law fits (after Rao *et al.*, 1982).

density is almost an order of magnitude longer than that for oceanic heat-flow density. This fact may provide some justification for an order of magnitude less certainty of age for continental crust. Chapman and Pollack (1975) proposed a 'predictor' of continental heat-flow, based on this age-dependence, to infer heat flow values where actual measurements do not exist. They used this 'predictor', together with the similar but better founded oceanic predictor, to map the world heat-flow density distribution and, hence, the global heat output. Figure 9.6 shows the relation for heat-flow density values averaged for all the continents. When broken down to different continents, the relation looks like Figure 9.7. It must be noted that the relation is quite obscure for Australia and Africa. Therefore, age dependence of heat flow for continents should be treated with reservation.

Because continental crust contains a significant amount of radioactive heat sources (Table 9.3, see also Table 4.7), it must contribute to continental heat-flow density. There is an empirical relation, between the heat-flow density and radiogenic contribution. Namely for a particular region, called a heat-flow density province, heat-flow density q and radiogenic heat generation of surface rocks H_0 are related as (Roy *et al.*, 1968), see also Equation (8.4):

$$q = q_r + D\,H_0. \tag{9.2}$$

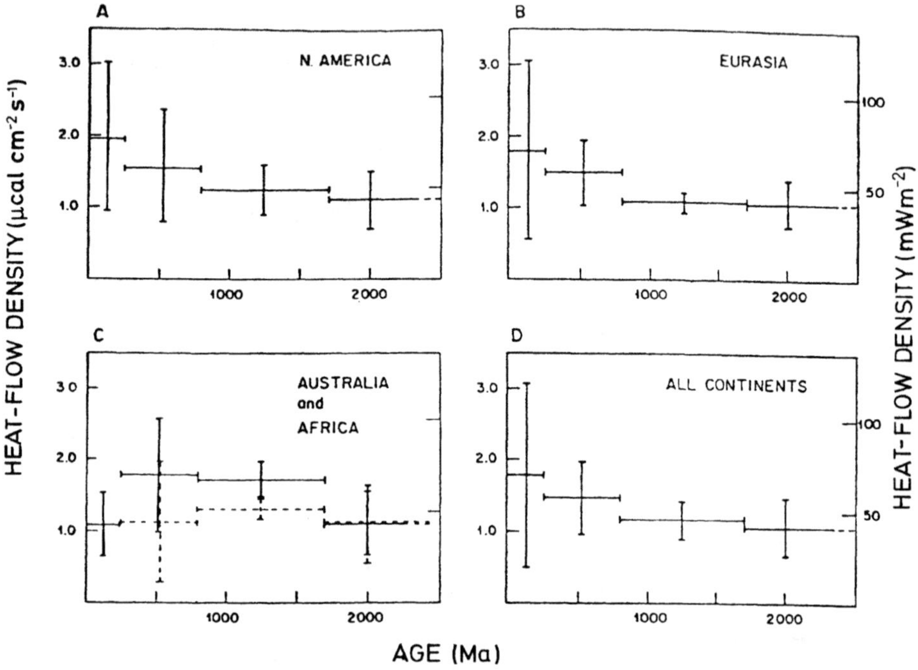

Fig. 9.7. Heat-flow density vs. age relation for each continent and all continents. Broken lines are for Africa (Sclater *et al.*, 1980a).

TABLE 9.3
Heat generation of common igneous rocks (mainly from Rybach, 1976, after Pollack, 1982).

Igneous rock type	U (ppm)	Th (ppm)	K (%)	Th/U	K/U	Density ($g\ cm^{-3}$)	Heat generation $10^{-6}\ W\ m^{-3}$
Granite/Rhyolite	3.9	16.0	3.6	4.1	0.9×10^4	2.67	2.5
Granodiorite/Dacite	2.3	9.0	2.6	3.9	1.1×10^4	2.72	1.5
Diorite/Andesite	1.7	7.0	1.1	4.1	0.7×10^4	2.82	1.1
Gabbro/Basalt	0.5	1.6	0.4	3.2	0.8×10^4	2.98	0.3
Peridotite	0.02	0.06	0.006	3.0	0.3×10^4	3.23	0.01
Dunite	0.003	0.01	0.0009	3.3	0.3×10^4	3.28	0.002
Continental crust	1.25	4.8	1.25	3.8	1.0×10^4	—	0.8

Figure 9.8 is the compilation for North America where it can be seen that the relation holds well for the eastern United States and the granitic terrain of the Sierra Nevada, but that there is a considerable scatter for the recently active area such as the Basin and Range Province. In the relation (9.2), $D\ H_0$ represents the contribution from radiogenic heat in the upper crust, while q_r is the heat coming from further below. q_r is called the reduced heat-flow density. D has the dimension

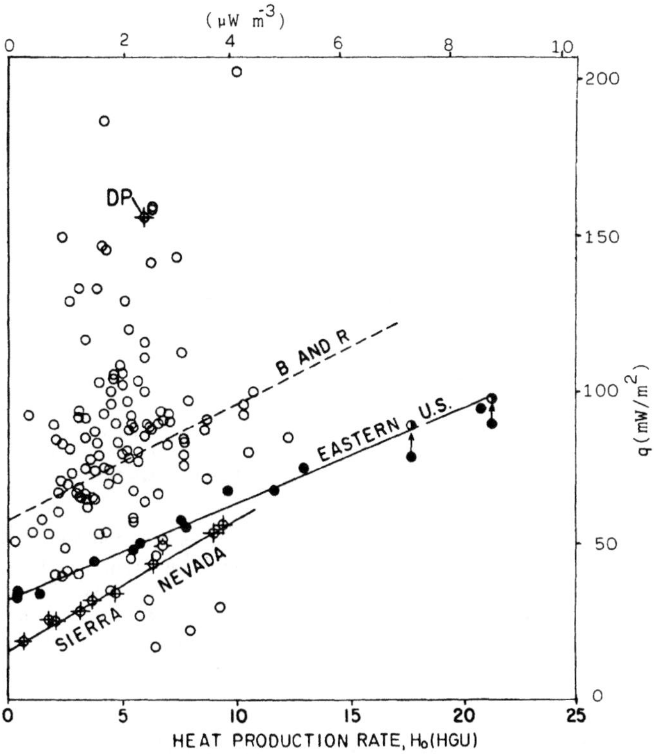

Fig. 9.8. Heat-flow density vs. heat production rate in crystalline rock for the United States. B and R represents Basin and Range province. Cricles represent all the data from the contiguous United States: solid circles represent points east of the Great Plains, crossed circles represent points interior to the Sierra Nevada province. Open circles are for points in other areas, but majority belongs to the Basin and Range province. DP is adjacent to the Long Valley volcanic center. (after Lachenbruch and Sass, 1977).

of length and represents the thickness of the near-surface radioactive layer, if its rate of heat generation is taken to be constant H_0. If, however, one assumes that the rate of heat generation H depends on the depth z as:

$$H(z) = H_0 \, e^{-z/D}. \tag{9.3}$$

D may be taken as the characteristic thickness of the heat generating layer. Relation (9.3) is favored over the constant rate model, because it appears more realistic and it can also explain the difference in radioactive heat generation of surface rocks in a province as due to differential erosion (Lachenbruch, 1970).

Up to present, ten or so heat-flow density provinces have been recognized. Their average heat-flow density $\bar{q}_0$ and q_r, are plotted as shown in Figure 9.9 (Pollack and Chapman, 1977). There is an intriguing relation:

$$q_r \cong 0.6 \, \bar{q}_0. \tag{9.4}$$

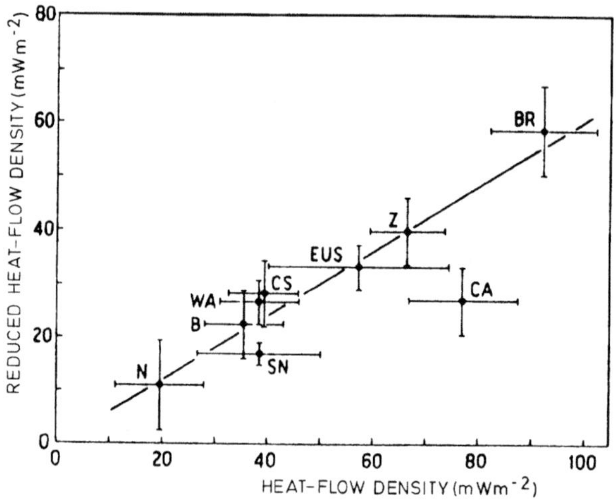

Fig. 9.9. Reduced heat-flow density (q_r) vs mean heat-low density ($\bar{q}_0$) in several heat-flow density provinces. Horizontal bars represent one standard deviation of the mean heat-flow density; vertical bars represent uncertainty of estimate of reduced heat-flow density. N = Niger, B = Baltic Shield, SN = Sierra Nevada, WA = Western Australia, CS = Canadian Shield, EUS = Eastern United States, Z = Zambia, CA = Central Australia, BR = Basin and Range (after Pollack and Chapman, 1977).

9.1.5. Heat-Flow Density and Lithospheric Structures

For oceanic regions, lithospheric plates are well interpreted as thermal boundary layer at least for the first approximation. For oceanic plates older than some 80 Ma, however, neither observations of heat-flow density nor water depth follow the prediction of the simple cooling half space model. They appear to approach asymptotically some constant values. The so called 'plate' model (e.g. McKenzie, 1967; Sclater *et al.*, 1980a) places importance on this deviation from the cooling half space model and postulates that the thickness of older plates is held constant in the real Earth. Parsons and McKenzie (1978) tried to defend the 'plate' model by assuming small scale mantle convection holds the temperature constant at a definite depth. This model has been criticized as conceptually inappropriate or quantitatively unreasonable (Davies, 1980a) but the above observations remained difficult to explain otherwise. Recently, however, the observation that the depth of old sea-floor deviates from the theoretical prediction has been questioned. Heestand and Crough (1981), for instance, suggest that the apparent deviation is the result of thermal disturbance due to hot spots and should be regarded as 'noise' to the cooling half space model. Hilde and Uyeda (1983) have also shown that the depth of oceanic trenches increases with age of the subducting oceanic plate and that this tendency holds even when subducting plates are older than 80 Ma, supporting the plate thickening model. Heat-flow density measurements in very old sea-floor are still too few to differentiate the models with confidence and more measurements are needed. Despite this problem, oceanic lithosphere can be regarded to be better understood in terms of plate tectonics than continental lithosphere.

For continental areas, the situations are less clear. As summarized in the previous section, the major observations are:

(1) Age dependence of heat-flow density.
(2) Linear relationship between heat-flow density and heat production rate of surface rocks.
(3) Proportionality between average surface heat-flow density and reduced heat-flow density.

With these relations and a certain set of further assumptions, i.e. on steady state conduction and constant thermal conductivity, one can estimate the geotherms for various regions, including continental areas; see also Section 8.1. From the intersections of such geotherms with estimated solidu curves, Chapman and Pollack (1977) tried to estimate the global distribution of lithospheric thickness. It is an interesting approach but needs further check on premises employed. Indeed, the large scatter of data does not warrant validity of above relationships except for (2). Furthermore, it has been pointed out that if relation (2) is valid, the data set for each heat-flow province should not necessarily be normally distributed around the average and therefore the average value is not meaningful in statistical sense (Rao *et al.*, 1982). This point needs to be seriously considered in future studies. Here, however, we will proceed assuming the above three relationships hold true, at least roughly.

The first problem is how to interpret observation (1), i.e. the age dependence of continental heat-flow density. The first interpretation to mention may be that the secular decrease of heat-flow density is mainly caused by removal of the surficial heat producing layer by erosion and its accumulation in orogenic areas (England and Richardson, 1980). Although this interpretation may be applied to such areas as the Alpine Mountains where the crust might have been thickened initially by orogenic movements (even for the Alps, apparent high heat-flow density may be ascribed to uplift and erosion, Bodmer *et al.*, 1979), it does not work for areas such as North American Basin and Range Province where the initial condition was probably set by some 'thermal event'. In the latter case, secular cooling of the lithosphere may also play an important role (Crough and Thompson, 1976; Kono and Amano, 1978).

Lithospheric cooling models have been criticized because:

(1) It is difficult to explain the vast difference in the time constants for the secular decrease of heat-flow density in oceanic and continental areas, in as much as the thermal properties of both lithospheres are not much different.
(2) Some continental lithosphere is as old as 10^9 a and simple cooling throughout such a long period of time would cause enormous subsidence, but such a subsidence has not been observed (Oxburgh and Parmentier, 1978; Jordan, 1981).

The calculations made by Crough and Thompson (1976) and Kono and Amano (1978), however, demonstrated that the time constant involved in the cooling of modeled continental lithosphere is in rough agreement with observed values and the thickness of the cool boundary layer (plate) does not exceed about 150 km. This is somewhat strange but may be due to two reasons: (1) the long time constant in the order of 10^9 a (Figure 9.7) is only apparent (cf. Vitorello and Pollack's arguments below) and the real time constant may be $2 - 5 \times 10^8$ a and (2) the modeled continental lithosphere at time = 0 already has a thickness of several tens of kilometers, so that in the continental case the 'tail end' (Crough and Thompson, 1976) of the cooling curve of oceanic lithosphere can be seen.

In these cooling plate models, old continental lithosphere has a greater thickness than oceanic lithosphere. But it is still about 150 km and the mantle below that depth is at essentially the same temperature beneath the oceans and the continents. Sclater *et al.* (1980a) emphasize this point. Jordan (1981), on the other hand, defies these models based on his long-held view that the difference in mechanical properties of continental and oceanic mantles, deduced from seismological observations, extends deeper than 250 km. Jordan postulates that, under the continents, the mantle is less attenuative and has higher seismic wave velocities than under the oceans. Normally this would mean lower temperature and higher density under the continents, which seems unlikely to be sustained for geological time of 10^9 a. To get around this difficulty, Jordan suggests that the coninental mantle consists of peridotite which has been more depleted in basaltic components, so that it, despite its higher seismic wave velocities, has more or less the same density as the less depleted oceanic mantle. This means continental mantle has a deep root the chemical composition of which is different from other parts of the mantle and that the root has been fixed to the overlying continental lithosphere throughout geologic history. Jordan distinguishes this structure from lithosphere by calling it the 'tectosphere'. The deep difference in the oceanic and continental mantles is, however, still a matter of debate among seismologists and undoubtedly is one of the most important problems in geodynamics today.

Coming back to the problem of continental heat-flow density, comment is warranted on the model proposed by Vitorello and Pollack (1980). Their model is akin to the cooling models mentioned earlier but may serve for re-considering the relationship between continental lithosphere and heat-flow density. First, surface heat-flow density (q) is split into three parts; (I) q_c: heat from the near-surface heat generating layer (D), (II) q_t: heat associated with the 'thermal event', and (III) q_b: background heat-flow density from deeper sources. Namely:

$$q = q_c + q_t + q_b = q_c + q_r. \tag{9.5}$$

These components of heat-flow density are illustrated in Figure 9.10, which is apparently based on the following ideas. First, q_b is estimated to be equal to that of old oceans, i.e. $q_b = 25$ mW m^{-2}. q in the recently active orogenic zones is estimated from observations to be $q = 90$ mW m^{-2}. Then, considering Equation (9.4), q_t $(t = 0) = 30$ mW m^{-2}, q_c $(t = 0) = 25$ mW m^{-2} and q_c $(t = \infty) = 17$ mW m^{-2} are inferred. Decrease of q_c from 25 mW m^{-2} at $t = 0$ to 17 mW m^{-2} at $t = \infty$ is due to the erosional effect. If this idealization is allowed, the problem of continental heat-flow density is reduced to how to interpret the secular variation of part (II), namely q_t (t). Vitorello and Pollack (1980) suggest that the time change of q_t depicted in Figure 9.10 can be explained if the average temperature of the upper 300 km was raised about 700 °C at the time of orogeny. This interpretation is by no means unique but may provide some clue. The origin of q_t is

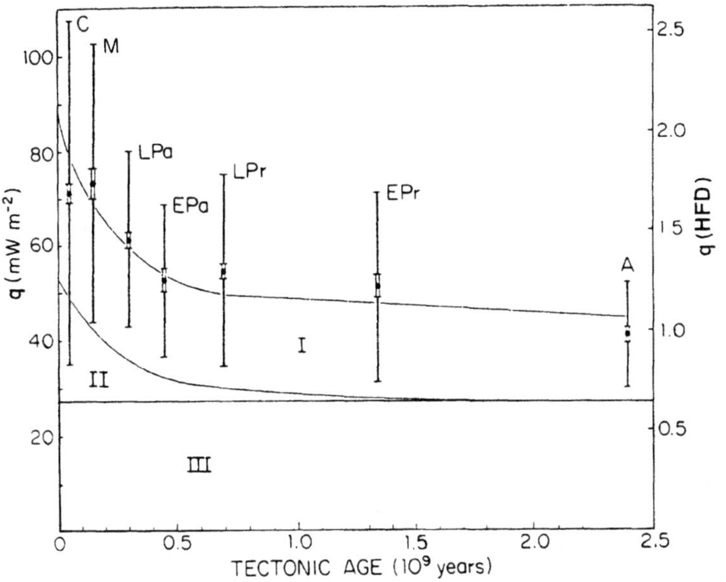

Fig. 9.10. Decrease of continental heat-flow density with age, and its three principal components (see text). C = Cenozoic, M = Mesozoic, LPa = Late Palaeozoic, EPa = Early Palaeozoic, LPr = Late Proterozoic, EPr = Early Proterozoic, A = Archean. Double bars represent the standard error of the mean and single bars standard deviation (after Vitorello and Pollack, 1980).

probably the igneous activity associated with orogenic processes. In the next section we consider this problem in the case of subduction type-orogenic zones. q_b is related to the thermal history of the entire Earth and will be discussed in the last section.

9.1.6. Thermal Processes in Orogenic Zones

In the model outlined in the previous section, it was assumed that the orogenic process involves 'thermal events' which reset the 'age' of the crust, and that the lithosphere is subsequently subjected to cooling and erosion. The nature of the 'thermal event' was, however, not considered.

Oxburgh and Parmentier (1978) proposed a model in which continental lithosphere grows by diapiric accretion from below. But, in view of our insufficient knowledge of the complex heat-flow density distribution in orogenic zones, as depicted, for example, in the European Heat Flow Monograph (Čermák and Rybach, 1979), it is at present difficult to check the validity of their model. In comparison, heat-flow density distributions in subduction zones are relatively clear. As an example of a well studied subduction zone, the distribution of heat-flow density in the Japanese area is illustrated in Figure 9.11 (Uyeda and Horai, 1964, Yasui *et al.*, 1967, 1968; Yamano *et al.*, 1982). In the northeast Japan Arc (Tohoku), heat-flow density is subnormal on the subducting Pacific plate and the

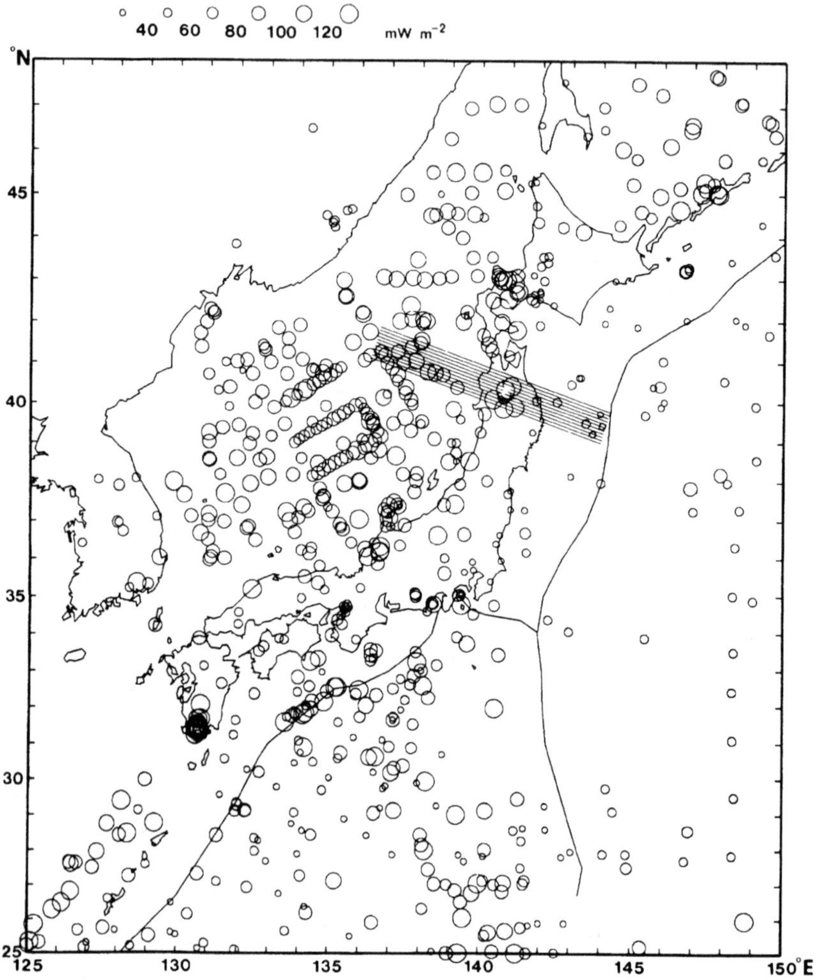

Fig. 9.11. Heat-flow density distribution in and around Japan. Stippled zone represents the cross-section for Figure 9.12.

trenchward side of the arc, whereas it is high in the area landward of the volcanic front, including the backarc basin (Japan Sea). This type of heat-flow density distribution is also observed in the areas of the Kurile, Ryukyu, Mariana and a few other subduction zones (e.g. Savostin and Vlasenko, 1974; Watanabe *et al.*, 1977; Hobart *et al.*, 1979; Anderson, 1980; Calvalho *et al.*, 1980; Blackwell *et al.*, 1982). On the other hand, in the southwestern Japan arc, heat-flow density is characteristically high in the trench (Nankai Trough). This anomalous situation may be caused by subduction of young Shikoku Basin lithosphere (Yamano *et al.*, 1984).

Discussions about the thermal processes associated with subduction, based on the observed heat-flow density distribution as shown in Figure 9.11 and other related data (e.g. Yoshii, 1979) have been presented by a number of investigators (e.g. Hasebe *et al.*, 1970; Minear and Toksöz, 1970; Oxburgh and Turcotte, 1970; Andrews and Sleep, 1974; Bodri and Bodri, 1978; Hsui and Toksöz, 1979; Furlong *et al.*, 1982; Honda and Uyeda, 1983). Here, brief mention will be made on the recent study by Honda (1985). Figure 9.12 shows an example of the cross section of the northeastern Japan Arc. The model is two dimensional and treats essentially the steady state solution. The medium has a temperature dependent viscosity. Most calculations were made on the assumptions of linear rheology. A few runs were made non-linear rheology but the results were not much different from linear rheology case. Radiogenic heat production and possible endothermic reactions, such as dehydration, are not included in the calculation. In this model, oceanward low heat-flow density is explained by the cooling effect of the cold oceanic plate and landward high heat-flow density by both the convective flow in the mantle wedge induced by the subducting slab and by shear heating associated with the flow. Major constraints imposed on the model were:

(1) Observed heat-flow density profile across the arc.
(2) Existence of volcanic chain in a zone where the depth of the Wadati-Benioff zone is approximately 120–150 km, defining the volcanic front.
(3) Temperatures as high as 1400 °C in the mantle wedge beneath the volcanic front (Tatsumi *et al.*, 1983).

In order to explain the low heat-flow density oceanward of aseismic front (Yoshii, 1979), it is required to assume that the upper seaward corner (hatched triangle in Figure 9.12) is solid, whereas to explain the high temperature of 1400 °C as required from petrological studies of the parent magma (Tatsumi *et al.*, 1983), the upper mantle beneath the Japan Sea also has to be assumed (as a boundary condition) to be fairly hot. The high temperature would be even more enhanced if the backarc basin were actively spreading, because the situation at a backarc spreading center should be similar to that at a normal oceanic spreading ridge.

Thus, heat-flow density in subduction zones has a characteristic distribution which has definite physical meaning. Therefore, simple 'average' of heat-flow density in such provinces do not have any meaning. Long after cessation of subduction, the heat-flow density distribution may be smeared. Even so, it is unclear if the simple 'average' becomes meaningful (Blackwell *et al.*, 1981). When orogenic processes take place repeatedly in a broad region, perhaps taking several hundred million years, the whole region may be heated more or less uniformly. In such a case, a vague initial condition for the 'age vs. heat-flow density' relation may be set for the region.

9.1.7 Thermal Process at Transform Faults

Extensive heat-flow density studies of transform fault plate boundaries have not been made except for the famous San Andreas Fault, California (e.g. Lachenbruch and Sass, 1980). Figure 9.13 shows the heat-flow density profile plotted against distance from the fault. The lower part of the figure indicates theoretically

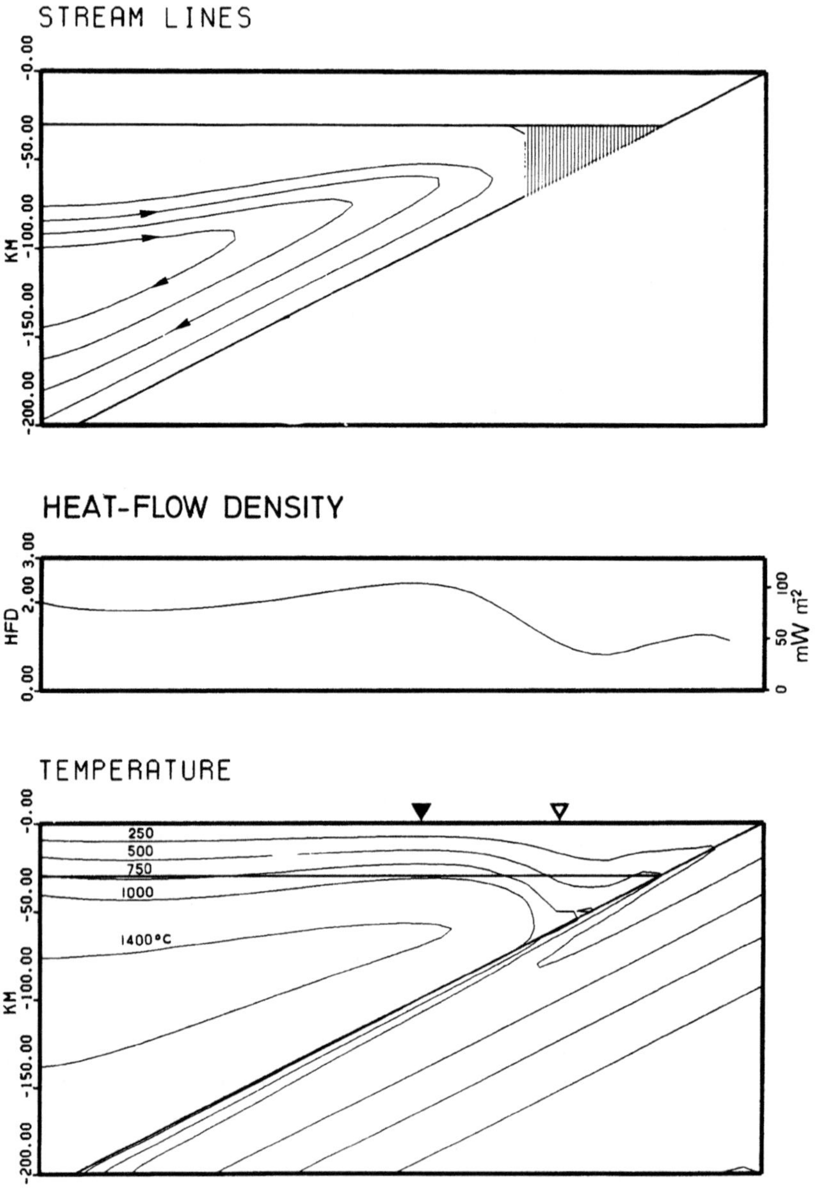

Fig. 9.12. Example of calculated thermal and dynamical state under Northeast Japan (Tohoku) Arc. Full triangle = volcanic front, open triangle = aseismic front. Stippled triangular zone = solid portion of the mantle wedge. Horizontal and vertical scales are same (after Honda, 1985).

expected heat-flow density anomalies on a frictional model in which a shear stress of 1 kbar is assumed for relative motion of 2 cm a^{-1}. Obviously, the observations do not show such an anomaly, indicating that shear stress across the fault should

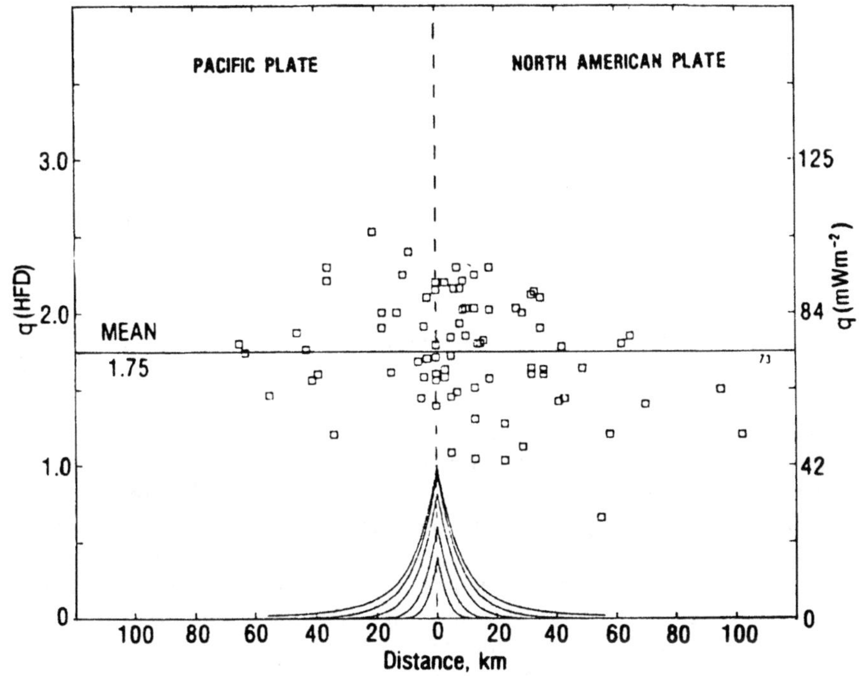

Fig. 9.13. Heat-flow density as a function of the distance from the main San Andreas fault zone. Pattern of curves shows theoretically expected heat-flow density anomaly if the slip velocity is 2 cm a^{-1} and frictional resistance is 1 kb for various times after initiation of slip. From inner to outer, curves correspond to 0.3, 0.8, 2.4, 8 and ∞ Ma (after Lachenbruch and Sass, 1980).

be no more than about one hundred bars. This result does not contradict the observed stress drops at earthquakes, but is contradictory with expectations from laboratory experiments on the friction of rocks. It is interpreted that along actual faults, some mechanism, either the development of mylonite or high pore pressure, may be operating to lower the friction.

Although no localized heat-flow density anomaly was observed along the fault, the general area (Coast Range) shows higher than normal heat-flow density. As to the origin of this widespread anomaly, one interpretation may be that the heat generation associated with fault movements is distributed over approximately a 100 km width. But in the case of the San Andreas Fault, the suggestion has been made that the high heat-flow density is the result of a northward migration of the Mendocino triple junction, which has altered the plate boundary from a subduction type to a transform fault type (Dickinson and Snyder, 1979). As the triple junction moved northward the North American plate was progressively put into contact with hot asthenosphere instead of a cold slab, and was heated.

The suggestion that fault motion is accompanied by low stress is important from tectonic point of view, in particular with regard to the observation that strike slip motion is very common at collisional/accretional plate boundaries (Coney *et al.*,

1980; Taira *et al.*, 1983). When a transform fault has some divergent component in its motion, a window or pullapart basin may develop. In such a case a notable heat-flow density anomaly will develop as in the Gulf of California (Von Herzen, 1963; Lawver *et al.*, 1975) and Andaman Sea (Burns, 1964).

9.1.8. Geothermics of Sedimentary Basins

There are numerous large sedimentary basins in the world underlain by continental or subcontinental crust, typical ones being the North Sea and Pannonian Basins. These basins have thick sedimentary sections, indicating that large scale subsidence has taken place. Many theories have been put forward to explain the origin of subsidence and geothermal anomalies associated with it: subcrustal erosion (e.g. Collette, 1971), dike intrusions (e.g. Burke, 1975), basalt-eclogite transformation due to thermal events (E. Arthushkov, personal communication) and lithospheric stretching (McKenzie 1978). Among these the stretching model is summarized here, because of its simplicity. In this model, it is assumed that continental lithosphere is suddenly extended by a factor of β (Figure 9.14) at time $t = 0$. This causes upwelling of hot asthenosphere that produces a thermal anomaly which decays through secular cooling, just as the oceanic lithosphere cools after its formation. Depending on the details of the models, one can calculate the history of subsidence and thermal state in this general scheme that can be tested with geological information. Subsidence histories thus predicted have been demonstrated to explain the data for some basins, including the North Sea (Sclater and Christie, 1980) and even the Bay of Biscay where extension went far enough to break the continental crust (Le Pichon and Sibuet, 1981). In fact, such a model can, in principle, be applied to backarc basins formed by extension and bounded by passive margins. Sclater *et al.*, (1980b) examined the intra-Carpathian basins and concluded that both stretching and an additional thermal input from below, probably associated with subduction, are required. Present heat-flow density, however, is not by itself very useful information to distinguish between various models because after some 10–15 Ma of basin formation, many models yield more or less the same heat-flow density. However, a very particular use may be derived from this approach in evaluating the degree of maturation of hydrocarbons in basins, because the hydrocarbon maturation is supposed to be controlled by the time-integrated temperature history of the basins (Sclater *et al.*, 1980a; McKenzie, 1981; Stegena *et al.*, 1981, see also Chapter 9.3).

9.1.9. Heat-Flow Density Anomaly Associated with Intraplate Heat Sources

Intraplate rifts, volcanoes and hot-spots are considered to produce some heat-flow density anomalies. The studies, however, have been limited to a few regions. Both the Baikal Lake rift (e.g. Lysak, 1978; Smirnov, 1980) and Rhine Graben (e.g. Bram, 1979; Haenel, 1983) indicate high heat-flow density anomalies coinciding the trend of depressions (Figures 9.15 and 9.16). In both areas, it has been noted, moreover, that heat-flow density varies locally, so that in addition to hot raised mantle, some near-surface disturbances are required to explain the heat-

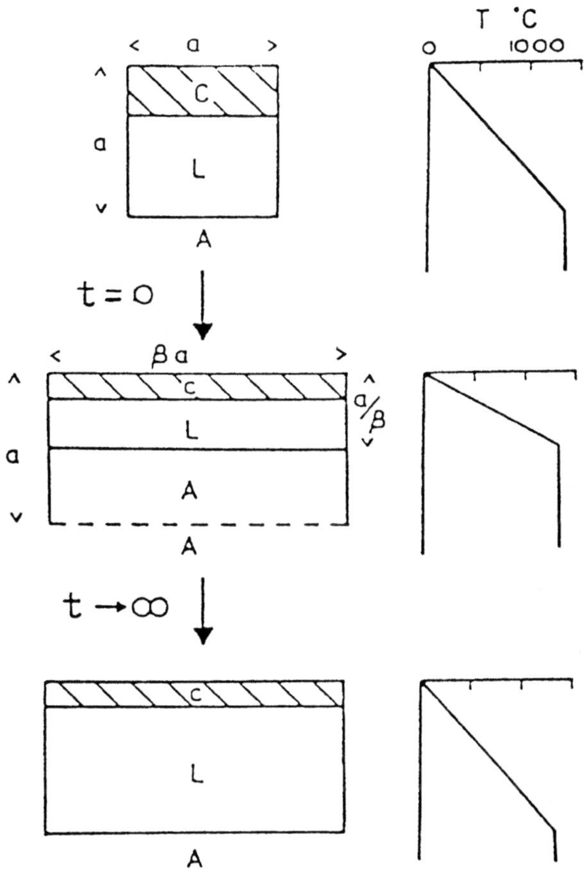

Fig. 9.14. Sketch to show the principal features of the stretching model. At time $t = 0$, a continental lithosphere is extended by β. Isostatic uplift causes upwelling of hot asthenosphere. Subsequent cooling produces subsidence. C: crust, L: lithosphere, A: asthenosphere (after McKenzie, 1978).

flow density distribution, such as hydrothermal convection (e.g. Hoffers 1981). Heat-flow density studies of the East African rift system have been conducted in the lakes (e.g. Lake Malawi, Von Herzen and Vacquier, 1967; Lake Tanganika, Degens *et al.*, 1971; Lake Kivu, Degens *et al.*, 1973). The results showed a few very high heat-flow density values up to some 180 mW m^{-2} locally but most of the values were low. Degens *et al.* (1971) argued that the predominantly low heat-flow density is consistent with a lack of active volcanism and spreading in Lake Tanganika. Chapman and Pollack (1977), on the other hand, inferred from relatively high heat-flow density in Zambia (54–76 mW m^{-2}) that incipient arm of the East African rift system extends into the Central African Plateau. It seems, to the present author, that there is a possibility that the low heat-flow density in lakes may be caused by hydrothermal circulation just like in young ocean floor.

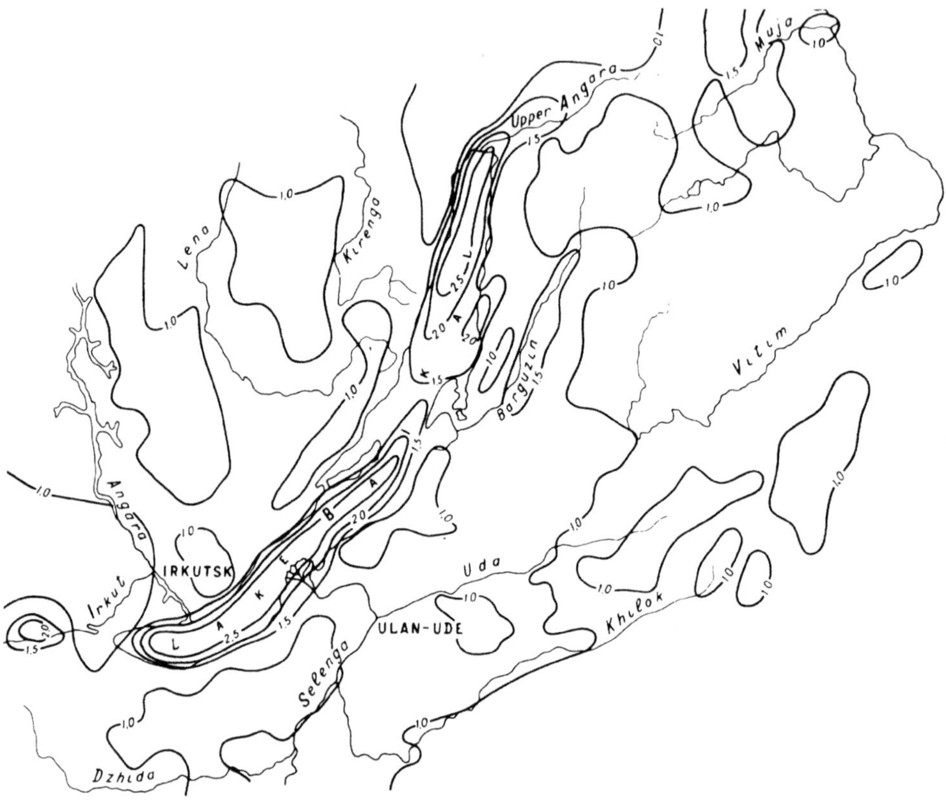

Fig. 9.15. Distribution of heat-flow density in the Baikal rift area (isolines in HFD). The map is based on about 70 measurements and inference from correlations between heat-flow density and topography, gravity and seismicity of the area (Lysak, 1978).

As to the possible effect of intra-plate volcanism on regional heat-flow density, studies in Rhenish Massif (e.g. Haenel, 1983), Germany, where active volcanism is recognized until the youngest Pleistocene time, show regionally high heat-flow density of 65–70 mW m^{-2} (Figure 9.16). Haenel (1983), however, suggests, on the basis of model calculations, that post-Miocene magma intrusion should not produce any observable heat-flow density anomalies.

Hot-spot in oceanic areas has been examined on the Hawaiian Swell by von Herzen *et al.* (1982). From 95 measurements on the swell, these authors recognized that heat-flow density increases systematically from near-normal values (52 mW m^{-2}) near Hawaii to values near Midway (59 mW m^{-2}) about 20–25% higher than normal for lithosphere of its age (110 Ma). This heat-flow density pattern was found to be consistent with the model that a lithospheric plate moved over a deep fixed hot spot and was reheated as shown in Figure 9.17. As seen in Figure 9.17, reheated lithosphere is thinned and the water depth is decreased to form a swell, and heat-flow density is increased. This effect, as mentioned earlier, may explain why the depth-age and heat flow density-age relations, predicted by plate cooling models, fail for older lithosphere.

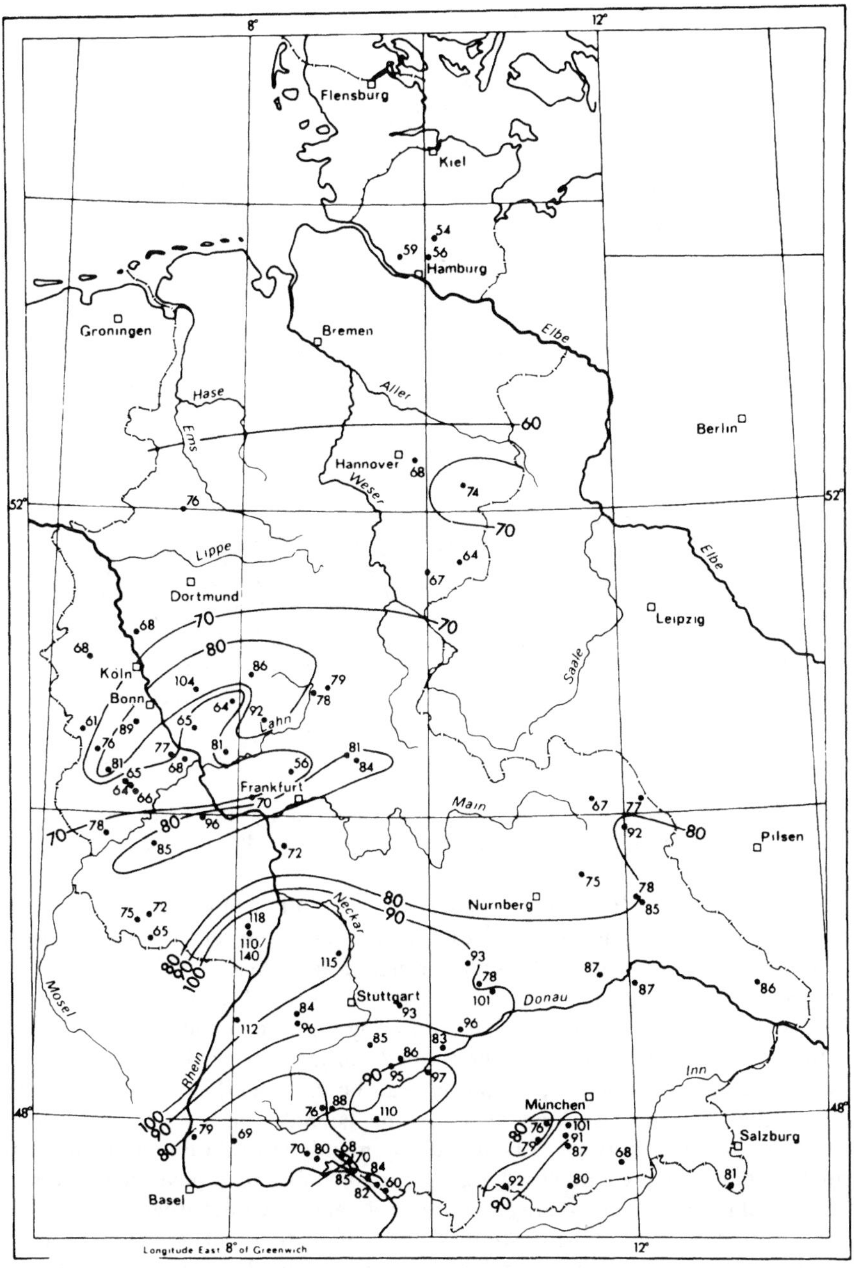

Fig. 9.16. Heat-flow density distribution in the Federal Republic of Germany (Haenel, 1983).

9.1.10. Temperature Distribution in the Earth

Despite its great importance, temperature in the Earth, especially at greater

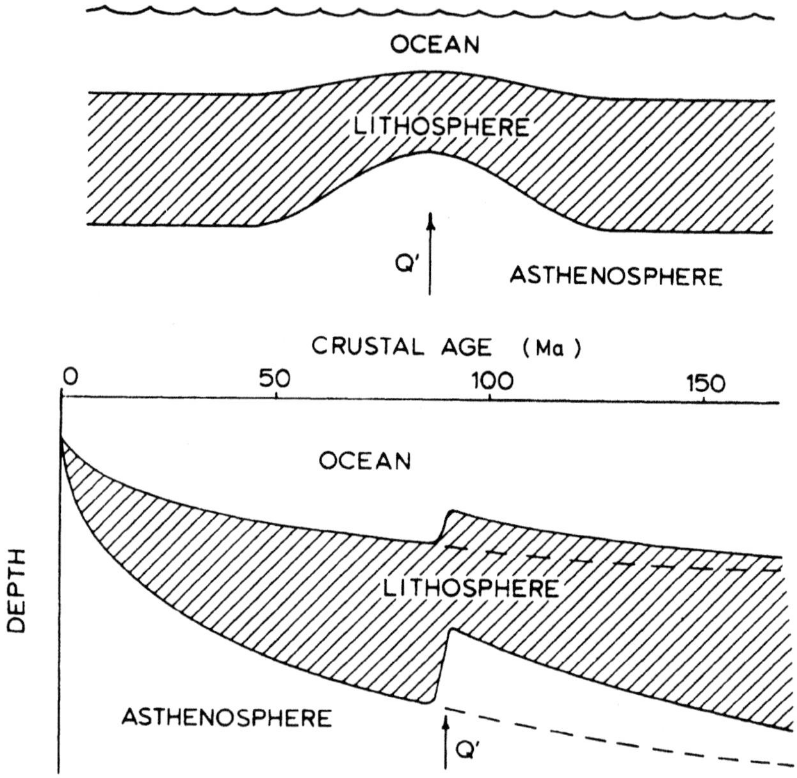

Fig. 9.17. Transverse (above) and parallel (below) cross section of the effect of a hot-spot beneath a moving lithospheric plate (von Herzen *et al.*, 1982).

depths, is much less well known compared to some other physical quantities such as the elastic constants and density. For example, even the recent estimates of the temperature at the core-mantle boundary range from 2500 K – 5000 K. Therefore, in this section, only a brief review on the basic ideas on which temperature profile in the Earth is assessed will be given without going into details.

In the shallow depth range, as mentioned earlier, where the heat transfer is solely due to solid conduction, geotherms can be calculated readily from surface heat-flow density data, as long as one dimensional steady state condition is assumed and the distributions of heat sources, mainly crustal radiogenic ones, and thermal conductivity are known. A number of such attempts have been made for both particular areas and average continental and oceanic areas, usually with simple assumptions about heat sources and conductivity. The validity of such geotherms fails as soon as non-conductive or convective heat transfer comes into the problem. Namely the depth range in which this method works is limited, at best, to within the 'solid' lithosphere. (Of course the method fails also in geothermal areas, where processes, such as hydrothermal circulation, are active, and other areas where underground water movement is significant.)

For the regions a little deeper in the upper mantle, to about 200 km depth, temperature can be assessed by so called geological thermometers, which are based on temperature-pressure dependent behaviors of elements like Ca and Al in pyroxene minerals contained in ultrabasic rocks brought up from the upper mantle (e.g. Boyd, 1973; Mercier and Carter, 1975). The temperature given by this method, strictly speaking, are for the times when the pyroxene minerals in question were in the upper mantle under particular localities.

Another method, for the upper mantle, which is less direct but more general and pertains to the present day temperature, uses both seismological data and results of high pressure-high temperature experiments on phase equilibria of mantle materials. For instance, olivine-modified spinel-spinel transformation of the Mg_2SiO_4-Fe_2SiO_4 system (Akaogi and Akimoto, 1979), when combined with a seismic model, gave 1400 °C at 380 km, 1550 °C at 520 km and 1610 °C at 610 km (Anderson, 1981).

Melting temperature is also useful. Although there are some contrary arguments, the upper mantle low velocity zone (LVZ), which underlies the lithosphere, may be in a partially molten state. It is thus customary in plate tectonics to identify the bottom of the lithosphere with the solidus of the upper mantle materials. Sclater *et al.* (1980a), for instance, assigns the temperature at the base of the 125 km thick plate to 1335 K, which is within the estimated range of melting point of peridotite under the mantle condition.

More important fixed point of temperature profile of the Earth's interior has long been based on the idea that the inner core boundary (5120 km depth) marks the melting point of the core material: the inner core boundary being defined as the boundary between the liquid outer core and the solid inner core. The question here is how to estimate the melting temperature of Fe or Fe plus some other elements like S at 3.2 Mbar (3.2×10^{11} Pa) pressure. For this purpose, relations such as Lindemann's law and Simmon's equation, which was shown to be equivalent to Lindemann's law (Gilvarry, 1956):

$$p = A\left\{\left(\frac{T_m}{T_{m0}}\right)^C - 1\right\} \tag{9.6}$$

have been used to extrapolate the experimental data. In (9.6), p is pressure, T_m and T_{m0} are the melting temperature at pressure p and zero respectively. A and C are constants. Recently, Lindemann's law has been given a sound thermodynamic basis by Stacey and Irvine (1977) for solids and by Stevenson (1980) for liquids. Stacey (1977) used Gilvarry's (1956) differential form of Lindemann equation:

$$\frac{1}{T_m}\frac{dT_m}{dp} = \frac{2(\gamma - 1/3)}{K_s}, \tag{9.7}$$

where γ is the Grüneisen parameter and K_s the incompressibility, both of which were assessed from seismic data. Temperature at the inner core boundary, thus, was estimated to be 4168 K. Anderson (1981) similarly obtained 4080 K for the inner core boundary temperature.

Temperature profile between above mentioned fixed points are usually estimated on the assumption that adiabatic temperature gradient prevails in the outer core and mantle below lithosphere. This approach is based on the idea that in a convecting system temperature gradient is held close to the adiabatic one. The adiabatic gradient dT/dz in the Earth is assessed from:

$$\frac{dT}{dz} = \frac{\alpha_v}{c_p} Tg \tag{9.8}$$

$$\frac{\alpha_v}{c_p} = \gamma \frac{\rho}{K_s}, \tag{9.9}$$

where

z = depth
c_p = specific heat capacity at constant pressure
g = gravitational acceleration
ρ = density
α_v = volumetric expansion coefficient.

Figure 9.18 shows the summary of results obtained by Stacey (1977) and Anderson (1981). At the present stage, these may be regarded as the best estimates of the temperature profile in the Earth.

9.1.11. Thermal History of the Earth

A knowledge of Earth's thermal history is essential to the understanding of the Earth's processes as a heat engine. As is well known, studies of the Earth's thermal history go back to Lord Kelvin (1862). He solved the thermal conduction equation of a cooling half space, assuming that the Earth started in an initially hot state (melting temperature). He obtained the time needed for the surface geothermal gradient to become as low as is observed and assigned this time (about 10^8 a) to the age of the Earth. This argument has only historical interest, because the Earth is known to contain heat sources and not simply be a cooling body. Kelvin's argument that the Earth has cooled in such a short time has, in fact, sometimes been quite misleading, because the part which was cooled in Kelvin's model was actually only a several tens of kilometer thick veneer below which the material is still molten; (it is interesting to note that the modern theory of plate cooling exactly follows Kelvin's calculation). Therefore, when a heat source is assumed within the Earth, it should heat up rather than cool down (e.g. Jeffreys, 1929; Slichter, 1941).

In the early 1940's, the cold origin theory of the Earth replaced the hot origin theory. The argument that the Earth, starting from a cold initial condition, can heat up was welcomed, because some degree of melting must have occurred at some stage of the Earth's history, regardless of the initial state, in order to form the core-mantle layering. With the advent of modern computers, investigations were carried out to demonstrate that an initially cold Earth can heat up sufficiently to allow core-mantle segregation (e.g. Lubimova, 1958; MacDonald, 1959). The Earth, however, had to be cooled to its present state. Thus the difficulty of

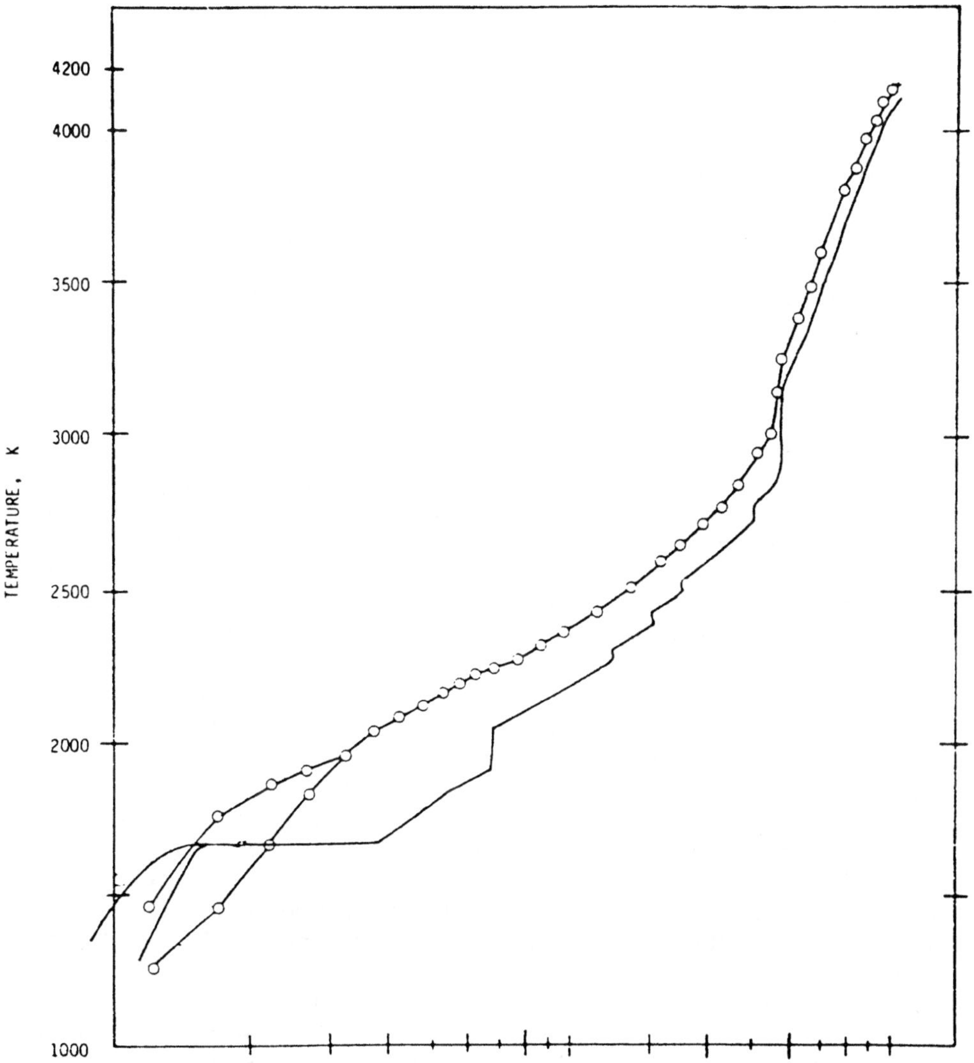

Fig. 9.18. Examples of temperature profile in the Earth's mantle and outer core (from Anderson, 1981). Both solid line (Anderson, 1981) and solid line with circles (Stacey, 1977) have branches in the upper mantle, corresponding to continental (upper branch) and oceanic (lower branch) areas.

cooling the radioactive Earth again emerged. In order to overcome this difficulty, various possibilities were introduced, e.g. introduction of radiative thermal conductivity (Clark, 1957), upward concentration of heat sources and so on. It was shown empirically, however, that radiative thermal conductivity of peridotite is not of much help (Fukao, 1969; Schatz and Simmons, 1972). Through these attempts, it became increasingly apparent that manipulation of solid thermal conduction equations would lead nowhere. It was almost obvious that such theories as continental drift, sea floor spreading and plate tectonics all require large scale material movements in the Earth's interior, and that a thermal history that does not take such movements into consideration is meaningless. It was felt, however, very difficult to solve the problem involving material motions if we were to solve hydrodynamic equations of the Earth's interior at the same time.

New light was shed on these difficulties by Tozer's (1970, 1972) concept of stabilization temperature. The viscosity (ν) of mantle materials follows the exponential dependence on temperature (T, in K) as:

$$\nu = \nu_0 T\, e^{E/kT}, \tag{9.10}$$

where E is activation energy and k is Boltzmann constant. Therefore, when the temperature is raised, the Rayleigh number Ra of the mantle must exceed the critical value, before melting takes place, and active thermal convection must start. Once convection current sets in, cooling will proceed much more efficiently than before and any further increase of T will cease. Because of this feedback mechanism, the internal temperature will be kept close to a certain value (stabilization temperature), regardless of the initial temperature. The only secular change would be gradual cooling of the whole system more or less dictated by thickening of surface plates. Thus, as far as the temperature history is concerned, one does not need a rigorous solution to the hydrodynamic equations. A number of investigators, who were engaged with mantle convection problems in relation to plate tectonics, developed the approach of parameterized convection (e.g. McKenzie and Weiss, 1975; Sharpe and Peltier, 1978, 1979; Turcotte, 1980; Schubert *et al.*, 1980; Davies, 1980b; Peltier and Javis, 1982; Honda 1983). Although there are differences among individual studies, the essential point of the approach may be summarized as follows: characteristics of a convective system can be represented by a certain number of non-dimensional quantities, namely Rayleigh number Ra, Reynolds number Re, Prandtl number Pr, and Nusselt number Nu (see Table 2.9 for their definitions). Reasonable estimates on the magnitudes of these quantities for the Earth's mantle suggest $\mathrm{Ra} \cong 10^5 - 10^7 \gg \mathrm{R}_c \cong 10^3$, $\mathrm{Re} \cong 10^{-20} - 10^{-21} \ll 1$, $\mathrm{Pr} \cong 10^{24} \gg 1$, where R_c is the critical Rayleigh number for onset of thermal convection. These values indicate that convection is occurring in the mantle and the flow is not turbulent but slow enough so that the inertia force is negligible. It is also known (e.g. Rossby, 1969) that, when Pr is large, Nu and Ra are related as:

$$\mathrm{Nu} \cong (\mathrm{Ra}/\mathrm{R}_c)^{1/3}. \tag{9.11}$$

For $\mathrm{Ra} \cong 10^6$, Nu in the mantle is estimated to be of the order of 10, suggesting that the efficiency of heat transfer there is about 10 times that of thermal conduction. The thermal time constant is accordingly small.

Another important aspect is that in a system with high Rayleigh number, temperature difference is concentrated within upper and lower boundary layers, in which heat is transferred only by conduction, and the bulk of the mantle has essentially an adiabatic temperature gradient as mentioned in the last section. The upper boundary layer corresponds to the lithosphere. In such a system mantle temperature can be approximately represented by a single value T_M. Therefore, the heat balance of the whole mantle may be described by:

$$mc_p \frac{\mathrm{d}T_M}{\mathrm{d}t} = -\mathrm{Nu}\lambda \frac{(T_M - T_\mathrm{o})}{D} A + Q_c + H_M, \tag{9.12}$$

where

m = total mass of the mantle
c_p = specific heat capacity
λ = thermal conductivity
T_o = surface temperature
D = thickness of the mantle
A = Earth's surface area
Q_c = heat from the core
H_M = total heat production rate of the mantle.

In the parameterized approach Equation (9.12) is to be solved, using the relations such as (9.11) and (9.14). Davies (1980b), through this type of approach, indicated that the heat generated in the mantle now is only about half of the present heat output. Although convection is an efficient heat remover, according to Davies, the Earth is still emitting its original heat. Sharpe and Peltier (1978, 1979), through a little less simplified model, arrived at similar conclusion, namely the portion of original heat in the present heat output may be as high as 50 % and the present temperature distribution in the mantle is not much different from the initial one. Originally at the melting point, the mantle now is at the stabilization temperature and transferring the heat confined in the core. Although the convective mantle has a time constant much shorter than that of a conductive mantle, its time constant is still comparable with the age of the Earth. The reason for this is probably the existence of thermal boundary layers that limit the rate of heat transfer. Surface boundary layer develops through the strong temperature dependence of viscosity. This way plate tectonics is playing an important role in the thermal history of the Earth.

Recently, suggestion has been made that the mantle may consists of two or more chemically distinct layers (e.g. O'Nions *et al.*, 1979; Wasserburg and De Paulo, 1979). McKenzie and Richter (1981), Spohn and Schubert (1982) and Honda (1982) applied the parameterized convection scheme to the heat transfer problem of a layered mantle and found that increased number of boundary layers lowers the efficiency of heat transfer.

9.1.12. Final Words

Plate tectonics has been remarkably successful in many aspects. It has, however, been essentially a kinematic theory. Heat-flow density and other thermal aspects must be fully incorporated with plate tectonics, if we are to achieve truly meaningful geodynamic understanding of the Earth. This has been just started and there is still much to be done in the future.

Acknowledgements

The author wishes to thank Editors of this monograph, Satoru Honda, Earthquake Research Institute, and T. W. C. Hilde, Texas A&M University for constructive comments on the manuscript. Mrs. N. Matsushima is acknowledged for her assistance in preparation of the manuscript and figures.

9.1.13. References

Akaogi, N. and Akimoto, S.: 1979, 'High Pressure Equilibria in a Garnet Lherzolite with Special Reference to $Mg^{2+} - Fe^{2+}$ Portioning among Constituent Minerals', *Phys. Earth Planet. Interiors* **19**, 31–51.

Anderson, R. N.: 1980, 'Update of Heat Flow in the East and Southeast Asian Seas', in D. Hayes (ed.), *Tectonic/Geologic Evolution of Southeast Asia*; Geophys. Monogr. **23**, Amer. Geophys. Un., Washington D.C., 319–326.

Anderson, R. N. and Skilbeck, J. N.: 1980, *Oceanic Heat Flow*, in C. Emiliani (ed.), *The Sea*, Wiley-Interscience, New York, **7**, 489–523.

Anderson, O. L.: 1981, 'Temperature Profiles in the Earth', in R. J. O'Connell and W. S. Fyfe (eds.), *Evolution of the Earth*, Geodynamics Series, **5**, AGU/GSA, 19–27.

Andrews, D. J. and Sleep, N. H.: 1974, 'Numerical Modeling of Tectonic Flow Behind Island Arcs', *Geophys. J. Roy. Astr. Soc.* **38**, 237–251.

Birch, F.: 1954, 'The Present State of Geothermal Investigation', *Geophys.* **19**, 645–659.

Birch, F.: 1958, 'Differentiation of the Mantle', *Geol. Soc. Amer. Bull.* **69**, 483–486.

Blackwell, D. D., Bowen, R. G., Hull, D. A. Riccio, J., and Steele, J. L.: 1982, 'Heat Flow, Arc Volcanism and Subduction in Northern Oregon', *J. Geophys. Res.* **87**, 8735–8754.

Bodmer, Ph., England, P. C., Kissling, E. and Rybach, L.: 1979, 'On the Correction of Subsurface Temperature Measurements for the Effects of Topographic Relief, Part II; Application to Temperature Measurements in the Central Alps', in V. Čermák and L. Rybach (eds.), *Terrestrial Heat Flow in Europe*, Springer, 78–87.

Bodri, L. and Bodri, B.: 1978, 'Numerical Investigation of Tectonic Flow in Island-Arc Areas', *Tectonophys.* **50**, 163–175.

Boyd, F. R.: 1973, 'A Pyroxene Geotherm', *Geochim. Cosmochim. Acta* **37**, 2533–2546.

Bram, K.: 1979, 'Heat Flow Measurements in the Federal Republic of Germany', in V. Čermák and L. Rybach (eds.), *Terrestrial Heat Flow in Europe*, Springer-Verlag, 191–196.

Bullard, E. C., Maxwell, A. E. and Revelle, R.: 1956, 'Heat Flow Through the Deep Sea Floor', *Adv. Geophys.* **3**, 153–181.

Burke, K.: 1975, 'Hot Spots and Aulacogens of the European Margin', Leicester/Shropshire, Oslo/Fen, *Geol. Soc. Amer. Abstr.* **7**, 34–35.

Burns, R. E.: 1964, 'Sea Bottom Heat-Flow Measurements in the Andaman Sea', *J. Geophys. Res.* **69**, 4918–4919.

Calvalho, H. da. S., Siswoyo, P., Thamrin, M. and Vacquier, V.: 1980, 'Terrestrial Heat Flow in the Tertiary Basin of Central Sumatra', *Tectonophys.* **69**, 163–188.

Čermák, V. and Rybach, L. (eds.): 1979, *Terrestrial Heat Flow in Europe*, Springer-Verlag, Berlin, Heidelberg, New York, 328 p.

Chapman, D. S. and Pollack, H. N.: 1975, 'Global Heat Flow: A New Look', *Earth Planet. Sci. Lett.* **28**, 23–32.

Chapman, D. S. and Pollack, H. N.: 1977, 'Heat Flow and Heat Production in Zambia: Evidence for Lithosphere Thinning in Central Africa', *Tectonophys.* **41**, 79–100.

Chapman, D. S. and Pollack, H. N.: 1980, 'Global Heat Flow: Spherical Harmonic Representation', *EOS, Trans. Am. Geophys. Un.* **61**, 383.

Clark, S. P.: 1957, 'Radiative Transfer in the Earth's Mantle', *Trans. Amer. Geophys. Un.* **38**, 931–938.

Collette, B. J.: 1971, 'Vertical Crustal Movements in the North Sea Area Through Geological Time', in F. M. Delany (ed.), *ICSU/SCOR Working Party 31, Symposium, Cambridge, 1970*, The Geology of the East Atlantic Margin, H.M.S.O., London.

Coney, P. J., Jones, D. L. and Monger, J. W. H.: 1980, 'Cordilleran Suspect Terranes', *Nature* **288**, 329–333.

Corliss, J. B., Dymond, J. R., Gordon, L. I., Edmond, J. M., Von Herzen R. P., Bullard, R., Green, K. E., Williams, D. L., Bainbridge, A., Crane, K. and van Andel, T. H.: 1979, 'Submarine Thermal Springs on the Galapagos Rift', *Science* **203**, 1073-1083.

Crough, S. T. and Thompson, G. A.: 1976, 'Thermal Model of Continental Lithosphere', *J. Geophys. Res.* **81**, 4857–4862.

Davies, G. F.: 1980a, 'Review of Oceanic and Global Heat Flow Estimates', *Rev. Geophys. Space, Phys.* **18**, 718–722.

Davies, G. F.: 1980b, 'Thermal Histories of Convective Earth Models and Constraints on Radiogenic Heat Production in the Earth', *J. Geophys. Res.* **85**, 2517–2530.

Decker, E. R. and Burcher, G. J.: 1977, 'Geothermal Studies in Antarctica', *U.S. Antarct. J.* **12**, 102–104.

Degens, E. T., Von Herzen R. P. and Wong How-Kim: 1971, 'Lake Tanganika: Water Chemistry, Sediments, Geological Structure', *Naturwiss.* **58**, 229–241.

Degens, E. T., Von Herzen, R. P., Wong How-Kim, Deuser, W. G. and Jannasch, H. W.: 1973, 'Lake Kivu: Structure, Chemistry and Biology of an East African Rift Lake', *Geol. Rundschau* **62**, 245–277.

Dickinson, W. R. and Snyder, W. S.: 1979, 'Geometry of Subducted Slabs Related to San Andreas Transform', *J. Geol.* **87**, 609–627.

England, P. C. and Richardson, S. W.: 1980, 'Erosion and the Age-Dependence of Continental Heat Flow', *Geophys. J. Roy. Astr. Soc.* **62**, 421–438.

Fukao, Y.: 1969, 'On the Radiative Heat Transfer and the Thermal Conductivity in the Upper Mantle', *Bull. Earthq. Res. Inst.* **47**, 549–569.

Furlong, K. P., Chapman, D. S. and Alfeld, P. W.: 1982, 'Thermal Modelling of the Geometry of Subduction with Implications for the Tectonics of the Overriding Plate', *J. Geophys. Res.* **87**, 1786–1802.

Gilvarry, J. J.: 1956, 'The Lindemann and Grüneisen Laws', *Phys. Rev.* **102**, 308–316.

Haenel, R.: 1983, 'Geothermal Investigations in the Rhenish Massif', in K. Fuchs, K. von Gehlen, H. Mälzer, H. Murawski and A. Semmel (eds.), *Plateau Uplift*, Springer-Verlag, 228–246.

Hamza, V. M. and Verma, R. K.: 1969, 'The Relationship of Heat Flow with Age of Basement Rock', *Bull. Volcanologique* **33**, 123–152.

Hasebe, K., Fujii, N. and Uyeda, S.: 1970, 'Thermal Processes under Island Arcs', *Tectonophys* **10**, 335–355.

Heestand, R. L. and Crough, S. T.: 1981, 'The Effect of Hot Spots on the Oceanic Age-Depth Relation', *J. Geophys. Res.* **86**, 6107–6114.

Hilde, T. W. C. and S. Uyeda: 1983, 'Trench Depth: Variation and Significance', in T. W. C. Hilde and S. Uyeda (eds.), *Geodynamics of the Western Pacific-Indonesian Region*, Geodynamics Series, **11**, Amer. Geophys. Un./Geol. Soc. Amer., 75–89.

Hobart, M., Anderson, R. N. and Uyeda, S.: 1979, 'Heat Transfer in the Mariana Trough', *EOS* **60**, 383.

Hoffers, B.: 1981, 'A Model for Hydrothermal Convection in the Rheingraben and its Tectonic Implications', *Tectonophys.* **73**, 141–149.

Honda, S.: 1982, 'Numerical Analysis of Layered Convection-Marginal Stability and Finite Amplitude Analysis', *Bull. Earthq. Res. Inst.* **57**, 273–302.

Honda, S.: 1983, 'Mantle Convection – A Review on Recent Topics', *Nagare* **2**, 341–354, (in Japanese).

Honda, S.: 1985, 'Thermal Structure Beneath the Northeast Japan', *Tectonophys.* 112, 69–102.

Honda, S. and Uyeda, S.: 1983, 'Thermal process in Subduction Zones', in D. Shimozuru and I. Yokoyama (eds.), *Arc Volcanism: Physics and Tectonics*, Terra Sci. Pub. Co., Tokyo, 117–140.

Hsui, A. and Toksöz, M. N.: 1979, 'The Evolution of Thermal Structures Beneath a Subduction Zone', *Tectonophys.* **60**, 43–60.

Jeffreys, H.: 1929, *The Earth*, 2nd ed., Cambridge.

Jessop, A. M., Hobart, M. A. and Sclater, J. G.: 1976, *The World Heat Flow Data Collection – 1975*. Geothermal Series 5, Earth Phys. Branch, Ottawa, 125 p.

Jordan, T. H.: 1981, 'Continents as a Chemical Boundary Layer', *Phil. Roy. Soc. Lond. A.* **301**, 359–373.

Kappelmeyer, O. and Haenel, R.: 1974, 'Geothermics with Special Reference to Application', *Geoexplor. Monogr. Ser.* **1**, 4, Gebr. Bornträger, Stuttgart, 238 p.

Kelvin, Lord: 1862, 'On the Secular Cooling of the Earth', *Roy. Soc. Edinburgh, Trans.* **23**, 157–169.

Kono, Y. and Amano, M.: 1978, 'Thickening Model of the Continental Lithosphere', *Geophys. J. Roy. Astr. Soc.* **54**, 405–416.

Lachenbruch, A. H.: 1970: 'Crustal Temperature and Heat Production: Implication of the Linear Heat Flow Relation', *J. Geophys. Res.* **75**, 3219–3300.

Lachenbruch, A. H. and Marshall, B. V.: 1966, 'Heat Flow Through the Arctic Ocean Floor – the Canada Basin-Alpha Rise Boundary', *J. Geophys. Res.* **71**, 1224–1248.

Lachenbruch, A. H. and Sass, J. S.: 1977, 'Heat Flow in the United States and Thermal Regime of the Crust', in J. G. Heacock (ed.), *The Earth's Crust*, Geophys. Monogr. **20**, Am. Geophys. Un. Washington D.C., 626–675.

Lachenbruch, A. H. and Sass, J. S.: 1980, 'Heat Flow and Energetics of the San Andreas Fault Zone', *J. Geophys. Res.* **85**, 6185–6222.

Langseth, M. G. and Anderson, R. N.: 1979, 'Correction', *J. Geophys. Res.* **84**, 1139–1140.

Langseth, M. G. and Von Herzen, R. P.: 1970, 'Heat Flow Through the Floor of the World Oceans', in A. E. Maxwell (ed.), *The Sea*, Instersci., New York, 299–352.

Lawver, L. A., Williams, D. L. and Von Herzen, R. P.: 1975, 'A Major Geothermal Anomaly in the Gulf of California', *Nature*, *257*, No. 5221, 23–28.

Lee, W. H. K.: 1970, 'On the Global Variations of Terrestrial Heat Flow', *Phys. Earth. Planet. Inter.* **2**, 332–341.

Lee, W. H. K. and MacDonald, G. J. F.: 1963, 'The Global Variation of Terrestrial Heat Flow', *J. Geophys. Res.* **68**, 6481–6492.

Lee, W. H. K. and Uyeda, S.: 1965, 'Review of Heat Flow Data', in W. H. K. Lee (ed.), *Terrestrial Heat Flow*, Geophys. Monograph. **8**, Amer. Geophys. Un. Washington, D.C., 87–190.

Le Pichon, X. and Sibuet, J. C.: 1981, 'Passive Margins: A Model of Formation', *J. Geophys. Res.* **86**, 3708–3720.

Lister, C. R. B.: 1972, 'On the Thermal Balance of a Midocean Ridge', *Geophys. J. Roy. Astr. Soc.* **26**, 515–535.

Lister, C. R. B.: 1980, 'Heat Flow and Hydrothermal Circulation', *Ann. Rev. Earth Planet. Sci.* **8**, 95–117.

Lubimova, E. A.: 1958, 'Thermal History of the Earth with Consideration of the Variable Thermal Conductivity of the Mantle', *Geophys. J. Roy. Astr. Soc.* **1**, 115–134.

Lysak, S. V.: 1978, 'The Baikal Heat Flow', *Tectonophys.* **45**, 87–93.

MacDonald, G. J. F.: 1959, 'Calculations on the Thermal History of the Earth', *J. Geophys. Res.* **64**, 1967–2000.

MacDonald, K. C., Becker, K., Spiess, F. N. and Ballard, R.: 1980, 'Hydrothermal Heat Flux of the "Black Smoker" Vents on the East Pacific Rise', *Earth. Planet Sci. Lett.* **48**, 1–7.

McKenzie, D. P.: 1967, 'Some Remarks on Heat Flow and Gravity Anomalies', *J. Geophys. Res.* **72**, 6261–6271.

McKenzie, D. P.: 1978, 'Some Remarks on the Development of Sedimentary Basins', *Earth Planet. Sci. Lett.* **40**, 25–32.

McKenzie, D. P.: 1981, 'The Variation of Temperature with Time and Hydrocarbon Maturation in Sedimentary Basins Formed by Extension', *Earth Planet. Sci. Lett.* **55**, 87–98.

McKenzie, D. P. and Richter, F. M.: 1981, 'Parameterized Thermal Convection in a Layered Region and the Thermal History of the Earth', *J. Geophys. Res.* **86**, 11.667–11.680.

McKenzie, D., and Weiss, N.: 1975, 'Speculations on the Thermal and Tectonic History of the Earth', *Geophys. J. Roy. Astr. Soc.* **42**, 131–174.

Mercier, J. C. and Carter, N. L.: 1975, 'Pyroxene Geotherms', *J. Geophys. Res.* **80**, 3349–3362.
Minear, J. W. and Toksöz, M. N.: 1970, 'Thermal Regime of a Downgoing Slab and new Global Tectonics', *J. Geophys. Res.* **75**, 1397–1419.
Oldenburg, D. W.: 1975, 'A Physical Model for the Creation of the Lithosphere', *Geophys. J. Roy. Astr. Soc.* **43**, 425–451.
O'Nions, R. K., Evenson, N. M. and Hamilton, P. J.: 1979, 'Geochemical Modelling of Mantle Differentiation and Crustal Growth', *J. Geophys. Res.* **84**, 6091–6101.
Oxburgh, E. R. and Parmentier, E. M.: 1978, 'Thermal Processes in the Formation of Continental Lithosphere', *Phil. Trans. R. Soc. Lond. A.* **288**, 415–429.
Oxburgh, E. R. and Turcotte, D. L.: 1970, 'Thermal Structure of Island Arcs', *Geol. Soc. Amer. Bull.* **81**, 1665–1688.
Parsons, B. and McKenzie, D. P.: 1978, 'Mantle Convection and the Thermal Structure of the Plates', *J. Geophys. Res.* **83**, 4485–4496.
Parsons, B. and Sclater, J. G.: 1977, 'An Analysis of the Variation of Ocean Floor Bathymetry and Heat Flow with Age', *J. Geophys. Res.* **82**, 803–827.
Peltier, W. R. and Javis, G. T.: 1982, 'Whole Mantle Convection and the Thermal Evolution of the Earth', *Phys. Earth Planet. Inter.* **29**, 281–304.
Pollack, H. N.: 1982, 'The Heat Flow from the Continents', *Ann. Rev. Earth Planet. Sci.* **10**, 459–481.
Pollack, H. N. and Chapman, D. S.: 1977, 'On the Regional Variation of Heat Flow, Geotherms, and the Thickness of Lithosphere', *Tectonophys.* **38**, 279–296.
Polyak, B. G. and Smirnov, Ya. B.: 1968, 'Relationships Between Terrestrial Heat Flow and Tectonics of Continents', *Geotectonics*, (Engl. Transl.), **4**, 205–213.
Rao, R. U. M., Rao, G. V. and Reddy, G. K.: 1982, 'Age Dependence of Continental Heat Flow – Fantasy and Facts', *Earth Planet. Sci., Lett.* **59**, 288–302.
Rossby, H. T.: 1969, 'A Study of Bénard Convection with and without Rotation', *J. Fluid, Mech.* **36**, 309–335.
Roy, R. F., Blackwell, D. D. and Birch, F.: 1968, 'Heat Generation of Plutonic Rocks and Continental Heat Flow Provinces', *Earth Planet. Sci. Lett.* **5**, 1–12.
Rybach, L.: 1976, 'Die Gesteinsradioaktivität und ihr Einfluss auf das Temperaturfeld in der Kontinentalen Kruste', *J. Geophys.* **42**, 93–101.
Savostin, L. A. and Vlasenko, V. I.: 1974, 'Results of Geothermal Investigations in the Sea of Okhotsk', *Geotermiya* **1–2**, 91–95 (in Russian).
Sharpe, H. N. and Peltier, W. R.: 1978, 'Parameterized Convection and Earth's Thermal History', *Geophys. Res. Lett.* **5**, 737–740.
Sharpe, H. N. and Peltier, W. R.: 1979, 'A Thermal History Model for the Earth with Parameterized Convection', *Geophys. J. Roy. Astr. Soc.* **59**, 171–203.
Schatz, J. F. and Simmons, G.: 1972, 'Thermal Conductivity of Earth's Materials at High Temperatures', *J. Geophys. Res.* **77**, 6966–6983.
Schubert, G., Stevenson, D. and Cassen, P.: 1980, 'Whole Planet Cooling and Radiogenic Heat Source Contents of the Earth and Moon', *J. Geophys. Res.* **85**, 2531–2538.
Sclater, J. G. and Christie, P. A. F.: 1980, 'Continental Stretching: An Explanation of the Post-Mid-Cretaceous Subsidence of the Central North Sea Basin', *J. Geophy. Res.* **85**, 3711–3739.
Sclater, J. G. and Francheteau, J.: 1970, 'The Implications of Terrestrial Heat Flow Observations on Current Tectonic and Geochemical Models of the Crust and Upper Mantle of the Earth', *Geophys. J. Roy. Astr. Soc.* **20**, 509–542.
Sclater, J. G., Jaupart, C. and Galson, D. A.: 1980a, 'The Heat Flow Through Oceanic and Continental Crust and the Heat Loss of the Earth', *Rev. Geophys. Space Phys.* **18**, 269–311.
Sclater, J. G., Royden, L., Horvàth, F., Burchfiel, B. C., Semken, S. and Stegena, L.: 1980b, 'The Formation of Intra-Carpathian Basins as Determined from Subsidence Data', *Earth Planet. Sci. Lett.* **51**, 139–162.
Simmons, G. and Horai, K.: 1968, 'Heat Flow Data 2', *J. Geophys. Res.* **73**, 6608–6629.
Slichter, L. B.: 1941, 'Cooling of the Earth', *Bull. Geol. Soc. Amer.* **52**, 561–600.
Smirnov, Ya. B.: 1980, 'Thermal Field of the Territory of the USSR, Main Adminst. Geodesy and Cartography, Council of Ministers, USSR, Moscow, 150 p.
Spiess, F. N., MacDonald, K. C., Atwater, T., Ballard, R., Carranza, A., Cordoba, D., Cox, C.,

Diaz Garcia, V. M., Francheteau, J., Guerrero, J., Hawkins, J., Haymaon, R., Hessler, R., Juteau, T., Kastner, M., Larson, R., Luyendyk, B., Macdougall, J. D., Miller, S. P., Normark, W., Orcutt, J. and Rangin, C.: 1980, 'East Pacific Rise: Hot Springs and Geophysical Experiments', *Science* **207**, 1421–1433.

Spohn, T. and Schubert, G.: 1982, 'Modes of Mantle Convection and Removal of Heat from the Earth's Interior', *J. Geophys. Res.* **87**, 4682–4696.

Stacey, F. D.: 1977, 'A Thermal Model of the Earth', *Phys. Earth Planet. Interiors* **15**, 341–348.

Stacey, F. D. and Irvine, R. D.: 1977, 'Theory of Melting: Thermodynamic Basis of Lindemann's Law', *Austr. J. Phys.* **30**, 631–640.

Stegena, L., Horvath, F., Sclater, J. G. and Royden, L.: 1981, 'Determination of Paleotemperature by Vitrinite Reflectance Data', *Earth Evolution Sciences* **3/4**, 292–297.

Stevenson, D.: 1980, 'Applications of Liquid State Physics to the Earth's Core', *Phys. Earth Planet. Interiors* **22**, 42–52.

Taira, A., Saito, Y. and Hashimoto, M.: 1983, 'The Role of Oblique Subduction and Strike-Slip Tectonics in the Evolution of Japan', in T. W. C. Hilde and S. Uyeda (eds.), *Geodynamics of the Western Pacific-Indonesian Region*, Geodynamics Series **11**, Amer. Geophys. Un./Geol. Soc. Amer., 303–316.

Tatsumi, Y., Sakuyama, M., Fukuyama, H. and Kushiro, I.: 1983, 'Generation of Arc Basalt Magmas and Thermal Structure of the Mantle Wedge in Subduction Zone', *J. Geophys. Res.* **88**, 5815–5825.

Tozer, D. C.: 1970, 'Factors Determining the Temperature Evolution of Thermally Convecting Earth Models', *Phys. Earth Planet. Inter.* **2**, No. 5, 393–398.

Tozer, D. C.: 1972, 'The Present Thermal State of the Terrestrial Planets', *Phys. Earth Planet. Inter.* **6**, 182–197.

Turcotte, D. L.: 1980, 'On the Thermal Evolution of the Earth', *Earth Planet. Sci., Lett.* **48**, 53–58.

Uyeda, S. and Horai, K.: 1964, 'Terrestrial Heat Flow in Japan', *J. Geophys. Res.* **69**, 2121–2141.

Uyeda, S.: 1980, 'Review of Heat Flow Studies in the Eastern Asia and Western Pacific Region', *UN ESCAP, CCOP/SOPAC, Tech. Bull.* **3**, 153–169.

Uyeda, S. and Watanabe, T.: 1982, 'Terrestrial Heat Flow in Western South America', *Tectonophys.* **83**, 63–70.

Verma, R. K.: 1969, 'Review of Heat Flow Studies in India', *Proc. 2nd Symp. Upp. Mantle Proj.*, Hyderabad, 147–174.

Verma, R. K., Hamza, V. M. and Panda, P. K.: 1970, 'Further Study of the Correlation of Heat Flow with Age of Basement Rocks', *Tectonophys.* **10**, 301–319.

Vitorello, I. Hamza, V. M. and Pollack, H. N.: 1980, 'Terrestiral Heat Flow in Brazilian Highlands', *J. Geophys. Res.* **85**, 3778–3788.

Vitorello, I. and Pollack, H. N.: 1980, 'On the Variation of Continental Heat Flow With Age and the Thermal Evolution of Continents', *J. Geophys. Res.* **85**, 983–995.

Von Herzen, R. P.: 1963, 'Geothermal Heat Flow in the Gulfs of California and Aden', *Science* **140**, 1207–1208.

Von Herzen, R. P., Detrick, R. S., Crough, S. T., Epp, D. and Fehn, V.: 1982, 'Thermal Origin of the Hawaiian Swell; Heat Flow Evidence and Thermal Models', *J. Geophys. Res.* **87**, 6711–6773.

Von Herzen, R. P. and Vacquier, V.: 1967, 'Terrestrial Heat Flow in Lake Malawi, Africa', *J. Geophys. Res.* **72**, 4221–4226.

Wasserburg, G. J. and De Paulo, D. J.: 1979, 'Models of Earth Structure Inferred from Neodymium and Strontium Isotope Abundances', *Proc. Natl. Acad. Sci, U.S.* **76**, 3594–3598.

Watanabe, T., Langseth, M. G. and Anderson, R. N.: 1977, 'Heat Flow in Back-Arc Basins of the Western Pacific', in M. Talwani and W. C. Pitman III (eds.), *Island Arcs, Deep Sea Trenches and Back-Arc Basins*, Amer. Geophys. Un., Washington, D.C., 137–161.

Williams, D. L., Von Herzen, R. P., Sclater, J. G. and Anderson, R. N.: 1974, 'The Galapagos Spreading Center: Lithosphere Cooling and Hydrothermal Circulation', *Geophys. J. Roy. Astr. Soc.* **38**, 587–608.

Yamano, M., Honda, S. and Uyeda, S.: 1984, 'Nankai Trough, A Hot Trench?', *J. Marine Geophys. Res.*, 6, 187-203.

Yamano, M., Uyeda, S., Aoki, Y. and Shipley, T. H.: 1982, 'Estimates of Heat Flow Derived from Gas Hydrates', *Geology* **10**, 339–343.

Yasui, M., Kishii, T. and Sudo, K.: 1967, 'Terrestrial Heat Flow in the Sea of Okhotsk (1)', *Oceangr. Mag.* **19**, 87–94.

Yasui, M., Kishii, T., Watanabe, T. and Uyeda, S.: 1968, 'Heat Flow in the Sea of Japan', in L. Knopoff, C. L. Drake and P. J. Hart (eds.), *The Crust and Upper Mantle of the Pacific Area*, Geophys. Monogr. **12**, Am. Geophys. Un., Washington, D.C., 3–16.

Yoshii, T.: 1973, 'Upper Mantle Structure Beneath the North Pacific and Marginal Seas', *J. Phys. Earth* **21**, 313–328.

Yoshii, T.: 1979, 'A Detailed Cross-Section of the Deep Seismic Zone Beneath Northeastern Honshu. Japan', *Tectonophys.* **55**, 349–360.

9.2. THERMAL EXPLORATION METHODS

R. HAENEL
Department of Geophysics, Geological Survey of Lower Saxony, Hannover, FRG

and

F. MONGELLI
Dipartimento di Geologia e Geofisica, Università di Bari, Italy

9.2.1. Introduction

Exploration and interpretation methods are based mainly on thermal changes in the subsurface and, therefore, also on changes of heat-flow density near the Earth's surface, caused by both heat sources and heat sinks. Furthermore, thermal conductivity and/or thermal diffusivity contrasts are often also of interest for exploration and interpretation.

The range of application of geothermal methods is given in Table 2.1 of Chapter 2. and can be also roughly subdivided into two branches: geothermal-energy exploration and non-energy exploration. Of course, there already exist many publications on geothermal exploration and interpretation, such as Carslaw and Jaeger (1959), Kappelmeyer and Haenel (1974), Mongelli (1981), Kestin *et al.* (1980), Rybach and Muffler (1982), Edwards *et al.* (1982), Armstead (1983).

With regard to the numerous already published textbooks and existing case histories, the intention is to demonstrate the possibilities of application as well as the importance of heat-flow density for exploration.

9.2.2. Energy Exploration

(1) GENERAL REMARKS

As defined in Chapter 2., the geothermal energy carriers may be subdivided into high-enthalpy fluids ($T > 150\,°C$) and low-enthalpy fluids ($T < 150\,°C$), or into fluids for electric power generation and for heating purposes. In both cases, the surface area above the heat reservoir is called geothermal area, and if the reservoir is under exploitation it is called geothermal field.

Commercially useful geothermal reservoirs are usually associated with upward moving materials, such as mantle diapirisms, mantle plumes, crustal intrusions, and migrating water and gases.

There are three essential requirements for a geothermal energy reservoir:
- a heat source, consisting mainly of a hot magma intrusion at greater depth,
- a permeable zone of hot fluid (water and/or steam),
- and an impermeable cover, which is usually characterised by fractures and fissures giving rise to the areas of 'surface activity'.

R. Haenel, L. Rybach, and L. Stegena (eds.)
Handbook of Terrestrial Heat-Flow Density Determination, 353—389.

In low-enthalpy areas water reservoirs mostly exist within the normal temperature field of the Earth. Because of its mostly low temperature change compared with the surroundings, geothermics generally does not constitute a very good tool in water exploration, except of areas where water is moving.

(2) MEASUREMENTS

An optimum solution is to cover the area under consideration with a grid system and drill a borehole at each node. The borehole depth can be some hundred, but mostly only 50–100 m, and the spacing ranges between 100 and 2000 m. The temperature is measured continuously or at specified depths. Temperature gradients and, if cores are retrieved and the thermal conductivities determined, heat-flow density can then be calculated. The instruments, methods, accuracy and sources of errors have already been discussed in Chapters 3. to 5.

In the case of a flat, homogeneous and impermeable area, a temperature survey using very shallows holes (1–2 m) and small distances can also provide useful results (see Kappelmeyer and Haenel, 1974; LeSchack and Lewis, 1983). Because of the annual temperature variation and of the near-surface water migration and evaporation, this kind of measurement must be carried out carefully, and the results must be interpreted with reservation. All measurements must be performed on the same day or must be reduced to the same day by means of fixed reference stations where the temperature is recorded throughout the entire survey; more details for correction procedures are given in Section 9.2.3.

Figure 9.19 shows a small area in Tuscany, Italy, as an example, where it was possible to delineate an anomaly of about 500 mW m^{-2} by a number of heat-flow density determinations. The depth of the shallow boreholes is about 35 m.

The observed heat-flow density q may be considered as the sum of a component of internal origin q_{in} and a component of external origin q_{ex} due to thermal disturbances which affect the Earth's surface:

$$q = q_{\text{in}} + q_{\text{ex}}. \tag{9.13}$$

Usually, the corrections are of second order in geothermal prospecting, especially in high-enthalpy areas. Nevertheless, correction for topography and sedimentation may be very important, particularly in rough and young areas, when high accuracy is required.

It should be noted that q_{ex} needs not always be eliminated in geothermal prospecting; for example, if we need to know the temperature at a certain depth we have to take into consideration the external thermal disturbances at that depth, too.

The heat-flow density of internal origin q_{in} may in some cases be the sum of the regional heat-flow density q_r and an additional anomalous term of heat-flow density q_a. The component q_r may be identified with the heat-flow density of the geothermal province. Therefore, it yields:

$$q_{\text{in}} = q_a + q_r. \tag{9.14}$$

For interpreting the measured thermal anomalies, the main task is to determine

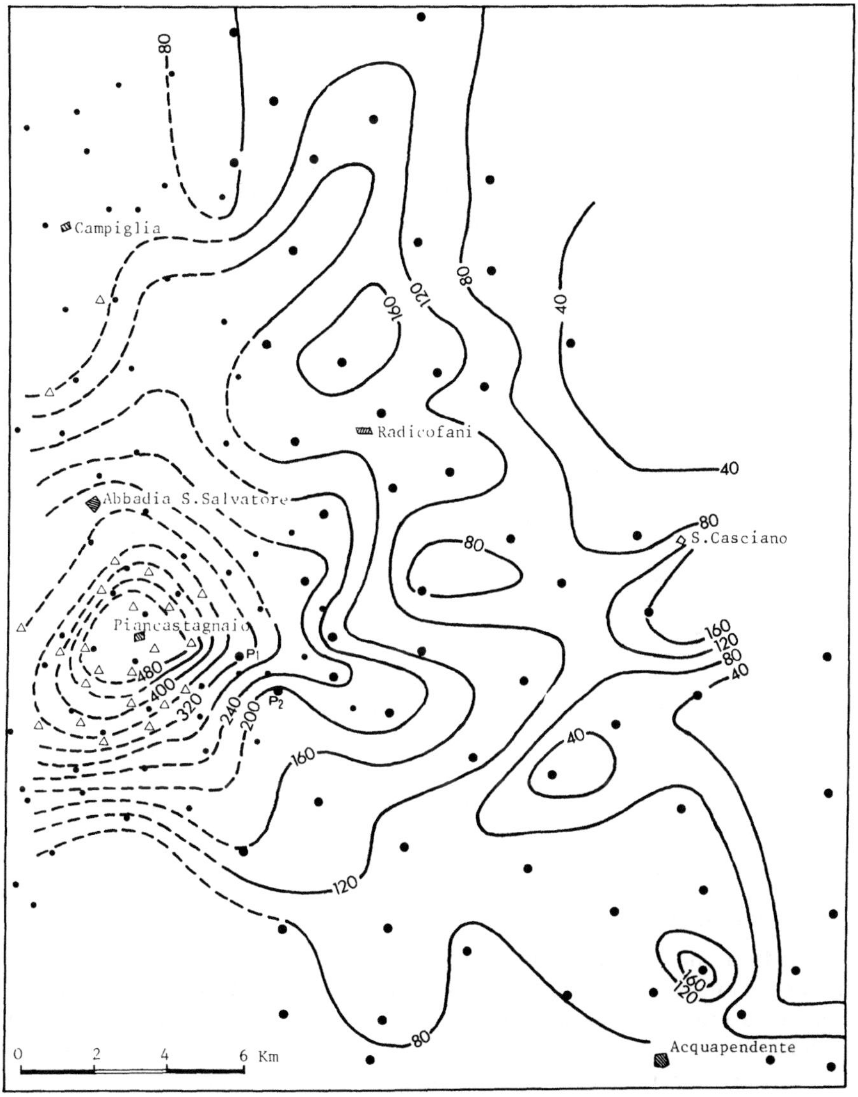

Fig. 9.19. Heat-flow density, in mW m^{-2}, in the Radicofani region (Italy). Dashed heat-flow density contour lines refer to part of Mt. Amiata prospection carried out in 1962. Burgassi *et al.* (1970). (●) shallow boreholes, (△) deep borehoes

depth and shape of the body which generates the anomaly. With $(\partial T/\partial z)_a$ a temperature gradient due to the body, we can write:

$$q_a = -\lambda\left(\frac{\partial T}{\partial z}\right)_a. \tag{9.15}$$

Interpretation is effected by fitting values of 1D, 2D, or 3D models to the measured heat-flow densities or temperature gradients, or temperatures at specified depths.

In most cases, the surplus of heat-flow density q_a is generated by hot magmatic bodies; these are the 'primary' sources of heat anomalies. When a confined fluid system lies between the primary source and the Earth's surface, convection generally occurs by heating. This aquifer may be considered as an effective heat source, which has been named 'secondary source'; see Figure 9.20 (Mongelli 1980a, b).

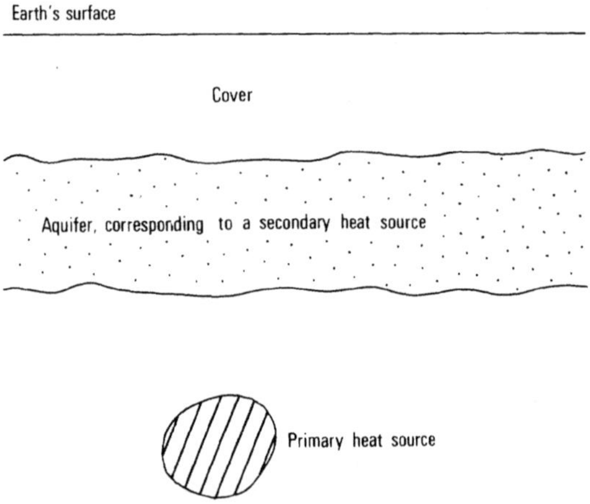

Fig. 9.20. Relative position of primary and secondary sources.

(3) INTERPRETATION OF PRIMARY SOURCE

The heat from an igneous intrusion rarely reaches the Earth surface by conduction alone. However, this mechanism is of interest for hot dry rock problems, as well as for a gross evaluation of the heat available in a promising region (Smith and Shaw, 1975).

The thermal effects of igneous intrusions have been reviewed by Jaeger (1964) and can be outlined as follows: at a certain time, a mass of magma at a known temperature and of known shape is injected into the rocks, which have a known temperature. Since many rocks and magmas have roughly the same thermal properties, by assuming these to be constant and equal, the problem reduces to the conduction of heat in an infinite medium with a prescribed initial distribution of temperature. We have to integrate Fourier's equation (see also Equation (2.5a)):

$$\frac{\partial^2 T}{\partial x^2} + \frac{\partial^2 T}{\partial y^2} + \frac{\partial^2 T}{\partial z^2} = \frac{1}{a}\frac{\partial T}{\partial t}, \tag{9.16}$$

or for steady state problems Laplace's equation:

$$\nabla^2 T = 0 \tag{9.17}$$

with adequate boundary conditions.

Mainly two methods of integration are used for the case of simple geometry: Laplace's solution and the 'sources' method. For more complicated shapes, numerical methods are preferable, but conformal mapping techniques (Schwarz and Christoffel theorem; Morse and Feshbach (1953)) can also be applied in the case of linear boundary conditions (see Hardee and Larson, 1980).

Laplace's Solution

The fundamental solution is used for the temperature in an infinite medium of constant thermal diffusivity, a, with a prescribed initial temperature $T(x', y', z')$ at time $t = 0$ (Carslaw and Jaeger, 1959, p. 56):

$$T(x, y, z, t) = \frac{1}{8}(\pi a t)^{-3/2} \int_{-\infty}^{\infty}\int_{-\infty}^{\infty}\int_{-\infty}^{\infty} T(x', y', z') \cdot \exp\left(-\frac{(x-x')^2+(y-y')^2+(z-z')^2}{4at}\right) \mathrm{d}x'\,\mathrm{d}y'\,\mathrm{d}z'. \tag{9.18}$$

In the present case, the temperature $T(x', y', z')$ has a constant value T_0 (the initial temperature difference between the magma and the country rock) over regions occupied by the magma and is zero elsewhere. If certain simple geometrical shapes are considered (as in other geophysical interpretation methods), the integral (9.18) can be reduced to one or two dimensions and restricted to some finite boundaries, and can be expressed in terms of the error function and related functions.

Infinite Horizontal Sheet (Dike). The simplest case is that of the infinite sheet (dike) of thickness 2D, lying at depth z, $b < z < c$, with constant initial temperature T_i and zero temperature outside. Starting from Equation (9.18) written in one dimension, the temperature at time t is (Carslaw and Jaeger, 1959, p. 54):

$$T(z, t) = \frac{T_i}{2}\left(\mathrm{erf}\left(\frac{z-b}{2\sqrt{at}}\right) - \mathrm{erf}\left(\frac{z-c}{2\sqrt{at}}\right)\right) \tag{9.19}$$

or

$$T(z, t) = \frac{T_i}{2} F(z, b, c), \tag{9.20}$$

where

$$F(z, b, c) = \mathrm{erf}\left(\frac{z-b}{2\sqrt{at}}\right) - \mathrm{erf}\left(\frac{z-c}{2\sqrt{at}}\right), \tag{9.21}$$

and

b = top of the dike, m
c = bottom of the dike, m
$D = (c - b)/2$; 2D = dike thickness, m.

In Figure 9.21, some numerical values of T/T_i are shown for various parameters Fo = at/D^2 = Fourier number, this gives an idea of the internal and external behaviour of a sheet in an infinite medium. Assuming, for example, $T_i = 1000\,°C$, sheet thickness 2D = 2000 m, a = 1×10^{-6} m^2 s^{-1}, the curve labeled 5 is attained after about 160000 a and T in the half space $z > 0$ equals about 200 °C.

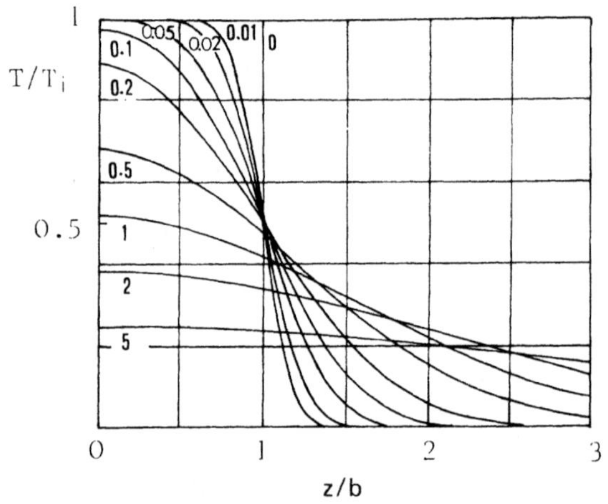

Fig. 9.21. Temperature field produced by an infinite sheet. The curve parameters are given in at/D^2; Carslaw and Jaeger (1959, p. 55).

In geothermal prospecting at the Earth's surface, $z = 0$, zero temperature is generally assumed. The solution for this case may be written by superimposing the solutions for the sheet (b, c) and an image sheet ($-b$, $-c$), (Carslaw and Jaeger, 1959, p. 62):

$$T(z, t) = \frac{T_i}{2}(F(z, b, c) + F(z, -b, -c)). \tag{9.22}$$

This solution, which better fits the geothermic situation is shown in Figure 9.22. The temperature gradient at $z = 0$ above the semi-infinite sheet is given by:

$$\left(\frac{\partial T}{\partial z}\right)_{z=0} = \frac{T_i}{\sqrt{\pi a t}}\left(\exp\left(-\frac{b^2}{4at}\right) - \exp\left(-\frac{c^2}{4at}\right)\right), \tag{9.23}$$

and the heat-flow density:

$$q = \lambda\left(\frac{\partial T}{\partial z}\right)_{z=0} = \frac{\lambda T_i}{\sqrt{at}}\left(\exp\left(-\frac{b^2}{4at}\right) - \exp\left(-\frac{c^2}{4at}\right)\right). \tag{9.24}$$

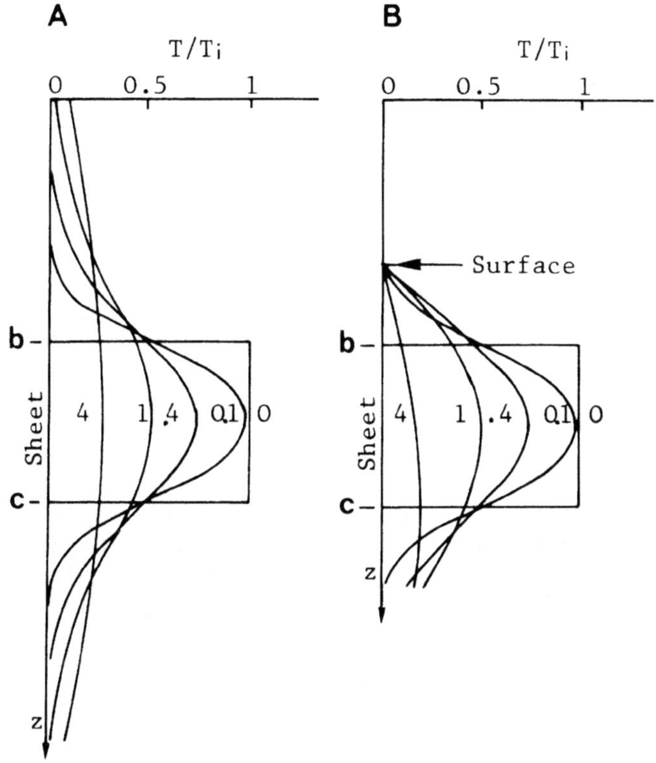

Fig. 9.22. Temperature distribution of an infinite sheet in an infinite (A) and a semi-infinite (B) medium ($T = 0$ at $z = 0$). The curve parameters are given in at/D^2; Jaeger (1964).

Horizontal Cylinder (Pipe Intrusion). A horizontal pipe-like intrusion can be represented by an infinitely long horizontal bar of rectangular cross section $b < z < c$, $d < x < e$ with constant initial temperature T_i and zero temperature outside. Starting from Equation (9.18), the two-dimensional solution is given by:

$$T(x, z, t) = \frac{T_i}{4} \cdot F(z, b, c) \cdot F(x, d, e), \tag{9.25}$$

and for a semi-infinite medium, it is:

$$T(x, z, t) = \frac{T_i}{4} \cdot (F(z, b, c) + F(z, -b, -c)) \cdot F(x, d, e), \tag{9.26}$$

from which the surface gradient is easily calculated.

As can be seen, the temperature given by Equation (9.26) is reduced by the factor $F(x, d, e)$ with respect to the temperature in Equation (9.22). The heat-flow density is simply given by multiplying Equation (9.23) by the thermal conductivity λ and $F(x, d, e)$.

Rectangular Parallelepiped (Batholith). The body $b < z < c$, $d < x < e$, $h < y < l$ having a wedge along the z direction at constant initial temperature T_i and zero initial temperature outside, may represent a batholith. From Equation (9.18) the solution for the infinite medium is:

$$T(x, y, z, t) = \frac{T_i}{8} \cdot F(z, b, c) \cdot F(x, d, e) \cdot F(y, h, l), \tag{9.27}$$

and for a semi-infinite medium:

$$T(x, y, z, t) = \frac{T_i}{8}(F(z, b, c) + F(z, -b, -c)) \cdot F(x, d, e) \cdot F(y, h, l). \tag{9.28}$$

The heat-flow density is given by multiplying Equation (9.23) by the thermal conductivity λ and $F(x, d, e) \cdot F(y, h, l)$. For practical application see Equations (9.44) to (9.47).

Sphere (Magma Chamber). A magma chamber (laccolith) may be represented by a sphere $r \leq b$ at constant initial temperature T_i and zero temperature outside. After rewriting the Fourier equation in spherical coordinates, the solution for an infinite medium in this case is also obtained by the Laplace's solution (Carslaw and Jaeger, 1959, p. 257):

$$T(r, t) = \frac{T_i}{2}\left\{\operatorname{erf}\left(\frac{r + b}{2\sqrt{at}}\right) - erf\left(\frac{r - b}{2\sqrt{at}}\right) - \frac{2\sqrt{at}}{r\sqrt{\pi}}\left[\exp\left(-\frac{(r - b)^2}{4at}\right) - \exp\left(-\frac{(r + b)^2}{4at}\right)\right]\right\}. \tag{9.29}$$

The heat-flow density or the heat loss of the sphere is then simply calculated as:

$$q = \lambda \frac{\partial T}{\partial r}. \tag{9.30}$$

Temperatures at the Centre of an Intrusive Body. The temperature T_m at the centre of an intrusive body can be found for some simple geometric forms of the body (Mundry, 1968):

$$\text{sphere: } T = \operatorname{erf}\left(\frac{1}{2\sqrt{\mathrm{Fo}}}\right) - \frac{1}{\pi \mathrm{Fo}} \exp\left(-\frac{1}{4\mathrm{Fo}}\right) \tag{9.31}$$

$$\text{infinite cylinder: } T = 1 - \exp\left(-\frac{1}{4\mathrm{Fo}}\right) \tag{9.32}$$

$$\text{plate: } T = \operatorname{erf}\left(\frac{1}{4\sqrt{\mathrm{Fo}}}\right) \tag{9.33}$$

where

$T = (T_m - T_0)/(T_2 - T_0)$
T_m = temperature at the centre of the instrusive body, °C
T_0 = temperature in the surroundings at $t = 0$, °C
T_2 = temperature of the instrusive body at $t = 0$, °C
Fo = Fourier number = $at/r^2 = at/d^2$, respectively
a = thermal diffusivity, $m^2 \, s^{-1}$
r = radius of sphere or cylinder, m
d = thickness of plate, m
t = time, s.

Source Method

In this method, which has been extensively treated by Simmons (1967), a magma body at the moment of intrusion is considered as an instantaneous heat source. The elementary solution used here is the temperature generated by a point source $P(x', y', z')$ which emits instantaneously at time $t = 0$ the quantity of heat $Q = S\rho c$ in an infinite medium initially at zero temperature (Carslaw and Jaeger, 1959, p. 256):

$$T(x, y, z, t) = S \exp\left(-\frac{(x - x')^2 + (y - y')^2 + (z - z')^2}{4at}\right)/(8(\pi at)^{3/2}), \quad (9.34)$$

where

S = strength of the source, K m^3
ρ = density, kg m^{-3}
c= specific heat capacity of the medium, J kg^{-1} K^{-1}.

From this solution, expressions are obtained for a finite line source, strip source and various solid sources by successive integrations. The results are formally equivalent to those obtained by the previous method for the same bodies.

As an elementary solution for steady-state problems, the temperature produced by a line source which emits heat at constant rate $\bar{Q}$ per unit time and unit length may be assumed to be:

$$T = \bar{Q} \ln\left(\frac{1}{r}\right)/(2\pi\lambda), \quad (9.35)$$

where r = distance from the line source.

Certain kind of heat sources are to be handled as parallel lying line source. In this case, the temperature at any point is obtained by integration of Equation (9.35). For bodies of simple shape the temperature distribution can be given by simple mathematical expressions (see Von Herzen and Uyeda, 1963).

The temperature field generated by a vertical dike may easily be obtained by starting from Equation (9.35). First, the temperature due to a steady line source lying parallel to the surface in a semi-infinite medium $z > 0$ at depth d, is;

$$T = \bar{Q} \ln\left(\frac{x^2 + (z - d)^2}{x^2 + (z + d)^2}\right)/(4\pi\lambda), \tag{9.36}$$

where x = horizontal distance from the line source.

Integration with respect to z for $d < z < \infty$, gives the temperature of an infinitely thin dike which generates the following gradient at the surface above the dike:

$$\left(\frac{\partial T}{\partial z}\right)_{z=0} = C\left(1 + \frac{x^2}{d^2}\right)^{-1/2}/d. \tag{9.37}$$

By integrating further with respect to x for $-L/2 < x < L/2$, we obtain the surface gradient due to a dike of thickness L (Von Herzen and Uyeda, 1963):

$$\left(\frac{\partial T}{\partial z}\right)_{z=0} = C \ln\left(\frac{b_1 + b_2 + \sqrt{(b_1 + b_2)^2 + 1}}{b_1 - b_2 + \sqrt{(b_1 - b_2)^2 + 1}}\right), \tag{9.38}$$

where $b_1 = x/d$, $b_2 = L/2d$ and C is a constant.

Again, the heat-flow density is given simply by multiplying $\partial T/\partial z$ by the thermal conductivity λ.

Some Comments

Critical Remarks. The simplifications used in the previous mathematical treatment neglect a number of complications, of which the most important are: latent heat and range of solidification of magma; mechanism of intrusion; metamorphism in the country rock; transport of heat by volatile substance produced by the magma; effect of convection in the magma; preheating of the magma. Some of these effects may be taken into account if the necessary parameters are specified quantitatively.

According to Jaeger (1964) the effect of latent heat can be considered approximately by replacing T_0 by $T_0 + L/c$, where L is the latent heat and c the specific heat capacity of the magma. This is only an approximation, because it does not take into account the ranges of solidification during the cooling of the magma. Nevertheless, Jaeger (1964) considers the case of a sheet and concludes that the temperature variation at distances above one-quarter of the thickness of the sheet from its surface is little affected; hence, if the sheet is quite deep, the use of the previous simple formula close to the Earth's surface is correct.

Two important phenomena may affect the life of the intrusion: internal convection of magma and subsequent injections. The first produces a uniform temperature in the magma as a whole, or in large portions thereof. Jaeger (1964) considers the magma as a well-stirred fluid until solidification is complete; Lachenbruch *et al.* (1976) retain the upper surface of the magma at constant temperature. For the case of subsequent intrusion, Jaeger (1964) considers the sum of separated effects with different zero time. With the assumption of continuous feeding, Lachenbruch *et al.* (1976) consider a body emitting constant heat flow, while Reiter and Clarkson (1983) suppose steady-state sources.

In spite of all these limitations, simple solutions are generally used; they are always useful for a more or less accurate interpretation of the anomalous heat-flow density in a region.

Information on Depth from Heat-Flow Density. In principle, two major targets exist in geothermics: the determination of temperatures and the determination of temperature changes, both by means of heat-flow density and/or direct temperature measurements. In the following, therefore an attempt will be made to assess the maximum depth from which the measured data may yield information.

Of course, the heat can be transported by convective as well as conductive mechanisms. Due to the complexity of convective heat transport (see Chapter 2.), only heat conduction will be considered. Because heat conduction is a very slow process, time also has to be taken into account.

To get information from the depth at which the transient heat-flow density originates, a sudden increase of temperature T_1 at an infinite surface at depth d will be assumed. Then, the temperature increase $T(z, t)$ at any level above that surface can be calculated by developing Equation (1) of Carslaw and Jaeger (1959, p. 100):

$$T(z, t) = T_0 + (T_1 - T_0)\frac{z}{d} + \frac{2}{\pi}\sum_{n=1}^{\infty}\frac{1}{n}[(-1)^n T_1 + (-1)^{n+1}(T_0 + d \text{ grad } T)] \sin\left(\frac{n\pi z}{d}\right)\exp\left(-\frac{an^2\pi^2 t}{d^2}\right), \tag{9.39}$$

where T_0 = mean annual temperature at the Earth's surface and grad T = undisturbed temperature gradient at $t \leq 0$. The temperature gradient is given by differentiating of Equation (9.39):

$$\frac{\partial T}{\partial z} = \frac{T_1 - T_0}{d} + \frac{2}{d}\sum_{n=1}^{\infty}[(-1)^n T_1 + (-1)^{n+1}(T_0 + d \text{ grad } T)] \cos\left(\frac{n\pi z}{d}\right)\exp\left(-\frac{an^2\pi^2 t}{d^2}\right). \tag{9.40}$$

The heat-flow density $q(z, t)$ is given by multiplying Equation (9.40) with thermal conductivity λ. Because the variation of heat-flow density with depth cannot at present be determined accurately enough, only the heat-flow density at the Earth's surface ($z = 0$) will be considered (this corresponds to heat-flow density determination near the Earth's surface):

$$q(0, t) = q_0(t) = \lambda\frac{\partial T}{\partial z} = \lambda\left[\frac{T_1 - T_0}{d} + \frac{2}{d}\sum_{n=1}^{\infty}[(-1)^n T_1 + (-1)^{n+1}(T_0 + d \text{ grad } T)]\exp\left(-\frac{an^2\pi^2 t}{d^2}\right)\right], \tag{9.41}$$

For the case where only the change $\Delta q(0, t)$ of heat-flow density is now considered, we take:

$$T_0 = 0, \text{ grad } T = 0, \ \Delta T = T_1 - T_0$$

then (see also Carslaw and Jaeger, 1959, Equation (10) on p. 313):

$$\Delta q_0(0, t) = \frac{\lambda \Delta T}{d}\left[1 + 2\sum_{n=1}^{\infty}(-1)^n \exp\left(-\frac{an^2\pi^2 t}{d^2}\right)\right], \tag{9.42}$$

and for steady state:

$$\Delta q_0(0, t = \infty) = \frac{\lambda \Delta T}{d}. \tag{9.43}$$

For more realistic conditions it is better to approximate a finite intrusive body by a rectangular parallelepiped. Again, the corresponding formulae can be derived as described by Carslaw and Jaeger (1959, pp. 56–62), see also Equation (9.28). At any point and time the temperature is:

$$\begin{aligned} T(x, y, z, t) = T_0 + z \text{ grad } T + \frac{\Delta T}{8}&\left[\operatorname{erf}\left(\frac{z-b}{2\sqrt{at}}\right) + \operatorname{erf}\left(\frac{z+b}{2\sqrt{at}}\right)\right. \\ &\left. - \operatorname{erf}\left(\frac{z-c}{2\sqrt{at}}\right) - \operatorname{erf}\left(\frac{z+c}{2\sqrt{at}}\right)\right]\left[\operatorname{erf}\left(\frac{h-y}{2\sqrt{at}}\right)\right. \\ &\left. + \operatorname{erf}\left(\frac{h+y}{2\sqrt{at}}\right)\right]\left[\operatorname{erf}\left(\frac{d-x}{2\sqrt{at}}\right) + \operatorname{erf}\left(\frac{d+x}{2\sqrt{at}}\right)\right], \end{aligned} \tag{9.44}$$

the gradient is:

$$\begin{aligned} \frac{\partial T(x, y, z, t)}{\partial z} = \text{grad } T + \frac{\Delta T}{8\sqrt{\pi a t}}&\left[\exp\left(-\left(\frac{z-b}{2\sqrt{at}}\right)^2\right) + \exp\left(-\left(\frac{z+b}{2\sqrt{at}}\right)^2\right)\right. \\ &\left. - \exp\left(-\left(\frac{z-c}{2\sqrt{at}}\right)^2\right) - \exp\left(-\left(\frac{z+c}{2\sqrt{at}}\right)^2\right)\right]\left[\operatorname{erf}\left(\frac{h-y}{2\sqrt{at}}\right)\right. \\ &\left. + \operatorname{erf}\left(\frac{h+y}{2\sqrt{at}}\right)\right]\left[\operatorname{erf}\left(\frac{d-x}{2\sqrt{at}}\right) + \operatorname{erf}\left(\frac{d+x}{2\sqrt{at}}\right)\right], \end{aligned} \tag{9.45}$$

and the heat-flow density at the Earth's surface. $z = 0$, is:

$$\begin{aligned} q_0(x, y, 0, t) = \lambda\frac{\partial T}{\partial z} = \lambda \text{ grad } T + \frac{\lambda \Delta T}{4\sqrt{\pi a t}}&\left[\exp\left(-\frac{b^2}{4at}\right) - \exp\left(-\frac{c^2}{4at}\right)\right] \cdot \\ &\left[\operatorname{erf}\left(\frac{h-y}{2\sqrt{at}}\right) + \operatorname{erf}\left(\frac{h+y}{2\sqrt{at}}\right)\right]\left[\operatorname{erf}\left(\frac{d-x}{2\sqrt{at}}\right) + \operatorname{erf}\left(\frac{d+x}{2\sqrt{at}}\right)\right]. \end{aligned} \tag{9.46}$$

At the centre above the body, and with only the change in heat-flow density (grad T

= 0) taken into consideration:

$$\Delta q_0(0, 0, 0, t) = \frac{\lambda \Delta T}{\sqrt{\pi a t}} \left[\exp\left(-\frac{b^2}{4at}\right) - \exp\left(-\frac{c^2}{4at}\right) \right] \cdot \left[\operatorname{erf}\left(\frac{d}{2\sqrt{at}}\right) \operatorname{erf}\left(\frac{h}{2\sqrt{at}}\right) \right] \tag{9.47}$$

where

a = thermal diffusivity, $m^2\ s^{-1}$
λ = thermal conductivity, $W\ m^{-1}\ K^{-1}$
b = depth to the top surface of the intrusion, m
c = depth to its base, m
d, h = half-axes of lateral extension, m
x, y = horizontal coordinates.
$\Delta T = T_i - 0$, where T_i is the intrusion temperature.

The effect of latent heat may be included by setting $\Delta T = \Delta T + L/c$, if crystallisation commences at $T(x, y, z, t)$ or $q_0(x, y, t)$ and $\Delta q_0(t)$, respectively; L = latent heat of crystallisation, and c = specific heat capacity of the magma.

In Figure 9.23 the sudden intrusion of a magma body is to have an infinite surface of the body in one case, and a finite, but large extension in the other case. The temperature difference ΔT at great depth is small; e.g., at a depth of 80 km, the temperature is about 1000 °C, and the intrusive body has temperatures between 1100 and 1400 °C.

From Figure 9.23 it is seen that temperature changes from 40–80 km depth can be recognized only if the body has a (nearly) infinite extension, and if geologically stable conditions prevail over more than 10^7 to 10^8 a. Information cannot be expected from depths below 100 km. On the other hand, if the intrusive body has a finite extension, the temperature change ΔT can be recognised only within a relatively small time span; the deeper the body, the smaller the time range. A body deeper than about 40–80 km cannot be recognised. Furthermore, the inaccuracy of the determined heat-flow density is about 10–20 %, corresponding to 4–20 mW m^{-2}. Therefore, the indication of an infinite body is limited to a depth of approximately 50 km, and in the case of a finite, extended body, to about 20 km.

Numerical Methods. The numerical methods are an important aid for calculating and simulating conditions which cannot be described by analytical solutions. The finite difference method as well as the finite element method are discussed extensively in Section 2.4.

As an example, we refer the readers to the results of Mundry (1970) for a two-dimensional irregular body in a transient state, taking into account the latent heat of solidification.

(4) INTERPRETATION OF SECONDARY SOURCES

Top Aquifer

In addition to the primary source, usually a hot-water system, and less frequently

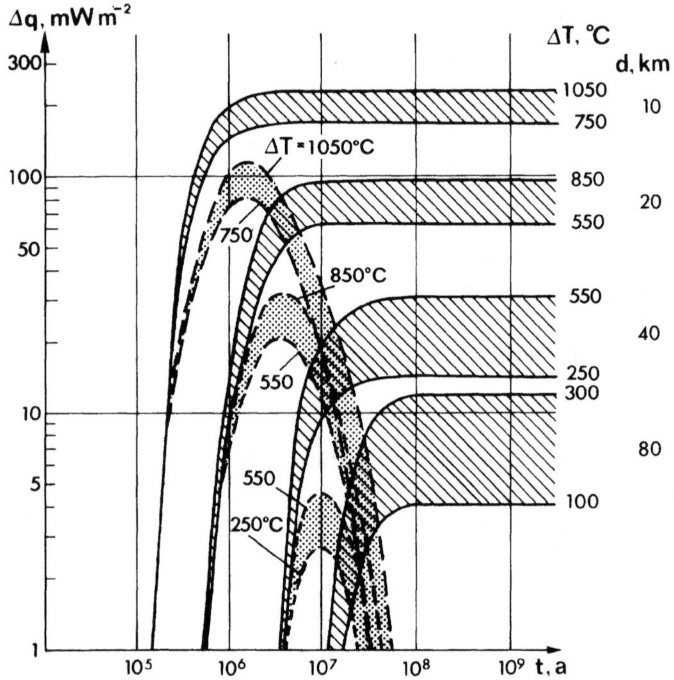

Fig. 9.23. The increase of heat-flow density at the Earth's surface, assuming an infinite horizontal plate of thickness d (———) with continuous heat supply at the lower surface using Equation (9.42), and an intrusive body of 50 × 50 × 30 km, length × width × height (– – –) with no continuous heat supply using Equation (9.47). In both cases the temperature increased suddenly by ΔT at depth d, and the parameters used are:

d = 80 km; $\lambda = 3.25$ Wm^{-1} K^{-1}, $c = 850$ J kg^{-1} K^{-1}, $\rho = 3.1 \times 10^3$ kg m^{-3}, $a = 39$ m^2 a^{-1},

d = 10, 20, 40 km: $\lambda = 2.25$ Wm^{-1} K^{-1}, $c = 850$ J kg^{-1} K^{-1}, $\rho = 2.85 \times 10^3$ kg m^{-3}, $a = 29$ m^2 a^{-1}.

a vapour-dominated system, exists (Figure 9.20). For a vapour-dominated system a deep water table can be expected (Renner *et al.*, 1975). Fluid motion in hot-water systems is usually controlled by the hydraulic pressure gradient and intrinsic permeability of the aquifers. The fluid system has a recharge and a discharge zone: meteoric water enters the aquifer mostly through permeable rocks or along faults, and after circulating in the aquifer it reaches the surface in a column of relatively small cross-section. In many cases, the aquifer also may be invaded from below or along deep faults. The convected heat accelerates the cooling of a magma intrusion, sometimes substantially.

In the simplest case, the Earth's surface and the upper surface of a confined aquifer are horizontal with constant temperature T_0 and T_2 respectively. Urban *et al.* (1976) have studied this one-dimensional problem and calculated the evolution of the temperature in the cover. The initial and boundary conditions are:

$T(0, t) = T_0$	at the surface
$T(d, t) = T_2$	at the reservoir top, d is the reservoir depth
$T(z, 0) = T_0 + z \cdot (T_1 - T_0)/d$	$(T_1 - T_0)/d$ is the undisturbed initial gradient.

The solution is obtained from Equation (9.39). Figure 9.24 shows the results obtained by taking $T_0 = 15\,°C$, $T_1 = 240\,°C$, the initial gradient grad $T = 32$ K km^{-1}, $d = 1.5$ km. It can be seen that the steady state is rapidly attained.

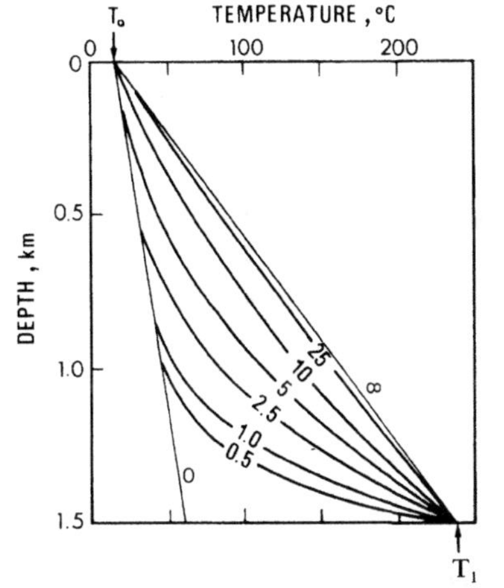

Fig. 9.24. Evolution of the temperature profile above a geothermic reservoir. Curve parameter: time in 10^3 a after sudden formation of a steam zone (240°C at depth of 1.5 km; from Urban *et al.*, 1976).

Fault, Step

Model 1. A surface anomaly can also be expected from moving water, which flows in the x direction at a depth $d + d_1$ within an aquifer of thickness h and which suddenly ascends along a fault (which lies in the y, z plane) into the uplifted aquifer at depth d (see Figure 9.25). The temperature in the positive direction x at the top of the aquifer, and the resulting heat-flow density are thereby sought.

Starting with Equation (2.17):

$$a\nabla^2 T = \frac{\partial T}{\partial t} + \mathbf{v}_f \nabla T,$$

the thermal effect of the water flowing through a porous rocks is given by:

$$\lambda_r \nabla^2 T = (\rho_r c_r (1 - P) + \rho_w c_w P)\frac{\partial T}{\partial t} + \rho_w c_w \mathbf{v} P \nabla T, \qquad (9.48)$$

where

$\mathbf{v}_f$ = Darcy velocity, filtration velocity, $\mathbf{v}_f = \mathbf{v}P$
$\mathbf{v}$ = pore velocity, distance velocity
c = specific heat capacity
ρ = density
P = porosity
r = refers to rock
w = refers to water.

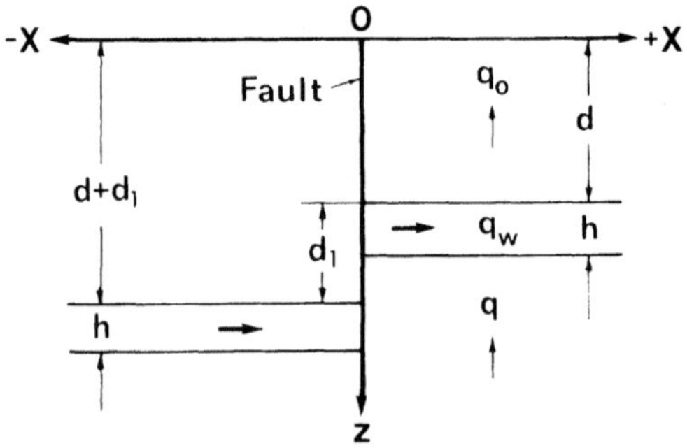

Fig. 9.25. Water flow at a higher level caused by a fault.

The heat transfer by conduction in direction (x, y) within the water can be neglected compared with the heat transport by fluid movement, and the water velocity v acts only in direction x; $v_x = v$. For the steady state case, Equation (9.48) yields:

$$\lambda_r \frac{\partial^2 T}{\partial z^2} = v\rho_w c_w P \frac{\partial T}{\partial x}. \tag{9.49}$$

Introducing the Fourier equation for the z component, $q_z = q$, it gives:

$$q = -\lambda_r \frac{\mathrm{d}T}{\mathrm{d}z}. \tag{9.50}$$

Differentiating Equation (9.50) and inserting in Equation (9.49), we get:

$$-\frac{\partial q}{\partial z} = v\rho_w c_w P \frac{\partial T}{\partial x}. \tag{9.51}$$

This equation describes the vertical change of heat-flow density within the water, caused by the water flow. The mass-flow rate of the water is $m = v\rho_w P$ in kg m^{-2} s^{-1}. As an approximation we can write:

$$\left| -\frac{\partial q}{\partial z} \right| = \frac{q_w}{h}, \tag{9.52}$$

and:

$$q_x = hmc_w \frac{\mathrm{d}T}{\mathrm{d}x}. \tag{9.53}$$

In the positive x-region the heat-flow density is given by:

$$q_0 = q_w + q \tag{9.54}$$

$$\lambda_r \frac{\Delta T}{d} = hmc_w \frac{\mathrm{d}T}{\mathrm{d}x} + \lambda_r \text{ grad } T, \tag{9.55}$$

where

grad T = undisturbed temperature gradient, K m^{-1}
$\Delta T = T_d - T_0$
T_d = temperature at depth d, °C
T_0 = temperature at the Earth's surface, °C.

Rearranging the last equation:

$$T_d = T_0 + d \text{ grad } T + \frac{hmc_w d}{\lambda_r}\frac{\mathrm{d}T}{\mathrm{d}x}, \tag{9.56}$$

substituting:

$$A = hmc_w d/\lambda_r \tag{9.57}$$

it follows:

$$\mathrm{d}x = -\frac{A}{T_0 - T_d + d \text{ grad } T}\mathrm{d}T; \tag{9.58}$$

integrating yields:

$$T_d(x) = T_0 + d \text{ grad } T - B \exp(-x/A). \tag{9.59}$$

The temperature at depth $d + d_1$ and at $x = 0$ is given by:

$$T_d(o) = T_0 + (d + d_1) \text{ grad } T; \tag{9.60}$$

from this equation and Equation (9.59) at $x = 0$ it also follows that:

$$B = - d_1 \text{ grad } T, \tag{9.61}$$

and for the positive region at the top of the aquifer:

$$T_d(x) = T_0 + d \text{ grad } T + d_1 \text{ grad } T \exp(-x\lambda_r/hmc_w d) \tag{9.62}$$

and the heat-flow density is given by:

$$q_0(x) = \lambda_r\left(\frac{T_d(x) - T_0}{d}\right). \tag{9.63}$$

The result allows an interpolation between T_0 ($z = 0$) and T_d ($z = d$), and can be compared with measurements in shallow boreholes.

EXAMPLE: Given $d = 500$ m, $d_1 = 100$ m, $h = 5$ m, $T_0 = 10$°C, grad $T = 5 \times 10^{-2}$ K m^{-1}, $c_w = 4200$ J kg^{-1} K^{-1}, $\lambda_r = 2.4$ W m^{-1} K^{-1}, and in direction x within a unit strip of 1 m width ($\pm y = \pm 0.5$ m) the mass flow rate per m^2 is given by:

$m = 1 \times 10^{-4}$ kg m^{-2} s^{-1}. The results are $1/A = 2.29 \times 10^{-3}$ m^{-1} and for $T_d(x)$ from Equation (9.62), $q_0(x)$ from Equation (9.63), and $q = \lambda_r$ grad T follows:

x in m	0	1	100	500	1000
T_d in °C	40	40	39	36.6	35.5
q_0/q	1.2	1.2	1.16	1.07	1.02

Model 2

In the case where the reservoir is affected by a fault, its upper surface represents a step. The Earth's surface temperature T_0 and the geometry of the structure are assumed to be known; the depths of the upper and lower sections of the reservoir top, respectively, are indicated by h_u and h_l. Let us consider the case of *constant temperature* T_s at the reservoir surface. Moreover, we assume the vertical temperature trend far from the step is linear:

$$T = T_0 + z \cdot (T_s - T_0)/h_u \quad 0 \leq z \leq h_u \tag{9.64}$$

$$T = T_0 + z \cdot (T_s - T_0)/h_l \quad 0 \leq z \leq h_l. \tag{9.65}$$

The solution to the steady-state two-dimensional Fourier equation for obtaining the temperature in the cover can be found by the finite difference or finite element method (see Section 2.4.). T_s is varied until the best fit between calculated and measured temperature gradients near the surface is obtained.

Figure 9.26 shows the results obtained by assuming T_0 = 15 °C and T_s = 220 °C

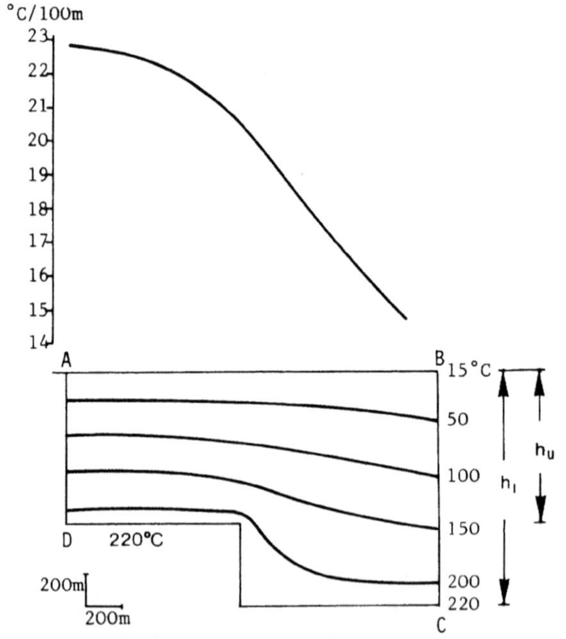

Fig. 9.26. Calculated temperatures and surface temperature gradient for the step.

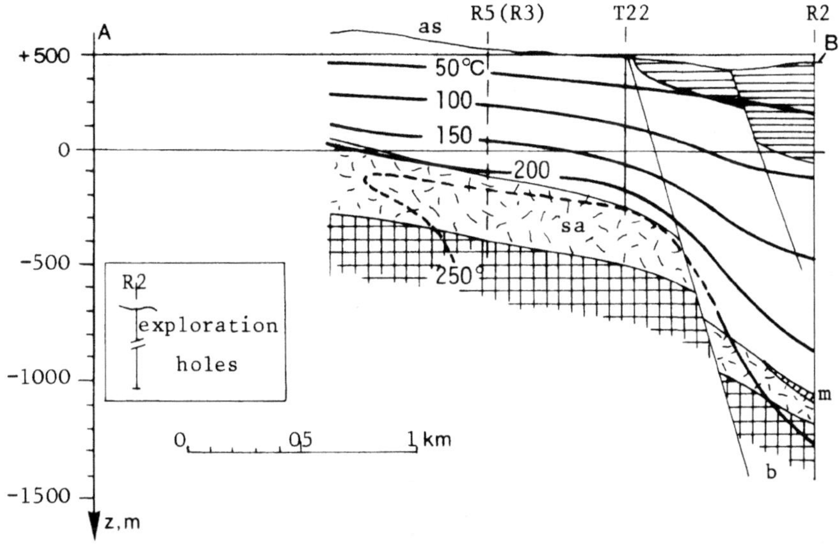

Fig. 9.27. Temperature distribution in °C around a fault of the Travale field; as = shaley clays, m = sandstone (macigno), sa = anhydritic series, b = Palaeozoic basement.

for the reservoir; they fit the temperatures observed over a fault of the Travale field, Italy, as shown in Figure 9.27 (Galeone and Mongelli, 1982). In correspondence with the observed or calculated gradient, with a conductivity of 1.68 W m^{-1} K^{-1}, the assumed value for the clays, the heat-flow density ranges between 250 and 380 mW m^{-2} with a local anomaly of only 125 mW m^{-2}.

This procedure can be easily extended to the cases of variable temperature on the reservoir top.

Dipping Surface (Wedge)

Consider the case where the reservoir top is a plane surface sloping from the recharge zone. We can assume a constant, known temperature on the Earth's surface and a temperature increasing exponentially along the sloping surface and asymptotically attaining a temperature of geothermal interest T_s. Moreover, to close the wedge of Figure 9.28A we assume a constant gradient on the side BC and denote by ξ the distance measured on AC, starting from A. In this way, we have the boundary conditions:

$$\text{on AB} \quad T = T_0 \tag{9.66}$$

$$\text{BC} \quad T = T_0 + z(T_s - T_0)/h \tag{9.67}$$

$$\text{AC} \quad T(\xi) = T_s - b \exp(-c\xi). \tag{9.68}$$

If the geometrical structure is known, changing the parameters T_s, b and c allows different near surface temperatures or gradients to be fitted to measured values.

Figure 9.28A shows the results obtained for the geometry sketched, assuming:

$$T_0 = 15\,°\mathrm{C},\ T_s = 250\,°\mathrm{C},\ T(\xi) = 250 - 235\, e^{-c\xi},\ c = 0.00121,\ \xi \text{ in km}.$$

This model was tested by comparison with the data observed on the Travale field (Galeone and Mongelli, 1981); see Figure 9.28B.

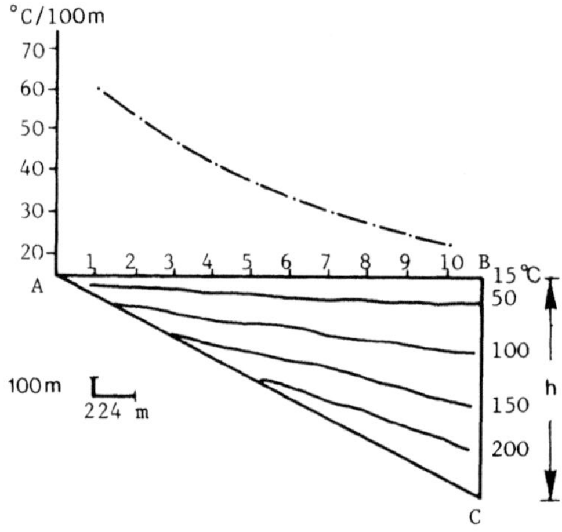

Fig. 9.28A. Calculated temperatures and surface temperature gradient for the edge.

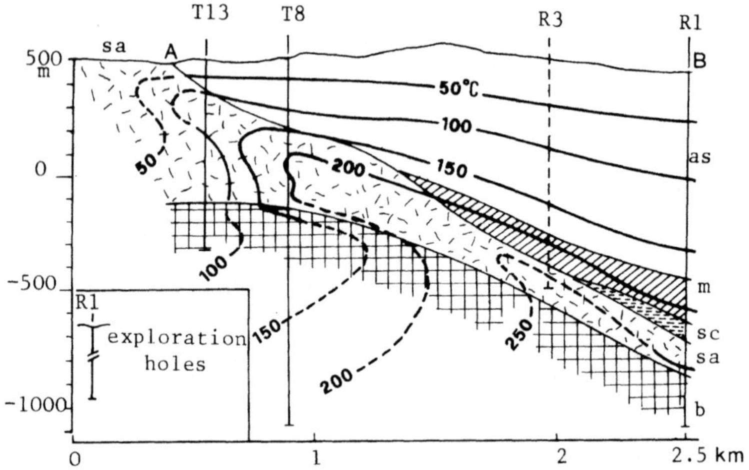

Fig. 9.28B. Temperature distribution in the cover and reservoir of the Travale field in °C; as = shale clays, m = sandstones (macigno), sc = shales, sa = anhydritic series b = Palaeozoic basement (Galeone and Mongelli, 1981).

(5) HOT-DRY ROCKS

The hot-dry-rock concept is based on the idea of independence of location for extracting heat.

The exploration of a hot-dry-rock region is quite similar to that for high- or low-enthalpy areas. In the case of high-enthalpy areas, the boreholes necessary for exploitation are shallower than those for exploitation in areas of low enthalpy or normal temperature distribution. Thus, the formulae already presented can be applied.

The heat exchange from rocks into the water-filled fracture also perturbs the heat conductance. Of interest are the temperature, pressure, and flow distributions within a fracture, as well as the temperature of the extracted water, as shown in Figure 9.29; see also Section 9.4.3.(2). Therefore, only references will be given here: Lawton (1974), Lawton *et al.* (1976), Milora and Tester (1976), Bodvarsson (1969) and Rodemann (1979, 1982).

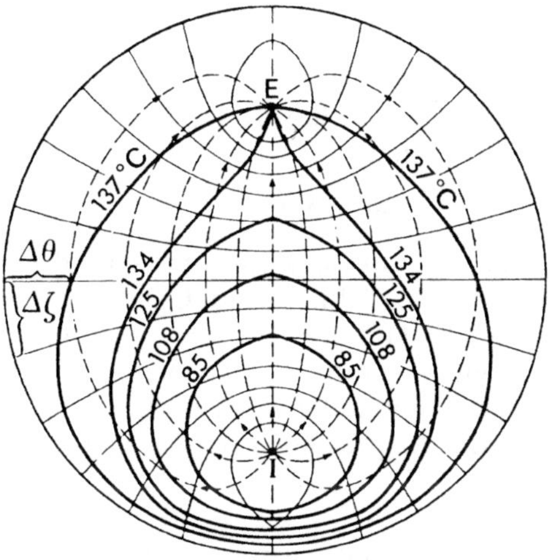

Fig. 9.29. The distribution of pressure (——), fluid flow (– – –) and temperature ——) for conditions as given in the Urach hot-dry-rock project after three days of injection after Rodemann (1982). The following model parameters have been used (I = point of injection, E = point of extraction):

injection temperature:	60 °C
rock temperature:	140 °C
fracture radius:	50 m
half-distance between boreholes:	31.25 m
flow rate:	$3\ l\ s^{-1}$
water viscosity:	$0.3\ 10^{-3}$ Pa s (100 °C)
rock density (gneiss):	$2700\ kg\ m^{-3}$
specific heat capacity of rock:	$840\ J\ kg^{-1}\ K^{-1}$
thermal conductivity:	$2.7\ W\ m^{-1}\ K^{-1}$
specific heat capacity of water:	$4180\ J\ kg^{-1}\ K^{-1}$
density of water:	$10^3\ kg\ m^{-3}$
fracture width:	10^{-3} m
$\Delta\xi$	0.01 bar
$\Delta\theta$	$0.25\ l\ s^{-1}$.

(6) REGIONAL HEAT-FLOW DENSITY ESTIMATES FROM SILICA GEOTHERMOMETRY

An empirical, linear relationship between heat-flow density (HFD) and the silica geothermometer value of groundwater T_s, established by Swanberg and Morgan (1979, 1980), enables to estimate the regional HFD q. The empirical relationship between q and T_s has the form:

$$T_s = mq + b, \qquad (9.69)$$

where the constants m and b are related to circulation depth of groundwater and to the mean annual surface temperature, respectively. Swanberg and Morgan (1981) obtained:

$$m = 675\,°\mathrm{C\ m^2\ W^{-1}} \quad \text{and} \quad b = 12.8\,°\mathrm{C}$$

by statistical analysis of data from the USA. For other areas of the world the b value should be selected according to the local mean surface temperature T_0.

The revised silica geothermometer equation (Fournier and Potter, 1982) is:

$$T_s = 1309/(5.19 - \log SiO_2) - 273.15, \qquad (9.70)$$

where SiO_2 is the concentration of dissolved silica in groundwater in ppm. Thus HFD can be calculated as follows:

$$q = 1.48(T_s - T_0), \text{ in mW m}^{-2} \qquad (9.71)$$

The method yields reliable results as long as the data density exceeds 1×10^{-2} km^{-2} (one data point per 10 × 10 km quadrangle).

Swanberg and Morgan (1981) compared 'conventional' and 'estimated' regional HFD patterns and reported satisfactory agreement between them. Another example of applying the groundwater silica geothermometer for regional HFD mapping is the Geothermal Map of Mexico (Prol and Juarez, 1986).

9.2.3. Non-Energy Exploration

(1) GENERAL REMARKS

Geothermal methods are suitable not only for exploration of geothermal energy sources, but also for:

- detection of subsurface bodies with different thermal conductivity;
- detection of subsurface bodies representing (positive or negative) heat sources, such as oxidising ores, radioactive deposits, and aquifers.

The change of temperature and heat-flow density caused by the geometric pattern of rocks with different conductivity is generally low, because of the small conductivity contrast occurring in nature. The elimination of disturbing factors (topography, vegetation differences etc.) is therefore necessary. Because of the high specific heat capacity of the water, its high mobility and the low thermal conductivity of the rocks, water conserves a part of its initial temperature distribution during its migration and, therefore, differs thermally from its environment.

(2) GEOLOGICAL STRUCTURES

Layers with different thermal conductivity cause a change in heat-flow density and temperature. In particular horst and graben structures, as well as salt domes exhibit thermal focusing effects (chimney effect) or anti-focusing effects (dispersion effect), see e.g. Giesel and Holz (1970) as well as Section 9.4.2.(2).

For a semi-infinite composite solid, as shown in Figure 9.30, a solution can be derived.

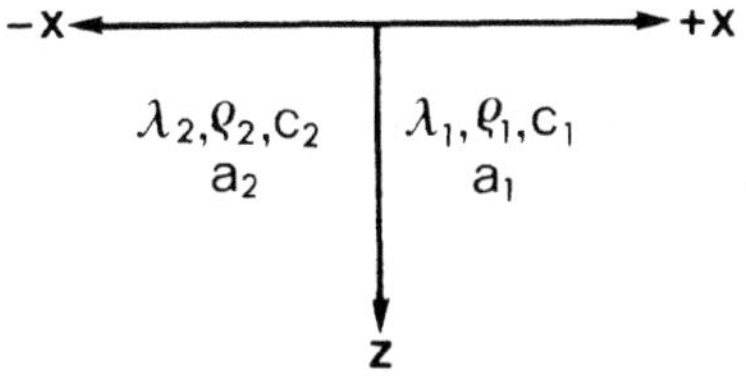

Fig. 9.30. A semi-infinite composite solid.

At the Earth's surface, the mean annual temperature T_0 is given, and with heat production neglected, the following equation is valid for $x \rightarrow \pm\infty$:

$$T(z) = T_0 + z \text{ grad } T, \tag{9.72}$$

with the boundary condition at the plane of separation, $x = 0$, $t > 0$:

$$T_1 = T_2 = T_m \tag{9.73}$$

$$\lambda_1 \left(\frac{\partial T_1}{\partial x}\right)_n = \lambda_2 \left(\frac{\partial T_2}{\partial x}\right)_n, \tag{9.74}$$

where n indicates the gradient perpendicular to the plane separating both solids. A general solution is given by Carslaw and Jaeger (1959, p. 88):

$$x > 0\text{: } T(x, t) = A_1 + B_1 \operatorname{erf}(x/(4a_1t)^{1/2}) \tag{9.75}$$

$$x < 0\text{: } T(x, t) = A_2 + B_2 \operatorname{erf}(x/(4a_2t)^{1/2}). \tag{9.76}$$

The Equations (9.75) and (9.76) must also be satisfied:

$$x = 0\text{: } T(x, z, t) = T_m, \quad \text{which leads to } A_1 = A_2 = T_m \tag{9.77}$$

$$x \rightarrow +\infty\text{: } T(x, z, t) = T_0 + z \text{ grad}_1\, T, \quad \text{which leads to } A_1 + B_1 = T_0 + z \text{ grad}_1\, T \tag{9.78}$$

$$x \rightarrow -\infty\text{: } T(x, z, t) = T_0 + z \text{ grad}_2\, T, \quad \text{which leads to } A_2 + B_2 = T_0 + z \text{ grad}_2\, T. \tag{9.79}$$

Substituting in Equations (9.73) and (9.74) yields:

$$T_m = \frac{b_1(T_0 + z \text{ grad}_1\, T) + b_2(T_0 + z \text{ grad}_2\, T)}{b_1 + b_2} \tag{9.80}$$

with the thermal inertia:

$$b_1 = (\lambda_1\rho_1c_1)^{1/2}, \quad b_2 = (\lambda_2\rho_2c_2)^{1/2}. \tag{9.81}$$

The final solution is found by inserting Equation (9.80) into Equations (9.75) and (9.76):

$$x > 0\colon T(x, z, t) = \frac{b_1 T_{a1} + b_2 T_{a2}}{b_1 + b_2} - \left(\frac{b_1 T_{a1} - b_2 T_{a2}}{b_1 + b_2} - T_{a1}\right) \mathrm{erf}\left(\frac{x}{\sqrt{4a_1 t}}\right) \tag{9.82}$$

$$\mathrm{grad}_{+x} T = \frac{\partial T(x, z, t)}{\partial z}$$

$$= \frac{b_1 T_{b1} + b_2 T_{b2}}{b_1 + b_2} - \left(\frac{b_1 T_{b1} + b_2 T_{b2}}{b_1 + b_2} - T_{b1}\right) \mathrm{erf}\left(\frac{x}{\sqrt{4a_1 t}}\right) \tag{9.83}$$

$$q_{+x} = \lambda_1 \,\mathrm{grad}_{+x} T \tag{9.84}$$

and

$$x < 0\colon T(x, z, t) = \frac{b_1 T_{a1} + b_2 T_{a2}}{b_1 + b_2} - \left(T_{a2} - \frac{b_1 T_{a1} + b_2 T_{a2}}{b_1 + b_2}\right) \mathrm{erf}\left(\frac{x}{\sqrt{4a_2 t}}\right) \tag{9.85}$$

$$\mathrm{grad}_{-x} T = \frac{\partial T(x, z, t)}{\partial z}$$

$$= \frac{b_1 T_{b1} + b_2 T_{b2}}{b_1 + b_2} - \left(T_{b2} - \frac{b_1 T_{b1} + b_2 T_{b2}}{b_1 + b_2}\right) \mathrm{erf}\left(\frac{x}{\sqrt{4a_2 t}}\right) \tag{9.86}$$

$$q_{-x} = \lambda_2 \,\mathrm{grad}_{-x} T \tag{9.87}$$

with the abbreviations:

$$\begin{aligned} T_{a1} &= T_0 + z\,\mathrm{grad}_1\, T, & T_{a2} &= T_0 + z\,\mathrm{grad}_2\, T \\ T_{b1} &= \mathrm{grad}_1\, T, & T_{b2} &= \mathrm{grad}_2\, T. \end{aligned} \tag{9.88}$$

If the initial temperature is $T(z) = T_0 + z \,\mathrm{grad}\, T$ for $x > 0$ and $T_c = \text{const.}$ for $x < 0$ (which could be approximately given along a coast line, where the water temperature will be constant), it leads to the same equations as Equations (9.82) to (9.87), and one has only to replace the abbreviations (9.88) by:

$$\begin{aligned} T_{a1} &= T_0 + z\,\mathrm{grad}_1\, T, & T_{a2} &= T_c \\ T_{b1} &= \mathrm{grad}_1\, T, & T_{b2} &= 0. \end{aligned} \tag{9.89}$$

A practical example is given in Figure 9.31 by Andl (1982).

(3) SUBSURFACE WATER

Local or semi-regional temperature and heat-flow differences are often caused by water percolation. The exploration methods, as well as the exploration sequences for subsurface water are in principle similar to those for high-enthalpy fluids, as described in this Section 9.2.3.

Aquifer Temperature Determination by Means of Springs

For planning the measurement depth and for getting a first idea of the anomalous behaviour of an aquifer, the existing springs may constitute a helpful tool. By means of geochemical thermometers (Fournier, 1981), an estimation of the aquifer

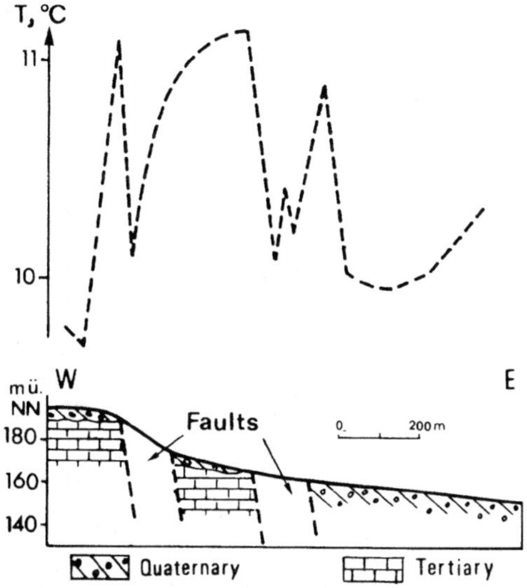

Fig. 9.31. Temperature change, as follows from the change in geothermal parameters; depth of boreholes: 2 m.

temperature is possible. Indications of the aquifer temperature can also be derived from the highest spring temperature, by assuming that this is a lower limit for the temperature of the water ascending from the depths. From the geological situation, an estimation of the aquifer depth d is usually possible. Moreover, in the case where the undisturbed temperature field (T_0, grad T) is known from outside the area under consideration, it follows from Equation (9.43), for steady-state conditions that:

$$\triangle q_0 = \lambda \triangle T/d,$$

where

$\triangle T$ = temperature contrast between aquifer temperature and temperature outside the anomalous area at the same depth d,
λ = thermal conductivity
$\triangle q_0$ = heat-flow density increase caused by the aquifer.

Area of Heated Spring Water

An estimate of the necessary extension of the surface F at which the heat transfer between the rocks and the aquifer takes place is given by Kappelmeyer (1957), see Figure 9.32:

$$F = r^2\pi = \frac{cQ_m(d + (\lambda/\alpha))}{\lambda} \ln\left[q/\left(q - \frac{\lambda}{d + (\lambda/\alpha)}(T_2 - T_1)\right)\right] \qquad (9.90)$$

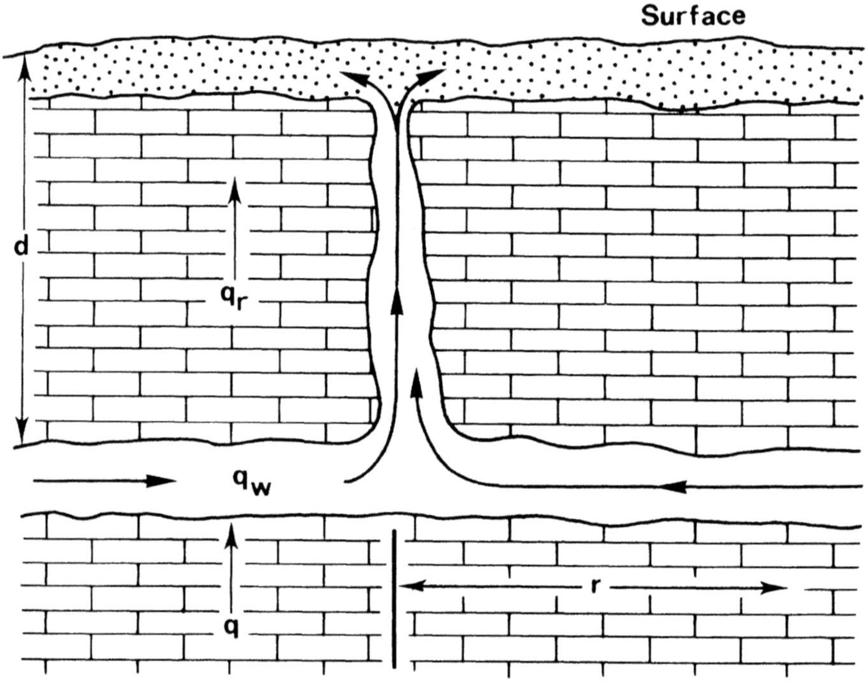

Fig. 9.32. Sketch illustrating the heating of subsurface water.

and because $\alpha \gg \lambda$, we can write:

$$r = \left\{\frac{cdQ_m}{\lambda\pi} \ln\left[q/\left(q - \frac{\lambda(T_2 - T_1)}{d}\right)\right]\right\}^{-2}, \quad q > \frac{\lambda(T_2 - T_1)}{d} \tag{9.91}$$

where

c = specific heat capacity of fluid, J kg^{-1} K^{-1}
d = depth to the top surface of the aquifer, m
Q_m = water yield, kg s^{-1}
λ = thermal conductivity of rocks above the aquifer, W m^{-1} K^{-1}
q = heat-flow density below the aquifer, W m^{-2}
q_w = heat-flow density absorbed from the aquifer, W m^{-2}
$q_r = q - q_w$
T_1 = temperature of the water when it enters the aquifer at distance r, °C
T_2 = temperature of the water leaving the aquifer for entering the Earth's surface, °C
r = radius of recharge area, m
α = heat transfer coefficient between rocks and fluid, W m^{-2} K^{-1}.

Corrections

A further step in exploration might be to conduct soil temperature measurements in shallow boreholes with depths of 1 to 10 m, usually arranged in a grid. The grid spacing depends on the objective, and ranges between 10 and 50 m; see Kappelmeyer and Haenel (1974), LeSchack and Lewis (1983). Temperatures are usually measured at the bottom of the boreholes only; sometimes they are also measured at half the borehole depth. The results from shallow boreholes are very sensitive to near-surface influences; these effects have to be considered, and if possible, eliminated.

Possible disturbances are: the annual (air) temperature (wave) ($\triangle T_t$), the variation with topographic height ($\triangle T_h$), the inclination of a slope ($\triangle T_i$), the position of a slope with regard to incoming solar radiation ($\triangle T_p$), the different thermal conductivities ($\triangle T_\lambda$), and the different vegetation ($\triangle T_v$). The influence of water moving by capillarity and evaporation will not be considered, because of the complicated heat transport mechanism.

In the extreme case all corrections are necessary:

$$\triangle T_c = \triangle T_t + \triangle T_h + \triangle T_i + \triangle T_p + \triangle T_\lambda + \triangle T_v. \tag{9.92}$$

If the exploration of an area requires several days or weeks, the soil temperatures will be influenced by the penetration of long period changes such as annual temperatures. Therefore, reference boreholes (reference stations) are necessary, in order to allow temperature registrations to be carried out at the depth of measurements in other boreholes. All measured temperatures must be mutually related for a comparison to a certain day on which the temperature T_{ref} at the reference station is measured. The correction amounts to (see Figure 9.33A):

$$\triangle T_t = T_m \pm T_{\text{ref}}. \tag{9.93}$$

The topographic correction according to Section 5.4. or obtained by the finite difference method is quite tedious for the numerous boreholes. Therefore, the problem is simplified and split into the influences of topographic height ($\triangle T_h$) and surface inclination ($\triangle T_i$). The correction for $\triangle T_h$ (see Figure 9.33B) for borehole 2 is given by:

$$\triangle T_h = -h \,\text{grad}_a\, T, \tag{9.94}$$

where

h = height with respect to a reference height, m
$\text{grad}_a\, T$ = temperature gradient of air above the Earth's surface (positive downwards), K m^{-1}

The effect of inclination of the Earth's surface is given by (see Figure 9.33C):

$$\triangle T_i = z(1 - \cos \alpha) \,\text{grad}\, T, \tag{9.95}$$

where

z = depth of measurement, m
α = inclination of slope, degree
grad T = undisturbed subsurface temperature gradient measured in the borehole, K m^{-1}.

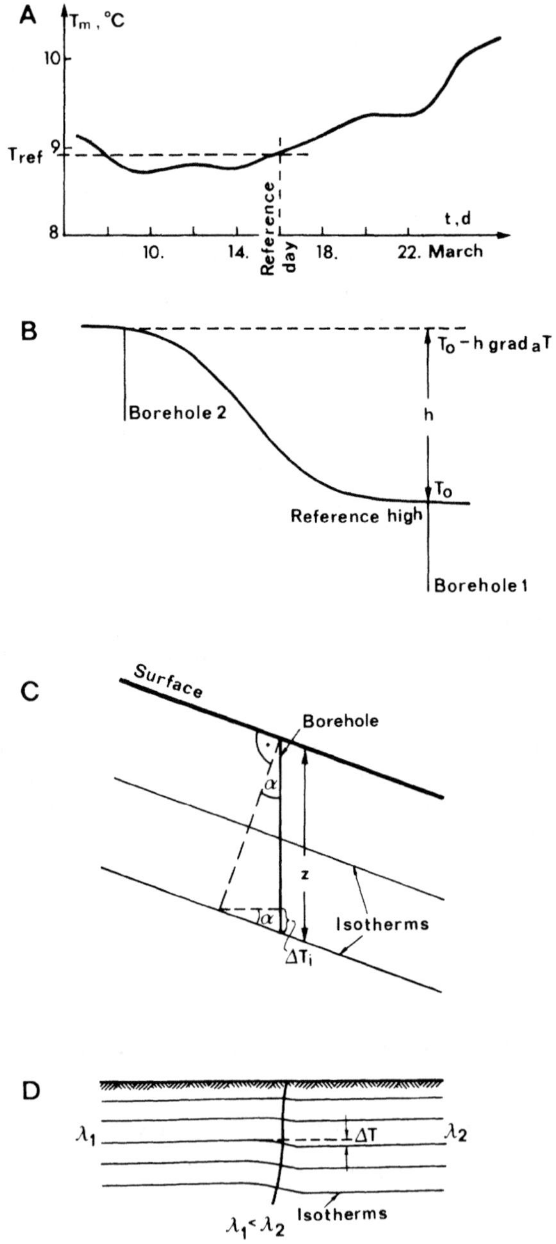

Fig. 9.33. Temperature corrections for shallow boreholes. (A) Determination of reference temperature T_{ref}. (B) Simplified correction for topographic height. (C) Correction for a sloping surface. (D) Effect of thermal conductivity caused by two different rocks or soils.

For an area of exploration with slopes in different orientations to the incident solar radiation, the effect of inclination has to be eliminated with respect to a reference area, usually a flat area. It must be taken into consideration that the flow density of radiation is proportional to the sine of the angle of incidence.

The effect of differences in the conductivity of soils can be estimated by means of:

$$T = T_0 + z \text{ grad } T = T_0 + qz/\lambda \tag{9.96}$$

and (see Figure 9.33D):

$$\triangle T_\lambda = \left(T_0 + \frac{qz}{\lambda_1}\right) - \left(\mathrm{T}_0 + \frac{qz}{\lambda_1}\right) = qz\left(\frac{1}{\lambda_1} - \frac{1}{\lambda_2}\right), \tag{9.97}$$

where

z = depth of measurement, m
q = heat-flow density in the area under exploration, W m^{-2}
$\lambda_{1/2}$ = thermal conductivity of the two different soils, W m^{-1} K^{-1}.

In forests, the temperatures at the Earth's surface are for example in Central Europe about 1 °C lower than in open areas, and in areas with bushes about 0.5 °C. This has to be taken into account for $\triangle T_v$.

The measured and corrected data can then be plotted on a map and contoured by isolines.

If the above-mentioned isoline map shows anomalies, the next step is to drill deeper boreholes, 30–100 m in depth to determine reliable temperature gradients and heat-flow density values. The correction of the data is again important and should not be neglected.

Separation of Warm and Cold Water

The water temperature of a spring or a producing borehole can change as a result of subsurface mixing of warm or cold water. A separation of both components is possible under the following assumptions:
- the water temperature from the deep aquifer is constant with time;
- water from the shallow aquifer has the same temperature as the air (rain temperature) and changes annually with time.

The principle is demonstrated in Figure 9.34: the measured water temperature T_w, the measured air temperature T_A and the yield Q_m, which change with time, are given.

The deep aquifer temperature must be higher than the measured maximum temperature $T_{w,m}$. Therefore, if the temperature gradient grad T and the mean annual air temperature T_0 of the area in question are known, the minimum depth of the deep aquifer can be estimated by means of:

$$z_{\min} = (T_{w,m} - T_0)/\text{grad } T. \tag{9.98}$$

The portion due to the warm water can be calculated by starting with:

$$Q = mc\triangle T. \tag{9.99}$$

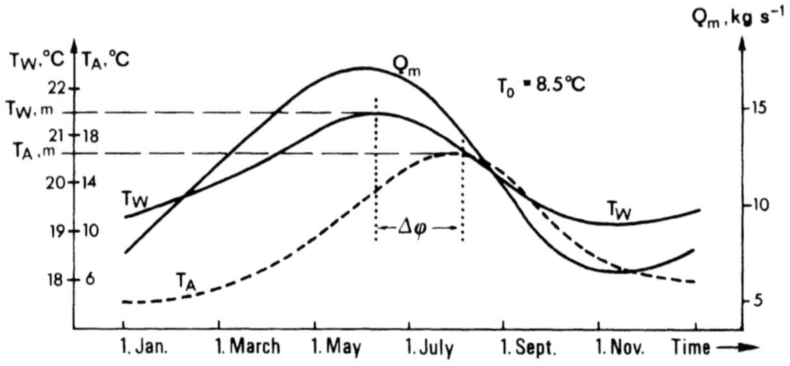

Fig. 9.34. Determination of the mixing rate of warm and cold water of a spring or borehole by using variations in time of outflow temperature T_w, air temperature T_A and yield Q_m.

Let the index 1 refer to the mixed water, index 2 to the warm water, and index 3 to the cold water with $\Delta T = T - 0$ and $c_1 \approx c_2 \approx c_3$, it follows that:

$$Q_1 = Q_2 + Q_3 \tag{9.100}$$

$$m_1 T_1 = m_2 T_2 + m_3 T_3 \tag{9.101}$$

where $m_1 = m_2 + m_3 = Q_m$, $T_1 = T_w$, $T_2 = T_{w,m}$, $T_3 = T_A$.

After rearranging, the equation leads to:

$$m_2 = Q_m(T_w - T_A)/(T_{w,m} - T_A); \tag{9.102}$$

and the amount of cold water can be easily calculated from:

$$m_3 = Q_m - m_2. \tag{9.103}$$

The phase shift $\Delta\varphi$ in days, is given by:

$$\Delta\varphi = T_{A,\max} - T_{w,m}. \tag{9.104}$$

This can provide an indication of the distance between the recharge area for cold water and the spring investigated, if the water velocity v (or its path) from the shallow aquifer is known:

$$\Delta x = \Delta r = v\Delta\varphi. \tag{9.105}$$

EXAMPLE: The following data are given: $T_{w,m} = 21.5\,°C$, $T_0 = 8.5\,°C$ and grad $T = 0.03$ K m^{-1}; and for the date of 5th May: $T_w = 21\,°C$, $T_{A,m} = 16.5\,°C$, $T_A = 9\,°C$, $Q_m = 17$ kg s^{-1}. Thus, it follows that:

Equation (9.98) : $z_{\min} = 433$ m
Equation (9.102): $m_2 = 16.3$ kg s^{-1}
Equation (9.103): $m_3 = 0.7$ kg s^{-1}.

The figures vary with time for m_2 and m_3.

(4) INTRUSIVE BODIES WITH HEAT PRODUCTION CONTRAST

In cases where the embedded body has a different heat production than the surrounding rocks, a heat-flow density anomaly at the Earth's surface can be expected. Simmons (1967) has derived equations analogous to those used in the gravity method for geometrically simple forms:

$$\text{Sphere: } \Delta q = \frac{2}{3}\frac{\Delta H r^3}{z^2}\left(1 + \left(\frac{x}{z}\right)^2\right)^{-3/2} \tag{9.106}$$

$$\text{Horizontal cylinder: } \Delta q = \frac{\Delta H r^2}{z}\left(1 + \left(\frac{x}{z}\right)^2\right)^{-1} \tag{9.107}$$

$$\text{Vertical thin cylinder: } \Delta q = \frac{\Delta H r^2}{2z}\left(1 + \left(\frac{x}{z}\right)^2\right)^{-1/2} \tag{9.108}$$

$$\text{Vertical sheet: } \Delta q = \frac{\Delta H h}{\pi}\left[\frac{1}{2}\ln\left(1 + \left(\frac{x}{z_2}\right)^{-2}\right) - \frac{1}{2}\ln\left(1 + \left(\frac{x}{z_1}\right)^{-2}\right)\right] \tag{9.109}$$

$$\text{Thin horizontal disc: } \Delta q = \frac{\Delta H h}{2\pi}\alpha, \quad \alpha = f\left(\frac{z}{r}, \frac{x}{z}\right). \tag{9.110}$$

where

Δq = heat-flow density anomaly at the Earth's surface at a horizontal distance, x from the centre of the body, W m^{-2}
ΔH = contrast in heat production, W m^{-3}
x = horizontal coordinate, m
z = top of vertical cylinder, centre of the sphere or the horizontal cylinder, m
r = radius of sphere or cylinder, m
z_1 = depth to the top surface of the vertical sheet, m
z_2 = bottom of the vertical sheet, m
h = thickness, m.

The parameters for the disc are shown in Figure 9.35.

The change in heat-flow density for a three-dimensional body can be calculated by replacing each contour of the body by a polygon of n sides; see Figure 9.36. This leads to:

$$\Delta q = \frac{\Delta H}{2\pi}\sum_{j=1}^{n}\sum_{i=1}^{n}\left[W\cos^{-1}\left(\frac{x_i}{r_i}\frac{x_{i+1}}{r_{i+1}} + \frac{y_i}{r_i}\frac{y_{i+1}}{r_{i+1}}\right) - \sin^{-1}\left(\frac{z_j q_i S}{\sqrt{p_i^2 + z_j^2}}\right) + \sin^{-1}\left(\frac{z_j f_i S}{\sqrt{p_i^2 + z_j^2}}\right)\right] \tag{9.111}$$

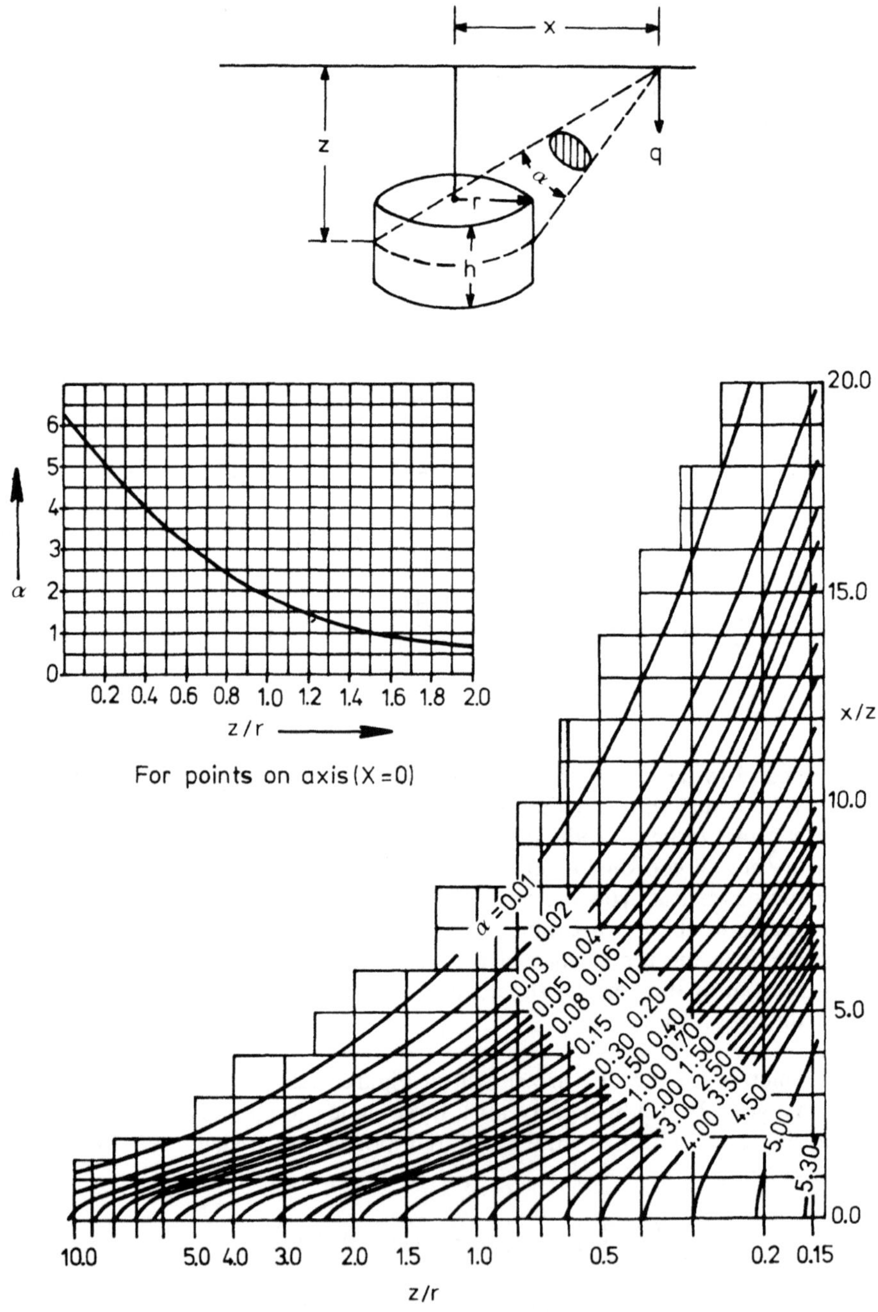

Fig. 9.35. Explanation of parameters in Equation (9.110).

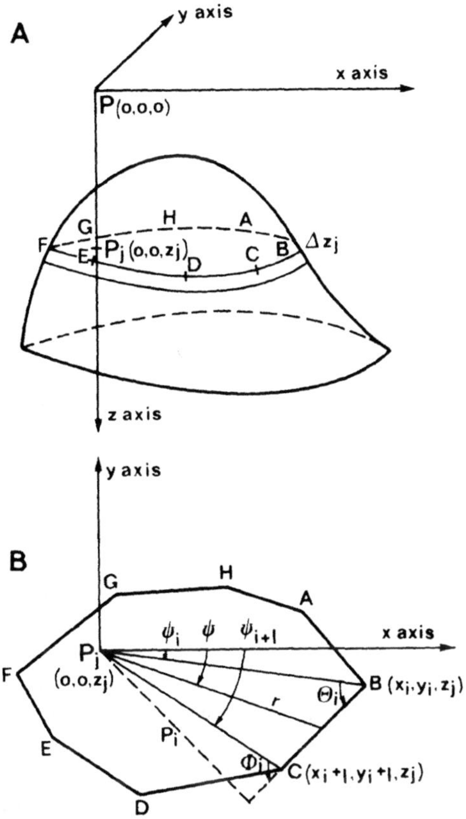

Fig. 9.36. (A) Geometrical relations for computation of heat-flow density due to fine three-dimensional heat source of arbitrary shape. (B) The actual contours at depths z_j are approximated as shown by the polygon.

where

$$S = \pm 1 \quad \text{if} \quad p_i \gtrless 0$$

$$W = \pm 1 \quad \text{if} \quad m_i \gtrless 0$$

$$p_i = \frac{y_i - y_{i+1}}{r_{i,i+1}} x_i - \frac{x_i - x_{i+1}}{r_{i,i+1}} y_i$$

$$q_i = \frac{x_i - x_{i+1}}{r_{i,i+1}} \cdot \frac{x_i}{r_i} + \frac{y_i - y_{i+1}}{r_{i,i+1}} \frac{y_i}{r_i}$$

$$f_i = \frac{x_i - x_{i+1}}{r_{i,i+1}} \frac{x_{i+1}}{r_{i+1}} + \frac{y_i - y_{i+1}}{r_{i,i+1}} \frac{y_{i+1}}{r_{i+1}}$$

$$m_i = \frac{y_i}{r_i} \frac{x_{i+1}}{r_{i+1}} - \frac{y_{i+1}}{r_{i+1}} \frac{x_i}{r_i}$$

$$r_i = (x_i^2 + y_i^2)^{1/2}$$
$$r_{i+1} = (x_{i+1}^2 + y_{i+1}^2)^{1/2}$$
$$r_{i \cdot i+1} = [(x_i - x_{i+1})^2 + (y_i - y_{i+1})^2]^{1/2}.$$

The application of Equations (9.106) to (9.110) assumes steady-state conditions, and therefore the time scale is of the same order as given in Figure 9.24.

(5) OTHERS

There are, of course, numerous other applications of geothermal methods in exploration. An idea of this is given in Table 2.1 of Chapter 2. Only brief comments and a few references will be given for the different topics here.

Borehole logging: By means of temperature logging, indications can be expected with regard to water influx, water loss, head of cementation as well as quality of cementation. Temperature measurements can also be used instead of a spinner flow meter. The process of temperature equilibration disturbance by drilling, may give information about the thermal conductivity of rocks. See Edwards *et al.* (1982), Michel and Haenel (1984), Murphy (1977, 1982).

Exploration for hydrocarbons: The temperature (liquid window) in which oil can be expected, as well as the correlation between temperature anomaly and oil/gas reservoirs are related directly to heat-flow density. Furthermore, there is considerable evidence that a certain temperature increase in the subsurface enhance the processes of alteration of the formation, of migration, and entrapment of oil and gas. See Cornelius (1975), Klemme (1975), Meinhold (1968, 1971).

Exploration for mineral deposits: Salt has a high thermal conductivity as compared with that of other rocks. Therefore, the heat is transported faster, and in the base of a salt dome a chimney effect occurs; Giesel and Holz (1970), Selig and Wallick (1966). In the sulphide ores and radioactive ores, an increased temperature and heat-flow density can be expected only under favourable conditions; Kappelmeyer and Haenel (1974).

Infrared survey: With regard to the geothermal noise due to microclimatic factors and other non-geothermal components, the minimum of detectable heat-flow density by infrared survey is 50–200 times the normal heat-flow density. Therefore, the method can be applied only to areas where geothermal surface manifestations or high heat loss into the surrounding space occurs. See Friedman *et al.* (1969), Grande (1976), Günther *et al.* (1977), Lorenz (1968), Reeves *et al.* (1975), Sabins (1978), Valle *et al.* (1970).

Mining: The creation and maintenance of bearable working conditions is one of the classical problems of geothermics; an updated report is given in Section 9.4.

Permafrost: The permanent frost causes serious problems in geological engineering. In glacier dynamics, geothermal measuring and modelling are of importance; see Section 9.4.

Radioactive wastes: Long-term predictions necessitate comprehensive modelling and continuous measuring of the heat loss from stored wastes. See Section 9.4.6.

Submarine investigation: Submarine springs, brine deposits, and the distinction between salt domes and volcanic intrusion below sea sediments can be detected by means of temperature measurements and/or heat-flow density determinations; see Kappelmeyer and Haenel (1974).

9.2.4. References

Andl, U.: 1982, *Geologische, geophysikalische und geochemische Untersuchungen einer aktiven Verwerfung bei Edenkoben (Pfalz)*, Oberrhein. geol. Abh., 31, Karlsruhe, 19–30.

Armstead, H. C. H.: 1983, *Geothermal Energy. Its Past, Present and Future Contribution to the Energy Needs of Man*, E. & F. N. Spon, 2, edition, New York, London, 404 p.

Bodvarsson, G.: 1969, 'On the Temperature of Water Flowing Through Fractures'. *J. Geophys. Res.* **74**, 1987–1992.

Burgassi, D., Ceron, P., Ferrara, G. C., Gestini, G., and Toro, B.: 1970, *Geothermal Gradient and Heat Flow in the Radicofani Region (East of Monte Amiata, Italy)*, Proceed. U.N. Symp. on the Development and Utilization of Geothermal Resources, Geothermics, Spec. Issue 2, Vol 2, Part 7, Pisa, 443–449.

Carslaw, H. A. and Jaeger, J. C.: 1959, *Conduction of Heat in Solids*, Clarendon Press, 2. Ed., Oxford, 510 p.

Cornelius, C. D.: 1975, 'Geothermal Aspects of Hydrocarbon Exploration in the North Sea Area', *Norges Geologiske Undersokelse, 316, Bull.* **29**, 29–68.

Edwards, L. M., Chilingar, G. V., Rieke III, H. H., and Fertl, W. H. (ed.): 1982, *Handbook of Geothermal Energy*, Gulf Publ. Comp. Book Div., Houston (USA), 613 p.

Fournier, R. O. and Potter, R. W.: 1982, 'A Revised and Expanded Silica (Quartz) Geothermometer', *Geotherm. Res. Conc. Bull.* **V**, 8–9.

Fournier, R. O.: 1981, 'Application of Water Geochemistry to Geothermal Exploration and Reservoir Engineering', in L. Rybach and L. J. P. Muffler (eds.), *Geothermal Systems*, J. Wiley & Sons, New York, 109–144.

Friedman, J. D., Williams, R. S., Palmason, G., and Müller, C. D.: 1969, *Infrared Survey in Iceland*, US Geological Survey Professional Paper 650-C, 89–105.

Galeone, L. and Mongelli, F.: 1981, 'Preliminary Geothermal Model of the Travale Field, Italy', *Geothermics* **10**, 39–44.

Galeone, L. and Mongelli, F.: 1982, 'Outlines for Interpreting Local Heat Flow Anomalies in the Tuscan-Latial pre-Apenninic Belt', in V. Čermák and R. Haenel (eds.), *Geothermics and Geothermal Energy*, E. Schweizerbartsche Verlagsbuchhandlung, Nägele n. Obermiller, Stuttgart, 181–190.

Giesel, W. and Holz, A.: 1970, 'Das anomale geothermische Feld in Salzstöcken – Quantitative Deutung an einem Beispiel', *Kali u. Steinsalz* **5**, 8, 272–274.

Grande, N. K. del: 1976, *An Advanced Airborne Infrared Method for Evaluating Geothermal Resources*, Proceed. 2. UN Symp. on the Development and Use of Geothermal Resources, San Francisco, May 1975, US Government Printing Office, Washington D.C. 29402, U.S.A., 947–953.

Günther, R., Kappelmeyer, O., and Kronberg, P.: 1977, *Zur Prospektion auf geothermale Anomalien, Erfahrungen einer Modelluntersuchung in Polichnitos, Lesbos (Griechenland)*, Geol. Rundschau Bd. 66, 10–33.

Hardee, H. C. and Larson, D. W.: 1980, 'Thermal Techniques for Characterizing Magma Body Geometries', *Geothermics* **3**, 237–250.

Jaeger, J. C.: 1964, 'Thermal Effects of Intrusions', *Rev. of Geophysics* **2**, 3, 443–466.

Kappelmeyer, O.: 1957, 'The Use of Near Surface Temperature Measurements for Discovering Anomalies Due to Causes at Depth', *Geophys. Props.* V, **3**, 239–258.

Kappelmeyer, O. and Haenel, R.: 1974, *Geothermics with Special Reference to Application*, Geoexplor. Monogr. Ser. 1, No. 4, Gebr. Borntraeger, Berlin, Stuttgart, 238 p.

Kestin, J. (ed. in Chief), Di Pippo, R., Khalifa, H. E., and Ryley, D. J. (ed.): 1980, *Sourcebook on the Production of Electricity from Geothermal Energy*, US Dep. of Energy, Washington, D.C. 20402, Contract, No. EY-76-S-4051. A 002, 997 p.

Klemme, H. G.: 1975, 'Geothermal Gradients, Heat Flow and Hydrocarbon Recovery', in G. Fischer and S. Judson (eds.), *Petroleum and Global Tectonics*, 251–304.

Lachenbruch, A. H., Sass, J. H., Munroe, R. J., and Moses, T. H. Jr: 1976, 'Geothermal Setting and Simple Heat Conduction Models for the long Valley Caldera', *J. Geophy. Res.* **81**, 5, 769–784.

Lawton, R. G.: 1974, *The AYER Heat Conduction Computer Programm*, Report LA-5613-MS, LASL, Los Alamos, USA.

Lawton, R. G., Murphy, H. D., Tester, J. W., and McFarland, R. D.: 1976, 'Geothermal Reservoir Modelling', in Blair, A. G., Tester, J. W., and Mortensen, J. J. (eds.), *LASL, Hot Dry Rock Geothermal Project*, progress report LA-6255-PR, LASL, Los Alamos, USA.

LeSchack, L. A. and Lewis, J. E.: 1983, 'Geothermal Prospecting with Shallow Temperature Surveys', *Geophysics* **48**, 975–996.

Lorenz, D.: 1968, 'Temperature Measurements of Natural Surfaces Using Infrared Radiometers', *Applied Optics* **7**, 1705–1710.

Meinhold, R.: 1968, *Der Zusammenhang zwischen Geothermie, Hydrodynamik, Geochemie und Erdöllagerstätten*, Erdöl-Erdgas-Erkundung und -Förderung, VEB-Verlag, Leipzig, 1, 422–433.

Meinhold, R.: 1971, *Hydrodynamic Control of Oil and Gas Accumulation as Indicated by Geothermal, Geochemical and Hydrological Distribution Patterns*, Eight world petroleum congress, London, 2, 55–66.

Michel, W. and Haenel, R.: 1984, *Quantitative Bestimmung von Wasserinjektionen und Extraktionen in Bohrungen mit Hilfe von Temperaturmessungen im Hot-Dry-Rock Projekt Urach, Teilprojekt Geothermik*, Report, NLfB Hannover, Archive No. 95 824.

Milora, S. L. and Tester, J. W.: 1976, *Geothermal Energy as a Source of Electric Power*, Thermodynamic and economic design criteria, The MIT Press, London, 186 p.

Mongelli, F.: 1980a, *Conductive Heat Flow Anomalies Interpretations-Lectures at 3rd Course 'Geophysics of Geothermal Areas'*, Ettore Majorana 'Intern. Centre for Scientif Culture', Erice, Italy.

Mongelli, F.: 1980b, 'Geothermal Prospecting for Geothermal Fields', in A. Rapolla, G. V. Keller, and D. J. Moore (eds.), *Geophysical Aspects of the Energy Problem*, Elsevier Sc. P. C., 83–99.

Mongelli, F.: 1981, *Elementi di Prospezione per l'Energia Geotermica*, Adriatica, Editrice, Bari, 150 p.

Morse, P. M. and Feshbach, H.: 1953, *Methods of Theoretical Physics*, McGraw Hill, New York, part I, 997 p.

Mundry, E.: 1968, *Über die Abkühlung magmatischer Körper*, Geol. Jb. 85, Hannover, 755–766.

Mundry, E.: 1970, *Mathematical Estimation Concerning the Cooling of a Magmatic Intrusion*, Proceed. UN Symp. on the Development and Utilization of Geothermal Resources. Geothermics, Spec. Issue 2, Vol. 2, Part I, Pisa, 661–668.

Murphy, H. D.: 1977, 'Fluid Injection Profiles – A Modern Analysis of Wellbore Temperature Surveys', – Soc. Petrol. Eng. AIME, Spe 6783, 1–8.

Murphy, H. D.: 1982, 'Enhanced Interpretation of Temperature Surveys taken during Injection or Production', *Journal of Petroleum Technology*, 1313–1326.

Prol, R. M. and Juarez, G.: 1986, 'Geothermal Map of Mexico', *J. Volc. Geotherm. Res.* **28**, 351–362.

Reeves, R. G., Anson, A., and Landen, D.: 1975, 'Manual of Remote Sensing', *Am. Soc. of Photogrammetry, Falls Church*, V. Vol. I (867 p.), Vol. II (2144 p).

Reiter, M. and Clarkson, G.: 1983, 'Geothermal Studies in the San Juan Basin and the Four Corners Area of the Colorado Plateau. II, Steady State Models of the Thermal Source of the San Juan Volcanic Field', *Tectonophysis* **91**, 253–270.

Renner, J. L., White, D. E., and Williams, D. L.: 1975, 'Hydrothermal Convection Systems', in D. F. White, and D. L. Williams (eds.), *Assessment of Geothermal Resources of the USA*, Geol. Survey Circ. 726, 5–57.

Rodemann, H.: 1979, *Modellrechnungen zum Wärmeaustausch in einem Frac*, Report, NLfB Hannover, Archive No. 81 990.

Rodemann, H.: 1980, 'Analytical Model Calculation on Heat Exchange in a Fracture', in R. Haenel (ed.), *The Urach Geothermal Project*, Schweizerbart'sche Verlagsbuchhandlung, Stuttgart, 351–353.

Rybach, L. and Muffler, L. J. P. (eds.): 1981, *Geothermal Systems – Principles and Case Histories*, Wiley & Sons, Ltd., 359 p.

Sabins, F. F. Jr.: 1978, *Remote Sensing, Principles and Interpretation*, W. H. Freeman and Cop., San Francisco, 1–426 p.

Selig, F. and Wallick, G. C.: 1966, 'Temperature Distribution in Salt Domes and Surrounding Sediments', *Geophys.* **31**, 346–361.

Simmons, F.: 1967, 'Interpretation of Heat Flow Anomalies, 1, Contrast in Heat Production', *Rev. of Geophys.* **5**, 42–52.

Smith, R. L. and Shaw, H. R.: 1975, 'Igneous-Related Geothermal Systems', in D. F. White and D. L. Williams (eds.), *Assessment of Geothermal Resources of USA*, Geol. Survey, Circ. 726, 58–83.

Swanberg, C. A. and Morgan, P.: 1979, 'The Linear Relation Between Temperatures Based on the Silica Content of Groundwater and Regional Heat Flow: A New Heat Flow Map of the United States', *Pure Appl. Geophys.* **117**, 227–241.

Swanberg, C. A. and Morgan, P.: 1980, 'The Silica Heat Flow Interpretation Technique: Assumptions and Applications', *J. Geophys. Res.* **85**, 7206–7214.

Swanberg, C. A. and Morgan, P.: 1981, 'Heat-Flow Map of the United States Based on Silica Geothermometry', in Y. S. Touloukian, W. R. Judd, and R. F. Roy (eds.), *Physical Properties of Rocks and Minerals*, McGraw-Hill Book Co., New York, 54–544.

Urban, T. C., Diment, W. H., Sass, J. H., and Jamieson, I. M.: 1976, *Heat flow at The Geysers, California, USA.*, Proceed. 2. UN Symp. on the Development and Use of Geothermal Resources, San Francisco, Vol. 2, 1241–1245.

Valle, R. G., Friedman, J. D., Gawarchi, S. J., and Banwell, C. J.: 1970, 'Photogeology and Thermal Infrared Reconnaissance Survey of the Los Megritos – Ixlan de Los Hervores Geothermal Area Michocan, Mexico', *Geothermics*, Special Issue No. 2, 381–398.

Von Herzen, R. P. and Uyeda, S.: 1963, 'Heat Flow Through the Eastern Pacific Floor', *J. Geophys. Res.* **68**, 14, 4234–4240.

9.3. PALAEOGEOTHERMICS

L. STEGENA
Department of Environmental Physics, Roland Eötvös University, Budapest Hungary

9.3.1. Introduction

Plate tectonics was based at first on palaeomagnetic evidences. In order to confirm our tectonophysical hypotheses and models, it would be of first order of importance to determine the variation of some other geophysical parameters like heat-flow density and geothermal gradient during geological time. Oil genesis in a layer is controlled by its content of organic material and by the temperature-time history of the layer (Lopatin, 1971, 1976). This idea raised great interest for palaeogeothermic studies in the petroleum exploration.

By solving the heat equation, only 'forward' solutions can be obtained. There are however many methods in geosciences, for the determination of temperatures in the geological past. Most of these methods are suitable to determine temperatures at which a (hydrothermal) mineral was formed or palaeotemperatures of suburface waters, and so on. These methods however are mostly not applicable to reach the mean goal of palaeogeothermics in general: i.e. to determine the (regional) heat-flow density and/or geothermal gradient variations during the geological past, for a larger geological unit, like a sedimentary basin. Till now, two methods produced important results on the field of palaeogeothermics in this sense: the studies on the maturation of organic materials in the sediments and the studies of fluid inclusions in sedimentary rocks (microthermometry). Other methods are mentioned only briefly in this chapter.

9.3.2. Palaeogeothermics Based on Maturation of Organic Materials in Sediments

It is widely recognized that temperature and time are the most important factors influencing the maturation of the organic materials of sediments; the influence of pressure seems to be negligible. Based on Karweil's (1956) work, Lopatin (1971) gave a simple scheme for describing the maturation level of organic materials. Supposing that the maturation process is to be handled as a first-order chemical reaction, the Arrhenius' equation is valid and the velocity of the reaction k depends exponentially on temperature:

$$k = \nu \exp(-E/RT), \tag{9.112}$$

where

ν = frequency-factor
E = activation energy
R = gas constant
T = temperature in Kelvin.

R. Haenel, L. Rybach, and L. Stegena (eds.)
Handbook of Terrestrial Heat-Flow Density Determination, 391—419.

Numerous chemical reactions double their reaction velocity for each 10 K temperature increase, not far from room temperatures, because their activation energy lies around 55 kJ mol^{-1}.

Lopatin (1971) accepted this value and suggested that the dependence of maturity on time is linear and the dependence on temperature has an exponential character. Therefore, the velocity of the 'maturation' reaction can be written as:

$$k \propto 2^{0.1T(t)} \tag{9.113}$$

and the parameter which describes the rank of maturity:

$$C \propto \int_0^t 2^{0.1T(t)}\,\mathrm{d}t, \tag{9.114}$$

where $T(t)$ is the temperature of the layer during the time interval dt and t is the age of the layer from the deposition till to present.

For practical reasons, Lopatin introduced the sum instead of integral, dividing the whole temperature history of the layer into 10 °C temperature intervals. He then chose arbitrarily the 100 to 110 °C temperature interval (which is the mean domain of oil generation) as the base interval and assigned to it an index value of $n = 0$, to the 120–130 °C interval $n = 2$, to the 90–100 °C interval $n = -1$, and so on. The maturity parameter calculated on this way he called as Time Temperature Index (TTI):

$$\mathrm{TTI} = \sum_{n\,\mathrm{max}}^{n\,\mathrm{min}} (\Delta t_n) 2^n, \tag{9.115}$$

where Δt_n is the time interval (im Ma) the layer spent in the n-th 10 °C temperature interval and $n$max and $n$min are the n-values of the highest and lowest temperature intervals occuring in the thermal history of the layer.

Figure 9.37 demonstrates the method of calculation of TTI, for a hypothetical layer 20 Ma old and lying at present at a depth of 2300 m. Let us suppose that the subsidence and burial history of the layer during geologic time was determined as shown by the curve of Figure 9.37. Let us suppose thereafter, that the present geothermal gradient is 50 mK m^{-1} and the gradient was constant during the whole sedimentary history, as shown on Figure 9.37, by the horizontal isotherms. In this case, for the layer of Figure 9.37, TTI = 15.2.

TTI's are calculated, theoretical values based on the burial history and on (recent) temperatures, characterizing the maturity of organic materials. There are direct methods too, to measure the maturity. The composition of organic matter in sediments is about (e.g. Ho, 1978) 90 % kerogen and 10 % bitumen (hydrocarbon, resin, asphaltene). One kind is soluble in organic solvents, this fraction is called bitumen. The other kind, insoluble in organic matter, is termed as kerogen. There are methods to estimate the maturity by examining the soluble organic matter: percent carbon in bitumen; carbon preference index (odd carbon number compounds to even carbon number), paraffin profile, percent wet gas. Other, more important methods examine the kerogen as a maturation index. Among the various kerogen methods (Kerogen Alteration Index KAI, thermal alteration

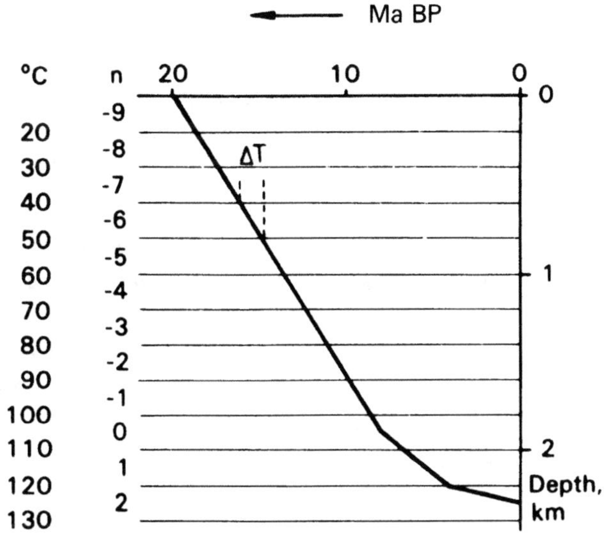

Time-Temperature Index (TTI)

$$TTI = \sum \Delta T_n 2^n$$

ΔT_n : in Ma

Fig. 9.37. Lopatin's (1971) method for the calculation of Time Temperature Index for a layer lying in the depth of 2300 m, aged 20 Ma. TTI is characteristic for the maturity of organic matter.

index TAI, pyrolysis, elementar CHO analysis, atomic H/C ratio, Electron Spin Resonance-ESR spectrometry), the vitrinite reflectance is the most commonly used in oil industry laboratories (Teichmüller, 1971).

Vitrinite reflectance is a ratio between the intensity of the reflected light and that of the source light, expressed in percent, using vitrinite (= woody kerogen) as the reflector. The value is often simply called R_0 or % R_0 (dimensionless, expressed in %), the subscript '0' designates that the measurement was made in oil. For a rock to be a good source rock of petroleum, it has to be rich in organic matter (at least 0.4 % organic carbon for shales and 0.3 % for carbonates), to contain the right kind of organic matter and to be mature. The diagenesis of organic matter, which results in the formation of kerogen, creates the source material for the generation of petroleum. Because of this, vitrinite reflectance measurements have a special importance in hydrocarbon exploration. Based on a lot of studies on various oil fields, it is suggested that the top of the liquid window (oil generation) is indicated by R_0 values of 0.5 for amorphous kerogen, 0.6 to 0.7 for herbaceous kerogen and 0.7 to 0.8 % for woody-coaly kerogen. The bottom of the generation window is indicated by 1.4 for amorphous, 1.8 for herbaceous-woody and 3.2 % for coaly kerogen. Some authors (Waples, 1980, 1983) give other but similar R_0 values.

Based on 402 thermal maturity (R_0) data from 31 worldwide wells, Waples (1980) determined a correlation between measured R_0 values and TTI values calculated using the burial histories of the boreholes and supposing the validity of present geothermal conditions during the geological past (Figure 9.38).

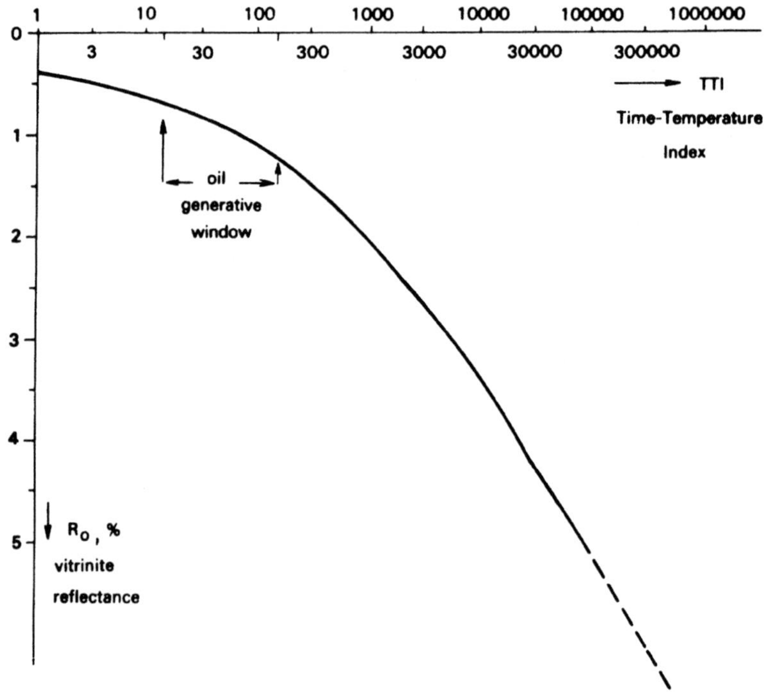

Fig. 9.38. Correlation between Time Temperature Index of maturity and vitrinite reflectance R_0 (after Waples, 1980).

The foregoings enables the palaeo heat-flow density estimation for a borehole, by the following steps:

- Based on known ages of some sedimentary layers in the borehole, the sedimentary history of these layers will first be determined (Figure 9.39, dashed lines).
- Using porosity-depth functions and/or other considerations, the sedimentation histories are corrected for the effect of compaction during the geological past (Figure 9.39, solid lines), as made by Du Rouchet (1980), Stegena *et al.* (1981) and Falvey and Deighton (1982).
- Based on present borehole temperatures, the geoisotherms for each 10 °C interval will be constructed in the time-depth section (Figure 9.40, left), thus assuming that present heat flow prevailed during the geological past. The

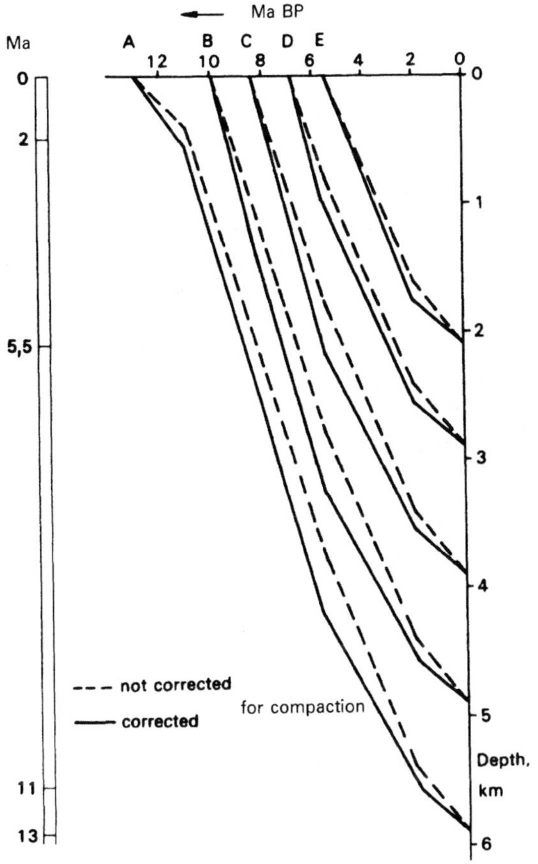

Fig. 9.39. Sedimentary history of a borehole (HOD) in the Pannonian basin, calculated on the basis of the ages at the left hand column, without and with correction of compaction (Stegena *et al.*, 1981).

constancy of heat flow during the past does not mean parallel and equidistant isotherms; it is possible to take into consideration the probable changes with time and depth, of thermal conductivity of the layers, with the aid of the lithology and burial history of the borehole.

- After calculating the TTI values for each layers of the borehole the TTI-s will transformed to R_0 values (Figure 9.40), using the correlation of Waples (1980) (Figure 9.38).
- These calculated R_0 values are then compared with the R_0 values measured in the borehole. The discrepancy between calculated and measured values is to be attributed to the variations of heat-flow density during the geological past. Using plausible hypotheses, one adjusts the past heat-flow density (Figure 9.40, right) and repeates the comparison of R_0 values calculated from TTI-s and measured R_0 values, until a good fit between calculated and measured vitrinite reflectances is reached (Figure 9.41).

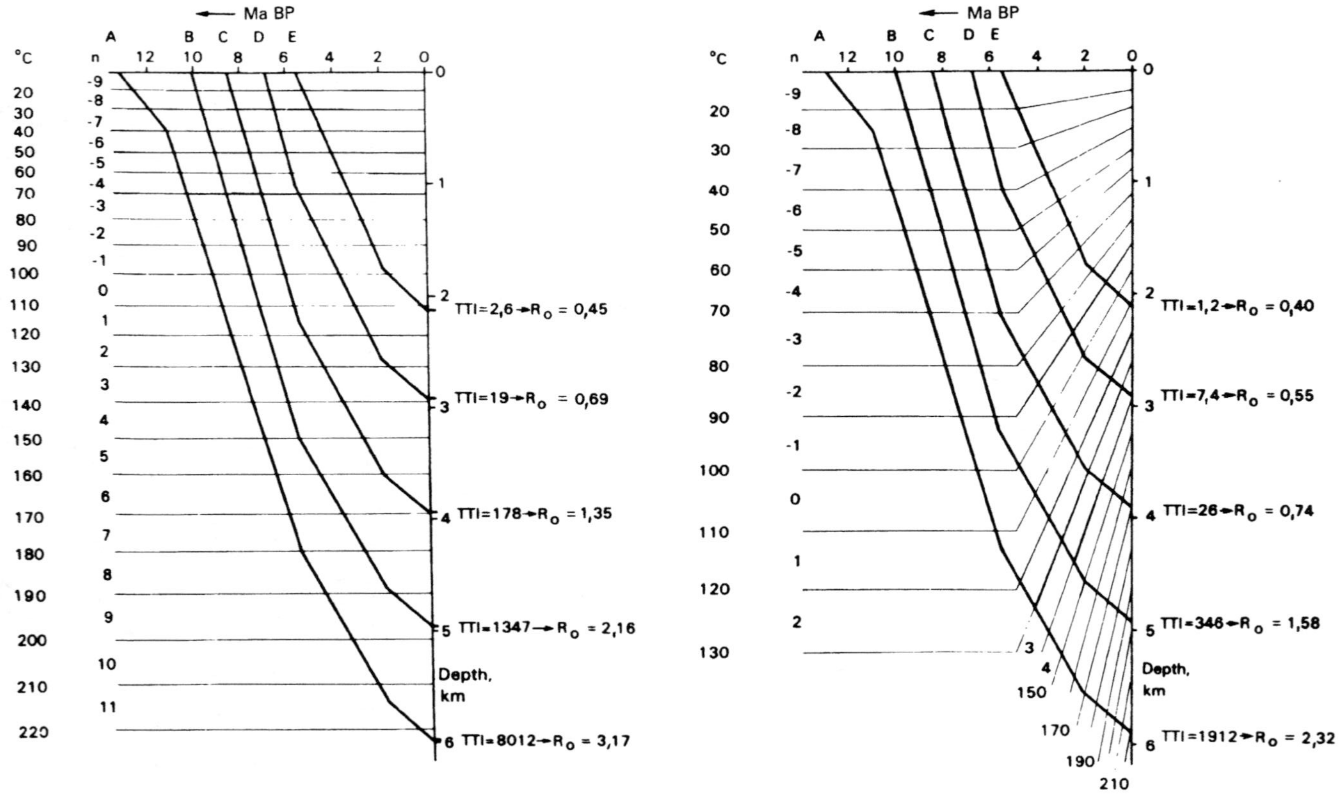

Fig. 9.40. Calculated TTI values for the borehole HOD assuming that the heat-flow density was constant through the sedimentary history (left), and that the borehole was heated up during the last 5 Ma to the present heat flow value (Stegena *et al.*, 1981).

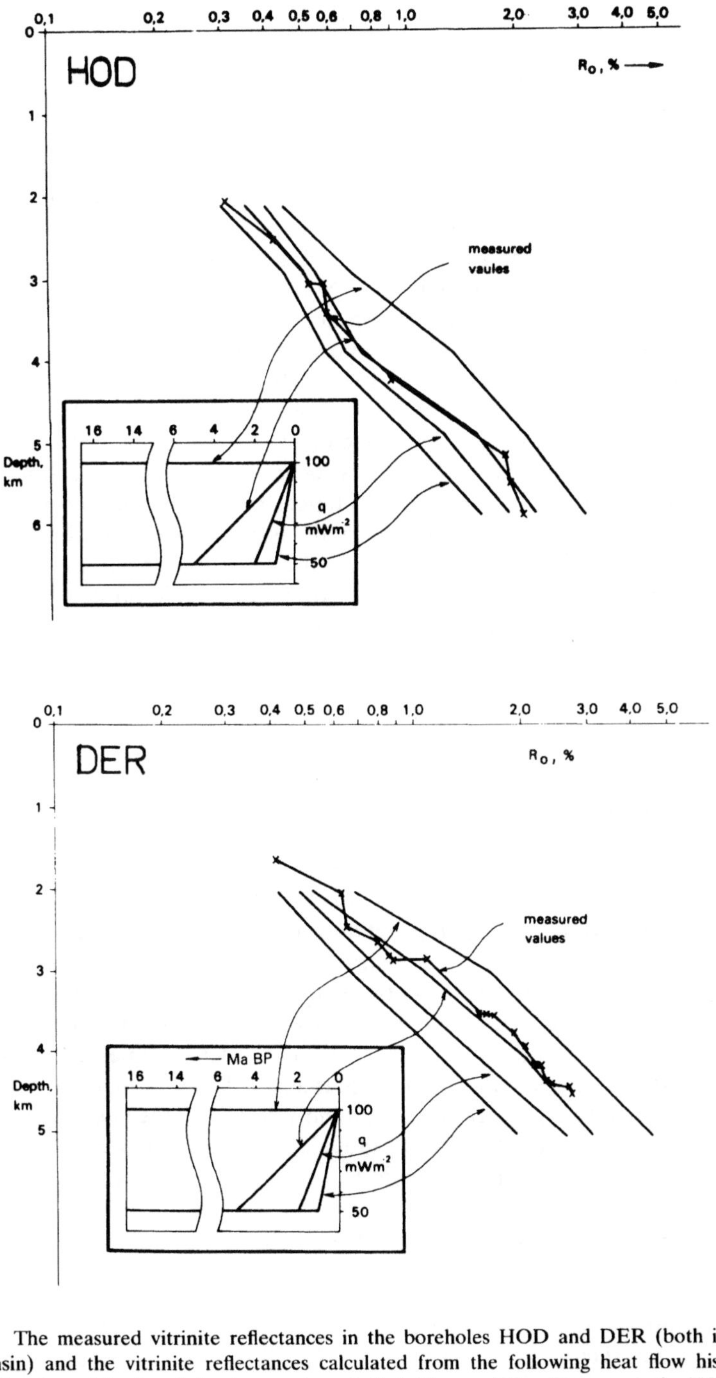

Fig. 9.41. The measured vitrinite reflectances in the boreholes HOD and DER (both in the Pannonian Basin) and the vitrinite reflectances calculated from the following heat flow histories: the upheating of the boreholes began at ∞, 5, 2, 1 Ma ago (after Stegena *et al.*, 1981).

Figure 9.41 shows two boreholes of the Pannonian Basin with heat flow histories calculated independently. Both boreholes gave the same result: the measured vitrinite reflectances are compatible with the assumption that the Pannonian Basin have had a low heat-flow density (~50 mW m^{-2}) before 5 Ma, and 5 Ma ago the heat-flow density began to increase (linearly?) to its present value (~100 mW m^{-2}).

The above scheme serves rather to illustrate the principles of the palaeogeothermal calculations but does not present a final solution of the question. There are some fundamental problems in oil geochemistry which are not solved satisfactorily and which can influence the above sketched model.

- It became common to assume that increases in vitrinite reflectance values were valid indicators of the extent to which organic matter maturated and oil generation had occurred (Waples, 1983). However there is an uncertainty in some R_0 measurements, because the results have a wide scatter, and sometimes it is hard to distinguish low reflecting resinite and high reflecting fusunite from vitrinites (Ho, 1978). During the beginning of oil generation, bitumen impregnations lower the vitrinite reflectance. In all red-coloured rocks organic matter is oxidized; in limestones vitrinite is very rarely preserved and if it occurs, the reflectance value differs from the value of vitrinite in the same rank in other lithologies. Reflectance, as any other electronic property of any solid, depends on three parameters: chemical composition, atomic structure and microstructure. The same value for reflectance can thus be measured for materials different in their microstructure and chemical composition which can be of various ranks or not. The use of transmission electron microscopy (TEM) by successive heating under an inert atmosphere to 1000 °C characterizes better the maturation of organic materials.
- It is generally supposed that pressure does not have a significant effect on the maturation of organic matter and on the amount of hydrocarbon generated. It is to be noted however that the role of pressure in oil generation has never been thoroughly examined (Waples, 1983).
- The maturation of organic matter exhibits a very complex process, with a lot of parallel chemical reactions, with various activation energies, and the whole process can hardly be described by a first-order kinetic expression (Siever, 1983), as showed by pyrolysis experiments (Cummings and Robinson, 1972). Lasage (1981) has compiled a table of activation energies for geochemical reactions that shows a range from less than 4 kJ mol^{-1} to more than 400 kJ mol^{-1}. Tissot (1969), Tissot and Espitalie (1975), Tissot and Espitalie (1975) and Jüntgen and Klein (1975) have modeled the thermal alteration of kerogen with a set of first-order rate equations:

$$\frac{dn_{ki}}{dt} = -n_{ki} a_i \exp\left(-\frac{E_i}{RT}\right), \quad i = 1, 2, \dots 6 \tag{9.116}$$

where

n_{ki} = mass function
a_i = frequency factor
E_i= activation energy of the i-th kerogen.

Integrating the equation over the thermal history of any horizon, the generated petroleum and the maturity of organic matter can be calculated. – This procedure however gives a rather theoretical treatment, it is hardly applicable for palaeo-geothermal calculations, because of the wrong or not known character of the quantities n, a, and E.

Lopatin (1971) tested his model on a very difficult well, Münsterland I. Recalibration of Lopatin's method with larger and more reliable data sets (Waples, 1980; Kettel, 1981) has verified the general validity of the model itself, but has modified Lopatin's original TTI-vitrinite reflectance correlation. Lopatin and Bostick (1973) and Lopatin (1976) later suggested some improvements in the original scheme. Lopatin (1976) used fewer and larger temperature intervals; instead of $\Delta T = 10$ K:

$$\Delta T = \frac{1.37T^2}{E - 1.37T}, \qquad (9.117)$$

where

E = activation energy = 42 kJ mol^{-1}, T in Kelvin.

This formula gives 15 K for ΔT at $T < 80\,°\mathrm{C}$, 20 K at $80\,°\mathrm{C} < T < 120\,°\mathrm{C}$, 25 K at $120\,°\mathrm{C} < T < 170\,°\mathrm{C}$ and 30 K for ΔT above 170 °C.

The maturation of organic matter accelerates exponentially with temperature. In the maturation process, the time on which the organic matter into consideration spent at maximum temperature, plays a decisive role. Hood *et al.* (1975) worked out a model, at which the period spent within 15 K of the rock's maximum palaeotemperature was taken into consideration for the maturation of organic matter and indirectly for the vitrinite reflectances. They created a scale of thermal maturity called 'level of organic metamorphism' (LOM), which is controlled only by the maximum temperature survived by the layer and by the 'effective heating time' spent by the layer within 15 K of the rock's maximum temperature (Figure 9.42). Staplin's (1969) similar scale (TAI, thermal alteration index) is based on microscopic structure variation and colouring of organic debris.

Pusey (1973) suggested that maximum palaeotemperatures can be obtained accurately from ESR (electron spin resonance) analysis of kerogen. The ESR is sensitive to free radicals; the number of free radicals increases as kerogen is subjected to increasing temperatures and kerogen free radicals are stable through geologic time. The ESR geothermometer was calibrated by obtaining data from cores of Tertiary basins believed to be actively subsiding and therefore very probably satisfying the assumption that samples from these basins now are at maximum temperature since burial, supposing constant heat flow. But ESR signals are dependent not only on temperature but also are subject to variations in kerogen type, diagenetic changes in kerogen, weathering and age of the rock.

Price (1982) improves on the idea that vitrinite reflectance depends first of all on maximum temperature. A plot of R_0 versus present temperature from a number of areas that have not undergone major geologic mutilation, increases in tight ($r = 0.97$) linear fashion, yet burial times for these different areas range from 0.3 to 240 Ma. He suggests that some geochemical postulates are in error

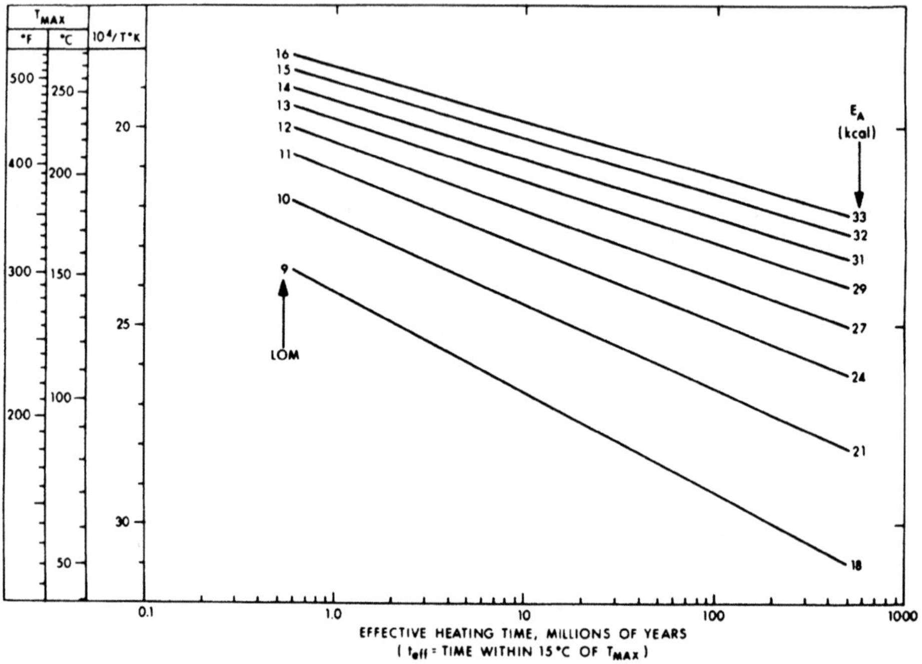

Fig. 9.42. Relation of LOM and R_0 to maximum temperature and effective heating time (after Hood *et al.*, 1975, modified).

and that time has little effect on organic maturation. It appears that vitrinite reflectance can be used as an absolute paleogeothermometer from 20 °C to at least 400 °C.

Middleton and Falvey (1983) propose, for simplicity, that maturation (C) and R_0 are related by the equation:

$$\ln R_0 = A + BC. \tag{9.118}$$

Empirical studies give $A = -2.275$ and $B = 0.177$. A, B, C (and also R_0) are dimensionless quantities.

For maturation C, they accept Lopatin's original idea with insignificant modification ($\Delta T = 10.2$ instead of 10 K) and for simplicity the logarithm of the previously given integral is used:

$$C = \ln \int_0^t 2^{T(t)/10.2}\,\mathrm{d}t \tag{9.119}$$

(as used by Royden *et al.*, 1980 and De Braemaeker, 1983). These equations combined give an equation relating R_0 to temperature as a function of time:

$$(R_0)^a = b \int_0^t \exp[cT(t)]\,\mathrm{d}t, \tag{9.120}$$

where

$a = 5.635$, $b = 2.7 \times 10^{-6}\ \mathrm{Ma}^{-1}$ and $c = 0.068\ \mathrm{K}^{-1}$.

Given the thermal history of an organic sediment $T(t)$, Equation (9.120) can be used to predict the reflectance of vitrinite in the sediment after a time, t. – Neither Welte and Yukler's (1981) formula gives more as Lopatin's (1971) and Waples' (1980) results rewritten in another form:

$$R_0\ \% = 1.301\ \lg \mathrm{TTI} - 0.5282. \tag{9.121}$$

Buntebarth (1979) tried to calculate palaeogeothermal gradients without any theory as far as possible. According to him a relationship exists between the coal rank, measured by the mean optical reflectance of vitrinite (R_m, mean value of the individual R_0 measurements) and the integral of depth and duration of burial:

$$R_m^2 \propto \int_0^{t_1} z(t)\,\mathrm{d}t, \tag{9.122}$$

where

z = depth
t = time
t_1 = upper time limit, which means that the calculation can be restricted to a part of the whole burial history.

Furthermore, that in this relationship the coal rank is proportional to a function of the geothermal gradient:

$$R_m^2 = f(\mathrm{grad}\ T) \int_0^{t_1} z(t)\,\mathrm{d}t. \tag{9.123}$$

Figure 9.43 shows measured R_m values in some boreholes in the Federal Republic of Germany, as a function of the burial history $\left(\int_0^{t_1} z(t)\,\mathrm{d}t\right)$. At $R_m = 1$,

$$f(\mathrm{grad}\ T) = \frac{1}{\left.\int_0^{t_1} z(t)\,\mathrm{d}t\right|_{R_m=1}} = \frac{1}{I}. \tag{9.124}$$

Geothermal gradients measured presently in the four boreholes on Figure 9.43 allowed the determination of a function between palaeogeothermal gradients and I values:

$$\mathrm{grad}\ T = 9.87 - 14.6\ \ln I\ (^\circ\mathrm{C\ km^{-1}}). \tag{9.125}$$

The applicability of this equation for other areas is questionable.

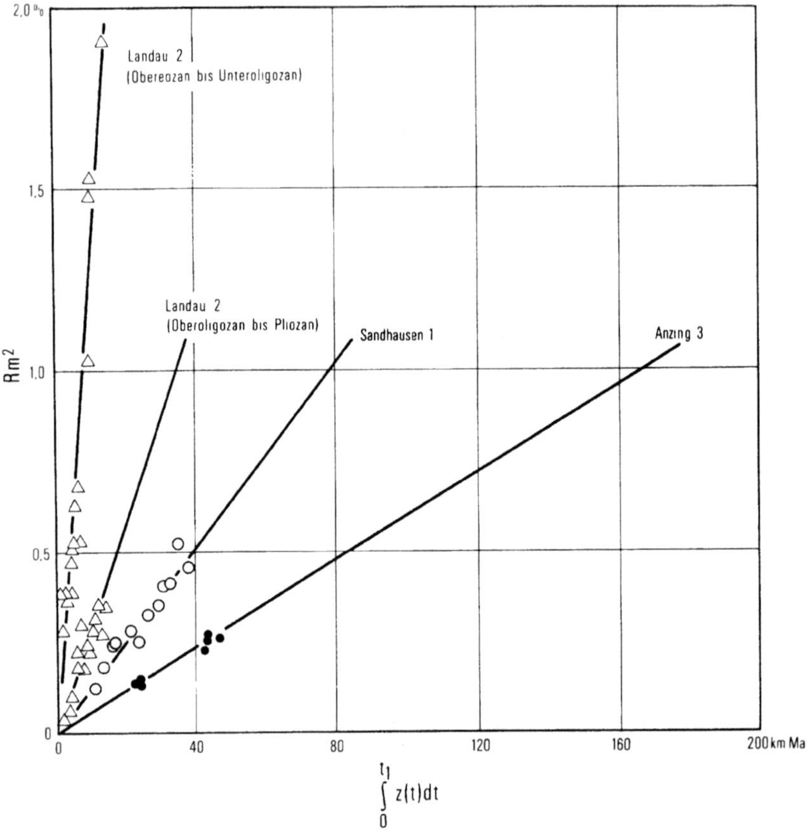

Fig. 9.43. Relation between the mean vitrinite reflectance (R_m) and the integral of depth and time, in four boreholes of the Rhine graben (Buntebarth, 1979).

Some case histories:
Based on maximum measured vitrinite reflectance data (Hacquebard 1977) and burial history of 28 wells lying in the Central Prairies Basin, Canada, Majorowicz and Jessop (1981) estimated a lower average palaeogeothermal gradient (27 mK m^{-1}) for the early Oligocene time as the present day one (30.6 mK m^{-1} in average) (Figure 9.44). For the calculation they used the method of Karweil (1956) with Bostick's (1973) modifications and the method proposed by Hood *et al.* (1975).

Eggen (1984) worked with a lot of vitrinite reflectance data but present heat-flow density estimations only. He stated that in the Viking Graben (Norwegian North Sea) the calculated palaeo heat-flow density (approx. 55 mW m^{-2}) fits well with the present heat-flow density estimation (50–60 mW m^{-2}); on the flank of the Viking Graben however an average palaeo heat-flow density close to 50 mW m^{-2} has to be assumed in order to obtain the observed maturity, while the present day estimation lies at 70 mW m^{-2}.

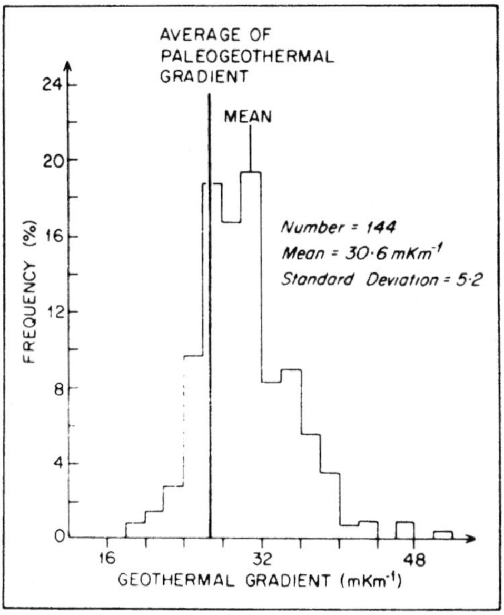

Fig. 9.44. Average of the palaeogeothermal gradients with the histogram of present geothermal gradients, in the Central Prairies basin, Canada (Majorowicz and Jessop, 1981).

Wang Ji-an *et al.* (1983) found that the coalification gradient increases from 0.25 to 0.65 (R_0, % km^{-1}) from mid to early Eocene, in the western part of Liaohe oil field, North China, and about 0.4 coalification gradient was determined in the early Tertiary sediments of the Central Hebei oil field. Karweil's (1956) and Lopatin's (1971) methods were used for palaeotemperature reconstructions. In contrast, Rybach (1984) gives 0.09–0.05 (R_0, % km^{-1}) coalification gradients for the Northern Alpine Foreland (Molasse basin).

Royden and Keen (1980) predict R_0 values for the sediments of Nova Scotia and Labrador shelfs, based on theoretically derived thermal evolution and on Lopatin's theory. A similar work was carried out by Royden *et al.* (1980), for the Falkland Plateau and for three places of the North Atlantic.

In contrast to the mentioned problems of vitrinite reflectance, these results seem to be very realistic.

To avoid the difficulties with vitrinite reflectance, McKenzie (1981) proposed a new idea:

Some of the problems relating to the empirical relations suggested by Lopatin (1971), Waples (1980) and others could be avoided if chemical reactions involving only one molecular type which occur during the maturation of the organic material were to be identified. MacKenzie and McKenzie (1983) have investigated the rates of three reactions which occur before and during the early stages of oil formation. Two of the reactions are isomerization reactions of hydrocarbon molecules, in a sterane with 20 carbon atoms and in a hopane with 22 carbon atoms; the third reaction converts C-ring monoaromatic to triaromatic steroid hydrocarbons. All three reactions were assumed to be first order and monomolecular; the isomerization reactions are reversible, with a conversion rate of the R to the S form 1.174 and 1.564 resp., while the aromatization was assumed to be irreversible.

This method excels by its clear theoretical (thermodynamic) principles; the Arrhenius equation is certainly valid for these reactions. The problem, however, is that the frequency factor and activation energy cannot be determined in laboratory, or only very inaccurately, because of the slowness of the reactions. Because of this, McKenzie's (1978) stretching theory for evolution of sedimentary basins was used to calibrate the reactions. This theory involves a thermal history, which can be derived from burial history, with some kind of accuracy. Based on chemical analyses of North Sea and Pannonian Basin cores, and using more or less determined or hypotethical stretching models, they determined the kinematics of the three reactions:

	Frequency factor (s^{-1})	Activation energy ($kJ\ mol^{-1}$)
Isomerization of steranes	6×10^{-3}	91
Isomerization of hopanes	16×10^{-3}	91
Aromatization of steroid HC-s	1.8×10^{14}	200

Figure 9.45 shows the results for the Pannonian basin, which are in a reasonable agreement with the theoretical curves derived from the assumption that the stretching rate $\beta = 2$ (Sclater *et al.*, 1980). A similar study was carried out by Hoffmann *et al.*, (1984) for the Malakam Delta, Kalimantan, Indonesia, and Sajgo and Lefler (1983) for the Pannonian Basin.

It seems that the molecular approach to the maturation of organic material will be applicable to decipher the thermal history of a basin. But more studies (e.g. catalythic effects, exact, non-hypothetical measure of the kimematics of the reactions) are needed before these suggestions can be used with confidence.

9.3.3. Homogenization Temperature Method (Microthermometry)

Fluid inclusions are cavities filled with liquid and/or other (gas, solid) phases usually observed in rock constituting transparent crystals. They are found in nearly all minerals; in sedimentary rocks they are studied mainly in quartz, carbonates and feldspar. They are formed by crystal growth when the advancing faces, edges and corners of the growing crystal are disturbed (primary inclusions) or by fracturing and healing of crystals during mechanical disturbances (secondary inclusions) (Poty and Pagel, 1984).

Rock or mineral samples, having some crystals with fluid inclusion, in a form of thin (~80 μm) section, are heated up and observed under a petrographic microscope by using a Leitz or similar heating stage. At a certain temperature, the bubble (the phase boundary) disappears in the inclusion. This temperature is

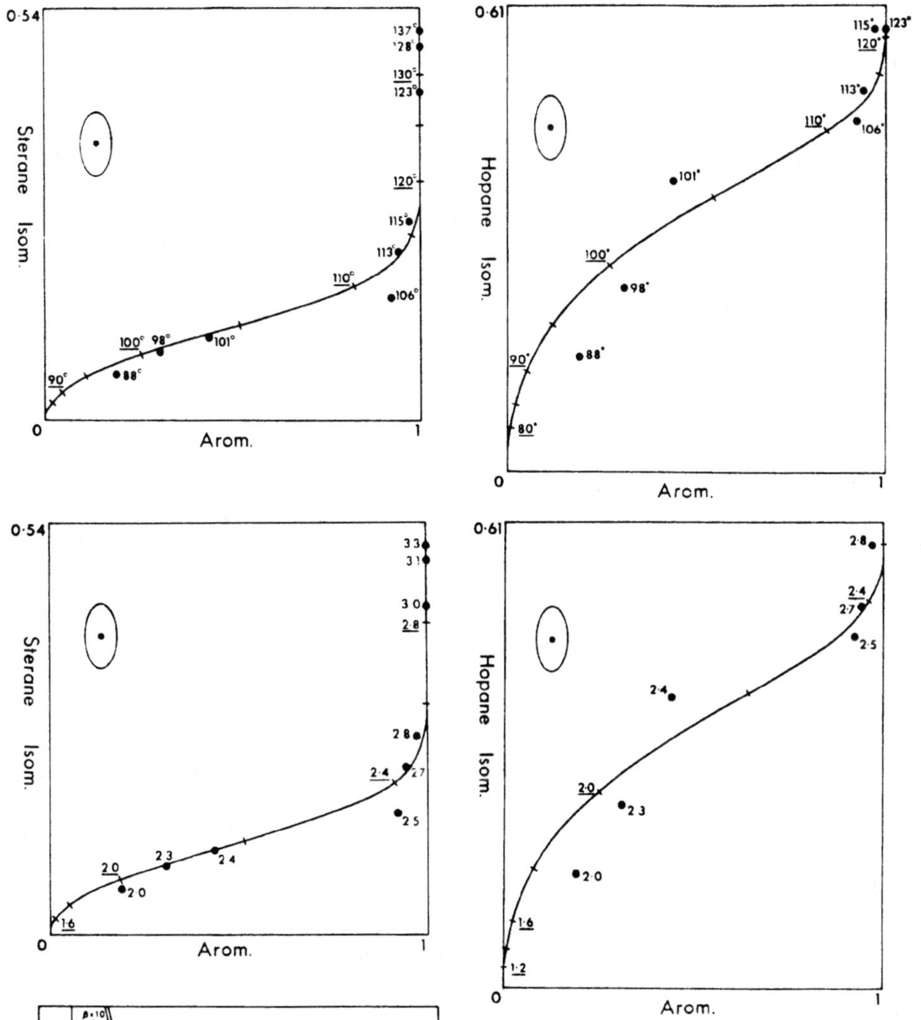

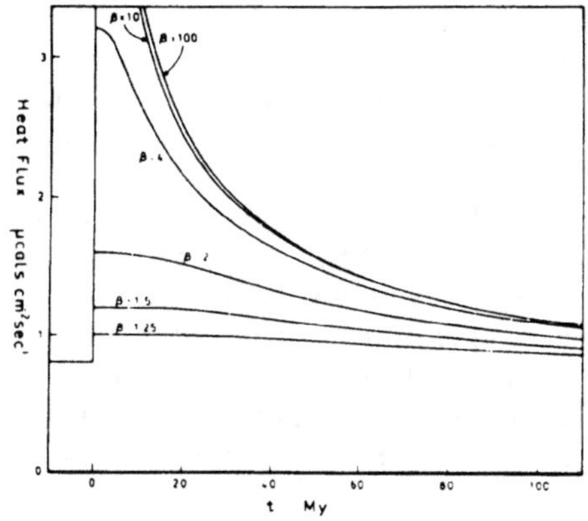

Fig. 9.45. The extent of sterane and of hopane isomerization as a function of steroid hydrocarbon aromatization for samples of the borehole HOD in the Pannonian Basin. The curves are calculated theoretically, based on the thermal history of the basin which comes out from McKenzies (1978) stretching theory, assumed the basin to have formed by sudden extension ($\beta = 2$), 15 Ma ago. The marks on the curves are present temperatures at 5 °C intervals (above) and present depths at 200 m intervals (below) (MacKenzie and McKenzie, 1983). The lowest left cartoon shows the approximate thermal history which belongs to different β values.

measured and called the homogenization temperature. The homogenization temperature, corrected by using the temperature-pressure-specific volume diagram for water (or salty water) gives the palaeo-temperature at which the mineral was formed.

Currie and Nwachukwu (1974) and Magara (1978) used this principle for determination of palaeo-geothermal gradients in Canadian Cardium sandstone as follows: Thin sections were made from fracture-filling materials (mainly quartz) of sandstone cores from 5 boreholes of a single reservoir. Those quartz fillings that contain fluid inclusions were heated and microscopically observed. The ranges of homogenization temperatures and calculated palaeo-geothermal and measured present geothermal gradients are shown on Table 9.4.

TABLE 9.4.
Ranges of homogenization temperatures and calculated geothermal gradients of Cardium sandstones (Currie and Nwachukwu, 1974; Magara, 1978).

Well	Homogenization temperature (°C)	Maximum burial depth geothermal gradient ($mK\ m^{-1}$)	Near present gradient ($mK\ m^{-1}$)	Present gradient ($mK\ m^{-1}$)
A	45–108	38	33	32
B	46–100	35	25	
C	50–85	33	33	
E	51–84	36	33	
F	49–88	33	31	

It was supposed that the highest homogenization temperature of a case refers to quartz fillings formed at maximum temperature and these of lowest homogenization temperature are originated from quartz fillings formed at lower ('near present') temperatures.

The palaeo-pressures were determined by sedimentation history, taking into consideration the rock compaction (using sonic logs). Figure 9.46 shows the diagram of interpretation: the measured maximum homogenization temperatures (A–F) mark out certain lines of specific volume of water. Following these lines till reaching the calculated palaeopressures we obtain the palaeotemperatures A′–F′. For the lowest homogenization-temperatures (a–f), the present pressure of cores (calculated by depth and density) were applied. The homogenization temperature method excels in its simplicity. There are however some points of consideration, as follow:

– Inclusions can contain not only water but a lot of other materials: salt, oil, CH_4, CO_2, etc. (Poty and Pagel, 1984). Roedder (1962, 1963) proposed to begin heating experiment at low (−35 °C) or even very low (−180 °C, Poty and Pagel, 1984) temperatures; the freezing temperatures are useful for discriminating among gas, liquid and supercritical fluid, and among liquid water, liquid oil and liquid CO_2; improving precision of the temperature determinations; estimating the minimum pressure at the time of deposition of certain samples etc.

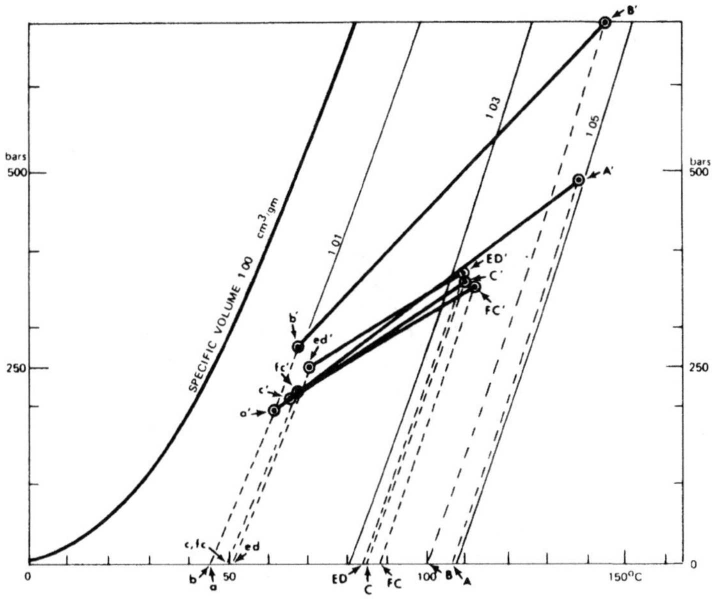

Fig. 9.46. Graphs showing ranges of homogenization temperature of Cardium sandstone in five (A–F) boreholes in western Canada and the interpreted paleotemperatures (a′–f′, A′–F′) (from Magara, 1978, modified). Lines with numbers 1.00 . . . 1.05 are specific volume of water.

– Fluid inclusions may also reequilibrate during burial. But experimental studies show that, in quartz, most fluid inclusions will not decrepitate at an internal overpressure lower than 8 MPa and that the smallest fluid inclusion can withstand overpressures as high as 40 MPa (Tugarinov and Naumov, 1970, Leroy, 1979).
– It is believed that fluid inclusion studies give minimal temperatures and only sometimes actual temperatures of inclusion formation. If the fluid was homogenous at the time of trapping, homogenization temperature will be a minimal temperature. The trapping temperature of the fluid in basins which are 5–7 km deep may be higher by a value up to 80°C, in comparison to the fluid inclusion temperature.

Poty and Pagel (1984) suggest that fluid inclusion techniques seem to give more detailed data than organic matter and illite crystallinity techniques.

Case Histories

Pagel (1975), using the fluid-inclusion microthermometry of detritic quartz grains of sandstones, received a 35 °C km^{-1} geothermal gradient for sedimentary basin in Athabasca (Canada) "which is characteristic for active sedimentary basins" (Figure 9.47). The fluids of inclusions exhibited a NaCl concentration of 30%.

Visser (1982) studied petroleum source rocks from Venezuela by fluid inclusion thermometry. Thin (~80 μm) sections were made, polished on both sides and mounted on coverslips with epoxy resin to give the necessary support. The

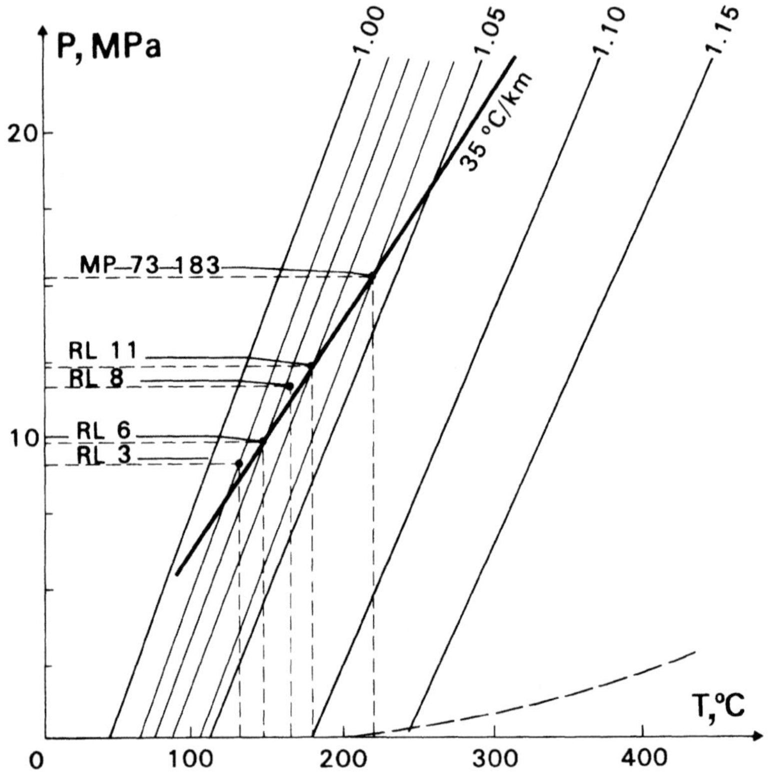

Fig. 9.47. PVT diagram for a 30 weight % NaCl solution and the calculated palaeotemperatures by microthermometry of detritic quartz grains from Athabasca sandstones of five (RL3 etc.) Canadian boreholes; after Pagel, 1975.

heating-freezing experiments were performed on a stage as described by Poty *et al.*, (1976). The fluid inclusion data of secondary inclusions in quartz- and calcite-filled veins showed that the maximum diagenetic temperature is in a good agreement with the actual measured formation temperature of 157 °C (Figure 9.48).

Based on primary inclusions – which were entrapped during the initial crystal growth – in diagenetic quartz overgrowths, Tillman and Barnes (1983) stated that the host rock (sandstone) temperatures in the Northern Appalachian Basin vary from 176 to 147 °C, with an average of 155 °C; the burial depth, at the end of the Palaeozoic or early in the Mesozoic, was about of 3.5 km (=~ 40 K km^{-1}). The average geothermal gradient measured today in central and western New York is about 25 K km^{-1}. The milky and clear calcite samples belong to the Oswego fault system and are mirroring a wider range of temperature of later hydrothermal events (Figure 9.49).

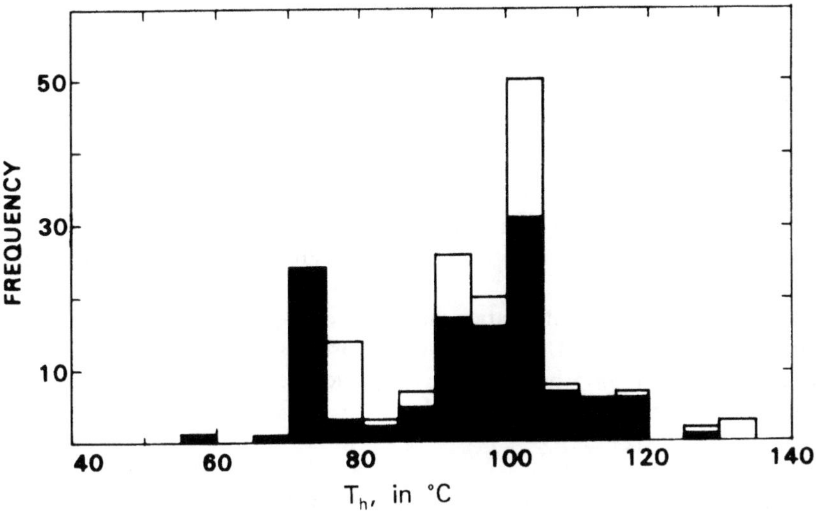

Fig. 9.48. Frequency diagram of homogenization temperatures in calcite (black) and quartz (white), in a petroleum source rock from Venezuela (Visser, 1982).

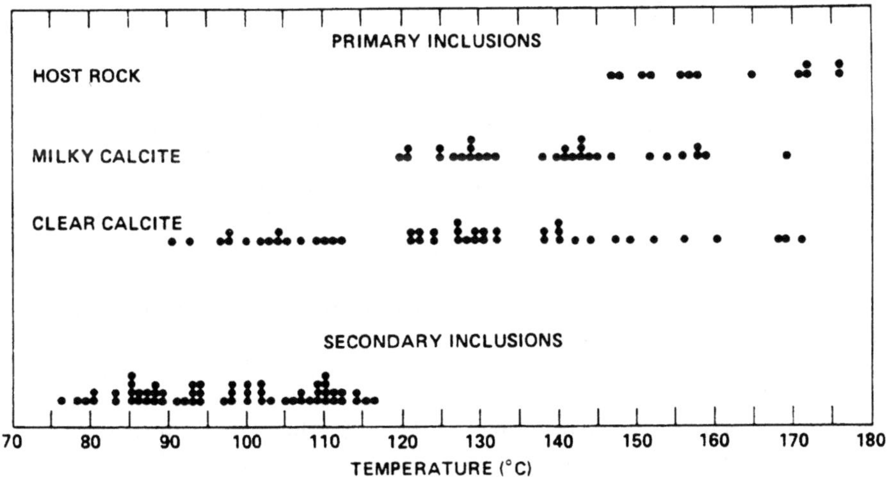

Fig. 9.49. Fluid inclusion temperature distribution in Northern Appachian basin sandstones. Tillman and Barnes (1983).

9.3.4. Other Methods

(1) CHEMICAL GEOTHERMOMETRY

Chemical geothermometry applied for fluid and vapor geothermal systems is based on the following requirements:

- Freely available elements (species) in the rock-water system
- Equilibrated reactions in the deep reservoirs
- Slow re-equilibration during upward migration

The following indicators are used (or mentioned in the literature):

1. Isotope exchange.

1.1. $HD+H_2O \rightleftharpoons H_2+HDO$ (Hulston, 1975; Arnorsson, 1975; Giggenbach and Lyon, 1977).

1.2. $CH_3D+H_2 \rightleftharpoons CH_4+HD$ (Hulston, 1976; Ellis and Mahon, 1977).

1.3. $H_2\,{}^{16}O+HS\,{}^{18}O^{16}O_3^- \rightleftharpoons H_2\,{}^{18}O+HS\,{}^{16}O_4^-$ (Hulston, 1976; Ellis and Mahon, 1977).

The time needed for re-equilibration does not exceed days or weeks for 1.1 and 1.2, and months for 1.3. Because of the shortness of this time, these isotope exchange reactions are poorly applicable for palaeo-geothermometry, sensu stricto.

2. Solute reactions.

2.1. $CO_2+4H_2 \rightleftharpoons CH_4+2H_2O$.

2.2. $2NH_3 \rightleftharpoons N_2+3H_2$.

The re-equilibration is too quick, also for these reactions.

3. Mineral solubility: Silica geothermometry (Fournier and Rowe, 1966; Mahon, 1966; Arnorsson, 1975; Ellis and Mahon, 1977).

The re-equilibration of an existing silica concentration in water is a very slow process, exists practically only above 600 K.

4. Ion exchange: Na/K ratio; for cooler waters Na−K−Ca. (White, 1965; Ellis and Mahon, 1977; Fournier and Truesdell, 1973; Paces, 1975).

The re-equilibration is also very slow, slower than for silica. As showed by this short enumeration, silica and Na/K thermometers are theoretically usable for palaeogeothermometry. The problem arises by the fact, that water, being mobile, migrates up- and downward through layers, and the dissolved silica or Na/K reflects first the temperature-differences caused by migration-processes; these methods are used indeed for the determination of the 'basic-temperature', the highest temperature encountered during migration. In special cases only, without water migration, silica and K/Na geothermometers can be used for palaeogeothermometry.

Kharaka (1980) have reported ^{18}O isotopic analyses from subsurface waters from a well of Brazoria County, Texas. The strata into consideration were deposited in marine or near-marine settings. During burial, formation water typically becomes enriched in ^{18}O as a result of reactions with the enclosing sediment. A plot of oxigen isotope values for calcium carbonate with depth shows a trend toward lighter values at greater depth (Milliken *et al.*, 1981), supporting the water isotope data, but with a big scatter. The results of Elders *et al.* (1984) from the ^{18}O content of calcites from sandstones in Cerro Prieto geothermal field boreholes are similarly scattering but meaningful (Figure 9.50). The four typical curves of Figure 9.50 reflect at first the cold and hot water migration processes and gave an important contribution to decipher the past fluid migration processes and thus, to estimate palaeogeotemperatures.

Isotope thermometry using *two minerals* has become well established since the classical paper of Urey (1947). The principle of this thermometry is that the

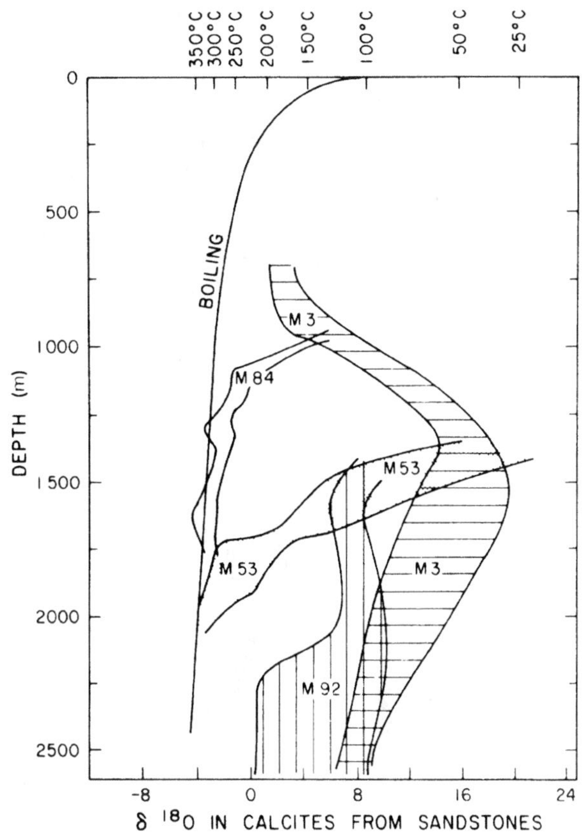

Fig. 9.50. Ranges of $\delta^{18}O$ in calcites from sandstone samples recovered from four different wells (M3, M53, etc.) of Cerro Prieto geothermal field. The shaded areas show the range of values measured. The 'boiling' curve shows the $\delta^{18}O$ of calcite in equilibrium with boiling ater with $\delta^{18}O$ = −8.33 ‰ (Elders *et al.*, 1984).

partitioning of two stable isotopes of an element between two minerals depends on the temperature. Isotopic compositions of two compounds (minerals) A and B are expressed as δ-values. The fractionation factor α of an isotope exchange reaction is related to the measured δ − values through the following approximation (Hoefs, 1983):

$$\delta_A - \delta_B = \Delta_{A-B} \approx 10^3 \ln \alpha_{A-B}. \tag{9.126}$$

Theoretical studies show that the fractionation factor, α, for isotope exchange between minerals is a linear function of $1/T^2$, where T is in degrees Kelvin, at crustal temperatures (Bottinga and Javoy, 1973):

$$1000 \ln \alpha = A/T^2. \tag{9.127}$$

For (palaeo) temperature determination, factor A has to be known. This is –

mostly poorly – known by theoretical calculations, by laboratory measurements or by empirical (geologic) considerations. Mostly oxigen, sulfur, hydrogen and carbon isotope fractionation reactions are used. The temperatures determined represent the last isotope equilibrium, after which no further isotope exchange took place. Temperatures obtained are often difficult to interpret: sometimes they indicate the peak metamorphic conditions, sometimes they indicate retrograde isotope exchange during the cooling period.

Figure 9.51 shows a comparison of temperatures obtained by fluid inclusions (microthermometry) and by sulfur isotope fractionations of the sphalerite-galena pair from two mines of Mexico and Central Peru (Hoefs, 1983).

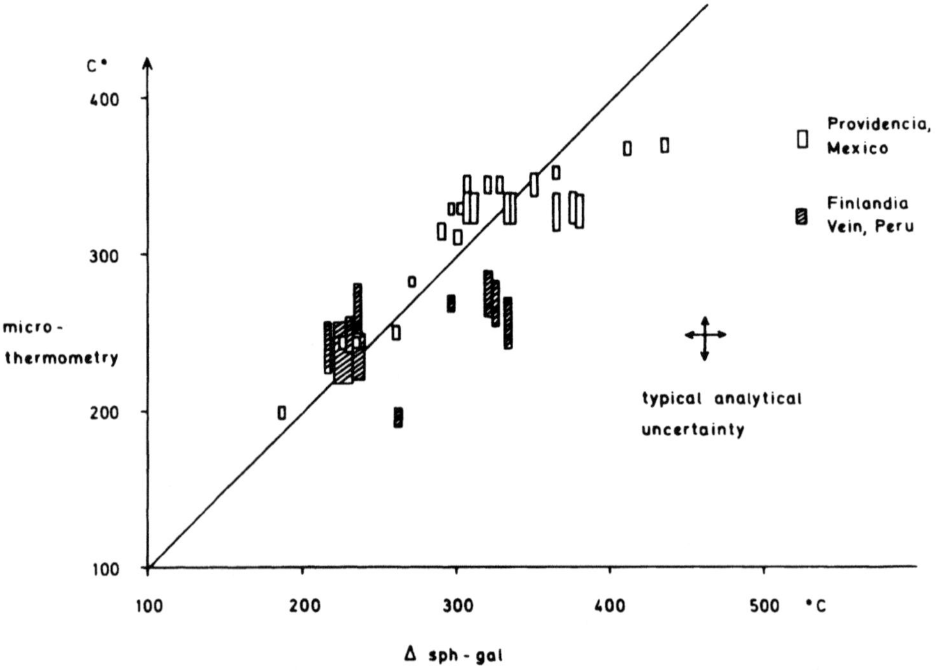

Fig. 9.51. A comparison of temperatures obtained by fluid inclusions (microthermometry) and by sulfur isotope fractionations of the sphalerite-galena pair, from a mine in Mexico and Peru (Hoefs, 1983).

However, it is suggested that most minerals, once formed in oxigen isotopic equilibrium with the ambient water, are extremely resistant to subsequent isotopic exchange with environmental waters at sedimentary and at most diagenetic temperatures except when they undergo chemical or mineralogical alteration. As a result, the $^{18}O/^{16}O$ ratios of minerals can provide information about their conditions of formation. If the $^{18}O/^{16}O$ ratios of two cogenetic mineral phases (e.g. illite and quartz, Figure 9.52) can be measured, then, both the temperature of formation and the $^{18}O/^{16}O$ ratio of the ambient water can be calculated.

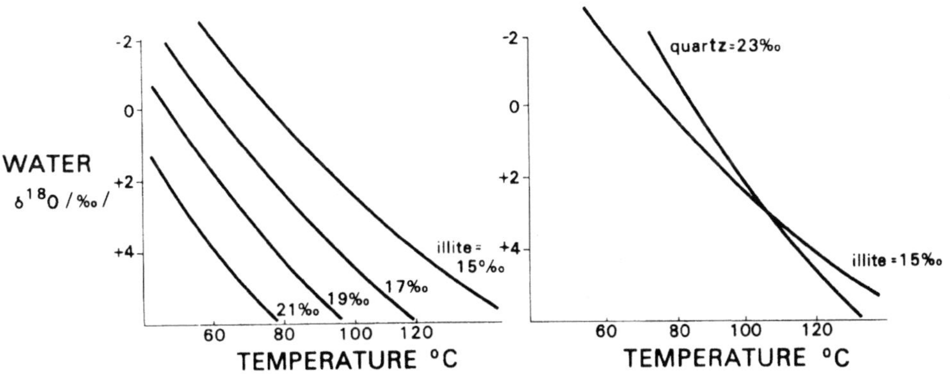

Fig. 9.52. Relationship between water and mineral $\delta^{18}O$ values (Savin and Lee, 1984). Left: lines indicating range of possible conditions of formation of illites with $\delta^{18}O$ values of +15, +17, +19 and +21 ‰, at various temperatures and $\delta^{18}O$ values of the ambient water. Right: $\delta^{18}O$ values of two cogenetic minerals uniquely define the temperature of formation and the isotopic composition of ambient water. This cogenetic mineral pair have been formed at about 106 °C in the presence of ambient water with $\delta^{18}O$ of about 3 ‰.

(2) HYDOTHERMAL ALTERATION OF MINERALS

The clay minerals of sediments, arriving at depths, are going through a diagenesis by influence of temperature and pressure. First the diagenesis of the expandable clay minerals, mostly of montmorillonite is of importance; the montmorillonite transforms, by means of water and CO_2 loss, to potassium-poor smectite/illite (nonexpandable clay) at growing temperature and pressure. The loss of the interstitial water of montmorillonite amounts to 10–15 % of the rock volume (Burst, 1969). This significant water mass can cause overpressures and probably plays an important role in the primary migration of hydrocarbons.

The processes resulting the diagenesis of montmorillonite begin at the depth corresponding to 80 °C geotemperature and generally end at 120 °C (Burst 1969; Jones, 1970). A part of montmorillonite remains below this depth because the absorbeable potassium being at disposal is not enough for the montmorillonite-illite transformation (Jones, 1970) – or transforms to a phase which is to be grouped to the pyrophyllites (Weiss and Roloff, 1965).

The above mentioned theoretical speculations do not reflect the applicability of montmorillonite/illite diagenesis for palaeogeothermics, regardless whether the sediment penetrated or not the 80–120 °C temperature limit, during its sedimentary history. It has been thought however that the montmorillonite-illite transformation may actually be a kinetically controlled process (Waples 1980), and the progress in transformation reflects the thermal history of the sediment. Waples (1980) stated a certain correlation between TTI and percent of expandable clay layers (Figure 9.53). Samples lying significantly left of the line (in thermally immature region), probably represent material which contained less than 100 % expandable layers when it was originally deposited. This is in agreement with the

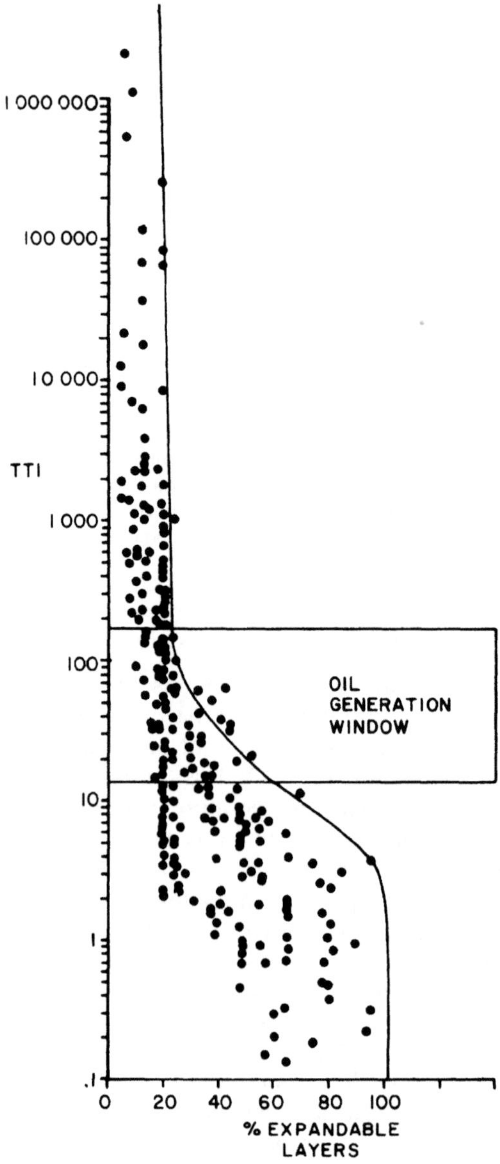

Fig. 9.53. Time Temperature Index of maturity versus % expandable layers in mixed-layer clays (Waples, 1980).

generally accepted view, that montmorillonite/illite ratio depends from the palaeogeographic environment too. It is believed that primary illite is associated with regressive, and montmorillonite with trangressive phases (Chapman, 1973). Because of this circumstance, palaeothermometry using clay mineral diagenesis will hardly be a quantitative method.

This statement is underlined by the studies of Elders *et al.*, (1984), on the Cerro Prieto geothermal system, Baja California, Mexico (Figure 9.54). Based on a lot of deep wells, they determined the progressive zones of hydrothermal alteration minerals in sandstones. The wide temperature-stripes and their overlap give some orientation referring to the applicability of this method for palaeogeothermal applications; for the Cerro Prieto field however, where very high temperature changes occurred as a consequence of a young (~50 000 a) hot (~1000 °C) shallow (~6 km) thermal plume, Elders *et al.*, (1984) constructed realistic thermal histories for the last 50 000 years.

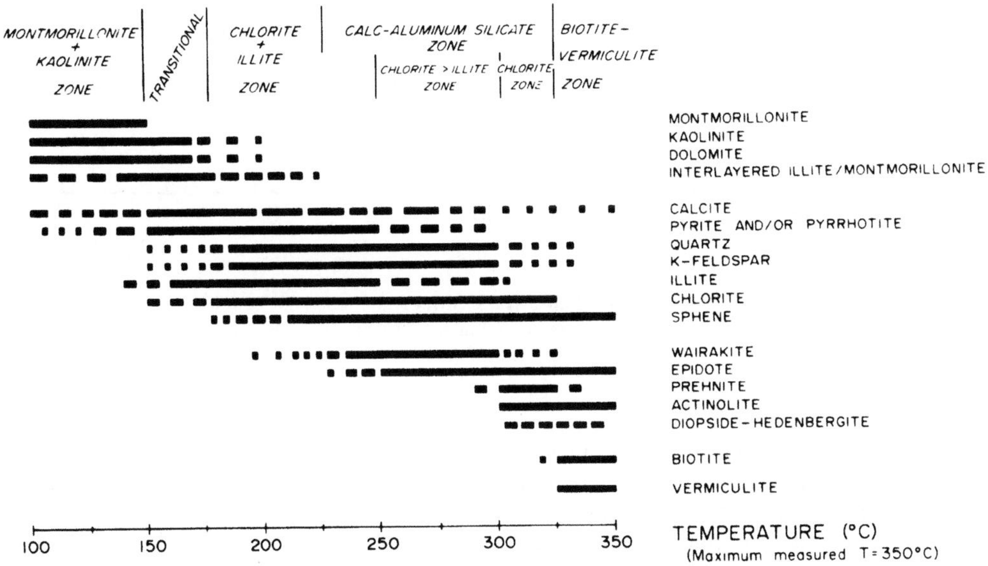

Fig. 9.54. Temperature ranges of zones of hydrothermal alteration minerals in the sandstones of Cerro Prieto geothermal field (Elders *et al.*, 1984).

(3) FISSION TRACK PALAEOGEOTHERMICS

It is supposed that apatite fission tracks are vanishing at 120 °C; similarly, the retention temperature for the Rb-Sr and K-Ar biotite systems lies at 300 °C, and for the K-Ar muscovite system at 350 °C. Based on these estimates, cooling histories can be derived by comparison of apatite fission track, mica Rb-Sr and mica K-Ar ages, as shown on Figure 9.55, on the example of the Bergell intrusive, Central Alps (Wagner *et al.*, 1979). Uranium-lead zircon data of the Bergell granite indicate that the intrusion is 30 Ma old. Muscovite K-Ar age determination of the granite gives 23.5 Ma, while the biotite Rb-Sr and K-Ar ages are 22 and 23 Ma respectively. Finally, the apatite fission track age was 13.5–14.0 Ma. From these data, cooling rates are calculated for the last 23.5 Ma, as shown on Figure 9.55.

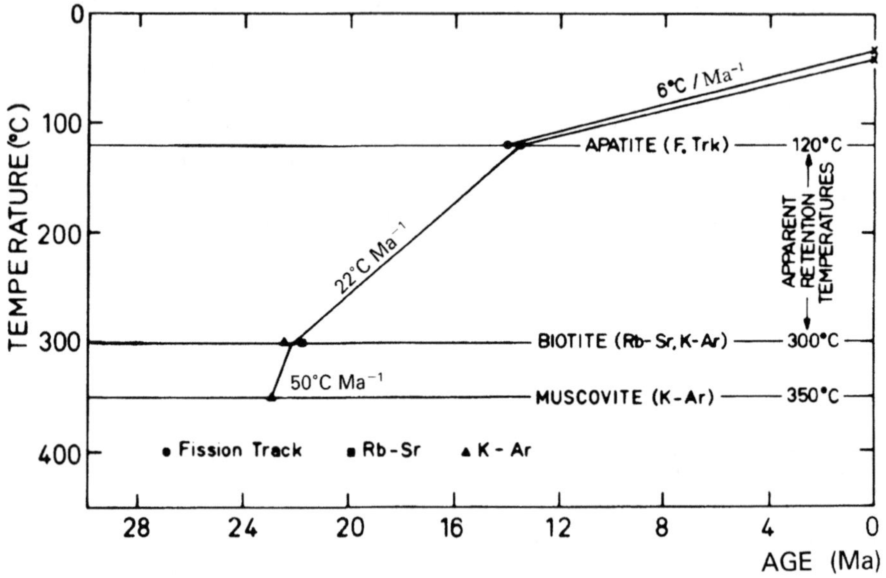

Fig. 9.55. Cooling history of the Bergell intrusive, Central Alps, based on the retention temperatures of Rb–Sr and K–Ar biotite, K–Ar muscovite and apatite fission track systems; after Wagner *et al.*, 1979, modified.

Except of certain uncertainities (Sanford and Elders, 1981, e.g. used 170 °C as the temperature annealing of ^{238}U fission tracks in detrital apatite of Cerro Prieto geothermal wells, corresponding to a duration of heating of 10^3-10^4 a), fission track palaeogeothermics seems to have some potential, to solve special problems.

9.3.5. References

Arnorsson, S.: 1975, 'Application of the Silica Geothermometer in Low-Temperature Hydrothermal Areas in Iceland', *Am. Jour. Sci.* **275**, 763–785.

Bostick, N. H.: 1973, 'Time as a Factor in Thermal Metamorphism of Phytoclasts (Coaly Particles)', *7th Congr. Int. Stratigr. Geol. Carbonif. Krefeld Compt. Rend.* **2**, 283–293.

Bottinga, Y. and Javoy, M.: 1973, 'Comments on Isotope Geothermometry', *Earth Plan. Sci. Lett.* **20**, 250–265.

Buntebarth, G.: 1979, 'Eine empirische Methode zur Berechnung von paleo-geothermischen Gradienten', in *Inkohlung und Geothermik, Fortschritte in der Geol.* Rheinl. Westfalen. 27, 97–108.

Burst, J. F.: 1969, 'Diagenesis of Gulf Coast, Clayey Sediments and its Possible Relation to Petroleum Migration', *A. A. P. G. Bull.* **53**, 73–77.

Chapman, R. E.: 1973, *Petroleum Geology*, Elsevier, 304 p.

Cummings, J. J. and Robinson W. E.: 1972, *Thermal Degradation of Green River Kerogen at 150 °C to 350 °C*, U. S. Bureau of Mines Rep. Invest. No. 7620.

Currie, J. B. and Nwachukwu, S. O.: 1974, 'Evidence of Incipient Fracture Porosity in Reservoir Rocks at Depth, *Bull. Can. Petr. Geol.* **22**, 42–58.

De Braemacker, J. C.; 1983, 'Temperature, Subsidence, and Hydrocarbon Maturation in Extensional Basins: A Finite Element Model', *A. A. P. G. Bull.* **67**, 1410–1414.

Du Rouchet, J.: 1980, 'The Program DIAGEN, Two Methods for Considering the Chemical Evolution of Organic Matter', *Bull. Centr. Rech. Expl.—Product., Elf-Aquit.* **4**, 813–831.

Eggen, S. S.: 1984; *Modelling of Subsidence, Hydrocarbon Generation and Heat Transport in the Norwegian North Sea*, IFP/Bernard Doronel.

Elders, W. A., Bird, D. K., Williams, A. E. and Schiffman, P.; 1984, 'Hydrothermal Flow Regime and Magmatic Heat Source of the Cerro Prieto Geothermal System, Baja California, Mexico', *Geothermics* **13**, 1/2, 27–47.

Ellis, A. J. and Mahon, W. A. J., 1977 *Chemistry and Geothermal Systems*, Acad. Press, New York, 392 p.

Falvey, D. A. and Deighton, L.: 1982, 'Recent Advances in Burial and Thermal Geohistory Analysis', *Journ. Austral. Petr. Expl. Assoc.* **22**, 65–81.

Fournier, R. O. and Rowe, J. J.: 1966, 'Estimation of Underground Temperatures from the Silica Content of Water from Hot Springs and Wet Steam Wells', *Am. Jour. Sci.* **264**, 658–697.

Fournier, R. O. and Truesdell, A. H.: 1973, 'An Empirical Na-K-Ca Geothermometer for Natural Waters', *Geoch. Cosmochim. Acta* **37**, 1255–1275.

Giggenbach, W. F. and Lyon, G. L.: 1977, *The Chemical and Isotopic Composition of Water and Gas from the Ngawha Geothermal Field*, Open File Rep. No. 30/555/7 Chem. Div., Petone.

Hacquebard, P. A.: 1977, 'Rank of Coal as an Index of Organic Metamorphism for Oil and Gas in Alberta', *Geol. Surv. Bull. Canada* **262**, 11–23.

Ho, T. T. Y.: 1978, *Manual of Organic Geochemical Interpretation*, Conoco, Ponca City, Oklahoma.

Hoefs, J.: 1986, *Isotope Geothermometers*, in G. Buntebarth and L. Stegena (eds.) Palaeogeothermics, Springer Verlag, 45–52.

Hoffmann, C. F.: MacKenzie, A. S., Lewis, C. A., Maxwell, J. R., Oudin, J. L., Durand, B. and Vandenbroucke, M.: 1984, 'A Biological Marker Study of Coals, Shales and Oils from the Mahakam delta, Kalimantan, Indonesia 1984', *Chem. Geol.* **42**, 1–23.

Hood, A., Gutjahr, C. C. M. and Heacock, R. L.: 1975, 'Organic Metamorphism and the Generation of Petroleum', *A.A.P.G. Bull.* **59**, 6, 986–996.

Hulston, J. R.: 1976, 'Isotope Work Applied to Geothermal Systems at the Institute of Nuclear Sciences', New Zealand, *Geothermics* **5**, 89–96.

Jones, P. H.: 1970, 'Geothermal Resources of the Northern Gulf of Mexico Basin', *UN Symp. Pisa Compt. Rend., Geothermics*, Spec. Issue **2**, 12–24.

Jüntgen, H. and Klein, J.: 1975, 'Formation of Natural Gas from Coaly Sediments', *Erdöl u. Kohle* **28**, 65–73.

Karweil, J.: 1956, 'Die Metamorphose der Kohlen vom Standpunkt der physikalischen Chemie', *Z. Deutscher Geol. Ges.* **107**, 132–139.

Kettel, D.: 1981, 'Maturity Calculations for the Upper Carboniferous of North-West Germany – A Test of Several Methods', *Erdöl Erdgas Ztschr.* **97**, 395–404.

Kharaka, Y. K.: 1980, *Geochemistry of Formation Waters from Pleasant Bayou No.2 Well*, Proc. 4th Geopressure-geothermal Energy Conf. Austin. Univ. Texas, 168–193.

Lasage, A.: 1981, 'Rate Laws of Chemical Reactions', in *Kinetics of Geochemical Processes, Reviews in Mineralogy* **8**, 1–68.

Leroy, J.: 1979, 'Contribution a l'etalonage de la pression interne des inclusions fluides lors de leur decrepitation', *Bull. Mineral.* **102**, 584–593.

Lopatin, N. V.: 1971, 'Temperature and Geologic Time as Factors in Coalification', *Izv. Acad. Sci. USSR, Ser. Geol.* **5**, 95–100.

Lopatin, N. V.: 1976, *Determination of the Influence of Temperature and Geologic Times on the Katagenic Processes of Coalification and Oil- and Gas-Formation*, Int. Geol. Congr. XXV. Session, Moscow, 361–366.

Lopatin, N. V. and Bostick, N. H.: 1973, 'Geologic Factors in Coalification', in *Priroda Organicheskogo Veshchestva Sovremennykh i Iskopaemykh Osadkov*, Nauka, Moscow, 79–90.

MacKenzie, A. S. and McKenzie, D. P.: 1983, 'Isomerization and Aromatization of Hydrocarbons in Sedimentary Basins Formed by Extension', *Geol. Mag.* **120**, 417–528.

Magara, K.: 1978, 'Compaction and Fluid Migration. Practical Petroleum Geology', *Developments in Petroleum Science* **9**, 319.

Mahon, W. A. J.: 1966, 'Silica in Hot Water Discharged from Drillholes at Wairakei, New Zealand', *N.Z. Jour. Sci.* **9**, 125–144.

Majorowicz, J. A. and Jessop, A. M.: 1981, 'Present Heat Flow and a Preliminary Palaeogeothermal History of the Central Prairies Basin, Canada', *Geothermics* **10**, 81–93.

McKenzie, D. P.: 1978, 'Some Remarks on the Development of Sedimentary Basins', *Earth Planet. Sci. Lett.* **40**, 25–32.

McKenzie, D. P.: 1981, 'The Variation of Temperature with Time and Hydrocarbon Maturation in Sedimentary Basins Formed by Extension', *Earth Planet. Sci. Lett.* **55**, 87–98.

Middleton, M. F. and Falvey, D. A.: 1983, 'Maturation Modeling in Otway Basin, Australia', *A.A.P.G. Bull.* **67**, 271–279.

Milliken, K. L., Land, L. S. and Loucks, R. G.: 1981, 'History of Burial Diagenesis Determined from Istopic Geochemistry, Frio Formation, Brazoria County, Texas', *A.A.P.G. Bull.* **65**, 1397–1413.

Paces, T.: 1975, 'A Systematic Deviation from Na-K-Ca Geothermometer below 75 °C and above 10^{-4} atm. P_{CO_2}', *Geoch. Cosmochim. Acta* **39**, 541–544.

Pagel, M.: 1975, 'Determination des conditions physico-chimiques de la silification diagenetique des gres Athabasca (Canada) au moyen des inclusions fluides', *C.R. Acad. Sc. Paris* **280**, D, 2301–2304.

Poty, B., Leroux, J. and Jachimowicz, L.: 1976, 'Un nouvel appareil pour la mesure des temperatures sous le microscope'. *Bull. Soc. Francais Mineral. Cristallogr.* **99**, 182–186.

Poty, B. and Pagel, M.: 1984, *Fluid Inclusion Techniques to the Study of Thermal Evolution of Sedimentary Basins*, Colloque International Phenomenes Thermiques dans les Bassins Sedimentaires, 6–10 Juin 1983, ADERA, Saint-Médard-en-Jalles/France, Paris.

Price, L. C.: 1982, 'Time as a Factor in Organic Metamorphism and Use of Vitrinite Reflectance as an Absolute Palaeogeothermometer. *A.A.P.G. Bull.* **66**, 619–620.

Pusey, C. W.: 1973, *How to Evaluate Potential Gas and Oil Source Rocks*, World Oil, April, 71–76.

Roedder, E.: 1962, 'Studies of Fluid Inclusions I', *Econ. Geol.* **57**, 1045–1061.

Roedder, E.: 1963, 'Studies of Fluid Inclusions II', *Econ. Geol.* **58**, 167–211.

Royden, L. and Keen, C. E.: 1980, 'Rifting Process and Thermal Evolution of the Continental Margin of Eastern Canada Determined from Subsidence Curves', *Earth Planet. Sci. Lett.* **51**, 343–361.

Royden, L., Sclater, J. G. and Von Herzen, R. P.: 1980, 'Continental Margin Subsidence and Heat Flow: Important Parameters in Formation of Petroleum Hydrocarbons', *A.A.P.G. Bull.* **64**, 2, 173–187.

Rybach, L.: 1984, 'The Palaeogeothermal Conditions of the Swiss Molasse Basin', *Inst. Français Petr. Rev.* **39**, 2, 143–146.

Sajgo, Cs. and Lefler, J.: 1983, 'A Reaction Kinetic Approach to the Temperature-Time History of Sedimentary Basins', in G. Buntebarth and L. Stegena (eds.) *Palaeogeothermics*, Springer Verlag, 119–152.

Sanford, S. J. and Elders, W. A.: 1981, *Dating Thermal Events at Cerro Prieto Using Fission Track Annealing*, Proc. Third Symp. Cerro Prieto Geothermal Field, San Francisco. Lawrence Berkeley Lab. Rep. LBL-11967, 123–133.

Savin S. K. and Lee M.: 1984. 'Estimation of subsurface temperatures from oxigen isotope ratios of minerals', in print.

Sclater, J. G., Royden, L., Horvàth, F., Burchfiel, B. C., Semken, S. and Stegena, L.: 1980, 'The Formation of the Intra-Carpathian Basins as Determined from Subsidence Data', *Earth Planet. Sci. Lett.* **51**, 139–162.

Siever, A.: 1983, 'Burial History and Diagenetic Reaction Kinetics', *A.A.P.G. Bull.* **67**, 4, 684–691.

Staplin, F. L.: 1969, 'Sedimentary Organic Matter, Organic Metamorphism and Oil and Gas Occurrence', *Can. Petr. Geol. Bull.* **17**, 47–66.

Stegena, L., Horvàth, F., Sclater, J. G. and Royden, L.: 1981, 'Determination of Palaeotemperature by Vitrinite Reflectance Data', *Earth Evol. Sci.* **3–4**, 292–300.

Teichmüller, M.: 1971, 'Application of Coal-Petrographic Methods in Petroleum and Natural Gas Prospecting', *Erdöl u. Kohle* **21**, 69–76.

Tillman, J. E. and Barnes, H. L.: 1983, 'Deciphering Fracturing and Fluid Migration Histories in Northern Appalachian Basin', *A.A.P.G. Bull.* **67**, 692–705.

Tissot, B.: 1969, 'First Data on the Mechanism and Kinetics of the Formation of Petroleum in Sediments', *Inst. Francais Petr. Rev.* **24**, 470–50.

Tissot, B., Deroo G. and Espitalie, J.: 1975, 'Etude comparée de l'epoque de formation et d'expulsion du petrole dans diverses provinces geologiques', *Ninth World Petr. Congr. Proc.* **2**, 159–169.

Tissot, B. and Espitalie, J.: 1975, 'L'evolution thermique de la materie organique des sediments: applications d'une simulation mathematique', *Inst. Francais Petr. Rev.* **30**, 743–777.

Tugarinov, A. I. and Naumov, V. B.: 1970, 'Dependence of the Decrepitation Temperature of Minerals on the Composition of their Gas-liquid Inclusions and Hardness', *Dokl. Acad. Sci. USSR* **195**, 112–114.

Urey, H.: 1947, 'The Thermodynamic Properties of Isotopic Substances', *J. Chem. Soc.*, pp. 562–682.

Visser, W.: 1982, 'Maximum Diagenetic Temperature in a Petroleum Source Rock from Venezuela by Fluid Inclusion Thermometry', *Chem. Geol.* **37**, 95–101.

Wagner, G. A., Miller, D. S. and Jäger, E.: 1979, 'Fission Track Ages on Apatite of Bergell Rocks from Central Alps and Bergell Boulders in Oligocene Sediments', *Earth Plan. Sci. Lett.* **45**, 355–360.

Wang, Ji-an, Wang Ji-yang, Wang Jun, Huang, Ge-shan, Yan Shu-zhen and Lu Xin-wen: 1983, 'Geothermal Studies in Oil Field District of North China', in G. Buntebarth and L. Stegena (ed.), *Palaeogeothermics*, Springer Verlag, 195–204.

Waples, D. W.: 1980, 'Time and Temperature in Petroleum Formation: Application of Lopatin's Method to Petroleum Exploration', *A.A.P.G. Bull.* **64**, 916–926.

Waples, D. W.: 1983, 'Physical-Chemical Models for Oil Generation', *Colorado School of Mines Quarterly* **78**, 1–4.

Weiss, A. and Roloff, G.: 1965, 'Über die Rolle glimmerartiger Schichtsilikate bei der Entstehung von Erdöl und Erdöllagerstätten', M.S. Thesis, Heidelberg.

Welte, D. H. and Yukler, M. A.: 1981, 'Petroleum Origin and Accumulation in Basin Evolution – A Quantitative Model', *A.A.P.G. Bull.* **65**, 1387–1396.

White, D. E.: 1965, 'Saline Waters of Sedimentary Rocks', in: *Fluids in Subsurface Environments, Symp., Mem. Am. Petr. Geol.* **4**, 342–366.

9.4. ENGINEERING APPLICATIONS

G. DELISLE
Department of Geophysics, Federal Institute for Geosciences and Natural Resources, Hannover, FRG

9.4.1. Introduction

The expansion of geothermics into the geotechnical field is the subject of this chapter. While the scope of this book does not allow a complete overview of the whole range of topics, a number of exemplary problems are discussed below to give an insight into the bandwidth of engineering geothermics.

9.4.2. Subsurface Temperature Distribution

Any mining or subsurface construction activity requires in the planning stage some knowledge of the subsurface temperature field. High rock temperatures at depth might well preclude economically worthwhile mining activities. Beyond a certain depth, the operating costs for ventilation become prohibitive. In Northern Germany this limiting depth lies at approximately 1200 m, which corresponds to a rock temperature level of close to 50 °C.

The prediction of the subsurface temperature distribution in an area of low topographic relief and flat or gently dipping sediments should be straightforward, if the pertinent boundary conditions – mean annual surface temperature T_0, thermal conductivity λ of the rock units in question and the regional terrestrial heat-flow density q_r – are known. The temperature–depth function can then be derived directly by applying the heat-flow density Equation (2.2).

A more complex situation arises, if significant topographic relief or substantial lateral thermal conductivity variations exist. A reliable prediction of the subsurface temperature distribution should then be based on a detailed mathematical analysis.

Again, T_0, q_r and a reasonable set of thermal conductivity values for the different rock types need to be known. With respect to the latter two points, we refer for a more detailed discussion to Chapters 4.1 and 5.

(1) MEAN ANNUAL SURFACE TEMPERATURE DISTRIBUTION

The mean annual temperatures at the Earth's surface are commonly available from climatological maps. Meteorological surveys usually register soil temperatures for agrometeorological purposes in depths of 0.2, 0.5, 1.0, 2.0 m and so on. The extrapolation to 0 m of these values averaged over many years exhibit very useful values for T_0.

T_0 depends to some extent on the type of vegetation at the surface. Bare surfaces tend to possess a higher T_0 than meadows and these again a higher value than forests. According to a field study over a period of one year (Kappelmeyer

R. Haenel, L. Rybach, and L. Stegena (eds.)
Handbook of Terrestrial Heat-Flow Density Determination, 421—448.

and Haenel, 1974), T_0 of a meadow exceeded T_0 of a forest by 1.1 °C. Moore and Fosberg (1971) and Haenel (1980) found an average difference of 2.0 ± 0.5 °C between forested and cultivated soils.

The T_0-value cannot be derived directly from temperature logs. The thermal influence of the daily (and annual) temperature variation controls the temperature distribution of the uppermost 10–50 cm (10 to 15 m) below surface. Therefore, in all likelyhood, measured near-surface T-values are not identical with the correct T_0-value.

Microclimatic variations may exert a significant influence on the subsurface temperature distribution in hilly terrain. Blackwell *et al.* (1980) based on field measurements on Wilbur Hill/Washington argue, that slope inclination and orientation may play an important role in defining the mean annual temperature distribution. Figure 9.56 shows the subsurface temperature distribution of Wilbur Hill/Washington as inferred by the authors. The 7 °C isotherm parallels the north slope, on the southern slope a continuous decrease of surface temperatures from 12 °C at the base to 7 °C at the 300 m higher top is inferred. The authors argue that the amount of effective solar radiation as function of slope inclination and orientation play an important part in controlling the surface temperatures.

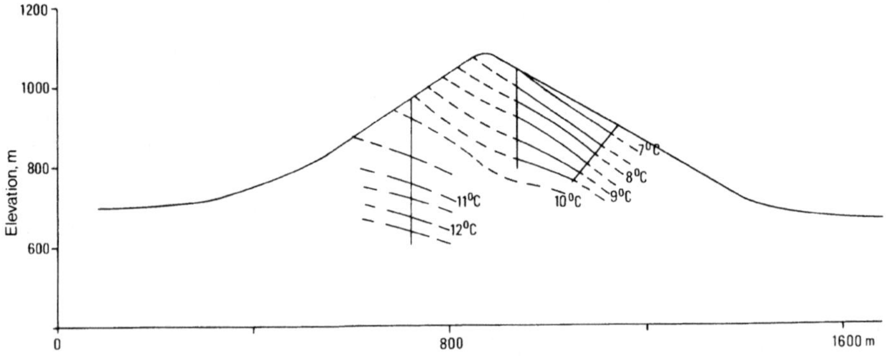

Fig. 9.56. Temperature cross section for a hill near Wilbur/Washington as inferred from measured temperatures in three drill holes shown on the section (after Blackwell *et al.*, 1980).

As a general rule it can be stated that the variation of ground temperature with elevation amounts to about −7 °C (−4 °C) km^{-1} in dry (wet) climate.

Unusual near-surface temperatures can also be found in the vicinity of settlements. Soil temperatures of 2 °C above normal under populated areas have been observed.

(2) CALCULATION OF THE NATURAL SUBSURFACE TEMPERATURE FIELD – AN EXAMPLE

The thermal conductivity of rock salt surpasses that of common sedimentary rocks by a factor of 2 to 2.5 at near surface temperatures. The contrast reduces with increasing rock temperatures. At temperatures higher than 200 °C the thermal conductivity of rock salt reaches the same level than other sedimentary rocks; see Figure 9.57 (and also Figure 10.1).

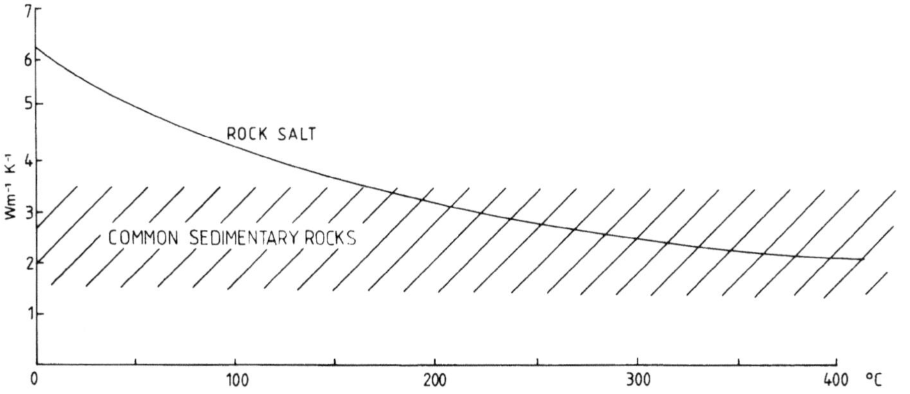

Fig. 9.57. The thermal conductivity of rock salt in W m^{-1} K^{-1} (data taken from Birch and Clark) surpasses that of other common sedimentary rocks by a factor of about 2 at low to moderate rock temperatures.

The lower thermal resistivity of rock salt at near surface rock temperatures causes a concentration of heat flow through the salt structure. Heat-flow density over a salt dome frequently surpasses the regional q_r-value by a factor of 2 or more. As a consequence, slightly reduced heat-flow density values at the salt dome flanks are observed in comparison to q_r.

The increase of heat-flow density over the salt dome depends on factors such as thickness of the sedimentary cover, depth, width and length of the salt structure and thermal conductivity of the surrounding rocks. Any calculation of the natural temperature field of a salt dome requires, therefore, foremost an approximate knowledge of its shape, its internal structure and the mineralogy of its salt units.

Equally important is the knowledge of the configuration and the thermal conductivity of the surrounding host rocks. The higher the λ-contrast between the salt structure and the host rock, the more pronounced will be the T-anomaly.

Several papers have treated the temperature field of a salt dome numerically. The papers of Selig and Wallick (1966) and Giesel and Holz (1970) took constant conductivity value for rock salt into consideration. Delisle (1980) has calculated the natural temperature field from an idealized model of a salt dome in Northern Germany (Figure 9.58A) using a polynomial function for $\lambda(T)$ of rock salt based on Birch and Clark (1940):

$$\lambda(T) = 5.734 - 1.83 \times 10^{-2} \times T + 2.86 \times 10^{-5} \times T^2 - 1.51 \times 10^{-8} \times T^3 \text{ (W m}^{-1}\text{ K}^{-1}\text{)}, T \text{ in °C.} \quad (9.128)$$

The near-surface T-increase over the salt dome and the negative T-anomaly at the root of the structure are recognizable (Figure 9.58B).

By calculating the temperature field of a salt dome, great care has to be taken in choosing the lateral and depth extent of the cross section, in which the temperature field will be computed (not shown in Figure 9.58B). The lateral and the depth extent of the thermal anomaly around and below the dome is at least of the same order as the lateral and vertical extent of the dome respectively. These

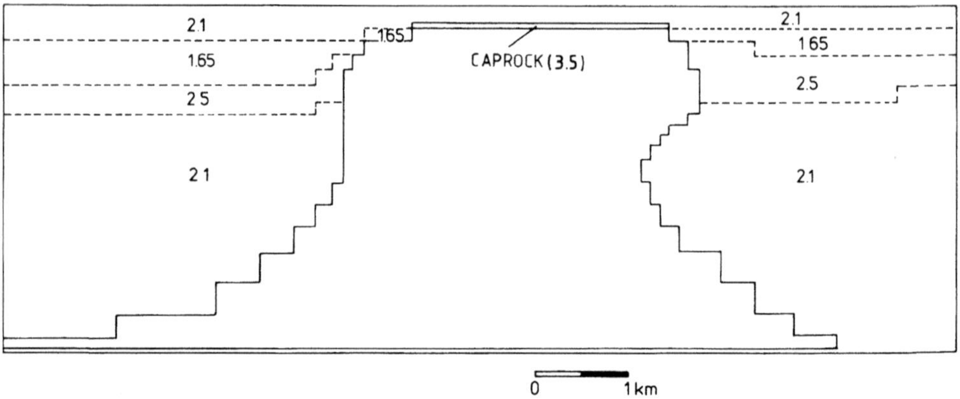

Fig. 9.58A. Calculation of the natural temperature field of a salt dome, the shape and the λ-distribution in the salt structure and the surrounding host rocks has to be defined (no vertical exaggeration). The thermal conductivity of rock salt is given by Equation (9.128). The temperature field below (Figure 9.58B) is based on λ-distribution as shown (λ in $Wm^{-1} K^{-1}$).

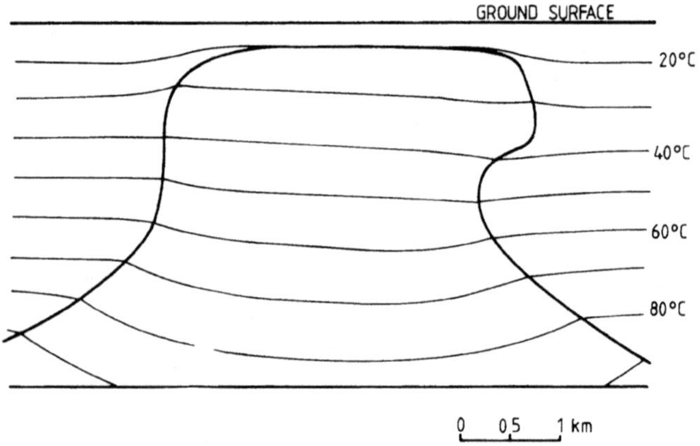

Fig. 9.58B. Calculated temperature field within the salt dome. At the top, an increase, at its base, a decrease of rock temperatures results from the given λ-distribution (after Delisle, 1980).

effects are well documented by Jensen (1983) on the natural temperature field of the Verum salt dome in Denmark.

The proper choice of a regional heat-flow density value for such a model might be difficult, since salt domes rarely occur in isolation. Large areas around it must be suspected to be within the radius of influence of such a structure. Overmore, drilling in such an area is often restricted to the flanks of a salt dome, where traps for oil and natural gas may exist. However, salt dome flanks are generally perturbed (see Figure 9.58B) and q_r cannot properly be determined in such boreholes.

The boundary condition on both sides of the cross section to be computed will usually be chosen to be $q_{\text{horizontal}} = 0$, indicating no heat flow through the lateral boundaries. At the top of the cross section the boundary condition will adequately be defined by fixing the temperature to be constant and equal to the mean annual surface temperature.

Given these boundary conditions, the natural temperature field can then be evaluated with the aid of Equation (2.4):

$$\rho c \frac{\partial T}{\partial t} = \mathrm{div}(\lambda \text{ grad } T).$$

The calculation requires the application of numerical methods (see discussion in Chapter 2.4).

9.4.3. Practical Aspects of Heat Transport by Convection

(1) HYDROGEOLOGICAL PROBLEMS

A significant redistribution of the terrestrial heat-flow density can be caused by ground water movement. By knowing the resulting subsurface temperature field, the underlying flow velocity field can be calculated. This approach is a widely used technique in ground water exploration, where T-logs from exploration bore holes are measured on a routine basis. A simple model to describe the thermal effect of vertical ground water flow was given by Bredehoeft and Papadopulos (1967). For details see Chapter 2.3.2.

Thermally effective convective heat transport is restricted to porous-permeable zones or fractures. The volume of the permeable rock zone in comparison to the adjacent rock volume being thermally influenced is frequently quite small. Heat-flow density measurements in or near to the zone of active subsurface water flow will yield useful information with respect to the definition of the boundary conditions required for any numerical model.

We restrict ourselves to the discussion of simplified approaches to describe heat transfer by convection.

A simple but often applicable mathematical expression describes the thermal effect (neglecting dispersion) of an advancing warm or cold water front in an aquifer of constant thickness. The thermal front, starting from $x = 0$ at $t = 0$, displaces with constant speed v the original aquifer fluid. Neglecting the heat flow from or to the less- or impermeable zones that bound the aquifer on both sides, the temperature in the aquifer as function of space and time is given by (Carslaw and Jaeger, 1959):

$$T(x, t) = \frac{\Delta T}{2}\left(1 - \mathrm{erf}\left(\frac{x - vt}{2\sqrt{at}}\right)\right) \tag{9.129}$$

where

ΔT = temperature difference at $t = 0$ between fluid and rock matrix
a = thermal diffusivity
x = distance in flow direction
v = velocity of fluid (Darcian flow)
t = time.

Homogeneous conditions – constant permeability, fluid flow and thermal diffusivity – are assumed. The attainment of thermal equilibrium between the fluids and the rock matrix is considered to be instantaneous.

Lauwerier (1955) has extended this approach by including an approximate solution for the temperature field in the bounding layers of the aquifer. The following simplifying assumptions were made:
- neglecting heat conduction in the flow direction
- thickness of aquifer being constant (= 2 h)

The resulting mathematical expression for the temperature as function of space and time is given by:

$$T(x, z, t) = \triangle T \operatorname{erfc}\left(\frac{\xi + /\eta/ - 1}{2\sqrt{\theta(\tau - \xi)}}\right) U(\tau - \xi) \qquad (9.130)$$

$$\xi = \frac{x\lambda_2}{h^2\rho_w c_w v}; \quad \tau = \frac{t \cdot \lambda_2}{h^2\rho_1 c_1} \quad \eta = \frac{z}{h} \quad \theta = \frac{\rho_1 c_1}{\rho_2 c_2}$$

U = unit function; $U(\xi) = 0$ for $\xi < 0$, $U(\xi) = 1$ for $\xi \geqslant 0$.

where

$\triangle T$ = temperature difference at $t = 0$ between fluid and rock matrix
x = coordinate in flow direction
z = coordinate perpendicular to flow direction; $z = 0$ at midpoint of aquifer
ρ = density
c = specific heat capacity
t = time
λ = thermal conductivity
v = velocity of fluid (Darcian flow)

Subscripts

1 = permeable zone
2 = bounding zone
w = water.

By reducing h to near zero values, the problem of heat transfer in a fracture and the adjacent rock matrix is touched upon.

(2) CONVECTIVE HEAT TRANSFER IN FRACTURED MEDIA

Fractures form often the dominant pathway for fluids in subsurface systems like geothermal reservoirs. Geothermal production wells are frequently fed exclusively by a small number of fractures they have penetrated. Two very basic cases of convective heat transfer in fractured media, for which analytical solutions have been worked out, are presented below.

A numerical approach to this type of problem is not necessarily advantageous over an analytical solution, since the accuracy of the solution depends critically on the accurate estimation of heat flow from the rock matrix into the fracture. Since the proper evaluation of this component requires irrespective of the chosen numerical method a high degree of discretization of the space around the fracture, the needed computer space and computing time will be substantial. In some cases, it might be in no relation to the gained additional information in comparison to the results from analytical solutions for somewhat simplified models.

PROBLEM 1. The reinjection of water (in most cases so-called waste waters, which result from the cooling of produced geothermal fluids) into the geothermal reservoir may aid in maintaining high reservoir pressures. On the other side, it might cause unwanted premature cooling of the geothermal reservoir. The temperature field around a fracture into which cold water is injected, is, therefore, from the point of view of developing a geothermal field of great interest.

The flow of water at a constant flow rate v into a fracture of constant width h is assumed. The host rock is considered as completely impermeable. According to the basic heat transport Equation (2.16), neglecting for the time being heat transport by conduction, one can write for any infinitesimal small fracture volume (see e.g. Lauwerier, 1955):

$$h\rho_f c_f \frac{\partial T_f}{\partial t} + h\rho_w c_w v \frac{\partial T_f}{\partial x} = 0 \tag{9.131}$$

where

h = fracture width
ρ = density
c = specific heat capacity
T = temperature
t = time
x = coordinate in fracture plane
v = velocity of fluid (Darcian flow).

Subscripts

f = fracture fillup (fluid and solids, e.g. sand)
w = fluid (water, oil)
r = rock.

We assume negligible heat conduction in the fluid phase as well as no heat conduction parallel to the fracture. The heat-flow density q across both fracture surfaces perpendicular to the fracture plane is given by the term:

$$q = 2\lambda \left.\frac{\partial T_r}{\partial z}\right|_{z=0}$$

where

λ = thermal conductivity of the rock bounding the fracture
z = coordinate perpendicular to the fracture plane.

No energy resistance at the interface rock – fluid in fracture is assumed (infinite heat transfer coefficient). Given these simplifying assumptions, the heat balance can be written as:

$$\rho_f c_f \frac{\partial T_f}{\partial t} + \rho_w c_w v \frac{\partial T_f}{\partial x} - \frac{2\lambda}{h} \frac{\partial T_r}{\partial z}\bigg|_{z=0} = 0 \tag{9.132}$$

for the case of a water front advancing in a fracture at constant speed.

If the injection of water occurs from a point source, then the fluid mass flow Q_m (m^3 s^{-1}) will decrease within the fracture radially (r) from the reinjection point (buoyancy forces not considered) by the factor:

$$\frac{1}{2\pi rh}$$

and the Equation (9.132) must be written as:

$$\rho_f c_f \frac{\partial T}{\partial t} + \frac{Q_m \rho_w c_w}{2\pi rh} \frac{\partial T_f}{\partial r} - \frac{2\lambda}{h} \frac{\partial T_r}{\partial z}\bigg|_{z=0} = 0 \tag{9.133}$$

(see also Lauwerier, 1955; Bodvarsson, 1969; Gringarten and Sauty, 1975).

The thermal gradient $\frac{\partial T_r}{\partial z}\Big|_{z=0}$ can be obtained by solving Equation (2.5) for the appropriate boundary conditions. Some solutions for this set of equations are discussed below:

Bodvarrson (1972) solved this problem for the further simplifying assumption of negligible width h. then, the advance of the radially propagating temperature front is given by:

$$T(r, t) = \triangle T \operatorname{erf}\left(\frac{\left(\frac{2\pi r^2 \lambda}{c_f \rho_f Q_v} + z\right)}{2\sqrt{at}}\right) \tag{9.134}$$

where

$\triangle T$ = temperature difference between initial rock temperature and temperature of injected water

Q_v = volume of injected water per unit time from a point source into the fracture.

If the width h is not neglected, the set of above equations can be solved analyticaly with the aid of the Laplace transform. The inversion of the solution in the Laplace domain must be carried out by numerical inversion.

Bodvarsson and Tsang (1982) have worked out solutions to the problem of heat transfer of equally spaced (=2D) horizontal fractures.

The presented solutions are based on a model as pictured in Figure 9.59. At $r = 0$, at times $t \geq 0$, water with the constant temperature T_i is injected into a fracture with a constant rate Q. Flow in the fracture occurs radially from the injection point. The initial rock temperature at $t = 0$ is uniformly at T_0, the rock matrix is impermeable for water.

If the temperature field for the case of a single fracture is to be considered, a large value for D must be chosen; otherwise, if a set of parallel and evenly spaced

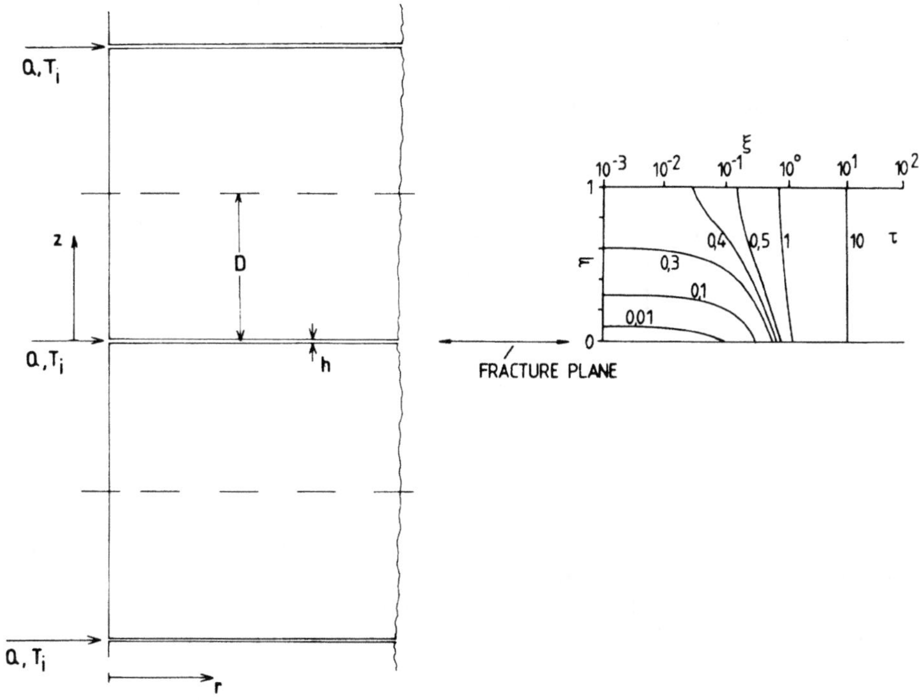

Fig. 9.59. A set of parallel, equally spaced (2D) fractures is given (left). A: Radially propagating volume flow Q_v into fractures occurs at $r = 0$ for $t > 0$. B: the position of the 'thermal front' (for explanation see text) for a given model geometry ($\theta < 0.01$) for various times;

$$\tau = \frac{\lambda t}{\rho_r c_r D^2}$$

as function of vertical distance from fracture:

$$\eta = \frac{z}{D}$$

and radial distance from injection point:

$$\xi = \frac{\lambda \pi r^2 (2 + \theta)}{\rho_w c_w Q D}$$

(after Bodvarrson and Tsang, 1982).

horizontal fractures is considered, in which equal flow rates for $t \geq 0$ exist, D must be equated with 1/2 of the fracture spacing).

To solve above problem, the following dimensionless parameters are defined:

$$\xi = \frac{\lambda \pi r^2 (2 + \theta)}{\rho_w c_w Q D}; \quad \theta = \frac{\rho_f c_f h}{\rho_r c_r D}; \quad \tau = \frac{\lambda t}{\rho_r c_r D^2}; \quad T_D = \frac{T - T_0}{T_i - T_0}$$

$$\eta = \frac{z}{D}.$$

These parameters are to be substituted into Equation (2.4) and (9.133). These equations, subject to above boundary conditions, yield in the Laplace domain solutions (see below), which must be evaluated by numerical methods.

Solution (in Laplace domain) for temperature field in fracture:

$$u_f = \frac{1}{p}\exp\left[-\frac{(\theta_p + 2\sqrt{p}\,\tanh\sqrt{p})\xi}{(2+\theta)}\right].$$

Solution (in Laplace domain) for temperature field in fractured rock:

$$u_r = \frac{1}{p}\exp\left[-\frac{(\theta_p + 2\sqrt{p}\,\tanh\sqrt{p})\xi}{(2+\theta)}\right]\left\{\cosh\sqrt{p}\,\eta - \sinh\sqrt{p}\,\eta\,\tanh\sqrt{p}\right\}$$

where

p = Laplace parameter.

Bodvarsson and Tsang (1982) have included in their paper a plot, showing the advancement of the 'thermal front' as function r, z, and t. In the case discussed here, the term 'thermal front' refers to the points at a given time, where the rock or fluid temperature $T = (T_i + T_0)/2$ has been reached. The plot, also shown in Figure 9.59, can be used, if the ratio of about $2h/D$ is smaller than 0.01.

At early times after the start of injection, the thermal front advances quite rapidly in the fracture. The heat transfer from the fracture walls into the fluid is not sufficient to change the temperature of the fluid to values near T_0. In this early period ($\tau \leq 10^{-4}$), the relation:

$$\tau = \frac{\theta}{(2+\theta)}\xi \tag{9.135}$$

holds (not shown in Figure 9.59).

At intermediate times ($10^{-4} \leq \tau \leq 1$), the position of the thermal front can be adequately described by the relation:

$$\tau = \frac{4.396}{(2+\theta)^2}\xi. \tag{9.136}$$

After sufficient long time ($\tau > 1.0$), the advancement of the thermal front is by and large inhibited by the heat flow into the fracture. A quasi-stationary situation has developed, any advancement of the thermal front occurs now very slowly. For this time period, the approximate relation describing the position of this particular boundary is given by:

$$\tau = \xi. \tag{9.137}$$

Substituting the real variables into this relation, Equation (9.138):

$$t \approx \left(\frac{2\pi\rho_r c_r D}{\rho_w c_w Q}\right) r^2 \tag{9.138}$$

demonstrates that the advancement of the thermal front is now dependent on $\sqrt{t}$.

PROBLEM 2. Another problem is the system of a fracture serving as a conduit for convecting fluids, capped by a low or impermeable rock formation. The convecting fluids are forced out of the fracture along the horizontal contact between the fractured rock and the overlying rock formation (Figure 9.60, left). This type of geologic situation is frequently found in graben structures, where crystalline or massive sedimentary rock masses were fractured during graben formation and blanketed subsequently by sedimentary rocks of very low permeability.

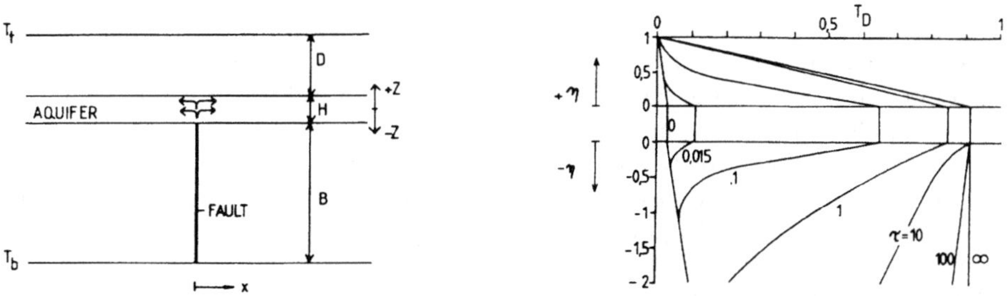

Fig. 9.60. Left: The case of a vertical fault entering an aquifer, covered by impermeable rock. Through the fault flows water with constant rate v and constant temperature T_f into the aquifer. There, water flow is diverted laterally to both sides in equal amounts. Rock parameters (ρ_a, c_a) in aquifer zone are constant. Above aquifer, impermeable rock with constant parameters ρ_1, c_1, λ_1 prevents any leakage of water to the surface. For definition of T_t and T_b see text (after Bodvarsson *et al.*, 1982).
Right: Evolution of the temperature field as function of time at point $\xi_1 = 0.1$ (number on curves are values for γ_1) for the following set of parameters (after Bodvarsson *et al.*, 1982):

$$\theta_1 = 0.1$$

$$\frac{\rho_2 c_2}{\rho_1 c_1} = 1.$$

$$\sigma = 30.$$

$$\kappa = 1.0$$

$$\omega = 1.0$$

$$\xi_1 = 0.1$$

$$T_g = 0.035.$$

This geothermal situation has been analyzed by Bodvarsson *et al.* (1982). As outlined in their paper, the governing equations, based on the following simplifying assumptions, are:

The energy balance for a given volume of the aquifer of thickness H (se also Equation (9.132)):

$$z = 0: \; \frac{\lambda_1}{H}\frac{\partial T_1}{\partial z}\bigg|_{z=0} - \frac{\lambda_2}{H}\frac{\partial T_2}{\partial z}\bigg|_{z=0} - \frac{\rho_w c_w Q}{H}\frac{\partial T_a}{\partial x} - \rho_a c_a \frac{\partial T_a}{\partial t} = 0$$

Subscripts

1 = rock material above aquifer
2 = rock material below aquifer
w = fluid
a = refers to aquifer zone.

Heat transport in the under- and overlying formations is assumed to be purely conductive. Therefore, (see also Equation (2.4)):

$$z > 0:\ \lambda_1 \frac{\partial^2 T_1}{\partial z^2} = \rho_1 c_1 \frac{\partial T_1}{\partial t}$$

$$z > 0:\ \lambda_2 \frac{\partial^2 T_2}{\partial z^2} = \rho_2 c_2 \frac{\partial T_2}{\partial t}.$$

The following simplifying assumptions are made:
- constant temperature at top (T_t) and base (T_b) of the cross section
- no heat losses from the fluid as it moves up the vertical fracture zone and into the aquifer with temperature T_f
- constant recharge rate Q_v($m^3\ s^{-1}\ m^{-1}$) in aquifer
- no horizontal conduction in the aquifer (convective heat transport dominant)
- instantaneous thermal equilibrium between solid and fluid phase in the aquifer
- uniform temperature in the aquifer in any vertical section.

The above equations, subject to the initial conditions:

$$T_a(x, 0, 0) = T_t + \Gamma_1 D$$

$$T_1(x, z, 0) = T_t - \Gamma_1(z - D)$$

$$T_2(x, z, 0) = T_t + \Gamma_1 D - \Gamma_2 z$$

B = thickness of rock formation from base of aquifer to base of cross section
D = thickness of rock formation above aquifer
Γ_1 = geothermal gradient above aquifer
Γ_2 = geothermal gradient below aquifer

and the following boundary conditions:

$$T_a(0, t) = T_f,\ t > 0$$

$$T_a(x, t) = T_1(x, 0, t) = T_2(x, 0, t)$$

$$T_1(x, D, t) = T_t$$

$$T_2(x, B, t) = T_b = T_t + \Gamma_1 D - \Gamma_2 B$$

(B in z direction has a negative value) can be solved for the overlying section, the aquifer itself and the section below in the Laplace domain. The following dimensionless parameters need to be defined:

$$\xi_1 = \frac{\lambda_1 x}{\rho_w c_w Q D};\quad \tau_1 = \frac{\lambda_1 t}{\rho_1 c_1 D^2};\quad \theta_1 = \frac{\rho_a c_a}{\rho_1 c_1}\frac{H}{D};$$

$$\eta = \frac{z}{D};\quad \kappa = \frac{\lambda_2}{\lambda_1};\quad T_D = \frac{T - T_t}{T_f - T_t}$$

$$T_g = \frac{\Gamma_1 D}{T_f - T_t};\quad \sigma = \frac{B}{D};\quad \omega = \frac{\Gamma_2}{\Gamma_1}.$$

For early ($\tau << 1$) and late times ($\tau > 1.0$) the solutions in the Laplace domain can be easily inverted into real space:

Early times:

$\eta = 0$: $T_D = [1 - T_g] \cdot U(\tau_1 - \theta_1 \xi_1) + T_g$; $U(\xi)$ = unit function (see above).

$\eta \neq 0$: T_D = unchanged.

Late times:

$$\eta = 0: T_D = (1 - T_g) \exp\left[-\left(1 + \frac{\kappa}{\sigma}\right)\xi_1\right] + T_g$$

$$\eta > 0: T_D = (1 - T_g)(1 - \eta) \exp\left[-\left(1 + \frac{\kappa}{\sigma}\right)\xi_1\right] - T_g(\eta - 1) \quad (9.139)$$

$$\eta < 0: T_D = (1 - T_g)\left(1 + \frac{\eta}{\sigma}\right) \exp\left[-\left(1 + \frac{\kappa}{\sigma}\right)\xi_1\right] - T_g(\omega\eta - 1).$$

For additional details the reader is referred to the original paper.

The resulting temperature regime in the subsurface as function of time for one set of parameters is shown in Figure 9.60, right. The example illustrates very well that deep seated circulating groundwater can substantially modify the near surface temperature regime. Areas, though characterized by high heat-flow density, show in cases like these an almost or even total absence of any surface manifestation of geothermal activity. Typical examples of this kind of geothermal systems exist today e.g. in the Rhinegraben (FRG) or in the Basin and Range Province (USA).

9.4.4. Permafrost

(1) GROWTH AND DECAY OF PERMAFROST

The expansion of human activity into Arctic territories has brought with it the need to construct technical installations and buildings on frozen ground. Consequently, a large number of scientific studies of the permafrost phenomenon were carried out. The knowledge of the most likely response of permafrost zones to manmade changes of the surface conditions is a key to the proper planning of necessary installations in this sensitive environment.

Growth and decay of permafrost zones are controlled by the mean annual surface temperature and by the terrestrial heat-flow density. The onset of permafrost occurs as a result of the drop of the regional mean annual temperature below the freezing point of the interstitial water of the soil. The velocity of penetration of the frozen zone into depth depends on the thermal properties of the subsurface materials and on the water content.

The primary resistance to permafrost growth is caused by the latent heat of fluids upon freezing. High porosity and water content of a subsurface stratum will slow down the growth rate of permafrost in contrast to low porosity material. On the other side, decay of a permafrost zone by melting will occur at a lower rate in rocks with high fluid content due to the high amount of latent heat to be absorbed. In addition, high salinity fluids with freezing points well below 0 °C can prevent the development of permafrost. Rivers and lakes, storing solar heat during the summer season act as local heat sources throughout the year and prevent permafrost development in the immediate underlying rock materials.

In total, a permafrost zone may have a very irregular lower boundary in the subsurface; even patches with no permafrost at all may dissect the otherwise frozen ground.

Figure 9.61 shows examples of measured temperature logs in deep bore holes penetrating permafrost zones at Prudhoe Bay in Alaska (Lachenbruch, 1982) and on the East Siberian Platform (Balobaev *et al.*, 1978). The data reflect the enormous thickness, permafrost zones can reach in areas of low mean annual surface temperatures.

In comparison to the Siberian data, the existing temperature logs from Prudhoe Bay (Lachenbruch, 1982) show a lesser depth of permafrost penetration. According to Lachenbruch, λ for the frozen section was found equal to 3.4 W m^{-1} K^{-1}, the resulting heat-flow density $q = 54$ m W m^{-2}.

A quantitative assessment of the rate of decay of any permafrost zone upon warming of the surface due to climatic changes can be made on the basis of the theoretical approach presented by Lachenbruch (1982). An instantaneous increase of the surface temperature from $T_0 (T_0 < 0\,°C)$ to T_s is assumed at $t = 0$. The temperature at the base of the frozen zone is given by T_b, the melting temperature of the frozen pore fluids. The initial thermal gradient is given by $(T_0 - T_b)/Z_1$, the final by $(T_s - T_b)/Z_2$;

where

Z_1 = thickness of frozen zone at $t = 0$
Z_2 = thickness of frozen zone after readjustment of the temperature field.

The transient temperature field can be calculated on the basis of the heat conduction Equation (2.4):

$$\frac{\partial T}{\partial t} = a \frac{\partial^2 T}{\partial z^2}$$

subject to the following boundary conditions:

$$T(z, 0) = (T_0 - T_s)\left(1 - \frac{z}{Z}\right)$$

$$T(z, t) = 0; \text{ at } z = 0 \text{ and } z = Z$$

where

T_0 = surface temperature before sudden temperature increase at $t = 0$
T_s = surface temperature after sudden temperature increase at $t = 0$
z = vertical coordinate
Z = permafrost thickness at $t = 0$.

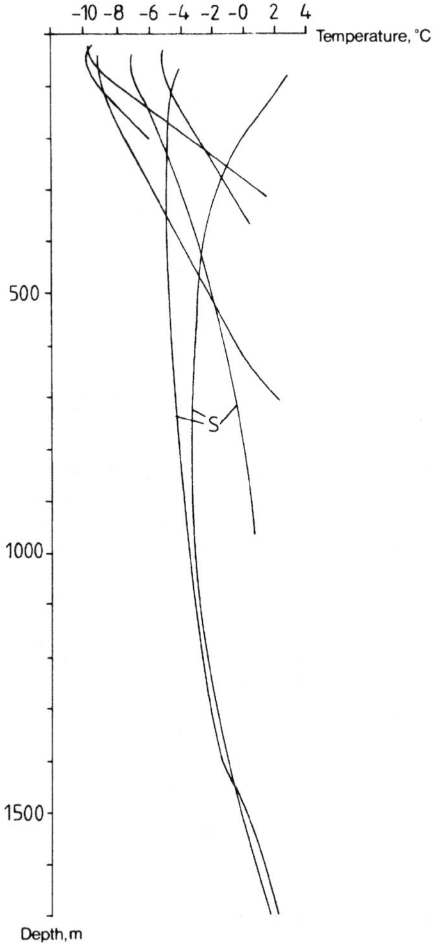

Fig. 9.61. Measured temperature logs in drill holes in permafrost near the Alaskan Arctic coast (after Lachenbruch, 1982) and – marked by S – to the west of the East Siberian Platform (after Balobaev *et al.*, 1978). All *T*-logs show the effect of recent climatic warming at shallow depth.

The solution to above equation is given by:

$$T(z, t) = (T_0 - T_s)\frac{2}{\pi}\sum_{n=1}^{\infty} \exp\left(-\frac{n^2\pi^2}{4}\frac{t}{\nu}\right) sin\left(n\pi\frac{z}{Z}\right) \tag{9.140}$$

ν = time constant.

By substitution of Equation (9.140) into Equation (2.4), we find $\nu = \frac{Z^2}{4a}$; ν is the so-called Fourier-criterion. The temporal change of the transient temperature field is shown in Figure 9.61. Lachenbruch have shown that given the parameters relevant for the situation in the Prudhoe Bay, Z = 600 m and $\lambda = 1.6\ 10^{-6}$ m^2 s^{-1}, ν = 1800 a. As can be seen from Figure 9.62, ν is a good approxima-

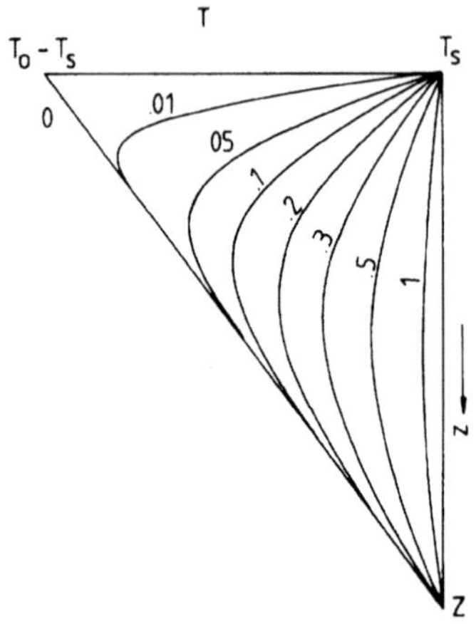

Fig. 9.62. Temporal adjustment of the subsurface temperature field to a change of the mean annual surface temperature, as based on Equation (9.140) is shown. Concurrent thinning of permafrost layer is neglected. Numbers on curves multiplied with the time constant v give the actual time period since the onset of the temporal change (after Lachenbruch, 1982).

tion for the time period required to raise the rock temperature at all levels to approximtely T_s. At the same time and subsequently a thinning of the permafrost zone is caused by the upward movement of T_b due to the heat supplied at the base by the heat-flow density q.

The energy balance at this point is given by:

$$LP\frac{\partial Z}{\partial t} = q - \lambda\frac{\partial T}{\partial z} - \lambda\frac{(0\,°\mathrm{C} - T_s)}{Z(t)}$$

where

L = latent heat of ice per unit volume;
P = porosity.

For $v > 1$, the approximate relation holds:

$$LP\frac{\partial Z}{\partial t} \approx q.$$

Given plausible rock parameters, the rate of thinning of permafrost is then in the order of cm a^{-1}.

(2) THERMAL EFFECTS OF PIPELINES IN PERMAFROST AREAS

The discovery of enormous oil reservoirs in Arctic regions and their exploitation led to the construction of pipelines that had to be routed across permafrost areas. For technical reasons, the oil to be pumped through the pipelines is kept at elevated temperatures (app. 80 °C). The heat losses from the pipeline into the ground might induce melting processes, which in turn under adverse circumstances might result in a destabilization of the foundations of the pipeline itself. Preventive measures can be designed, if the thermal influence of the pipeline on the underground can be predicted by model calculations, based on plausible material parameters.

The theoretical basis to treat this problem was presented by Lachenbruch (1970). The basic equations, on which his calculations are based, are as follows:

Conductive heat transport outside the thawed region:

$$\frac{\partial^2 T_1}{\partial x^2} + \frac{\partial^2 T_1}{\partial y^2} = \frac{1}{a_1}\frac{\partial T_1}{\partial t} \tag{9.141a}$$

Conductive heat transport inside the thawed region:

$$\frac{\partial^2 T_2}{\partial x^2} + \frac{\partial^2 T_2}{\partial y^2} = \frac{1}{a_2}\frac{\partial T_2}{\partial t} \tag{9.141b}$$

with T kept constant at surface of pipeline.

The movement of the boundary (with $x = X$, $y = Y$ as coordinates and $T_1 = T_2 = 0$) between the thawed and unthawed region is given by:

$$\lambda_1 \frac{\partial T_1}{\partial x} - \lambda_2 \frac{\partial T_2}{\partial x} = LP\frac{\partial X}{\partial t}$$

$$\lambda_1 \frac{\partial T_1}{\partial y} - \lambda_2 \frac{\partial T_2}{\partial y} = LP\frac{\partial Y}{\partial t}.$$

The annual temperature variation:

$$T(0, t) = T_0 + A \exp\left(-y\sqrt{\frac{\omega}{2a_2}}\right) \sin\left(\omega t - y\sqrt{\frac{\omega}{2a_2}}\right), \tag{9.141c}$$

where

T_0 = mean annual surface temperature
A = annual temperature amplitude
ω = angular velocity for the period of 1 a
y = vertical coordinate

has to be superimposed on the temperature field induced by the artificial heat source (pipeline).

This set of equations need to be solved numerically in an iternative fashion. After each new calculation of the temperature field, the new boundaries of the thawed and unthawed region are evaluated taking advantage of the Equations (9.141a) and (9.141b).

(3) CONSTRUCTION OF BUILDINGS IN PERMAFROST REGIONS

The construction of buildings on permafrost has opened a wide field for application of geothermal concepts. The primary objective in constructing these buildings must be the preservation of permafrost conditions around the building foundations. After the piles of the foundations have been sunk into the frozen subsoil, the ground space is covered by a plate allowing for a ventilation zone between the frozen surface and the base of the building. The resulting effect is twofold:

- The ground is no longer exposed to heat flow from the building.
- The thermal effect from the radiational heat from the sun is eliminated.

In some cases, where the natural mean annual temperature at the surface approaches melting conditions, an additional circulation of cold air or carbondioxid in the piles has been successfully attempted (i.e. Maksimov *et al.*, 1973).

An example of permafrost at successive stages of artificial cooling of the underground is shown in Figure 9.63.

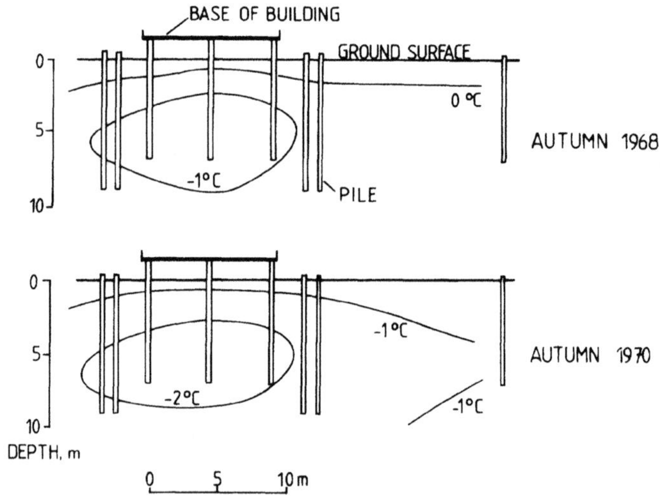

Fig. 9.63. Building in permafrost regions built on piles, to allow air circulation under the base. In addition, the subsurface was cooled by circulating carbondioxid in the piles. The temporal change of the resulting subsurface temperatures is shown (after Maksimov *et al.*, 1973).

9.4.5. Manipulation of Subsurface Temperatures by Technical Installations

(1) VENTILATION OF MINES

The generation and maintainance of bearable working conditions in deep mines by ventilation is one of the classic geothermal problems. The following heat exchange generation processes have to be dealt with:

- adiabatic compression of air flowing down into the mine
- absorption (desorption) during evaporation (condensation) of water, originating from inflow of formation waters or from condensation of inflowing wet air
- heat release by machines, electrical circuitry or personnel
- heat release by oxidation processes in the mine
- heat release by the mined material
- heat exchange between the rock wall and ventilated air.

The air temperature gradient due to adiabatic compression amounts to approximately 0.05 °C/100 m. The magnitude of heat production by various sources in the mine and the resulting air temperature increase can be determined from the volume of flowing air per unit time in the gallery considered and the thermal parameters of air. The calculation of the heat transfer gallery wall/air and secondly the influence of evaporation/condensation processes in the mine on air temperature are of complex nature. The theoretical basis for the mathematical treatment of both problems are discussed in detail by Mundry (1963, 1964).

(2) HEAT TRANSFER AT THE INTERFACE ROCK WALL – AIR

New galleries are commonly driven into thermally unperturbed country rock. After initiation of ventilation, heat will be removed from the rock masses and, with time, a growing cooling mantle around the gallery will develop. Assuming a cylindrical shape of the gallery and negligible thermal influence of convecting fluids on the temperature field in the rock formation – as is the case in dry mines, i.e. potassium mines – the developing temperature field around the gallery is given by equation:

$$T_a(\mathrm{Fo}, \xi, \mathrm{Nu}) = \frac{2}{\pi}\int_0^\infty \left[\sin\left(\frac{\frac{2\mathrm{Nu}\xi}{\pi x^2}}{A^2 + B^2}\right) \cdot \exp\left(-\mathrm{Fo}x^2 - \mathrm{Nu}\xi\frac{I_1(x)\cdot A + N_1(x)\cdot B}{A^2 + B^2}\right)\right]\frac{\mathrm{d}x}{x} \qquad (9.142)$$

$$\text{Fourier number: } \mathrm{Fo} = \frac{at}{r^2}$$

$$\text{Nusselt number: } \mathrm{Nu} = \frac{\alpha r}{\lambda}$$

$$\xi = \frac{2\lambda z}{c_a \rho_a v r^2}$$

$$A = I_1(x) + \frac{\mathrm{Nu}}{x}T_0(x)$$

$$B = N_1(x) + \frac{\mathrm{Nu}}{x}N_0(x)$$

where

α = rock surface – air heat transfer coefficient
r = radius of gallery
T = air temperature
T_0 = air temperature at entrance of gallery
T_r = rock temperature at $t = 0$
$T_a = \dfrac{T - T_0}{T_r - T_0}$

λ = thermal conductivity
a = thermal diffusivity
v = air stream velocity through gallery
c_a = specific heat content of air
ρ_a = density of air.

I_0, I_1 and N_0 and N_1 denote the Bessel and Neumann function of zero and first order.

The above expression will best be evaluated numerically. The results for a number of relevant sets of T_a, Fo, ξ, Nu – values were published by Mundry (1964) in graphical form. Some examples are shown in Figure 9.64.

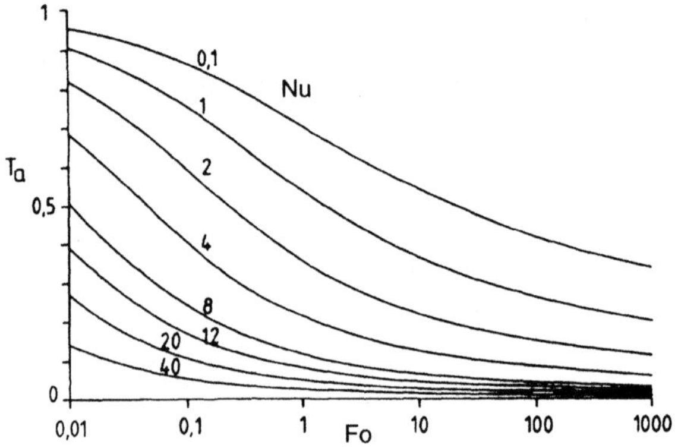

Fig. 9.64. Relative temperatures (T_a) at rock wall as function of time (Fo) at entrance of a gallery ($\xi = 0$) after onset of air ventilation in a dry mine. The Nu-value is proportional to the radius of the gallery as well as the heat transfer coefficient at the interface rock air (after Mundry, 1963).

The same theory holds also for cases, where the volume of air passing through the gallery changes at gallery cross points. It can be shown, that e.g. in the case of a gallery with two cross points, whereby at the first the air flow is being reduced by a factor of 3, at the second by a factor of 2, the following relations hold (see also Figure 9.65):

$$\xi_D = \xi; \text{ for } 0 < \xi < \xi_1$$

$$\xi_D = \xi_1 + \frac{v}{v_1}(\xi - \xi_1); \ \xi_1 < \xi < \xi_2$$

$$\xi_D = \xi_1 + \frac{v}{v_1}(\xi_2 - \xi_1) + \frac{v}{v_2}(\xi - \xi_2)\, \xi_2 < \xi$$

$$v_1 = v/3$$

$$v_2 = v_1/2$$

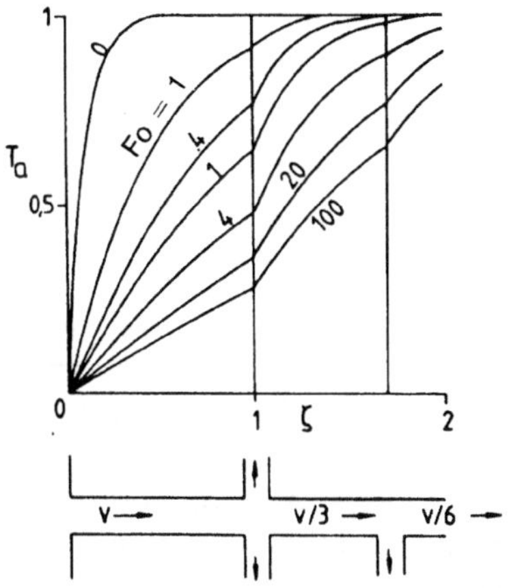

Fig. 9.65. Relative temperatures (T_a) in a dry mine at rock wall as function of time (Fo), the distance from the injection of air stream v and reduction of the air stream v by deviating galleries for a constant geometry (Nu = 10) of all galleries considered (after Mundry, 1963).

(3) WET MINES

In most geologic formations some inflow of water due to the high hydraulic gradient from the country rock toward the galleries and shafts is common. In that case the above calculation needs to be modified to allow for the thermal influence of evaporation/condensation processes to be considered.

Under wet conditions the air temperature commonly drops upon entering the gallery. Until the air humidity has reached the saturation point evaporation occurs and water absorbs heat from the air. Near this point the air temperature starts to raise as soon as the heat gained from heat exchange at the rock wall surpasses the heat losses. In accordance with the rate of temperature increase, additional water can be taken up by the ventilated air. This complex can be treated mathematically in the following fashion (Mundry, 1964):

$$\frac{dT}{dz} + \frac{s}{c_a}\frac{dx}{dz} = \frac{2q}{c_a \rho_a r v} \tag{9.143}$$

where

q = heat-flow density entering the gallery across the rock wall, J m^{-1} h^{-1}
s = heat of evaporation, J kg^{-1}
T_a = temperature of air stream, °C
c_a = specific heat capacity of wet air, J kg^{-1} °C^{-1}
ρ_a = density of air, kg m^{-1}
r = radius of gallery, m
x = steam content in air per air volume, kg^{-1}
z = distance along long axis of gallery, m
v = air stream velocity m s^{-1}.

The value of q can be approximated – assuming the temporal change of air temperature with time at any given point to be small – by the expression:

$$q = \lambda \frac{(T_g - T)}{r} P \tag{9.144}$$

with T_g = initial rock temperature, and:

$$P = \frac{4}{\pi^2} \int_0^\infty \frac{e^{-\tau i^2}}{\left[J_0(i) + \frac{i}{\mathrm{Nu}} J_1(i) \right]^2 + \left[N_0(i) + \frac{i}{\mathrm{Nu}} N_1(i) \right]^2} \frac{\mathrm{d}i}{i^3}.$$

The amount of evaporating water W is proportional to $x' - x$ (x' being the amount of water steam in air at point at saturation), the evaporation surface $2\,r\,\mathrm{d}z$ and the heat transfer coefficient α:

$$\beta = k_1 \frac{\alpha}{c_a} \quad 0 < k_1 < 1$$

k_1 being proportional to relative humidity at the gallery wall.

It follows:

$$\frac{\mathrm{d}x}{\mathrm{d}z} = \frac{2\pi r \beta}{\rho_a v \pi r^2}(x' - x) = \frac{2\beta}{\rho_a v r}(x' - x)$$

Equation (9.143) can now be written as:

$$\frac{\mathrm{d}T}{\mathrm{d}z} = \frac{2\lambda}{c_a \rho_a v r^2} P(T_g - T) - \frac{2 s k_1 \alpha}{c_a^2 \rho_a v r}(x' - x).$$

By linearisation of the temperature dependence of x' over sufficiently small temperature ranges (thereby assuring a good fit with the $x'-T$-function):

$$x' = K + LT$$

(K and L are coefficients);
we arrive at

$$\frac{\mathrm{d}T}{\mathrm{d}z} = AT_a + BX + C$$

$$\frac{\mathrm{d}x}{\mathrm{d}z} = E \cdot T_a + FX + G$$

The solution can be found to be:

$$T = X \cdot B\, e^{\lambda_1 z} - YB\, e^{\lambda_2 z} - \alpha_0$$

$$x = X(A - \lambda_1)\, e^{\lambda_1 z} + Y(A - \lambda_2)\, e^{\lambda_2 z} - \beta_0 \tag{9.145}$$

where

$$X = \frac{(T_0+\alpha_0)(A-\lambda_2)+B(x_0+\beta_0)}{B(\lambda_2-\lambda_1)} \qquad \alpha_0 = \frac{CF-BG}{AF-EB}$$

$$Y = -\frac{(T_0+\alpha_0)(A-\lambda_1)+B(x_0+\beta_0)}{B(\lambda_2-\lambda_1)} \qquad \beta_0 = \frac{AG-EC}{AF-EB}$$

$$A = -M - \frac{s}{c_a}N \cdot L \qquad M = \frac{2\lambda P}{c_a r^2 \rho_a v}$$

$$B = \frac{s}{c_a}N \qquad N = \frac{2k_1\alpha}{c_a r \rho_a v}$$

$$C = M \cdot T_0 - \frac{s}{c_a}N \cdot K$$

$$E = N \cdot L$$

$$F = -N \qquad \lambda_1 = \frac{A+F}{2} + \sqrt{(A+F)^2 - 4(AF-EB)}$$

$$G = N \cdot K \qquad \lambda_2 = \frac{A+F}{2} + \sqrt{(A+F)^2 - 4(AF-EB)}.$$

The dependence of air temperature along the gallery for a selected set of parameters was calculated by Mundry (1964; see also Figure 9.66). A low humidity level of the ventilated air upon entering the gallery causes a sudden drop of the air temperature due to the evaporation of mine water. The uptake of water along z depending on the amount of heat gained at the interface rock wall/air reduces the rate of temperature increase in the downstream direction. For reference, the resulting air temperature, given no evaporation process, is shown as well.

9.4.6. Storage of Heat Producing Wastes in the Underground

The need of burial of heat producing nuclear wastes (HAW = *h*igh level radio*a*ctive *w*astes) in geologic formations has opened up a whole range of questions of geothermal and geochemical nature. The calculation of the temperature field around a burial field (repository) in space and time forms the basis for the discussion of the thermomechanical as well as geochemical-mineralogical consequences of the heat-up of the host rock. The feasibility of construction of such facilities in geologic media such as basalt, granite, clay or rock salt is currently under intense investigation by various countries.

The construction of a repository in a salt structure appears currently as the most promising method to isolate nuclear wastes from the biosphere. The development of the temperature field in space and time for an idealized model of a proposed HAW burial site in a salt dome of Northern Germany is discussed below (Delisle, 1980). The pertinent calculation consists of three steps

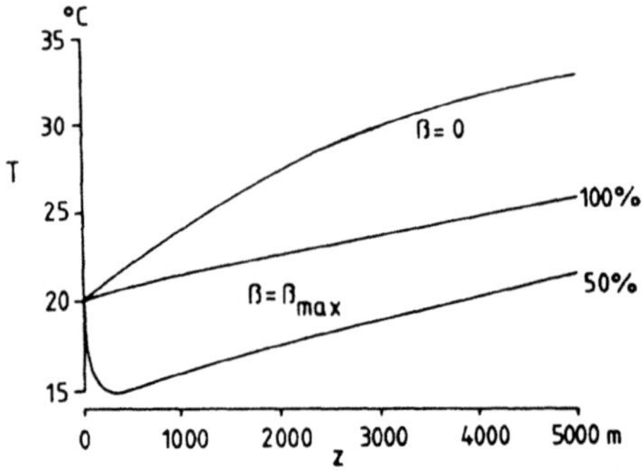

Fig. 9.66. Temperature of ventilated air in a wet mine as function of distance from the air injection point into gallery after one year of continuous ventilation (after Mundry, 1964). The following constant parameters were chosen:

T_0 = 20°C
T_g = 40°C
t = 1 a
λ = 2.3 W m^{-1} K^{-1}
a = 1.4 10^{-6} m^2 s^{-1}
α = 11.6 W m^{-2} K^{-1}
v = 2 m s^{-1}
r = 2 m

Considered are three cases
- no evaporation, $\beta = 0$
- relative humidity of air equal 100 %, $\beta = \frac{\alpha}{c_a}$
- relatively humidity of air equal 50 %, $\beta = 0.5 \times \frac{\alpha}{c_a}$

- determination of the unperturbed temperature field of the host formation
- definition of the thermal parameters of the repository, that is:
  heat generation as function of time by the wastes
  loading density of heat sources in repository
  geometry of repository
- calculation of the resulting temperature field in time and space.

The undisturbed rock temperatures of the salt dome in question have been discussed in 9.4.2.(2). The calculations were based on a tentative repository design.

Drifts, forming a rectangular grid, oriented in direction of the long axis of the salt dome, serve as accessory pathways to numerous disposal boreholes. Disposal fields for HAW consist of six drifts, spaced 50 m apart from each other. From each drift nine boreholes, 300 m deep, 50 m apart, are drilled into rock salt for the emplacement of wastes amounts to 720 W m^{-1} borehole length. The heat generation per cannister as function of time is shown in Figure 9.67.

Numerous HAW disposal fields with 6 × 9 disposal boreholes are stringed in line oriented parallel to the long axis of the salt dome. The drifts of each disposal field will be backfilled with rock salt as soon as emplacement of wastes in all boreholes has been carried out. Assuming the same material parameters as have been used in the calculation of the natural temperature field of the salt dome (for

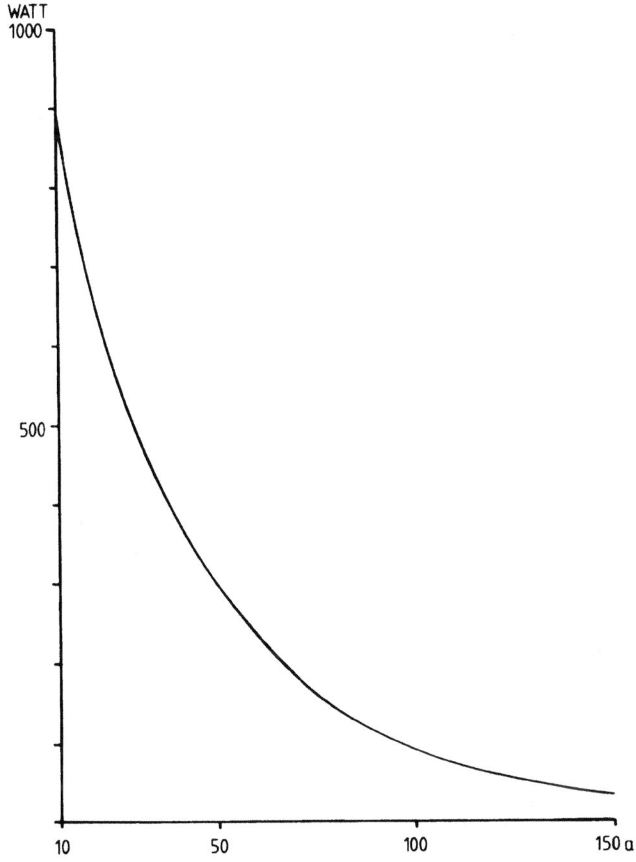

Fig. 9.67. Heat production of a 1.25 m long HAW-container as function of time. The highly radioactive elements, products of burnt-up fuel rods from nuclear reactors, had already decayed for a period of 10 a prior to packaging in the container (after Delisle, 1980).

$\lambda(T)$ see Equation (9.128) above, the near and far field of the rock temperature distribution around the emplaced wastes has been calculated in 3-D.

Figure 9.68 shows the temperature distribution along a line across midpoints of 4 disposal boreholes, which were filled with HAW-waste cannisters in subsequent years. The data are taken from a numerical model, which encompasses several disposal fields; the shown graphs, therefore, include the thermal effect of a number of adjacent boreholes. The outermost borehole shows 10 a after start of loading of the pertinent disposal field the highest rock temperature at its borehole surface, while the steepest temperature gradient is found around the innermost borehole. In later years, the average rock temperature T_{av} around each borehole rises continuously to a level of approximately 150 °C, while the difference between the peak temperature at the interface cannister to rock salt and T_{av} decreases due to the decrease of heat production by the wastes with time. In the far field

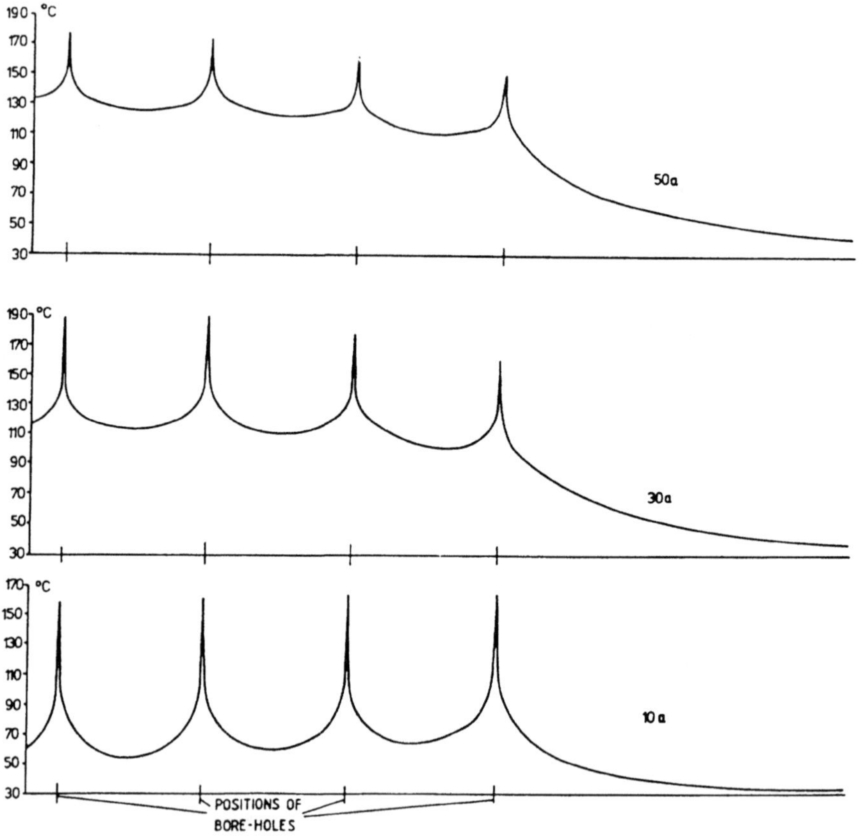

Fig. 9.68. Rock temperatures in a nuclear waste disposal field 10, 30, and 50 a after initiation of disposal. The rock temperatures are shown along a line connecting the mid-points of 4 disposal bore holes that have been filled with HAW-waste containers in successive years. (The calculation was actually performed in 3-D) (after Delisle, 1980).

calculations, where the heat production from one or several filled boreholes is averaged over one element of the numerical grid, the temperatures, obtained at each grid point, will represent T_{av} as discussed above.

The need to include the whole structure of the salt dome and the surrounding formations precludes the detailed consideration of the specific geometry of the repository. A vertical cut through the central disposal field is shown in Figure 9.69 and Figure 9.70. The resulting temperature field after emplacement of the HAW-waste 1000 a and 5000 a are shown.

As the calculation shows, the zone of thermal perturbation even 100 a after emplacement is essentially restricted to the repository and the immediate vicinity. A slow increase of ground water temperatures at the base of the sedimentary cover overlying the salt structure will commence 300 a after loading, the maximum thermal disturbance, however, will be reached only 1400 a later. A maxi-

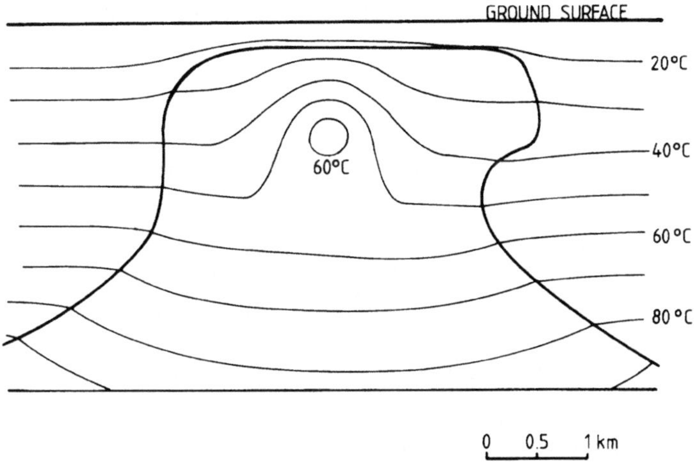

Fig. 9.69. Rock temperatures in a salt dome (same as in Figure 9.58) 1000 a after disposal of HAW-waste has stopped. Assumed is the burial of 1640 waste cannisters per year over a 50 a period in a disposal field, at a depth range from 850 m to 1150 m with a lateral extent of 420 m in the center of the salt dome. Other parameters are the same as the ones shown in Figure 9.58 (after Delisle, 1980).

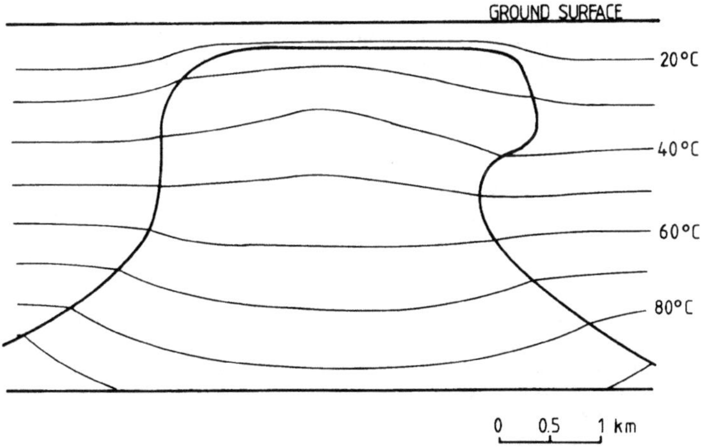

Fig. 9.70. Same as Figure 9.69, but for an elapsed time period of 5000 a after end of waste burial (after Delisle, 1980).

mum increase of 7 °C at the base of the sedimentary cover of the salt dome is expected.

The maximum T_{av} – values in the salt dome are reached approximately 120 a after loading. From then onwards, a slow, but continuous decrease of T_{av} is accompanied by a continuous enlargement of the rock volume affected by increased heat flow from the repository. 5000 a later, most portions of the salt dome are affected by a change of rock temperature by a fraction to several °C, depending on the distance from the repository. Even 20 000 a after loading we must expect slightly anomalous rock temperatures in the vicinity of the burial site, as the pertinent calculation has shown.

9.4.7. References

Balobaev, V. T., Devyatkin, V. N., and Kutasov, I. M.: 1978, 'Contemporary Geothermal Conditions of the Existence and Development of Permafrost', in F. J. Sanger and P. J. Hyde (eds.), *Proc. of the Second International Conference on Permafrost-USSR contributions*, Washington D.C. (National Academy of Sciences), 8–12.

Birch, F. and Clark, H.: 1940, 'The Thermal Conductivity of Rocks and its Dependence upon Temperature and Composition', *Am. J. Sc.*, **238**, 529–558.

Blackwell, D. D., Steele, J. L., and Brott, C. A.: 1980, 'The Terrain Effect on Terrestrial Heat Flow', *J. Geophys. Res.* **85**, B9, 4757–4772.

Bodvarrson, G.: 1969, 'On the Temperature of Water Flowing Through Fractures', *J. Geophys. Res.* **74**, 1987–1992.

Bodvarrson, G.: 1972, 'Thermal Problems in the Siting of Reinjection Wells', *Geothermics* **1**, 63–66.

Bodvarrson, G. and Tsang, C. F.: 1982, 'Injection and Thermal Breakthrough in Fractured Geothermal Reservoirs', *J. Geophys. Res.* **87**, B2, 1031–1048.

Bodvarrson, G., Benson, S. M. and Witherspoon, P. A.: 1982, 'Theory of the Development of Geothermal Systems Charged by Vertical Faults', *J. Geophys. Res.* **87**, B11, 9317–9328.

Bredehoeft, J. D. and Papadopulos, I. S.: 1967, 'Rates of Vertical Groundwater Movement Estimated from the Earth's Thermal Profile', *Water Resources Research* **1**, 325–329.

Carslaw, H. S. and Jaeger, J. C.: 1959, *Conduction of Heat in Solids*, 2nd Edit., Clarendon, Oxford, 510 p.

Delisle, G.: 1980, 'Berechnungen zur raumzeitlichen Entwicklung des Temperaturfeldes um ein Endlager für mittel- und hochradioaktive Abfälle in einer Salzformation', *Z. dt. geol. Ges., Hannover* **131**, 461–482.

Giesel, W. and Holz, A.: 1970, 'Das anomale geothermische Feld in Salzstöcken', *Kali und Steinsalz, Essen*. **5**, 272–274.

Gringarten, A. C. and Sauty, J. P.: 1975, 'A Theoretical Study of Heat Extraction from Aquifers with Uniform Regional Flow', *J. Geophys. Res.* **80**, B5, 4956–4962.

Haenel, R., (ed.): 1980, *Atlas of Subsurface Temperatures in the European Community*, (Th. Schäfer Druckerei GmbH), Hannover, 36 p and 43 maps.

Jensen, P. K.: 1983, 'Calculation on the Thermal Conditions around a Salt Diapir', *Geophys. Prosp.* **31**, No. 3.

Kappelmeyer, O., and Haenel, R.: 1974, *Geothermics with Special Reference to Application*, Geoexplor. Monogr. **1**, No. 4, (Borntraeger), Berlin, 238 p.

Lachenbruch, A. H.: 1970, *Some Estimates of the Thermal Effects of a Heated Pipeline in Permafrost*, Geological Survey Circular 632, Washington D.C., 23 p.

Lachenbruch, A. H.:1982, 'Permafrost, Heat Flow, and the Geothermal Regime at Prudhoe Bay, Alaska', *J. Geophys. Res.* **87**, B11, 9301–9316.

Lauwerier, H. A.: 1955, 'The Transport of Heat in an Oil Layer Caused by the Injection of Hot Fluid', *Appl. Sci. Res.* **Sec. A**, 145–150.

Maksimov, G., Zanyatin, S., and Konovaloz, S.: 1973, 'Methods of Cooling Plastically Frozen Ground', in F. J. Sanger and P. J. Hyde (eds.), *Proc. of the Second International Conference on Permafrost – USSR Contributions*, Washington D.C. (National Academy of Sciences), 553–560.

Moore, E., and Fosberg, M. A.: 1971, 'Temperatures and Classification of Soils in Latah, Benewah, and Kootenai Counties, Idaho', *Northwest Sci.* **45**, 137–144.

Mundry, E.: 1963, 'Vorausberechnungen von Wettertemperaturen in trockenen Gruben. Mathematische Behandlung des Problems', *Kali und Steinsalz, Essen*, **Bd.3**, 363–371.

Mundry, E.: 1964, 'Ein Verfahren zur Berechnung der Wettertemperatur in feuchten Gruben', *Kali und Steinsalz, Essen*, **Bd.2**, 37–41.

Selig, F., and Wallick, G. C.: 1966, 'Temperature Distribution in Salt and Surrounding Sediments', *Geophys.* **31**, 346–361.

10. APPENDIX

G. ZOTH and R. HAENEL
Department of Geophysics, Geological Survey of Lower Saxony, Hannover, FRG

10.1. THERMAL CONDUCTIVITY

The determination of heat-flow density requires not only the temperature gradient, but also the thermal conductivity from boreholes taken at that depth, where the temperature gradient is determined. In so far, the published thermal conductivity values are without interest. Anyhow, there exist a few standard books, which have compiled all existing values, such as Birch (1942), Kappelmeyer and Haenel (1974), Roy *et al.* (1981), Čermák and Rybach (1982). On the other hand, thermal modelling down to great depth, e.g. down to the Mohorivićić discontinuity, needs representative values at appropriate temperatures and pressures (depth). Therefore, an attempt is made to group a few of them.

In Table 10.1 a summarized compilation is given of thermal conductivity at room temperature. This table shows the large scattering.

The Figures 10.1–10.6 show salt rocks, limestone, metamorphic rocks, acid rocks, basic and ultrabasic rocks as a function of temperature. All values repre-

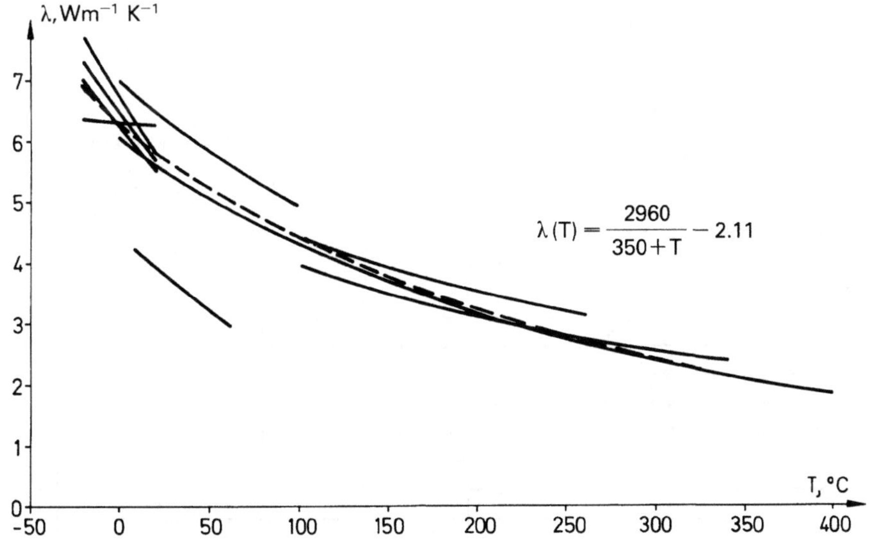

Fig. 10.1. Salt rocks. Birch and Clark (1940), Creutzburg (1965), Kopietz (1978), Zoth (1979).

R. Haenel, L. Rybach, and L. Stegena (eds.)
Handbook of Terrestrial Heat-Flow Density Determination, 449–466.

TABLE 10.1
Thermal conductivity of different rocks at room temperature Čermák and Rybach (1982), Kappelmeyer and Haenel (1974).

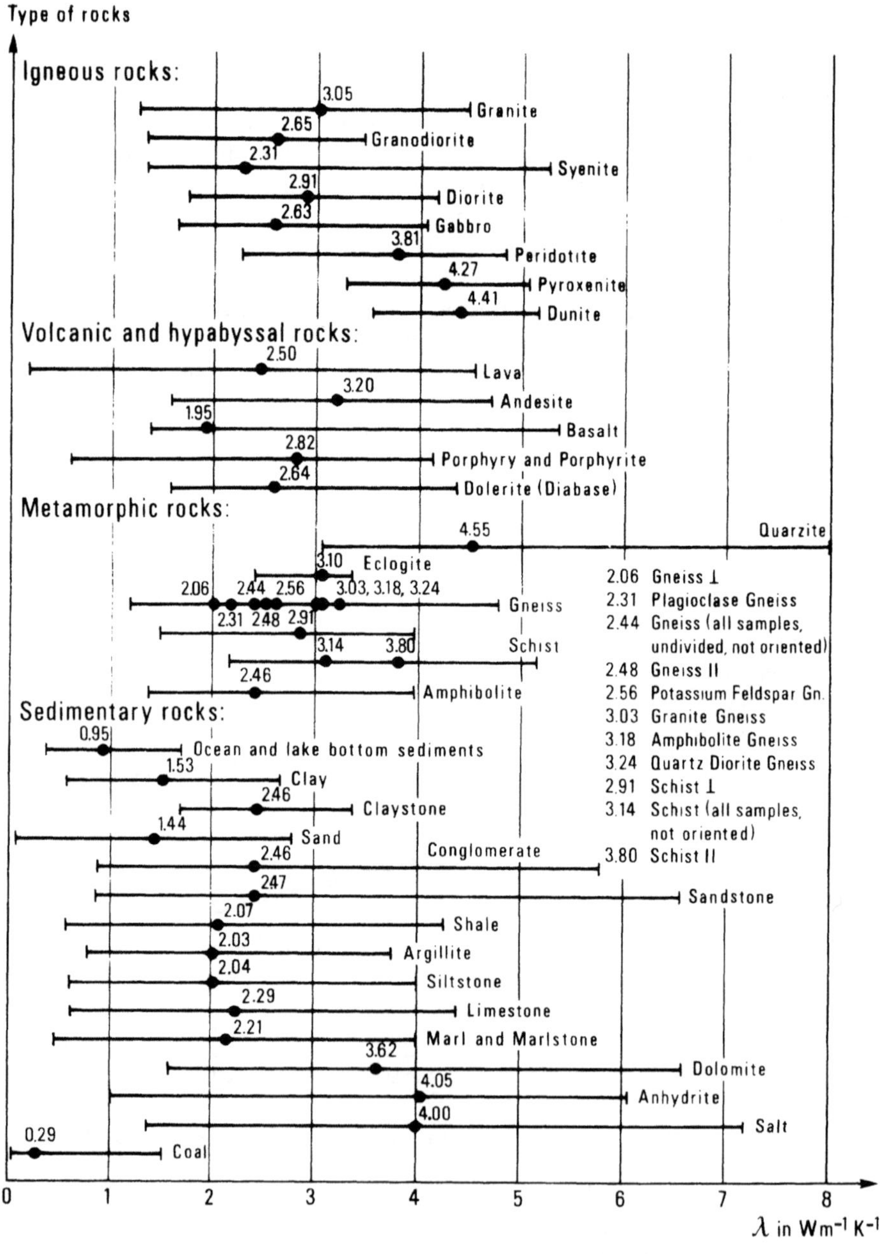

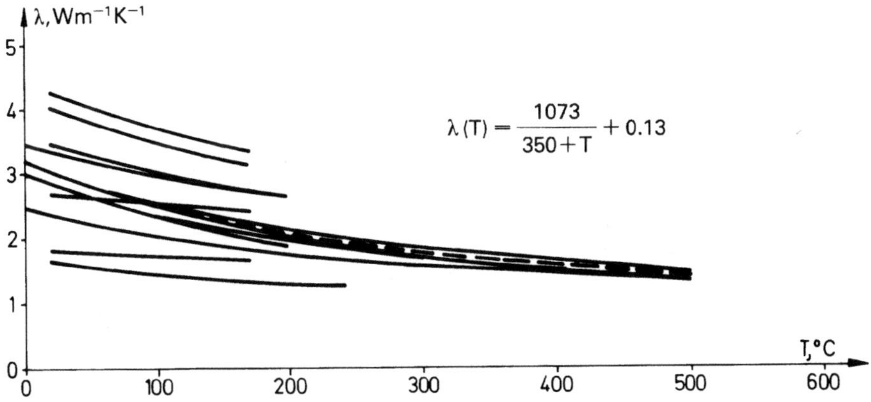

Fig. 10.2. Limestones. Čermák and Rybach (1982), Haenel *et al.* (1985), Mongelli *et al.* (1982).

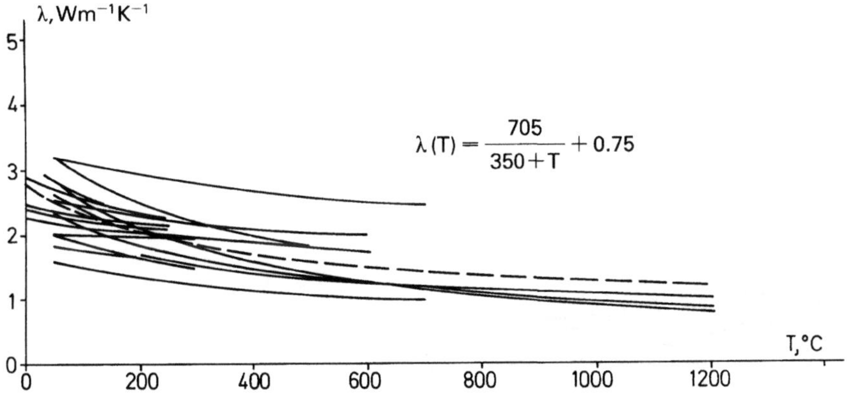

Fig. 10.3. Metamorphic rocks; albitite, amphibolite, biotite amphibolite, eclogite, hornblende gabbro, marble, phyllite, slate. Čermák and Rybach (1982), Haenel (1983), Sibbitt (1979).

sent published data. The mean curves (dashed lines) and the corresponding formulas are calculated according to the method of least squares.

In Figure 10.7 a compilation of these mean curves is shown. From that it follows that, excluding salt rocks and ultrabasic rocks, one formula can be used:

$$\lambda(T) = \frac{770}{350 + T} + 0.70 \text{ for } 0 < T < 800\ °\text{C}, \tag{10.1}$$

where λ is given in W m^{-1} K^{-1} and T in °C. In the case of ultrabasic rocks, the

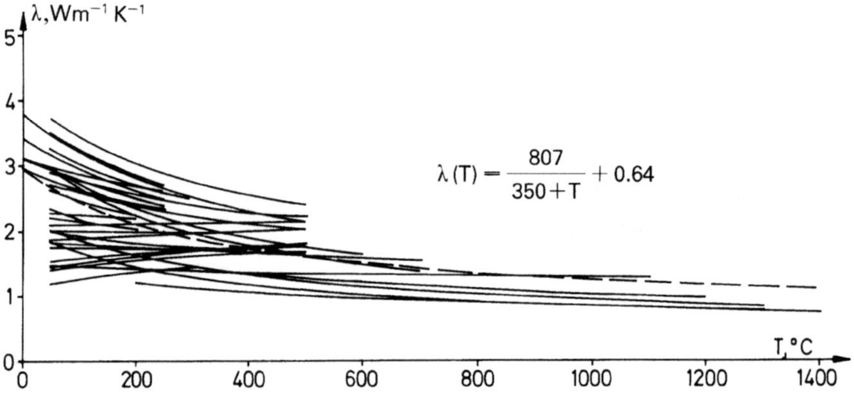

Fig. 10.4. Acid rocks; diabase, granite, granodiorite, granodiorite gneiss, monzogranite gneiss, porphyrite, quartz diorite, quartz porphyry. Čermák and Rybach (1982), Haenel (1983), Sibbitt (1979).

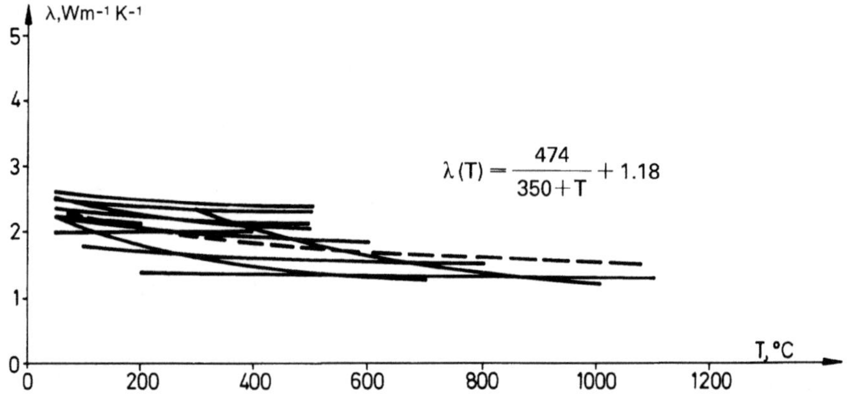

Fig. 10.5. Basic rocks; basalt, gabbro. Čermák and Rybach (1982), Haenel (1983), Touloukian *et al.* (1980).

formula is valid also for the temperature range of about $T > 800\,°C$. For temperatures above 1000 °C, see also the outline of Chapter 4.3. One has, of course, to be careful about applying this equation, and should keep in mind that the error range at low temperatures can be up to ± 1.5 W m^{-1} K^{-1}. Finally, the pressure effect is not considered due to lack of data.

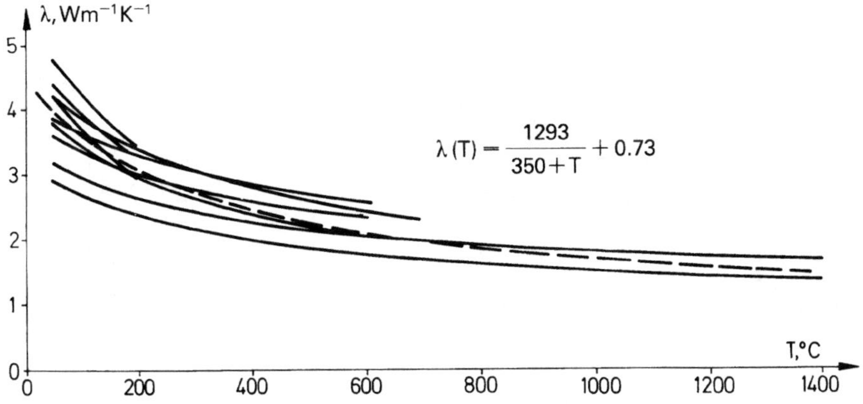

Fig. 10.6. Ultrabasic rocks; dunite, olivinite, peridotite. Čermák and Rybach (1982).

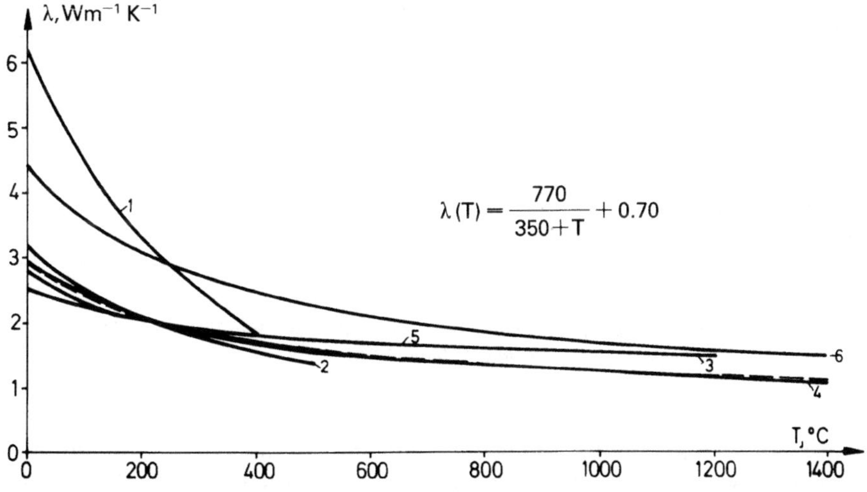

Fig. 10.7. Temperature dependency of various rocks; mean curves of Figures 10.1–10.6. The formula is calculated excluding salt rocks and ultrabasic rocks. (1) Salt rocks; (2) Limestones; (3) Metamorphic rocks; (4) Acid rocks; (5) Basic rocks; (6) Ultrabasic rocks.

10.2. PALAEOTEMPERATURES

The existence of surface palaeotemperature variations is well known (Figure

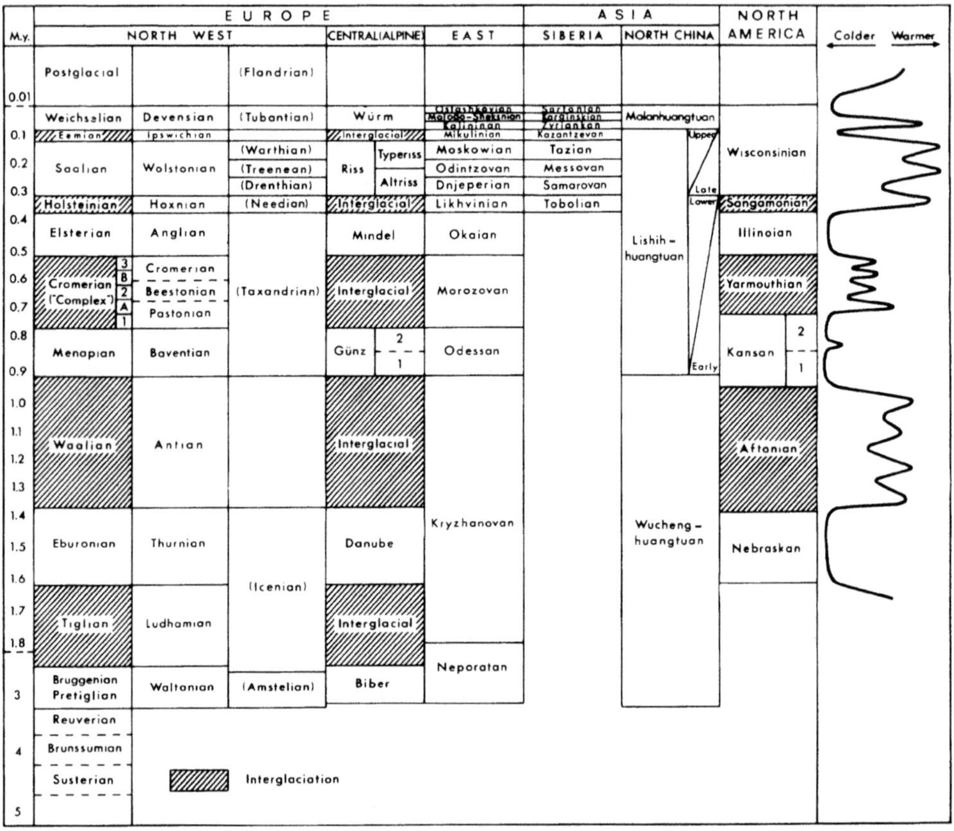

Fig. 10.8. The well known warm and cold periods during the past; Van Eysinga (1975) and Ericson and Welling (1970).

10.8), but due to inexact knowledge of the details the influence on the subsurface temperature gradient, and therefore on the heat-flow density, may be subject to large uncertainties. Due to the possible strong effect, the heat-flow density should be determined only from boreholes deeper than about 100 m (Haenel, 1983), 300 m (Jessop, 1983) or even deeper (see also Chapter 3.9.,13.). Anyhow, one of the first steps is to reconstruct the palaeotemperatures. There exist already few efforts. The results will be represented here, to encourage others to do similar. For more details see Chapter 5.4.3.

Haenel (1983) found for Germany that if the correction from the palaeotemperatures for 100000 a ago are taken as being 100 %, then the correction based on the last 380 a contributes about 60 % at 50 m depth and is reliable due

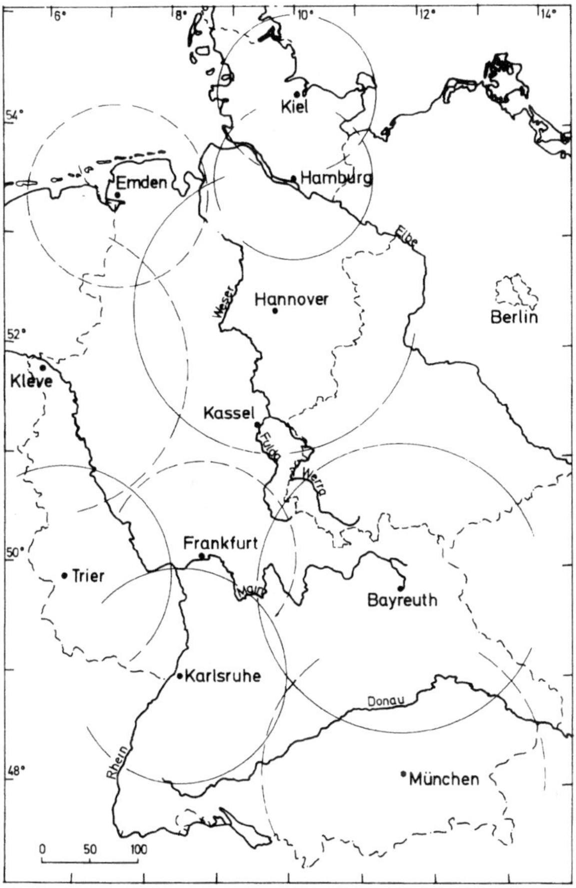

Fig. 10.9. Meteorological stations where palaeotemperature registrations exist (see Table 10.4) in the Federal Republic of Germany. Data for the dashed circles are not yet available.

to reasonably reliable temperature data. Assuming an accuracy in palaeotemperatures of only 60 % between 380 and 100 000 a, the total error of the correction is about 25 %. Therefore, Germany has been subdivided in several areas from which reliable temperatures exist; see Figure 10.9 and Tables 10.4 and 10.5.

In the Table 10.2–10.10 are shown the available data in hand, deduced from different sources. The influence on heat-flow density depending on latitude and thermal conductivity of rocks is shown in Figure 10.10.

TABLE 10.2.
Estimated surface palaeotemperatures with respect to present surface temperature ($\triangle T$) in North-eastern Ontario, Canada (Čermák, 1971). The time interval 0–2000 a was derived from the measured present-day temperature-depth profiles in Kapuskasing and/or Hearst borehole, resp. using statistical analysis. The time >2000 a is based on Prest's (1957) Pleistocene chronology, the detailed local geological study by Bennett *et al.* (1967) and the Holocene history by Brysson and Wendland (1967) and is generally similar to Birch's (1948) 1B model. The key figures are: ice-base temperature − 1.5 °C during ice ages (i.e. −4 relative to 'present' surface temperature being 2.5 °C – extrapolated straight part of the temperature-depth profile), and interglacials + 1.0 warmer than 'present' temperature.

Time, *a*	Kapuskasing		Time, *a*	Hearst	
from	to	$\triangle T$, °C	from	to	$\triangle T$, °C
0	40	2.2	0–	100	3.2
40	160	0.7	100	175	0
160	175	0			
175	475	−0.9	175	475	−0.9
475	670	0	475	670	0
670	1070	0.9	670	1070	1.3
1070	2000	0	1070	2000	0

Time, *a* from	to	$\triangle T$, °C
2 000	3 000	−1.0
3 000	3 500	0
3 500	4 000	1.0
4 000	6 000	2.0
6 000	6 500	0
6 500	7 250	−2.0
7 250	9 500	0
9500	45 000	−4.0
45 000	60 000	1.0
60 000	100 000	−4.0
100 000	225 000	1.0
225 000	325 000	−4.0
325 000	525 000	1.0
525 000	625 000	−4.0
625 000	925 000	1.0
925 000	1 025 000	−4.0

TABLE 10.3.
Surface palaeotemperatures in Denmark (Balling *et al.*, 1981). Reference time $t = 0$ corresponds to about 1980 for which ΔT is set zero.

Time, *a* from	to	ΔT, °C
	0	0.0
0	350	1.0
350	700	− 2.0
700	1000	1.0
1000	1300	0.5
1300	1600	− 1.0
1600	5000	− 1.0
5000	8000	2.0
8000	10000	10.0
10000	70000	−10.0
70000	130000	10.0
130000	190000	−10.0

TABLE 10.4.
Surface palaeotemperatures in the Federal Republic of Germany, which have been recorded during the past. Reference time $t = 0$ corresponds to 1980 for which ΔT is set zero. The value in bracket gives the temperature for 1980.

Time, *a*		Temperature difference ΔT, °C, for the station:					
from	to	Bayreuth	Hamburg	Hannover	Karlsruhe	Kiel	Trier
−2	−1						
−1	0						
0	1(1980)	0(7.4°C)	0(8.7°C)	0(8.3°C)	0(9.7°C)	0(7.8°C)	0(9.5°C)
1	2	0.2	0.4	−0.3	0.5	−0.5	0.0
2	3	0.0	1.1	0.2	0.0	0.3	−0.2
3	4	0.4	0.9	1.0	1.0	0.9	0.7
4	5	0.0	1.6	0.9	1.1	0.7	1.0
5	6	0.6	1.3	1.4	0.9	1.5	0.9
6	7	1.2	0.8	1.3	1.2	1.3	0.9
7	8	0.5	0.4	0.6	0.4	1.2	0.5
8	9	0.3	1.1	0.1	−0.2	0.6	−0.2
9	10	0.9	0.0	0.8	0.6	1.2	0.3
10	11	0.3	0.2	−0.1	0.2	0.0	0.1
11	12	0.3	0.8	0.0	0.0	0.3	0.0
12	13	0.5	1.5	0.4	0.2	0.9	0.0
13	14	1.3	0.6	1.3	1.0	1.4	0.6
14	15	1.2	−0.1	0.6	1.0	0.7	0.6
15	16	0.0	1.3	0.7	−0.1	0.2	−0.3
16	17	0.4	−0.4	0.2	0.9	0.6	0.6
17	18	−0.6	−0.5	−0.9	−0.6	0.3	−0.8
18	19	−0.8	1.2	−0.8	−0.3	−0.1	−0.7
19	20	0.7	0.6	1.1	1.3	1.5	1.0
20	30	0.2	0.5	0.4	0.4	0.8	0.3
30	40	0.6	0.2	0.6	0.4	0.7	0.4
40	50	0.6	0.4	0.9	0.4	1.2	0.2
50	60	0.6	−0.1	0.6	0.5	0.2	0.1
60	70	0.8	0.1	1.1	0.4	0.2	−0.2
70	80	0.5	−0.3	0.5	0.1	−0.1	−0.3
80	90	0.2	−0.4	0.4	0.9	−0.3	0.0
90	100	−0.3	−0.7	0.2	−0.1	−0.6	0.1
100	110	0.0	−0.3	0.9	−0.2	−0.4	0.3
110	120	0.3	0.0	1.0	0.8	−0.3	0.5
120	130	−0.2	−0.7	0.9	0.7	−0.5	−0.1

TABLE 10.5.
Surface palaeotemperatures in the Federal Republic of Germany (Grubbe, 1981). Data mainly based on Rudloff (1967) for 150–280 a, Lamb (1977) for 280–10 000 a, and Frenzel (1980) for 10 000–100 000 a.

Time, *a* from	to	ΔT, °C	Time, *a* from	to	ΔT, °C	Time, *a* from	to	ΔT, °C
130	180	−0.7	1 580	2 000	1.5	10 000	15 000	− 7.0
180	230	−0.6	2 000	2 480	1.5	15 000	26 000	−10.3
230	280	−0.3	2 480	3 500	−0.5	26 000	35 000	− 6.2
280	330	−0.7	3 500	4 500	1.0	35 000	55 000	−10.3
330	380	−0.6	4 500	6 700	2.0	55 000	65 000	− 4.9
380	580	−0.5	6 700	7 200	−1.0	65 000	70 000	− 8.0
580	680	0.3	7 200	8 000	2.0	70 000	75 000	− 4.9
680	830	0.8	8 000	9 000	−2.0	75 000	100 000	0
830	1 580	0	9 000	10 000	1.5			

TABLE 10.6.
Surface palaeotemperatures in France (Vasseur and Lucazeau, 1983). Reference time $t = 0$ is about 1980 for which ΔT is set zero.

Time, *a* from	to	ΔT, °C
	0	0.0
0	50	0.4
50	140	0.0
140	190	−0.2
190	270	0.0
270	320	−0.2
320	410	−0.3
410	470	0.1
470	540	−0.4
540	700	0.0
700	1 000	0.8
1 000	1 600	0.0
1 600	2 000	−0.4
2 000	3 250	−0.4
3 250	5 000	0.5
5 000	8 000	1.8
8 000	9 000	0.5
9 000	10 500	0.0
10 500	30 000	−8.0
30 000	40 000	−4.0
40 000	65 000	−6.0

TABLE 10.7.
Surface palaeotemperatures in Hungary (Stegena, pers. communication). Reference time $t = 0$ is about 1980 for which ΔT is set zero. The value in bracket gives the temperature for 1980.

Time, *a* from	to	ΔT, °C	Time, *a* from	to	ΔT, °C
	0(1980)	(10.3)			
0	150	0.3	7700	8000	− 2.2
150	200	−1.2	8000	8800	− 2.2
200	300	−1.0	8800	9500	0.0
300	400	−1.2	9500	10000	0.0
400	500	−1.0	10000	10250	− 5.5
500	650	−0.3	10250	10400	− 5.2
650	750	−1.1	10400	10550	− 5.7
750	900	0.0	10550	10750	− 5.0
900	1000	1.0	10750	10900	− 5.8
1000	1200	1.1	10900	11100	− 2.8
1200	1400	1.0	11100	11600	− 2.8
1400	1450	−0.2	11600	11800	− 6.4
1450	1550	−0.2	11800	12200	− 2.4
1550	1700	0.5	12200	12500	− 7.0
1700	2100	0.5	12500	12800	− 4.5
2100	2300	−1.5	12800	13000	− 8.0
2300	3100	−1.5	13000	13300	− 5.4
3100	3600	0.9	13300	13800	− 5.4
3600	3800	1.3	13800	14200	− 8.3
3800	4700	1.3	14200	14700	− 8.4
4700	5000	−1.5	14700	15000	− 8.0
5000	5900	+1.5	15000	15250	− 7.6
5900	6000	1.8	15250	15500	− 8.2
6000	6700	1.8	15500	16200	−11.0
6700	7100	1.0	16200	17000	−11.4
7100	7700	0.0	17000	17500	−11.6

TABLE 10.8.
Surface palaeotemperatures in Scandinavia (Swanberg *et al.*, 1974). Reference time $t = 0$ corresponds to about 1970 for which ΔT is set zero. The pollen diagram after Hammen *et al.* (1971) and mean annual ground temperatures (Norsk Meteoroligsk Arbok, 1970) were used. The above model is partly similar to Birch's (1948) 2B model.

Time, *a* from	to	ΔT, °C
	0	0.0
0	2000	−0.4
2000	8000	1.5
8000	13000	0.0
13000	27000	−4.0
27000	48000	0.0
48000	60000	−4.0
60000	90000	0.0
90000	165000	1.5

TABLE 10.9.
Surface palaeotemperatures in Switzerland (Rellstab, 1981, 1982). Reference time $t = 0$ corresponds to about 1980 for which ΔT is set zero.

Time, *a* from	to	ΔT, °C
0	50	0.0
50	80	− 0.4
80	170	− 0.8
170	200	− 0.6
200	230	− 0.7
230	280	− 0.8
280	400	− 1.0
400	650	− 0.7
650	850	0.4
850	1 000	− 0.3
1 000	2 000	0.0
2 000	3 200	− 0.5
3 200	5 000	0.5
5 000	8 000	1.5
8 000	10 000	0.0
10 000	14 000	− 6.0
14 000	25 000	−13.0 Würm
25 000	40 000	− 4.0
40 000	70 000	−11.0 Würm
70 000	125 000	− 2.0
125 000	200 000	−13.0 Riss
200 000	250 000	− 2.0
250 000	300 000	−13.0 Mindel
300 000	450 000	− 2.0
450 000	600 000	− 7.0
600 000	800 000	−10.0 Günz
800 000	900 000	− 5.0
900 000	1 000 000	−10.0 Günz
1 000 000	∞	− 0.0

TABLE 10.10.
Surface palaeotemperatures in the United Kingdom (Wheildon, 1977). Reference time $t = 0$ corresponds to about 1980 for which ΔT is set zero.

Time, *a* from	to	ΔT, °C	Time, *a* from	to	ΔT, °C
	0	0.0	10 200	10 500	−6.0
0	50	+0.5	10 500	40 000	−9.0
50	100	0.0	40 000	60 000	−7.0
100	550	−0.5	60 000	80 000	−9.0
550	600	0.0	80 000	100 000	+2.0
600	900	+1.4	100 000	130 000	−9.0
900	1 400	0.0	130 000	140 000	+2.0
1 400	2 100	−0.6	140 000	160 000	−7.0
2 100	3 400	+0.5	160 000	180 000	+2.0
3 400	3 900	+1.0	180 000	200 000	−7.0
3 900	5 000	+0.5	200 000	220 000	+2.0
5 000	7 000	+2.5	220 000	240 000	−4.0
7 000	8 500	+0.5	240 000	260 000	+2.0
8 500	9 500	+1.5	260 000	300 000	−9.0
9 500	10 200	0.0			

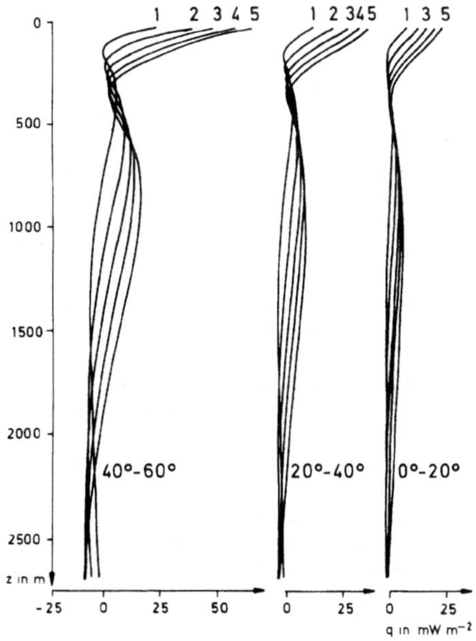

Fig. 10.10. Global averages of the influence of surface palaeoclimatic temperature changes on the heat-flow density. The values have to be added for the northern latitudes of 0 to 20°, 20 to 40°, and 40 to 60° respectively. The curves are calculated for different thermal conductivities (Beck, 1977). (1) 1.26 W m^{-1} K^{-1}; (2) 2.51 W m^{-1} K^{-1}; (3) 3.77 W m^{-1} K^{-1}; (4) 5.02 W m^{-1} K^{-1}; (5) 6.28 W m^{-1} K^{-1}; see also Table 5.5.

10.3. REFERENCES

Balling, N., Kristiansen, J. I., Breiner, N., Poulsen, K. D., Rasmussen, R., and Saxov, S.: 1981, *Geothermal Measurements and Subsurface Temperature Modelling in Denmark*, GeoSkrifter 16, Aarhus University, 172 p.

Beck, A. E.: 1977, 'Climatically Perturbed Temperature Gradients and Their Effects on Regional and Continental Heat Flow Means', *Technophysics* **41**, 17–39.

Bennet, G., Brown, D. D., George, P. T., and Leahy, E. J: 1967, *Operation Kapuskasing*, Ontario Dept. Mines, Misc. Paper 10, 98 p.

Birch, F. and Clark, H.: 1940, 'The Thermal Conductivity of Rocks and its Dependence upon Temperature and Composition', *Amer. J. Sci.* **238** (I), 529–558.

Birch, F.: 1942, 'Thermal Conductivity and Diffusity', in F. Birch, J. F. Schairer, and G. Cecil Spicer (eds.), *Handbook of Physical Constants*, Geological Society of America, Special Papers No. 36. Baltimore, 243–266.

Birch, F.: 1948, 'The Effect of Pleistocene Climatic Variations upon Geothermal Gradients', *Amer. J. Sci.* **246**, 729–760.

Bryson, R. A. and Wendland, W. M.: 1967, 'Tentative Climatic Patterns for some Late Glacial and Post-Glacial Episodes in Central North America', in W. J. Mayer-Oakes (ed.), *Life, Land and Water*, Univ. of Manitoba Press, Winnipeg, Man., 271–298.

Čermák, V.: 1971, 'Underground Temperature and Inferred Climatic Temperature of the Past Millenium', *Palaeogeography, Palaeoclimatology, Palaeoecology* **10**, 1–19.

Čermák, V. and Rybach, L.: 1982, 'Thermal Conductivity and Specific Heat of Minerals and Rocks', in G. Angenheister (ed.), *Landolt-Börnstein; Zahlenwerte und Funktionen aus Naturwissenschaften und Technik.* (Hrsg.): Bd. 1, Physikalische Eigenschaften der Gesteine, Teilbd. a.,Springer-Verlag Berlin, Heidelberg, New York, 305–343.

Creutzburg, H.: 1965, 'Bestimmung thermischer Stoffwerte von Salzgesteinen und Nebengesteinen', *Kali und Steinsalz* **5**, 170–172.

Ericson, D. R. and Welling, G.: 1970, *Pleistocene Climates in the Atlantic and Pacific Oceans: A Comparison Based on Deep-Sea Sediments*, Science 167, New York, pp. 1483–1485.

Frenzel, B.: 1980, 'Das Klima der letzten Eiszeit in Europa', in *Das Klima, Oeschger, Messerli Svilar*, Springer Verlag, 45–63.

Grubbe, K.: 1981, *Vertikalbewegungen und ihre Ursachen am Beispiel des Rheinischen Schildes*, Teilprojekt Geothermik (1.4.80–30.4.81). Report BGR/NLfB Hannover, Archive No. 89 157, 36 p.

Haenel, R.: 1983, 'Geothermal Investigation in the Rhenish Massif', in K. Fuchs, K. von Gehlen, H. Malzer, H. Murawski, and A. Semmel (eds.), *Plateau Uplift*, Springer-Verlag Berlin, Heidelberg, 228–246.

Haenel, R., Michel, W., Schulz, R., and Zoth, G.: 1985, *Hydrogeothermische Untersuchungen zum Geothermischen Demonstrationsprojekt Saulgau*, Report BGR/NLfB Hannover, Archive No. 97 839, 108 p.

Hammen, T. van der, Hijmstra, T. A., and Zagwijn, W. H.: 1971, 'The Flora Record of the Late Cenozoic of Europe', in Karl, K. Turekian (ed.), *The Late Cenozoic Glacial Ages*, New Haven and London, 391–424.

Jessop, A. M.: 1983, *The Essential Ingredients of a Continental Heat Flow Determination*, Zbl. Geol. Paläont. Teil I, H. 1/2 Stuttgart, 70–79.

Kappelmeyer, O. and Haenel, R.: 1974, *Geothermics with Special Reference of Application*, Geoexpl. Monographs, S. 1, No. 4, Gebr. Borntraeger, Berlin, 238 p.

Kohlrausch, F.: 1968, *Praktische Physik, Tafeln*, 22. Aufl., Stuttgart (Teubner), 671 p.

Kopietz, J.: 1978, *Bericht zur Bestimmung der Wärmeleitfähigkeit von Asse-Salzgesteinen*, Report BGR/NLfB Hannover, Archive No. 80 108, 43 p.

Lamb, H. H.: 1977, 'Climate Present, Past and Future', Vol. 2, *Climatic History and the Future*, Verlag Methuen & Co. Ltd., London, 835 p.

Mongelli, F., Loddo, M., and Tramacere, A.: 1982, 'Thermal Conductivity, Diffusivity and Specific Heat Variation of some Travale Field (Tuscany) Rocks versus Temperature', in M. L. Gupta (ed.) *Terrestrial Heat Flow, Tectonophysics*, **83**, 33–43.

Norsk Meteoroligsk Arbok: 1970, Mean annual ground Temperatures, Oslo.

Prest, V. K.: 1957, 'Pleistocene Geology and Surficial Deposits', in C. H. Stockwell (ed.), *Geology and Economic Minerals of Canada*, Economic Geology Series No. 1, Geol. Surv. Canada, Ottawa, 443–495.

Rellstab, W.: 1981, *Der Einfluss des Paläoklimas auf das Temperaturfeld in der Schweiz*, Institut für Geophysik ETH – Zürich, Archive No 90, 63 p.

Rellstab, W.: 1982, *Der Einfluss des Paläoklimas auf das Temperaturfeld in der Schweiz*, Institut für Geophysik ETH – Zürich. Revised version.

Roy, R. F., Beck, A. E., and Touloukian, Y. S.: 1981, 'Thermophysical Properties of Rocks', in Y. S., Touloukian, R. R. Judd, and Roy, R. F. (eds.), *Physical Properties of Rocks and Minerals*, McGraw Hill, New York, 409–502.

Rudloff, H. von: 1967, *Die Schwankungen und Pendelungen des Klimas in Europa seit dem Beginn der regelmäßigen Instrumenten – Beobachtungen 1670*, Die Wissenschaften. Bd. 122, Vieweg; Braunschweig, 370 p.

Sibbitt, W. L., Dodson, J. G., and Tester, J. W.: 1979, 'Thermal Conductivity of Crystalline Rocks associated with Energy Extraction from Hot Dry Rock Geothermal Systems', *Journal of Geophysical Research* **74**, 1117–1124.

Swanberg C. A., Chesman, M. D., Simmons, G, Smithson, S. B, Grönlie, G., and Heier, K. S.: 1974, 'Heat Flow – Heat generation Studies in Norway', *Tectonophysics* **23**, 31–48.

Touloukian, Y. S., Judd, W. R., and Roy, R. F.: 1980, *Physical Properties of Rocks and Minerals*, in: McGraw Hill/CINDAS Data Series on Material Properties, Vol. II-2. New York, London, 548 p.

Van Eysinga, F. W. B.: 1975, *Geological Time Table*, Elsevier, Amsterdam.

Vasseur, G. and Lucazeau, F.: 1983, 'Bounds on Palaeotemperatures and Palaeoclimatic Correction', in R. Haenel and M. Gupta (eds.), *Results of the First Workshop on Standardization in Geothermics*, Zentralblatt für Geologie u. Paläontologie, Teil 1, Heft 1/2, Stuttgart, 17–24.

Wheildon, J., Francis M. F., and Thomas-Betts, A.: 1977, 'Inverstigation of the S. W. England Thermal Anomaly Zone', in *Seminar on Geothermal Energy* (Ed. CEE), Bruxelles, 175–188.

Zoth, G.: 1979, *Temperaturmessungen in der Bohrung Hänigsen sowie Wärmeleitfähigkeitsbestimmungen von Gesteinsproben*, Report BGR/NLfB Hannover, Archive No. 81 828, 15 p.

10.4. GLOSSARY OF SYMBOLS

List of commonly used symbols. In order to avoid undue repetition the same symbols have always been used consistently as defined below and their units as strongly recommended by the International Organization for Standardization.

Symbol	Unit	Meaning
A	m^2	area
a	$m^2\ s^{-1}$	thermal diffusivity
b	–	constant, factor, abbreviation
C	$J\ K^{-1}$	heat capacity
C	–	coefficient, constant, factor
c	$J\ kg^{-1}\ K^{-1}$	specific heat capacity
c_p	$J\ kg^{-1}\ K^{-1}$	specific heat capacity at constant pressure
c_t	$J\ m^{-3}\ K^{-1}$	specific thermal capacity
c_v	$J\ kg^{-1}\ K^{-1}$	specific heat capacity at constant volume
D	m	thickness, distance
d	m	diameter
E	J	energy, radiant energy
Fo	–	Fourier number, $\mathrm{Fo} = at/r^2$
g	$m\ s^{-2}$	gravitational acceleration
H	J	enthalpy
H	$W\ m^{-3}$	heat production rate
h	m	height, thickness, distance

Continuation

Symbol	Unit	Meaning
h	J kg^{-1}	specific enthalpy
I	A	electrical current
k	m^2, D	permeability
k_f	m s^{-1}	hydraulic conductivity
L	J kg^{-1}	latent heat
l	m	length
M	m^2 K W^{-1}	thermal insulation
m	kg	mass
P	W, J s^{-1}	power
P	%	porosity
p	Pa, bar	pressure, 1 bar = 10^5 Pa
Q	J, W s	heat
$\tilde{Q}$	J m^{-1} s^{-1} or J m^{-1}	source term
Q_m	kg s^{-1}	mass-flow rate
Q_v	m^3 s^{-1}	volume-flow rate
q	W m^{-2}, mW m^{-2}	heat-flow density
R	m	Earth's radius
R	K W^{-1}	thermal resistance
R	Ω	electrical resistance
r	m K W^{-1}	thermal resistivity
r	m	radius
S	J K^{-1}	entropy
s	J K^{-1} kg^{-1}	specific entropy
T	°C, K	temperature
T_b	Dm	transmissibility
T_0	°C, K	mean temperature at the Earth's surface
T_v	m^2 s^{-1}	transmissivity
t	s	time; other units are: min (minute), h (hour), a (year)
V	m^3	volume
v	m s^{-1}	velocity
x, y, z	m	Cartesian coordinates
z	m	depth
α	rad	angle
α	W m^{-2} K^{-1}	coefficient of surface heat transfer
α_1	K^{-1}	coefficient of linear thermal expansion
α_v	K^{-1}	coefficient of volumetric thermal expansion
β	m^{-1}	scattering coefficient
Γ	K m^{-1}	temperature gradient
ε	–	emissivity
ε	m^{-1}	opacity
η	Pa s	dynamic viscosity
η	%	efficiency
κ	Pa^{-1}	compressibility, coefficient of compressibility
Λ	m	wavelength
λ	W m^{-1} K^{-1}	thermal conductivity
ν	m^2 s^{-1}	kinematic viscosity
ν	Hz	frequency
ρ	kg m^{-3}	density
ρ	Ω m	electrical resistivity

Continuation

Symbol	Unit	Meaning
σ	$S\ m^{-1}$	electrical conductivity
τ	s	period
τ	–	Fourier number
ϕ	$J\ s^{-1}$, W	heat flow, rediant energy flow
φ, ψ	rad	angle
ω	s^{-1}	angular frequency

$k = 1.380662 \cdot 10^{-23}\ J\ K^{-1}$ = Boltzmann constant
$\sigma = 5.67032 \cdot 10^{-8}\ W\ m^{-2}\ K^{-4}$ = Stefan-Boltzmann constant
$h = 6.626176 \cdot 10^{34}\ W$ = Planck constant
$c = 2.997925 \cdot 10^{8}\ m\ s^{-1}$ = velocity of light in vacuum

$$\mathrm{erf}(x) = \frac{2}{\sqrt{\pi}} \int_0^x e^{-u^2} du$$

$$\mathrm{erfc}(x) = 1 - \mathrm{erf}(x)$$

$$\mathrm{erf}(-x) = -\mathrm{erf}(x)$$

$$[\mathrm{erf}(x)]' = \frac{x'}{\sqrt{\pi}} \exp(-x^2)$$

$$\mathrm{erf}(\infty) = 1$$

$$\mathrm{erf}(x) = \frac{2x}{\sqrt{\pi}} \text{ for } x < 0.05$$

The error function and complementary error function

x	erf(x)	erfc(x)
0	0	1.0
0.05	0.056372	0.943628
0.1	0.112463	0.887537
0.15	0.167996	0.832004
0.2	0.222703	0.777297
0.25	0.276326	0.723674
0.3	0.328627	0.671373
0.35	0.379382	0.620618
0.4	0.428392	0.571608
0.45	0.475482	0.524518
0.5	0.520500	0.479500
0.55	0.563323	0.436677
0.6	0.603856	0.396144
0.65	0.642029	0.357971
0.7	0.677801	0.322199
0.75	0.711156	0.288844
0.8	0.742101	0.257899
0.85	0.770668	0.229332
0.9	0.796908	0.203092
0.95	0.820891	0.179109
1.0	0.842701	0.157299

Continuation

x	erf(x)	erfc(x)
1.1	0.880205	0.119795
1.2	0.910314	0.089686
1.3	0.934008	0.065992
1.4	0.952285	0.047715
1.5	0.966105	0.033895
1.6	0.976348	0.023652
1.7	0.983790	0.016210
1.8	0.989091	0.010909
1.9	0.992790	0.007210
2.0	0.995322	0.004678
2.1	0.997021	0.002979
2.2	0.998137	0.001863
2.3	0.998857	0.001143
2.4	0.999311	0.000689
2.5	0.999593	0.000407
2.6	0.999764	0.000236
2.7	0.999866	0.000134
2.8	0.999925	0.000075
2.9	0.999959	0.000041
3.0	0.999978	0.000022

10.5. COMPUTER PROGRAMME TO SECTION 4.33

```
PROGRAM RADCON      74/73    OPT=3                              FTN 5.1+552         13/03/86  18.56.28

      PROGRAM RADCON
C=======================================================================
C CHRISTOPH CLAUSER                                 FORTRAN 77 / CDC CYBER 825
C TU BERLIN
C ANGEWANDTE GEOPHYSIK
C ACKERSTR. 71-76
C D-1000 BERLIN 65
C FED. REP. GERMANY
C
C RADIATIVE THERMAL CONDUCTIVITIES AND OPACITIES ARE CALCULATED USING
C THE DATA OF:
C FUKAO, Y., MIZUTANI, H. UND UYEDA, S., 1968. OPTICAL ABSORPTION
C    SPECTRA AND RADIATIVE THERMAL CONDUCTIVITY, PHYS. EARTH PLANET.
C    INTERIORS, 1: 57-62.
C SOME SPARSE DATA GAPS WERE FILLED "BY HAND".
C
C A MEAN, BLACKBODY WEIGHTED OPACITY IS CALCULATED FOR EACH WAVELENGTH
C INTERVAL LAMBDA(J)-LAMBDA(J+1). THE ABSORPTION COEFFICIENT TAKEN FOR
C EACH INTERVAL IS THE ONE FOR THE LOWER WAVELENGTH. THE NUMERICAL
C INTEGRATION IS DONE FOR EACH INTERVAL AND OPACITIES, AND RADIATIVE
C THERMAL CONDUCTIVITIES ARE CALCULATED FROM EQUATIONS 3C AND 3B (1),
C 3D AND 3E (2), AND 3F (3) OF THIS CHAPTER IN THE IHFC GUIDELINE BOOK.
C FOR EQUATION 3F THE WAVELENGTH- AND TEMPERATURE DEPENDENCE OF THE
C REFRACTIVE INDEX WERE TAKEN FROM SHANKLAND ET AL. (1979). SCATTERING
C EFFECTS ARE NEGLECTED.
C
C THE NAG LIBRARY ROUTINE DO1AHF WAS USED FOR THE NUMERICAL INTEGRATION.
C
C INPUT PARAMETERS (PARAMETER AND DATA STATEMENTS):
C -----------------------------------------------------------------
C
C
C KK     : NUMBER OF TEMPERATURES FOR WHICH ABSORPTION SPECTRA EXIST
C MM     : NUMBER OF DISCRETE WAVELENGTHS IN EACH ABSORPTION SPECTRUM
C T      : FIELD CONTAINING THE TEMPERATURES IN K
C LAMBDA : FIELD CONTAINING THE WAVELENGTHS IN M*1E-6
C ALFA   : FIELD CONTAINING THE ABSORPTION COEFFICIENTS IN 1/CM FOR
C          ALL TEMPERATURES (1ST INDEX) AND WAVELENGTHS (2ND INDEX)
C=======================================================================
      PARAMETER (KK=11, MM=11)
      EXTERNAL F1,F2,REF
      REAL LAMBDA, KR
      DIMENSION T(KK), LAMBDA(MM), ALFA(KK,MM), OPA1(KK), OPA2(KK),
     +          RAD3(KK), KR(KK,MM-1)
      COMMON /A/ BRECH, TEMP
C
C - - PLANCK'S CONSTANT, BOLTZMANN'S CONSTANT, LIGHT-VELOCITY IN - - - -
C - - VACUUM, PI, REAL PART OF REFRACTIVE INDEX AT 5900 NM AND - - - - -
C - - 22 CELSIUS, AND STEFAN-BOLTZMANN CONSTANT  - - - - - - - - - - - -
C
      DATA PLANCK, BOLTZ, C, PI, BRECH, STEBOL / 6.626176E-34,
     +1.380662E-23, 2.997925E+8, 3.141492654, 1.67, 5.67032E-8/
      DATA T /
     + 300., 400., 500., 600., 700., 800., 900.,1000.,1100.,1200.,1300./
      DATA LAMBDA /
     +  0.4,  0.5,  0.6,  1.0,  1.4,  1.8,  2.2,  2.5,  3.0,  4.0,  5.0/
      DATA ALFA /
     +  9.8, 10.7, 11.9, 13.8, 16.5, 21.2, 22.8, 30.0, 40.0, 50.0, 60.0,
     +  4.1,  4.3,  4.8,  5.1,  5.7,  6.4,  7.4,  8.2,  9.0, 10.3, 10.3,
     +  2.2,  2.3,  2.5,  2.8,  1.9,  3.4,  3.9,  4.2,  4.5,  5.4,  5.2,
     + 19.4, 21.2, 17.7,  9.1,  4.5,  3.6,  3.9,  5.9,  7.4,  9.0, 10.2,
     +  8.8, 12.5, 11.1,  8.2,  4.8,  3.4,  3.8,  5.7,  6.7,  8.5, 10.1,
     + 0.38, 0.75,  1.3,  2.2,  2.2,  2.2,  3.1,  4.2,  6.2,  7.8,  9.2,
     + 0.04, 0.08, 0.17, 0.23, 0.65,  1.0,  1.4,  1.9,  2.8,  3.9,  4.1,
     + 0.20, 0.25, 0.30, 0.40, 0.70,  1.1,  1.4,  1.6,  2.3,  3.2,  3.5,
     +  2.5,  3.0,  3.3,  4.0,  4.5,  4.9,  5.4,  5.6,  7.4,  7.6,  7.8,
     +  6.9,  7.9,  8.3,  9.2,  9.6, 10.0, 10.5, 11.5, 13.0, 15.0, 17.0/
      WRITE (*,10)
 10   FORMAT('1','TABLE 1'//
     +' - A:  ABSORPTION COEFFICIENT (1/CM) BY FUKAO ET AL. (1968)'/)
      WRITE (*,11)
 11   FORMAT(7X,'T (K)',51X,'WAVELENGTH (10**-6 M)',/)
      WRITE (*,12) (LAMBDA(J),J=1,MM)
 12   FORMAT(11X,11F11.1/)
      WRITE (*,13) (T(I),(ALFA(I,J),J=1,MM),I=1,KK)
 13   FORMAT(5X,F5.0,1X,11F11.2)
      DO 20 J=1,MM
         LAMBDA(J) = LAMBDA(J) * 1.E-6
         DO 20 I=1,KK
            ALFA(I,J) = ALFA(I,J) * 100.
 20   CONTINUE
      FAKTOR = PLANCK*C/BOLTZ
C
C - - LOOP FOR ALL TEMPERATURES - - - - - - - - - - - - - - - - - - - -
C
      DO 100 I=1,KK
         TEMP = T(I)
         FAK  = FAKTOR / TEMP
         N = 0
         EPSR = 1.E-9
C
C - - NUMERICAL INTEGRATION BY NAG-ROUTINE DO1AHF OF FUNCTIONS F1 - - -
C - - AND F2 BETWEEN LOWER AND UPPER BOUNDS BNDLO AND BNDUP; MINIMUM- -
C - - ACCURACY IS EPSR, N IS A DEFAULT PARAMETER AND NPTS, RELERR - - -
C - - AND IFAIL ARE OUTPUT PARAMETERS - - - - - - - - - - - - - - - - -
C
         EPS1 = 0.
         EPS2 = 0.
         EPS3 = 0.
```

R. Haenel, L. Rybach, and L. Stegena (eds.)
Handbook of Terrestrial Heat-Flow Density Determination, 467—468

Continuation:

```
                DO 70 J=1,KK-1
                   JWO    = J
                   DIF    = (LAMBDA(J+1)-LAMBDA(J)) * 1E+6
                   BNDLO  = FAK/LAMBDA(J)
                   BNDUP  = FAK/LAMBDA(J+1)
                   EPSI   = ALFA(I,JWO)
                   AIGRA  = D01AHF(BNDLO,BNDUP,EPSR,NPTS,RELERR,F1,N,IFAIL)
                   EPS1   = EPS1 + AIGRA / EPSI
                   HILF2  = AIGRA * (REF(LAMBDA(JWO))**2) / EPSI
                   EPS2   = EPS2 + HILF2
                   AIGRA  = D01AHF(BNDLO,BNDUP,EPSR,NPTS,RELERR,F2,N,IFAIL)
                   HILF3  = AIGRA * REF(LAMBDA(JWO))/ EPSI
                   EPS3   = EPS3 + HILF3
                   KR(I,J) = (HILF2 + HILF3)/DIF
  70            CONTINUE
                FAK1     = 4. * (PI**4) / 15.
                FAK2     = FAK1 * BRECH**2
                FAK3     = 8.*PI*(BOLTZ**4)*(T(I)**3)/(3.*(C**2)*(PLANCK**3))
                OPA1(I) = FAK1 / EPS1
                OPA2(I) = FAK2 / EPS2
                RAD3(I) = FAK3 * (EPS2 + EPS3)
                DO 80 J=1,KK-1
                      KR(I,J) = KR(I,J)*FAK3
  80            CONTINUE
 100      CONTINUE
          WRITE(*,200)
          WRITE(*,210) (T(I),OPA1(I)/100.,OPA2(I)/100.,I=1,KK)
          WRITE(*,220)
          DO 150 I=1,KK
             RADFAK = 16. * STEBOL * (BRECH**2) * (T(I)**3) / 3.
             RAD1 = RADFAK / OPA1(I)
             RAD2 = RADFAK / OPA2(I)
             WRITE(*,230) T(I),RAD1,RAD2,RAD3(I)
 150      CONTINUE
          WRITE(*,240)
          WRITE (*,250) (T(I),(KR(I,J),J=1,MM-1),I=1,KK)
 200      FORMAT(//' - B:  T (K)',3X,'OPACITY (1/CM)'//18X,'(1)',7X,'(2)'/)
 210      FORMAT(5X,F5.0,1X,2F11.5)
 220      FORMAT(//' - C:  T (K)',3X,'RADIATIVE THERMAL CONDUCTIVITY (W/(M*K
         +))'//18X,'(1)',8X,'(2)',8X,'(3)'/)
 230      FORMAT(5X,F5.0,1X,(3F11.4))
 240      FORMAT(//' - D:  CONTRIBUTION TO (3) PER 10**-6 M (10**-6 W/(M*K))
         +'//,7X,'T (K)',46X,'WAVELENGTH INTERVALS (10**-6 M)'//,12X,
         +'    0.4-0.5 ','    0.5-0.6 ','    0.6-1.0 ','    1.0-1.4 ',
         +'    1.4-1.8 ','    1.8-2.2 ','    2.2-2.5 ','    2.5-3.0 ',
         +'    3.0-4.0 ','    4.0-5.0 '/)
 250      FORMAT(5X,F5.0,1X,10F11.7)
          WRITE (*,'(''1'')')
          STOP
          END
```

FUNCTION F1   74/73   OPT=3   FTN 5.1+552

```
          REAL FUNCTION F1(X)
          F1 = -(EXP(X) * (X**4)) / ((EXP(X)-1)**2)
          RETURN
          END
```

FUNCTION F2   74/73   OPT=3   FTN 5.1+552

```
          REAL FUNCTION F2(X)
          COMMON /A/ BRECH, TEMP
          F2 = -(2.*TEMP*(X**3)*1.8E-5) / (EXP(X)-1)
          RETURN
          END
```

FUNCTION REF   74/73   OPT=3   FTN 5.1+552

```
          REAL FUNCTION REF(LAM)
          REAL LAM
          COMMON /A/ BRECH, TEMP
          REF = BRECH - 2.05E+4 * (LAM-5.9E-6)
          RETURN
          END
```

AUTHOR INDEX

SUBJECT INDEX

GPSR Compliance

The European Union's (EU) General Product Safety Regulation (GPSR) is a set of rules that requires consumer products to be safe and our obligations to ensure this.

If you have any concerns about our products, you can contact us on ProductSafety@springernature.com

In case Publisher is established outside the EU, the EU authorized representative is:

Springer Nature Customer Service Center GmbH
Europaplatz 3
69115 Heidelberg, Germany

Batch number: 10815974

Printed by Printforce, the Netherlands